The Evolution
of
Genome Size

The Evolution
of
Genome Size

Edited by

T. CAVALIER-SMITH

Department of Biophysics
King's College London
University of London

A Wiley-Interscience Publication

JOHN WILEY & SONS

Chichester · New York · Brisbane · Toronto · Singapore

Library of Congress Cataloging in Publication Data:
Main entry under title:

The Evolution of genome size.

  'A Wiley–Interscience publication.'
  Includes bibliographies and index.
  1. Genomes. 2. Evolution. I. Cavalier-Smith, T.
QH447.E96 1985   574.87'3282   84–25659

ISBN 0 471 10272 5

British Library Cataloguing in Publication Data:

The Evolution of genome size.
  1. Chromosomes
  I. Cavalier-Smith, T.
  574.87'322   QH600

ISBN 0 471 10272 5

Typeset by Input Typesetting Ltd., London SW19 8DR.
Printed and bound in Great Britain by The Bath Press, Bath, Avon.

Contents

Contributors

AMMERMANN, D. Lehrstuhl Zoologie, Abteilung Zellbiologie, Institut fur Biologie III, Universität Tübingen, D 7400 Tübingen 1, Auf der Morgenstelle 28, Federal Republic of Germany.

BACHMANN, K. Hugo de Vries Laboratorium University of Amsterdam, Kruislaan 318, 1098 SM Amsterdam, The Netherlands.

BOYNTON, J. E. Department of Botany, Duke University, Durham, North Carolina 27706, USA.

CAVALIER-SMITH, T. Department of Biophysics, King's College London, 26–29 Drury Lane, London WC2B 5RL, England.

CHAMBERS, K. L. Herbarium, Department of Botany, Oregon State University, Corvallis, Oregon 97331, USA.

CHARLESWORTH, B. Department of Biology, University of Chicago, 1103 E. 57th Street, Chicago, Illinois 60637, USA.

CLARK-WALKER, D. Research School of Biological Sciences, Australian National University, Box 475, P. O., Canberra City, A. C. T. 2601, Australia.

CULLIS, C. A. John Innes Institute, Colney Lane, Norwich NR4 7UH, England.

DOOLITTLE, W. F. Department of Biochemistry, Dalhousie University, Halifax, Nova Scotia, Canada B3H 4H7.

DYSON, P. Department of Genetics, University of Glasgow, Glasgow G11 5JS, Scotland.

GILLHAM, N. W. Department of Zoology, Duke University, Durham, North Carolina 27706, USA.

HARRIS, E. H. Department of Botany, Duke University, Durham, North Carolina 27706, USA.

HERDMAN, M. Unité de Physiologie Microbienne, Département de Biochimie et Génétique Microbienne, Institut Pasteur, 28 rue du Docteur Roux, 75724 Paris, France.

JONES, R. N. Department of Agricultural Botany, School of Agricultural Sciences, Penglais, Aberystwyth, Dyfed SY23 3DD, Wales.

NURSE, P. Imperial Cancer Research Fund, Lincoln's Inn Fields, London WC2, England.

PRICE, H. J. Genetics, Department of Plant Sciences, Texas A and M University, College Station, Texas 77843, USA.

SHERRATT, D. Department of Genetics, University of Glasgow, Glasgow G11 5JS, Scotland.

Preface

The greatest advance in evolutionary biology since Darwin and Wallace's 1858 papers came in the period 1922–1932 with the formulation of the genetical theory of evolution by Muller, Haldane, Chetverikov, Fisher and Wright. Since then molecular genetics has progressed so dramatically that major new insights into evolution are probable during the next decade, as the structure of genes, and mechanisms of mutation, recombination and gene expression, become well understood. New discoveries are coming thick and fast but they will increase evolutionary understanding only if they are synthesized with existing knowledge and with new findings in all areas of biology.

The purpose of this book is to contribute to this synthesis by focusing on one relatively neglected but fundamental problem: the evolution of genome size. Bacterial genome size is probably directly related to the number of genes, and is a good measure of their genetic and organismic complexity. But in eukaryotes this is not so, and the existence of vast amounts of apparently non-genic DNA has long been a major puzzle, now known as the DNA C-value paradox. Does all this DNA, sometimes as much as 50 000-fold in excess of protein-coding requirements, have a function? Or is it merely useless 'junk' or a 'selfish' parasite? Attempts to answer these questions soon lead into a great variety of fascinating questions: the mechanisms of DNA replication, duplication and transposition; the significance of repetitive DNA, split genes, heterochromatin, and chromatin elimination; the genetic control and evolutionary significance of cell and nuclear volume and cell growth rates; and many others. Therefore a number of cell and molecular biologists and molecular, population, and cyto-geneticists, were invited to make their distinctive contributions to this book, which is the first to concentrate on the evolution of genome size. It presents important new arguments in favour of the skeletal DNA theory of the evolution of genome size, and against the junk and selfish DNA explanations of the C-value paradox.

Nearly half a century ago Darlington very clearly expressed the problems inherent in a study of this kind in his preface to the second edition of his *Recent Advances in Cytology* (1937, Churchill, London): 'Finding out *why* things happen in the cell is an entirely different matter from finding out *how* they happen. The one problem is a matter of skill and common sense. The other takes us into a new element. We are plunged into inferences, often speculative inferences, which connect mechanics, physiology and genetics. We find that the cell is part of an interlocking system of growth and reproduction, heredity and variation. Everything that happens in the cell is related to everything else that happens in the organism, or indeed has happened in its ancestors. It is impossible at one and the same time to deal with all these dialectical relationships. I can describe only those that seem to me most important at the moment.' Too often such evolutionary aspects of cell biology have been relegated to brief and often superficial speculations at the ends of papers primarily concerned with reporting new facts. To avoid such superficiality and to carry further the integrative approach pioneered by Darlington, and which must be central to evolutionary studies, it is best to consider in detail a single major evolutionary problem, and explore many different aspects of it. This book therefore does *not* attempt to deal with all aspects of genome evolution. It should interest not only evolutionary biologists but also geneticists and molecular and cell biologists concerned primarily with basic phenomena such as DNA replication and the cell cycle, since comparative evolutionary studies can also give valuable clues about fundamental mechanisms.

I warmly thank all the contributors, and apologise to them for the varying delays between receipt of their manuscripts and going to press. I much appreciate the encouragement given to my work on genome size by Professors H. G. Callan and M. H. F. Wilkins and thank Bridget Sarsby and Patricia Collins for the typing.

T. Cavalier-Smith

Units of measurement of genome size

Nuclear genome size of eukaryotes is usually measured in picograms (pg) of DNA (1 pg = 10^{-12}g).

The smaller prokaryotic genomes, consisting of single molecules, are more commonly measured in daltons – the unit of relative atomic and molecular mass.

$$1 \text{ dalton} \simeq 1.66 \times 10^{-12} \text{pg}$$
$$10^9 \text{ daltons} \simeq 0.00166 \text{ pg}$$
$$1 \text{ pg} \simeq 6.02 \times 10^{11} \text{ daltons}$$

Plasmid genome size is more often expressed as the number of base pairs per molecule or as the number of kilobases or kilobase pairs (abbreviated kb or kbp).

For double stranded DNA 1 kb = 1000 base pairs $\simeq 6.18 \times 10^5$ daltons.
1 pg $\simeq 0.98 \times 10^6$ kb.

The Evolution of Genome Size
Edited by T. Cavalier-Smith
© 1985 John Wiley & Sons Ltd

CHAPTER 1

Introduction: the Evolutionary Significance of Genome Size

T. Cavalier-Smith

THE DNA C-VALUE PARADOX

Molecular genetics began with the discovery of DNA (Miescher, 1871), and the formulation of four key ideas:

1. The idea of the exact replication of molecular genetic determinants (Roux, 1883; Nägeli, 1884).
2. The idea that these consist of chromosomal DNA (Sachs, 1882; Hertwig, 1884).
3. The idea that genes may be active or dormant (Darwin, 1868).
4. The idea that genes act by the passage of short-lived copies into the cytoplasm where they control biosynthesis and cell structure (de Vries, 1889).

A century of work has firmly established these basic qualitative ideas and clothed them with remarkable molecular detail, but a fundamental quantitative problem remains: why should the cellular DNA content of differing species vary by over a million-fold? This book attempts to answer this question.

The mass of DNA in an unreplicated haploid genome, such as that of sperm nucleus, is called the genome size (Hinegardner, 1976) or, alternatively, the C-value (Swift, 1950) because it is usually constant in any one species. Hardly had this constancy been established when Mirsky and Ris (1951) demonstrated huge interspecific variation in C-values that bore no relationship to differences in organismic complexity or to the likely number of different genes in the species studied. Molecular biologists immediately saw the constancy of the C-value within a species, and the fact that unreplicated diploid nuclei contained twice this amount of DNA (Boivin et al., 1948; Vendrely and Vendrely, 1948), as support for the idea that genes consisted

1

of DNA. But for years the problem posed by the great variability in genome size in different species was widely ignored.

There has been a recurrent tendency to regard DNA C-values as related simply to the number of genes, and to equate evolutionary accumulation of DNA with increases in gene number (e.g. Kimura, 1961). This view of the significance of genome size appears to be true only for bacteria (see Chapter 2) and viruses. But ever since the first measurements (Mirsky and Ris, 1951) the evidence has been strongly against it for eukaryotes. In Chapter 3 I show that variation in eukaryote genome size is over a thousand-fold greater than can be accounted for by variations in gene number and that variation in eukaryote genome size bears no relationship to organismic complexity.

As it became increasingly firmly established that genes consist of DNA this puzzling lack of relationship between estimates of the number of genes and the DNA C-values of different eukaryotes became more and more of a problem, now usually called the C-value paradox (Thomas, 1971).

Six quite different kinds of solution have been proposed:

1. The whole idea of particulate heredity and of a definite number of discrete genes is false (Goldschmidt, 1955).
2. Genes do not consist of DNA, which must have some other role, perhaps structural (Darlington, 1956).
3. Mendel's idea that there is only one copy of each gene per gamete is false; instead the number of copies vary with C-value, either because each chromosome contains a variable number of longitudinal DNA strands (the multistrand theory: Darlington, 1955; Martin and Shanks, 1966; Rothfels *et al.*, 1966) or because of tandem duplications along its length (Callan and Lloyd, 1960; Callan, 1967; Ohno, 1970). To reconcile his postulate of many copies of a gene per genome with the genetic evidence for a single copy, Callan (1967) proposed his ingenious master-slave theory.
4. Much DNA is genetically or physiologically inert (Muller and Painter, 1932; Darlington, 1937) and is useless junk (Ohno, 1972) carried passively by the chromosome merely because of its linkage to functional genes (Rees, 1972).
5. Much DNA is a functionless parasite (Östergren, 1945; 'selfish DNA' *sensu* Doolittle and Sapienza, 1980) or 'genetic symbiont' (Cavalier-Smith, 1983a) that accumulates and is actively maintained by intragenomic selection (Cavalier-Smith, 1980a).
6. DNA has quantitative non-genic, or 'nucleotypic' (Bennett, 1971) functions, in addition to its qualitative genic functions (Commoner, 1964; Stebbins, 1966; Bennett, 1971, 1972, 1973; Rees, 1972; Cavalier-Smith, 1978, 1980a, 1982a).

The first two possibilities were soon decisively rejected by the elucidation of the genetic code (Crick, 1967). But at about the same time the powerful new method of DNA renaturation kinetics (Bolton *et al.*, 1965) led to the discovery that most eukaryotes, unlike bacteria, contained large amounts of repetitive DNA (Britten and Kohne, 1968). This at first appeared to support the third idea that there might be several copies of protein-coding genes in each genome. However, further studies using renaturation kinetics soon showed that eukaryote chromosomes were not multistranded (Laird, 1971) and that protein-coding genes were mostly present in only one copy per genome, even in organisms with very large C-values (Rosbash *et al.*, 1974). This led to the rejection of the third possibility, and to the present position where we have to accept not only that genes do consist of DNA but also that in many eukaryotes the major part of the DNA does not consist of functional cellular genes. A convenient non-committal name for this extra DNA, which occurs in very large amounts in high C-value eukaryotes, is secondary DNA (Hinegardner, 1976). Why this DNA should be there, and the relative merit of the last three possibilities, is the fundamental problem considered in this book. Secondary DNA must either have a non-genic function or be the incidental result of mutational and intragenomic selective forces of no positive significance to the organism. Since this book is intended not only for molecular biologists but also for population geneticists, cell biologists and general evolutionary biologists, who may be less familiar with the arguments from DNA renaturation kinetics that have led to this conclusion, I shall start with a brief outline of them; a more detailed and documented explanation of renaturation kinetics is given by Lewin (1980, ch. 18, 19) and Bouchard (1982).

RENATURATION KINETICS, CHROMOSOME UNINEMY AND THE NUMBER OF COPIES OF EACH GENE

To measure the kinetics of renaturation, a sample of purified DNA broken into fragments a few hundred nucleotides in length (i.e. shorter than most genes) is first denatured by heating the DNA solution to force apart the complementary strands. If the denatured sample is placed at a temperature about 20°C less than was needed to cause complete denaturation, the separated single-stranded fragments will in time meet and pair with complementary molecules; the rate of this renaturation is determined by measuring the fraction of fragments that has become double-stranded as time progresses. Under standard conditions this rate depends only on the concentration and nature of the DNA. If the concentration and base composition of the DNA is kept constant the rate of renaturation under standard conditions depends only on the number of copies of each DNA sequence in the solution. The larger the number of copies of each sequence the greater the chance that complementary sequences will meet and reanneal by zipping together, so the faster will be the observed rate of renaturation.

Nuclear DNA from most eukaryotes proved to be a mixture of DNA that renatures rapidly and DNA that renatures much more slowly (Britten and Kohne, 1968). The rapidly renaturing DNA fragments therefore consist of sequences that are repeated many more times in the genome than the slowly renaturing ones. Almost all the rapidly renaturing sequences become double-stranded before a detectable fraction of the slowly renaturing sequences have begun to reanneal: therefore, by separating double and single-stranded DNA from each other at this stage of the renaturation reaction by hydroxyapatite chromatography, one can obtain relatively pure samples of the rapidly and slowly renaturing DNA for separate study. Because both the multistrand and the master-slave theories predicted many copies per genome of each gene, and that the average number of copies should be greater in high C-value species than in low C-value species, it was important to test this prediction.

This was first done by Laird (1971) who compared the renaturation rates of the slowly renaturing DNA fraction in mice, the fruit fly *Drosophila* and the sea squirt *Ciona*, with each other and with the renaturation rates of DNA from the bacteria *Escherichia coli* and *Bacillus subtilis* and the phage T$_4$. Unlike eukaryote DNA, neither bacterial nor viral DNA could be subdivided into fractions renaturing at different rates. However, the viral DNA renatured 18 times faster than the *E. coli* DNA. This was expected because the viral genome is 18-fold smaller than the bacterial genome; so long as both kinds of DNA contain only one copy of each gene per genome a solution of viral DNA would contain 18 times as many copies of each sequence per unit volume as a solution containing the same concentration of bacterial DNA (where concentration is measured in moles of nucleotide per litre) and will therefore renature proportionally faster. Laird found that the time taken for renaturation of the slowly renaturing DNA from the three eukaryotes was also directly proportional to the mass of this DNA per genome. Moreover, when the eukaryote and prokaryote data were plotted on the same graph (Fig. 1.1) a single straight line passed through all the experimental points. Thus, if phage T$_4$ and *E. coli* had only one copy of each sequence per genome, the same must be true for the slowly annealing DNA of each of the three eukaryotes even though they differed from each other by 22-fold in C-value. The slowly annealing fraction of eukaryote DNA is therefore now referred to as non-repeated, unique or single-copy DNA. Since single-copy DNA makes up well over half the genome in these and many other species with vastly differing C-values, it is clear that the bulk of the variation in genome size reflects variation in the amount of single-copy DNA and therefore cannot be explained either by a multistrand model, the master-slave hypothesis, or any modifications of them.

Three additional independent methods: relaxation viscometry (Lauer and Klotz, 1975; Kavenoff and Zim, 1973), ultracentrifugation (Petes and Fangman, 1973) and electron microscopy (Petes *et al.*, 1973) have now shown

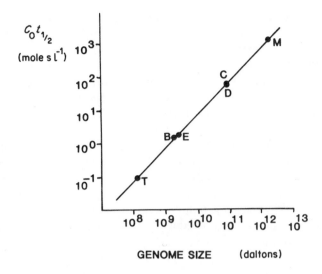

GENOME SIZE (daltons)

Fig. 1.1. Correlation between genome size and the time taken for the DNA to renature. The time taken for the renaturation of half the DNA at a concentration C_o was measured for the total DNA of phage T4 (T) and the bacteria *Bacillus subtilis* (B) and *Escherichia coli* (E); and for the major slowly-renaturing fraction of the DNA of three animals (the fruit fly *Drosophila melanogaster* (D); the seasquirt *Ciona intestinalis* (C) and the mouse *Mus musculus* (M)). The product of the half-time for renaturation ($t_{1/2}$) and the DNA concentration, C_o, is referred to as the C_ot value and is constant for any one species; the linearity and unitary slope of the curve shows that it depends only on the (haploid) genome size. Data from Laird (1971).

that the chromosomes of the yeast *Saccharomyces cerevisiae* and *Drosophila melanogaster*, which differ 20-fold in C-value, each consist of one single DNA molecule. Furthermore, the chromatids of the lampbrush chromosomes of *Triturus cristatus*, whose C-value is 3142 times that of yeast and 157 times that of *Drosophila*, showed the same kinetics of digestion by DNase as a single double-helical DNA molecule (Gall, 1963), and therefore could not be multistranded. Experiments stretching *Chironomus* polytene chromosomes to breakpoint (Gruzder and Reznik, 1981) confirm the uninemy of chromatids. There is therefore no longer reason to doubt that all unreplicated eukaryotic chromosomes consist of single linear DNA molecules, to which are attached numerous protein molecules.

Further evidence against explanations for the C-value paradox in terms of variable numbers of identical copies of genes lies in the demonstration by renaturation kinetics that most messenger RNA molecules are transcribed from unique sequences, present only once per haploid genome. This is true in adult tissues and for cells in tissue culture (Greenberg and Perry, 1971; Lewin, 1980: chapter 24) and in amphibian ovaries in which all the messengers required for early embryonic development are being synthesized and stored in oocytes for later use (Rosbash *et al.*, 1974). In the latter case it was shown that the kinetic complexity (a measure by renaturation kinetics of the number of different sequences) of the messenger RNA fraction was the same in the amphibians *Xenopus* and *Triturus* which differ 7-fold in genome size. If this could also be established for every stage of the life history, it would prove that most of the extra DNA in organisms with larger genomes does not code for proteins. Unfortunately, this is not strictly possible using present techniques which could not detect extremely rare messengers that might be present only transiently in a rare somatic cell type. But even in the absence of definitive proof of this the evidence for it outlined in Chapter 3 is very strong. The extra DNA in organisms with high C-values must therefore either have a non-coding (perhaps structural) function or be non-functional and simply produced as the net result of non-adaptive evolutionary forces.

CORRELATIONS BETWEEN EUKARYOTE GENOME SIZE AND QUANTITATIVE CHARACTERS

Many authors (e.g. Lewin, 1980) who have discussed the C-value paradox have appeared to be unaware of the extensive evidence for positive correlations between eukaryote genome size and a variety of quantitative characters, notably:

1. the total volume and mass of the metaphase chromosomes;
2. interphase nuclear volume and dry mass;
3. cell volume and weight;
4. nucleolar volume and dry mass;
5. centromere volume;
6. nuclear RNA content;
7. nuclear protein content;
8. seed weight in herbaceous flowering plants;
9. length of the cell cycle;
10. the duration of meiosis;
11. the length of the DNA synthesis or S-phase of the cell cycle;
12. pollen maturation time;
13. minimum generation time in flowering plants;
14. time taken for embryogenesis from fertilization to the hatching tadpole in amphibians.

Since many of these quantitative characters are of considerable adaptive and functional significance to the organisms possessing them, many biologists (e.g. Stebbins, 1966; Szarski, 1970, 1983; Bennett, 1973; Cavalier-Smith,

1978, 1982a) have reasonably assumed that natural selection acting on such characters has played a major role in the evolution of different genome sizes.

In flowering plants there is also clear evidence for an association between genome size and climate (Bennett, 1976; Levin and Funderberg, 1979), and for herbaceous perennials in the temperate zone between genome size and temperature at the season of maximal growth (Grime, 1983). Similar correlations between climate and chromosome size in plants have long been known to botanists (Heitz, 1926, 1927; Avdulov, 1931; Stebbins, 1971, 1976), which may be why they have been less ready than zoologists to regard variations in genome size as non-adaptive.

Those aware of these facts have tended to assume, like Bachmann *et al.* in Chapter 9, that they are incompatible with the idea that extra DNA in larger genomes is merely useless 'junk' DNA or 'selfish' DNA and that they show instead that genome size *itself* genetically controls most or all of these quantitative characters and is subject to selection so as to optimize them. Certainly the mere postulate of the existence of junk DNA and parasitic DNA does not solve the problem of the evolution of genome size, since it does not explain why these correlations exist and why they have such numerical values. But on the other hand the mere existence of these correlations does not prove that genome size exerts a direct nucleotypic control over these phenotypic characters, or that selection acting on such characters directly favours an increase in genome size; while this could be the case, it is in principle possible that none of these characters except chromosome volume is nucleotypically determined, and that secondary DNA has no positive function. My own view is that the majority of the quantitative characters that correlate with genome size are not directly and mechanistically determined by genome size, and that the observed correlations are mostly the result of indirect developmental and/or evolutionary correlations; but two phenotypic characters, namely chromosome volume and nuclear volume, do seem to be directly and causally determined by a combination of the sheer amount of DNA together with the tightness or looseness of its folding. Chapter 4 argues that this joint nucleotypic and genic control of nuclear volume is the fundamental reason for the observed correlations, but that the crucial evolutionary factor is cell volume, not genome size itself; evolutionary variations in cell volume probably lead to the selection of correlated variations in the other characters, including nuclear volume and therefore genome size which partly determines it. In order to substantiate or refute this thesis we need a better understanding of the genetic control of the cellular and organismic characters in question; several aspects of the problem are reviewed in Chapters 4 and 5.

Two other features of genome size variation that have been ignored until recently (Cavalier-Smith, 1978) require explanation and also hold important clues to the solution of the C-value paradox. One is the fact that the three amniote classes (mammals, birds, reptiles) are exceptional among eukaryotes in their small variation in genome size (2 to 4-fold). Most extensively studied eukaryotes, whether plant, animal or microorganism, show a much greater variation in genome size. Protozoa show a 1000 and unicellular algae a 3000-fold range, whereas most classes of multicellular animals and plants show a variation of the order of 10 to 100-fold. Great variation in C-value is therefore not the exception, but the general rule in most eukaryote groups: it is the near constancy in mammals and in birds and reptiles that is the exception and which requires a special explanation. The second neglected fact is that unicellular eukaryotes in general show a greater range in C-values than multicellular ones. Both the smallest, *and* the largest, eukaryote genome sizes are found in unicellular species; from yeast to Amoeba the range is 80 000-fold, but from sponges to humans only 55-fold! Multicellular genomes also cluster in the middle range (from 0.5–10 pg).

Therefore, whatever are the evolutionary forces that prevent large reductions or increases in genome size, they appear to be stronger in multicellular eukaryotes than in unicellular ones, stronger in animals than in plants and strongest of all in amniotes; one might argue conversely that the evolutionary forces causing large changes in genome size are strongest in unicellular eukaryotes.

An ideal solution to the C-value paradox would be a theory that could simultaneously explain:

1. these differences in the range of C-values in different eukaryote groups;
2. the positive correlations between genome size and so many quantitative characters;
3. the absence of any correlation between genome size and gene numbers.

My hypothesis that genic and non-genic DNA together have a universally important function as a nuclear skeleton that determines the volume of nuclei (Cavalier-Smith, 1978, 1982a) still seems to achieve this more satisfactorily than any competing hypothesis.

THE SKELETAL DNA HYPOTHESIS

'My basic argument is that natural selection acts powerfully on organisms to determine their cell size and developmental rates (which are inversely related). The mean cell volume of an organism is the result of an evolutionary compromise between conflicting selection for large cell size and for rapid developmental rates: the particular compromise reached for a particular species will depend on its ecological niche and organismic properties. Since larger cells require larger nuclei, selection for a particular cell volume will secondarily select for a corresponding nuclear volume, producing a close correlation between cell and nuclear volumes in different organisms. I suggest that the basic nucleotypic function of DNA is to act as a nucleoskeleton which determines the nuclear volume; small C-values are therefore required by small cells with small nuclei, and large C-values by large cells needing large nuclei. The DNA C-value of an organism is therefore simply the secondary result of selection for a given nuclear volume, which in turn is the secondary result of the evolutionary compromise between selection for cell size and for developmental rates.'

(Cavalier-Smith, 1978)

Chapter 4 reviews evidence that strongly supports the fundamental postulates of the preceding quotation. But there was also a variety of secondary ideas (Cavalier-Smith, 1978), some of which now require modification or reconsideration. Most important of these is the reason I gave as to why larger cells require larger nuclei. I originally suggested that this was because nucleocytoplasmic RNA transport would become rate-limiting to growth if large cells had only the same nuclear surface area as small cells. Though the comparative evidence strongly indicates that in many large cells, such as nerve cells or pollen mother cells, the need to increase nuclear pore numbers and nuclear surface areas actually is an important evolutionary factor (Cavalier-Smith, 1978, 1982a), and therefore that the basic idea underlying my proposal is valid, in Chapter 4 I give reasons for doubting that it can be the *universal* explanation for the evolution of larger nuclei by larger cells; instead I propose that nuclear volume itself, rather than nuclear surface area, is the more generally important factor, and that the reason why larger cells need a larger nuclear volume is that they need more transcriptional and RNA-processing machinery, which, for optimal efficiency and growth rate, must always vary in direct proportion to the number of ribosomes; this I suggest is the fundamental reason for the approximate constancy of the nucleocytoplasmic ratio first emphasized by Strasburger (1893).

On this hypothesis the basic nucleotypic function of DNA is to determine nuclear volume; cell volume is considered to be determined independently by specific genes. However, since it is possible to construct cell cycle models whereby the total DNA content could causally determine cell volume, it is necessary to consider the evidence concerning the genetic control of cell volume in some detail in order to decide whether such nucleotypic control really occurs, as various authors have suggested, or whether the correlation between cell volume and genome size arises simply because large cells evolve larger genomes, either because they need them as a nucleoskeleton (Cavalier-Smith, 1978) or simply because they can tolerate more useless DNA (Doolittle and Sapienza, 1980). In Chapter 5 Nurse reviews the direct genetic evidence and considers that the issue is still open. Though much comparative evidence is *consistent* with nucleotypic control of cell volume, none appears to *require* it; and, as pointed out in Chapter 4, only some nucleotypic models could directly explain the C-value/cell volume correlation.

JUNK DNA AND GENOME SIZE

'Why then are so many parts of chromosomes inert? Is it merely that they have been accumulated by mutation to inertness and have not been got rid of quickly enough by selection?' (Darlington and La Cour, 1941.)

Though, for reasons given on page 7, the junk DNA hypothesis alone is insufficient to explain the C-value paradox, it can be combined with a simple selective argument to provide a potential explanation (Cavalier-Smith, 1980a, 1982a). Think of the genome size of an organism as the result of an evolutionary balance between forces tending to increase it and those tending to reduce it. Mutation can in principle either increase or decrease genome size; if deletions and duplications are exactly in balance it will have no net effect, but if the rate and/or size of duplications exceed the rate and/or size of deletions then genomes would tend steadily to increase as a result of this mutation pressure (Cavalier-Smith, 1982a) even if the extra DNA has no function. If it is mildly deleterious the extra DNA will be selected against; a mutation-selection equilibrium will be set up when the rate of addition of extra bits of DNA by duplication is balanced by the rate of removal of genomes containing it by organismic selection resulting from its harmful effects.

In principle the observed direct correlation between genome size and nuclear volume (see Chapter 4) could result if either the strength of the mutation pressure, or the strength of counter-selection against extra junk DNA, were a function of nuclear volume of exactly the correct form to ensure that the mutation–selection equilibrium always produced a genome size in a constant ratio to nuclear volume. It is plausible that some types of selection against extra non-coding DNA might be relatively weaker in cells with larger nuclei; but others, such as counter-selection against larger genomes because of their greater sensitivity to mutagens (see Chapter 3), would appear not to be less for larger nuclei. Nor is it clear that there would necessarily be an inherent *mutational* bias towards DNA increase. In view of these considerations, and the apparently purely *ad hoc* nature of the quantitative assumptions about mutation rates and selective forces that would be necessary to reconcile the junk DNA hypothesis with the universal correlation between genome size and cell and nuclear volume, it is hard to view it as a serious scientific explanation of the C-value paradox.

This is not to say that junk DNA does not exist at all; if there is any truth in the suggestion that there is a net upward mutation pressure on genome size then this will ensure that genomes at mutation–selection equilibrium will tend to contain at least a little non-functional junk DNA even if it is continually being selected against. The greater tendency of large deletions compared with large duplications to be lethal would also cause a slight (selective) bias towards somewhat larger genomes. Moreover, fluctuations in the relative rates of duplication, deletion and strength of selection against extra DNA would also mean that genomes may come to contain a bit more DNA at any particular time than they really need. Such a slightly greater than optimal amount of DNA may well be the general rule and might be considered as a form off mutational load.

Another, possibly more potent, mode of spreading for junk DNA would be by 'hitch-hiking' with useful genes (Maynard Smith and Haigh, 1974). Some support for this arises from the important studies of experimentally induced changes in genome size reviewed by Cullis (Chapter 6); the evidence suggests that the extra DNA often consists of sizeable chunks of non-functional or 'neutral' DNA in which are embedded relatively small segments of DNA that are subject to strong positive natural selection. If natural selection favoured an increase in the number of copies of the embedded segment, then, if the molecular mechanism that generated extra copies did so by multiplying the whole chunk, the neutral DNA would also increase. Unless the rate of deletion of neutral DNA were comparable to its rate of generation it would accumulate as a useless 'hitch-hiker'. The selectable segment could either be part of a transposon and be selected intragenomically or else be selected by ordinary organismic selection.

This hitch-hiking effect would lead to a significant overall increase in genome size only if there was a corresponding increase in gene number; but the evidence reviewed in Chapter 3 indicates that most categories of gene, including protein-coding genes, do not increase in concert with genome size and therefore could not cause a corresponding increase in junk DNA. However, one category of gene could in principle do so: that is replicon origins which are interspersed throughout the genome. The evidence reviewed in Chapters 3 and 7 shows that their number must vary in direct proportion to genome size; this is consistent with, but does not prove, my hypothesis (Cavalier-Smith, 1978) that increase in replicon number is the major mode of evolutionary increase in genome size.

It is important to establish whether the direct proportionality between replicon numbers, genome size and cell size is purely the result of an evolutionary equilibrium or whether there is a more direct causal connection between them. In the case of bacteria there is a direct relationship between the number of copies of the unique replicon origin and cell volume; the mode of cell cycle regulation appears to be such that growth in cell volume to a critical level triggers the initiation of DNA replication and therefore the doubling of the number of replicon origins. It is possible (Cavalier-Smith, 1978) that in eukaryotes also there might be a similar direct causal relationship between cell volume and the number of replicon origins, such that the number of origins per unit volume remained constant in evolution and such that evolutionary changes in replicon number would directly cause corresponding changes in cell volume (Cavalier-Smith, 1978, 1980b,c, 1983b). If this were the case, selection for different cell volumes, for which the evidence is strong (Chapter 4), would automatically also select for corresponding changes in the number of replicon origins. Such a selective force would not lead to proportional increases in total DNA, so long as selection against extra DNA were strong, as I assumed (Cavalier-Smith, 1978) to be the case, since replicons would simply tend to get smaller as a result of deletion of redundant DNA linked to the extra replicon origins; therefore the conclusion was that selection for increased numbers of replicon origins would be an insufficient explanation for the quantitative relationship between cell volume and genome size. But this conclusion would not be valid if selection against the extra junk DNA were so weak that there had not been enough time since it originated for it to be eliminated. Under these conditions it is possible that selection for increased numbers of replicon origins could lead to a proportional increase in genome size (Cavalier-Smith, 1983b). To evaluate this possibility further it is necessary to understand the mechanism whereby the initiation of eukaryote DNA replication is controlled by cell volume. Chapter 7 therefore discusses the volume-dependent control of DNA-replication; the conclusion is that, although the mechanisms are still poorly understood, there are reasons for doubting that they necessarily entail the observed isometric relationship between eukaryote cell volume and the number of replicon origins. Therefore, if larger cells do not necessarily require more replicon origins irrespective of their genome size one cannot explain the larger amounts of non-genic DNA in large cells simply in terms of 'hitchhiking' of neutral junk DNA by virtue of its linkage to positively selected replicon origins.

The correlation between replicon numbers and cell volume can instead simply be explained as the result of (1) a necessarily isometric proportionality between genome size and cell volume (Chapter 4) and (2) evolutionary forces maintaining a rough constancy of replicon size (Chapter 7).

THE SIGNIFICANCE OF CONSTITUTIVE HETEROCHROMATIN

Evidence for 'genetically or physiologically inert' chromosomal regions first came from studies of heterochromatin (Muller and Painter, 1932), that is of chromatin that unlike typical 'euchromatin' does not normally decondense during interphase (Heitz, 1928). The frequent statement that constitutive heterochromatin is condensed in all cells is untrue. I have proposed (Cavalier-Smith, 1978, 1980b, 1982a) that the primary function of constitutive heterochromatin is to allow marked changes in nuclear volume during the life cycle: it is permanently folded up in cells that need relatively small nuclei, but unfolded for specialized phases, e.g. growing oocytes, where the nuclei need to be exceptionally large. It therefore should not exist in organisms in which nuclei do not undergo large volume changes during the life cycle, which appears to be the case for most, but *not* all, unicells. On this hypothesis, which appears consistent with the known distribution of heterochromatin, it is physiologically inert, but not genetically so, since it is important for the genetic control of the facultative increase in nuclear volume of certain cell types (Chapter 4).

Other important aspects of constitutive heterochromatin are discussed later in the book: its possible roles at centromeres and telomeres (Chapter 3), its frequently late replication (Chapter 7), and its frequent content of highly repeated DNA that probably evolves by *intragenomic drift* (Chapters 3, 8 and 12).

GENOME SIZE OF VIRUSES AND PLASMIDS

Plasmid DNA and virus chromosomes are always much smaller than cellular chromosomes. As they lack significant amounts of non-coding DNA reasonably strong selective forces must exist to prevent its accumulation. For viruses the nature of this force is fairly obvious. The closed geometry of a viral capsid cannot contain more than a limited amount of DNA. Moreover capsid size, unlike cell size, cannot increase gradually but would have to evolve in major discontinuous steps that are inherently improbable (Joklik, 1974). I have proposed (Cavalier-Smith, 1983a) that the strong selection for small genomes in viruses is so great as to lead to the evolution of single-stranded genomes and of ways of coding for more proteins without increasing their genome size, namely:

1. overlapping genes using different reading frames;
2. overlapping genes using alternative promoters in the same reading frame;
3. differential splicing of a single premessenger to produce a variety of polypeptides in adenovirus.

For plasmids the selective forces limiting genome size are less obvious. Since they are so small compared with the main chromosomes the time taken for replication might appear unimportant. However, if there are several copies per cell then intracellular competition might occur and more rapidly replicating plasmids would displace more slowly replicating ones. However, control of the frequency of replication of plasmids appears to involve only the initiation mechanism (Chapter 7). It is therefore doubtful whether increasing the length of a plasmid would affect its replication frequency and copy number. For transmissible plasmids increased length of the DNA would increase the transfer time during conjugation; if there is a reasonable chance of longer transfers being interrupted this would impose some selection against increase. However, the most important factor is probably the extra metabolic burden imposed on the cell by extra plasmid DNA which could lower its growth rate, for which there is good evidence in the case of longer plasmids

(Zund and Lebek, 1980), and so lead to the selection of cells containing shorter plasmids. This could also be a selective factor tending to keep the copy number of plasmids as low as possible. Plasmid-free cells have also been shown to be at a competitive advantage in the absence of positive selection for the plasmid-coded characters (Dale and Smith, 1979).

'SELFISH DNA' AND GENOME SIZE

Though plasmids and viruses are not usually thought of as part of the genome, many examples are known where they can become inserted into the host chromosome and inherited like normal chromosomal genes. Viral and plasmid integration is therefore a possible mode of increase in genome size additional to classical duplications of chromosomal genes. Transposable genetic elements capable of duplicative transposition seem to occur in all well-studied cellular genomes. Since, like plasmids and viral genomes, they have the potential to replicate independently of the chromosome, mutations that increase or decrease their multiplication rate relative to the rest of the chromosome will tend to alter their copy number within the genome and therefore also overall genome size; I called such differential intragenomic multiplication 'intragenomic selection' (Cavalier-Smith, 1980a). The frequency of such 'genetic symbionts' (Cavalier-Smith, 1983a) in natural populations will depend on the relative strength and direction of intragenomic and organismic selection, as well as the frequency with which they originate or become defective by mutation.

Such genetic symbionts certainly exist. But how important are they as agents causing changes in genome size compared with conventional duplications and deletions? Doolittle and Sapienza (1980) suggested that they are the major force causing genomes to increase and referred to them as 'selfish DNA'. Orgel and Crick (1980) made a similar proposal, but confusingly used 'selfish DNA' to include not only genetic symbionts but also junk DNA. I argue in Chapter 8 that the latter usage is too broad to be useful and that the use of the phrase 'selfish DNA' in two differing senses is confusing. Bachmann *et al*, in Chapter 9 point out the potential confusion between 'selfish DNA' and Dawkins' earlier phrase 'selfish genes': to Dawkins (1976) *all* DNA was 'selfish' even when useful to the organism. The concepts of 'selfish genes' and 'selfish DNA' are totally different. Such confusions are avoided by the terms genetic symbiont (Cavalier-Smith, 1983a) and intragenomic selection (Cavalier-Smith, 1980a), though even the latter is perhaps more animistic than is ideal—'differential intragenomic multiplication' would be less potentially misleading though more cumbersome.

In Chapter 8 I argue that genetic symbionts usually make up no more than a small fraction of the genome; that the balance between intragenomic selection and organismic selection lies strongly towards the latter as far as genome *size* is concerned; and that the undoubted existence of genetic symbionts and intragenomic selection does not explain the C-value paradox.

GENOME SIZE OF CHLOROPLASTS AND MITOCHONDRIA

The special cases of mitochondrial DNA and chloroplast DNA are discussed respectively by Clark-Walker (Chapter 10) and Gillham *et al.* (Chapter 11). The genome size of chloroplasts is remarkably uniform. Since it is generally accepted (Cavalier-Smith, 1982b; Gray and Doolittle, 1982) that chloroplasts evolved from symbiotic cyanophytes (= cyanobacteria, blue-green algae), having at least a 10-fold larger genome than chloroplasts, the origin of the chloroplast genome provides the most convincing example of a major evolutionary reduction in genome size; no-one would dispute that such reduction occurred in the case of *Cyanophora*'s blue-green chloroplast discussed by Herdman (Chapter 2). Some of the DNA lost by the symbiont during its conversion into an organelle was undoubtedly transferred to the nucleus but most was probably simply deleted. This cautions one against the assumption (Orgel and Crick, 1980) that it is difficult to delete non-functional DNA. Moreover, the uniformity of chloroplast genome size, in marked contrast to the roughly 80 000-fold variation in the (much larger) nuclear genome size strongly indicates that intragenomic selection does not *necessarily* lead to a vast increase in the DNA content of a genome. What force prevents such increase in chloroplast DNA genome size? One possibility is intracellular competition between different copies of the chloroplast DNA in the same cell. This could occur between the many copies in a single chloroplast as in *Chlamydomonas* or between copies in different chloroplasts in plants having many per cell. Cosmides and Tooby (1981) have discussed the possible role of organelle competition in evolution. Evaluation of this possibility requires more knowledge of the controls of plastid DNA replication.

Another possibility is selection for efficiency in use of energy and nutrients. Although chloroplast DNA contains introns (see note 1, p. 32) it probably does not have large amounts of non-coding DNA. In my view the uniformly low chloroplast genome size is a sign of strong selection against wasteful non-coding DNA.

Mitochondrial genomes are much more varied. Animal mitochondrial DNA (mt DNA) is extremely compact and uniform and altogether lacks any secondary DNA, so some strong selective force must prevent any accumulation of extra DNA. But mt DNA of both plants and fungi varies 6-fold in size, and it also shows some variation in protozoa. In fungi, at least, it seems that this variation is caused by variations in secondary DNA rather than genic DNA. Though it is not easy to see what function the extra DNA has, one can hardly maintain that it has a complete lack of function at the same time as proposing that in other plasmids intracellular competition prevents the accumulation of secondary DNA. Perhaps, however, the contrast with chloroplast DNA is more apparent than real: because of the small size of mt DNA in fungi compared with chloroplast DNA the differences in mitochondrial genome size between different species in absolute terms (e.g. number of nucleotides) is of the same order of magnitude as for chloroplasts even though in relative terms it is greater. Since any metabolic burden on the cell will depend on the absolute amount of secondary DNA involved per cell, rather than the relative amount per mitochondrial genome, it could be that the selective force against secondary DNA is fairly similar for chloroplasts and fungal mitochondria. The greater amount of secondary DNA in some species could simply be a sign that selection against it is not absolute. As Clark-Walker points out, the presence of some secondary DNA would make it easier for mutation or transposition to insert more without causing lethal mutations: once the animal ancestor lost all its secondary DNA it became resistant to further invasion of secondary DNA. Flowering plant mitochondria have genomes even larger than do chloroplasts; they seem rather prone to take up extra DNA, as shown by the presence in maize of sequences derived from chloroplast DNA (Stern and Lonsdale, 1982); they probably consist in *Brassica* of three different mt DNAs (Palmer and Shields, 1984), whereas in maize the number of different molecules varies from strain to strain (Kemble and Bedbrook, 1980).

The much larger size of mt DNA in flowering plants compared with fungi, protozoa and animals suggests that selection against secondary mt DNA is weaker in plants; could this be because the mt DNA forms a much smaller fraction of total DNA in photosynthetic plants (0.03–0.5%: Suyama and Bonner, 1966) than in heterotrophs (11.28% in yeast: Williamson, 1970) and up to 67% in animal oocytes (Dawid, 1966) and therefore that equal absolute increases in its genome size will impose a *relatively* smaller energy burden on the cell? Or does plant mt DNA simply have more genes?

Clark-Walker discusses other striking differences between the mitochondrial genomes of the different eukaryote kingdoms. Some authors interpret them in terms of the independent origin of mitochondria by symbiosis in each kingdom. However, this diversity could also have resulted from diversification *during* or *following* a single origin for mitochondria; much diversification *during* the gradual and divergent conversion of a single symbiont into a mitochondrion (Cavalier-Smith, 1983c,d,) is to be expected. Additional diversity must have arisen later.

As far as genome size is concerned, the important point is that, on either theory, the remarkably compact animal mitochondrial genome appears to be advanced rather than primitive, which supports the thesis that non-functional DNA can readily be lost given a suitable selective advantage. Furthermore, we are so ignorant of the manner of packaging and segregation of mt DNA that we really have no idea how it would be affected by different genome sizes; it is therefore premature to rule out the possibility of a structural function for the secondary DNA of some fungal and plant mt DNA. Such a role seems probable for the kinetoplast minicircular DNA of trypanosome mitochondria, which has been so remarkably conserved in the evolution of the Kinetoplastida that it is unlikely to be merely 'junk' or 'selfish' DNA.

SEX AND GENOME SIZE

It has been proposed that mitochondria and chloroplasts fail to accumulate 'selfish' secondary DNA simply because, unlike nuclei, they lack sex. But although it is true in principle, as briefly discussed by Charlesworth (Chapter 16), that sex provides a means whereby transposable elements can pass from one individual to another and thereby spread through the population in a way impossible in an asexual species, comparative evidence does not support the theories that the presence or absence of sex is ever a decisive factor in the evolution of genome size. For example, *Chlamydomonas* chloroplasts, which undergo fusion and recombination of chloroplast DNA, have no more DNA than do higher plant chloroplasts that do not undergo these two processes. In protists asexual species like *Euglena* and some amoebae can have very large nuclear genomes whereas sexual species like the malarial parasites and yeasts have very small genomes. Since the full 80 000-fold variation in eukaryote nuclear genome size can be found in asexual species the presence or absence of sex must be irrelevant to the C-value paradox. Finally, just because transposable elements in asexual species cannot spread through the population by sexual cell fusion followed by transposition, there is nothing about asexuality *per se* that prevents their originating and spreading within individual genomes and increasing even a 1000-fold. That this has not happened in *any* mitochondria or chloroplasts indicates that a positive selective force, and not merely the absence of sexuality, prevents it (see note 1, p. 32).

This emphasizes that understanding all the possible mechanisms whereby genomes can increase or decrease, as reviewed in Chapter 12 by Dyson and Sherratt, is only the first step in understanding the evolution of genome size; equally important are data on their relative frequency and on the actual strengths of selection affecting genome size. In the absence of such direct quantitative evidence comparative arguments may do much to distinguish probabilities from mere theoretical possibilities.

B-CHROMOSOMES AND CHROMATIN DIMINUTION

These two topics are important for understanding the C-value paradox because they are examples of the facultative increase and decrease, respectively, of genome size. B-chromosomes are chromosomes supernumerary to the normal set which, unlike ordinary A-chromosomes, can be present or absent in the genomes of different individuals in some plant and animal species without markedly affecting their phenotypic characteristics. They do, however, have quantitative effects on the phenotype of such characters as cell and nuclear volume, cell growth rates and sometimes fertility. Since B-chromosomes seem to consist mainly or entirely of secondary DNA this gives some support to the idea that such DNA might be positively selected because of its effect on such quantitative characters. However, Östergren (1945) suggested that B-chromosomes are really parasites maintained by intra-genomic selection. Jones (Chapter 13) reviews recent evidence, some of which gives increased support to Östergren's proposal.

Chromatin diminution and chromosome elimination, reviewed by Ammerman in Chapter 14, occurs only in certain, mostly minute, invert-ebrate animals and certainly heterokaryote protozoa (= Ciliophora); many chromosome segments or whole chromosomes are totally eliminated from their somatic nuclei but remain in the germ line. Since this eliminated secondary DNA, which may be over 90% of the genome, clearly contains no genes essential for ordinary bodily structure and function, why should it be retained in the germ line, and why is it eliminated in the somatic cells?

When I first realized the key importance of this phenomenon for under-standing the C-value paradox (Cavalier-Smith, 1978) it seemed obvious that the secondary DNA must be positively beneficial to the animal in the germ line and positively harmful in the somatic cells. Yet it is at least conceivable (Orgel *et al.*, 1980) that all this DNA is a benign parasite that allows itself to be removed (or even removes itself) from the somatic cells, lest it harm the host, but prevents the host from eliminating it in the germ line. Though the idea is far fetched, the evolutionary considerations discussed by Dawkins (1982) concerning 'evolutionary arms races' and gene 'outlaws' prevent me from dismissing it without careful consideration. I return to this question in Chapter 4, where I conclude that the eliminated DNA is probably not selfish DNA, but is maintained by classical individual selection because of its nucleo-skeletal role in the germ line; on the other hand, B-chromosomes probably often are 'selfish chromosomes' of no use to the organism carrying them.

GENETIC APPROACHES TO THE C-VALUE PARADOX

Another important approach to the C-value paradox is the study of genome size variation by conventional genetics. Since genome size is usually constant within a species this requires interspecific crosses, which are seldom fertile. The few studies that have been made are reviewed in Chapter 9 by Bachmann *et al*. However, none of these studied the cell volume in the F_2 populations which showed marked segregation for genome size. This is a pity since the nucleotypic and replicon origin hypotheses of the control of cell volume would predict a correlation between cell volume and genome size in the segregants, whereas if genome size and cell volume were merely correlated evolutionarily and not by a direct genetic mechanism they should segregate independently. Though accurate measurements of cell volume can be tedious it might be possible to obtain sufficiently precise data from electronic cell volume determination of pollen grains if the experiments were repeated. Though Bachmann *et al.* assume that genome size itself directly nucleotypically controls certain quantitative aspects of the phenotype, there is no direct evidence for this; in Chapter 4 I discuss some evidence against it.

EVOLUTION OF DNA SEQUENCES AND CHROMOSOME NUMBERS ARE SEPARATE PROBLEMS FROM GENOME SIZE EVOLUTION

Though, as argued in Chapter 8, it is unlikely that intragenomic selection is the decisive factor explaining.the evolution of genome size, evidence is growing that transposable elements may play a major role in the evolution of certain eukaryote genome *sequences*, notably the middle repetitive sequences whose evolution is discussed by Bouchard (1982) and by Doolittle in Chapter 15.

Just as the evolution of DNA sequences is a separate problem from the evolution of genome size, so is the evolution of chromosome numbers. This is because eukaryotes having the same numbers of chromosomes often have very different genome sizes (Rees and Jones, 1972) and eukaryotes with the same genome size often have very different chromosome numbers (Bennett and Smith, 1976). An extensive survey in flowering plants shows no significant correlation between chromosome numbers and genome size (Levin and Funderburg, 1979). This raises the question as to what evolutionary forces do determine chromosome numbers. Are they selectively neutral, or subject to selection because of their influence on overall levels of recombination (Darlington, 1937) or selectively constrained for other reasons? Charlesworth discusses this problem in Chapter 16. As he makes clear, there has been much loose discussion of the presumed long-term evolutionary value of recombination and variation. It has been suggested that organisms with large genomes show greater variability (Pierce and Mitton, 1980), and that large genomes evolve for this reason, but the factual basis for this proposal does not stand close investigation (Larson, 1981), and the rationale for it was poor.

For a given genome size variations in chromosome number change the size of the chromosome; therefore if selection on chromosome size, because of a need to fit the dimensions of the mitotic spindle (Darlington, 1937) or to pack into the nucleus (Chapter 4), was more powerful than selection on numbers, then chromosome numbers might evolve incidentally as a result of selection for particular genome sizes and for particular chromosome sizes. Though there are bound to be some constraints on chromosome size (see Chapter 4) they are probably less than for genome size, which would allow a major role for random drift in the evolution of chromosome numbers; they might depend mainly on the mutational processes that increase or decrease the number of centromeres per genome (Darlington, 1939).

GENOME SIZE AND QUALITATIVE EVOLUTION

It is often said that there has been a progressive increase in genome size during the course of evolution (Kimura, 1961), but how far is this true?

Viral versus cellular genome size

The very small size of viral genomes compared with those of cellular organisms is commonly cited in this connection. But viruses can hardly be more primitive than cells because they are obligate parasites on cells that are unable to make their own proteins. They probably originated by mutation and intragenomic selection from autonomous fragments of cellular DNA (or RNA) having (or acquiring) their own replicative controls (Cavalier-Smith, 1983a). Moreover, in contrast to cellular organisms which form a monophyletic unit (Cavalier-Smith, 1981, 1983c) viruses probably evolved polyphyletically on several different occasions from cellular nucleic acids (Joklik, 1974; Cavalier-Smith, 1983a,c). They appear to be a form of life that has secondarily become partially independent of the cell from which they originated and which may subsequently secondarily invade other species, especially if suitable vectors are available. If they did indeed originate in this way then the smaller size of viral genomes compared with cellular ones is an example not of a progressive increase in genome size but of decrease and simplification. Several features of some viral genomes such as overlapping genes and single-strandedness are indicative of intense selection on them to reduce genome size (Cavalier-Smith, 1983a).

Genome size during precellular evolution: from RNA to DNA

Many authors now accept Haldane's (1964) suggestion that genes originally consisted of RNA rather than DNA, and Eigen and Schuster (1979) and Schuster (1983) have cogently argued that self-replicating tRNA molecules must have been among the first genes. It is usual now to identify the origin of life with the origin of RNA replication and of the genetic code. I have suggested (Cavalier-Smith, 1980d) that the simplest assemblage of molecules capable of both replication and coding and therefore of evolution into more complex organisms must have had at least three genes: one for an enzyme that combined primitive RNA replicase, aminoacyl tRNA synthetase and peptidyl transferase activities, and two tRNA genes, one with a specificity for hydrophilic and one for hydrophobic amino acids. Eigen and Schuster (1979) pointed out that the absence of repair mechanisms for RNA severely limits the genome size that is possible with RNA as the genetic material. Since the number of different proteins that would be needed even for the simplest conceivable cell would probably be considerably in excess of this limit, I propose that DNA replication must have evolved well before the first cell.

Evolution of DNA replication and DNA repair would have had three major selective advantages:

1. greater genomic stability, and the consequent evolution of more precise and efficient macromolecular mechanisms;
2. greater number of different useful RNA and protein molecules;
3. linking all the genes into a single chromosome would ensure that all necessary genes were inherited together, and that essential genes were not readily lost.

But these properties would be of real advantage only if the gene products, as well as the genes, were linked together so as to form a discrete organism capable of evolving as a unit distinct from its competitors in the prebiotic soup. I therefore suggest that DNA replication evolved only after, or in conjunction with, a symbiosis between genes and membranes.

Blobel (1980) postulated that, before cells evolved, genes and ribosomes became attached to the *outside* of lipid vesicles and that cotranslational insertion of amphipathic proteins by means of hydrophobic 'signal' sequences evolved at this early stage of evolution. He does not explain why genes, ribosomes and membranes should become associated in this way; I suggest that this was because the membranes had evolved a capacity to synthesize ATP and other ribonucleoside triphosphates (NTPs) by evolving a light-driven proton pump that pumped protons into the lumen of the vesicle and a proton-driven NTP synthetase that made NTPs on the outside of the vesicle* where they could immediately be used for the replication of the genes (still RNA at this stage), and to provide ATP and GTP to power protein synthesis by the membrane-bound ribosomes (Fig. 1.2). The pigment that trapped the light may either have been a carotenoid, as in the light-driven proton pump of *Halobacterium* (Stoeckenius and Bogomolni, 1982), or more likely a crude mixture of carotenoids and bacteriochlorophylls, as in the green bacterium *Chlorobium*. (*See note 2, p. 32.)

Such an 'inside-out-cell' would have a copious supply of energy and there-fore be at a tremendous advantage compared with genes and ribosomes in

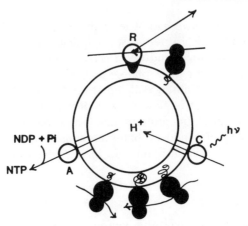

Fig. 1.2. Possible nature of the first symbiosis between RNA genes and energy-trapping membranes to form an 'inside-out-cell'.

Ribosomes are attached to the outside of a vesicular lipid bilayer and synthesize proteins that are inserted directly into the membrane; such integral membrane proteins include four key proteins: an RNA replicase (R); ribosome receptor proteins, such as ribophorins, for the attachment of the ribosomes to the membrane; the light-driven proton pump with associated pigments (possibly arranged in a primitive chlorosome (C) comparable to that of the Chlorobiaceae) that pumps protons into the vesicle; the proton driven NTP synthetase (A) that regenerates NTPs in the medium to drive protein and RNA synthesis. The genes replicated by the replicase would be tRNAs which would form an extracellular pool, ribosomal RNA genes, mRNAs for the four types of membrane proteins and mRNAs for the ribosomal proteins. The latter would be soluble proteins that could begin to attach to the rRNA while it and they were still being synthesized and thus help to attach the ribosomes indirectly to the vesicle. At this stage there would have been no RNA processing and only a single rRNA molecule embracing all the functions of 23s, 16s and 5s rRNA. The inside-out-cell would grow by random fusions with other vesicles and by incorporation of individual lipids from the prebiotic soup and proteins made by its ribosomes; division would be by random breakage of larger vesicles into smaller ones.

free solution unassociated with energy-transducing membranes. It would be able to grow by the spontaneous insertion into it of extra lipid molecules from the environment. Since such growth would tend to make it increasingly unstable its chances of being broken into two or more pieces by shearing forces in the liquid environment would increase, so an equilibrium would be set up between spontaneous growth and division. Spontaneous fusions would also inevitably occur which could bring together a variety of genes and their products, some of which might be far more efficient than earlier combinations. In this way the number of different genes and gene products associated with a particular vesicle (i.e. 'genome' size) could rapidly increase. However, random fission and fusion of inside-out cells would also have two major disadvantages: random fission, coupled with random segregation of the unlinked RNA molecules would lead to many daughter inside-out-cells lacking essential genes and therefore being less viable; random fusion could also lead to disruptive and less viable combinations. Further increases in complexity and efficiency would therefore only be possible if segregation became less random and fusion rarer.

Evolution of DNA replication and a single linkage group would allow non-random segregation only if it were combined with a regular segregation mechanism able to ensure that each daughter inside-out-cell got one DNA molecule. I suggest that this was the original function of the peptidoglycan murein, and that murein evolved simultaneously with DNA replication in an inside-out-cell. Secretion of murein into the interior of the vesicle and cross-linking it to the membrane would cause the latter to flatten so as to resemble a rough endoplasmic reticulum (RER) cisterna; since the murein is covalently cross-linked it would prevent spontaneous fission at random points; fission could occur only at places weakened by murein hydrolases; simply locating these enzymes and the murein synthesis enzymes specifically between the attachment sites of daughter DNA molecules would ensure that growth and division always occurred between them and that each daughter inside-out-cell acquired one DNA molecule as in *E. coli* today (Fig. 1.3, p. 24).

Until some such segregation mechanism evolved the number of different genes must have been relatively low. The presence in murein of equal amounts of D- and L-alanine, the second commonest amino acid produced in the Miller-Urey spark-discharge experiments, as well as of D-glutamic acid, also produced in high yield in these experiments, supports my proposal that peptidoglycan evolved at a very early stage in precellular evolution, either before the prebiotic soup had all been eaten or before possible competing life forms based on D-amino acids had become extinct.

If the peptidoglycan layer did not always grow in a plane but tended to become curved, then the margins of the inside-out-cell would tend to meet and enclose a volume of medium. It would then be advantageous if the proton-driven NTP synthetase, the DNA and the ribosomes all became concentrated on the inner, rather than the outer, membrane of this structure, since the NTPs could be used directly and more efficiently with less loss to the exterior and to competitors. As soon as DNA replication and repair had evolved, and more genes became possible, active-transport proteins and nutrient-binding proteins would evolve so as to pump soluble nutrients into the interior of the inside-out-cell so as to sequester them for future use. Once an inside-out-cell evolved the capacity to concentrate all essential nutrients in this way, accidental fusion of its incurved margins to cut it off totally from the exterior would no longer lead to starvation but to an even greater efficiency in use of soluble metabolites (Fig. 1.4). Such fusion, as first proposed by Blobel (1980), provides a much more plausible origin for the

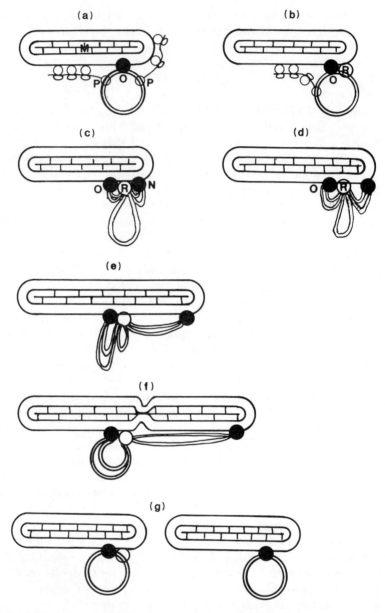

Fig. 1.3. Model for precellular DNA replication and segregation in a flattened inside-out-cell. (a) The key innovations compared with Fig. 1.2 are the evolution of separate DNA and RNA polymerases and of the peptidoglycan murein (M) which is cross-linked to both sides of the membrane vesicle thereby keeping it flat and rigid. The unreplicated circular DNA is attached directly by its origin of replication to the outside of the membrane by means of a specific integral membrane protein (O) and

first cell than the conventional assumption that it came about by the accidental trapping of DNA and ribosomes inside a lipid vesicle, with which they had not previously been associated, since such encapsulation would cut them off from their food supply and simply lead to starvation not evolutionary progress. The origin of the first cell might have led to a further increase in genome size because of the evolution of soluble enzymes, notably those of intermediary metabolism, as a result of the formation of a cytoplasm topologically separate from the environment. However, it is possible that many such enzymes had already evolved either inside the inside-out-cell or attached to its surface, and that the origin of the cytoplasm only involved a change in their cellular location from periplasm to cytoplasm, which could simply occur by the deletion of their signal sequences which might even involve a slight reduction in genome size! (See note 2, p. 32.)

This first cell would not only have a plasma membrane with attached ribosomes and DNA but also a separate outer membrane separated from it by a thin murein layer as in *E. coli* and other gram-negative bacteria (i.e. Gracilicutes); Blobel's suggestion, which is topologically similar to my earlier autogenous proposals for the origin of the double-layered mitochondrial envelope (Cavalier-Smith, 1975, 1980c), provides the only satisfactory explanation known to me for the origin of the outer membrane of gram-negative bacteria. As I see no plausible way (see note 3, p. 32) of adding such a membrane to prokaryotes that lack it (i.e. gram-positive bacteria (Firmacutes, mycoplasmas (Mollicutes) and archaebacteria (Mendocutes)), I cannot

---

also indirectly by the RNA polymerases (P), mRNA and ribosomes. (b) Replication starts by the synthesis of membrane-bound replisome proteins (R). (c) They replicate the origin sequences, one sister copy of which remains attached to the old origin attachment protein (O) while the other becomes attached to a newly made origin attachment protein (N) located on the other side of the replisome. (d) Murein growth occurs specifically at the end of the flattened inside-out-cell to which the new origin protein is attached, thus pushing it and the membrane to which it is attached away from the old origin protein (O) and the replisome. (e) The replisome continues to replicate the DNA bidirectionally as the murein and membrane grow. (f) The arrival of both chromosome termini at the replisome and their complete replication triggers division. Division is mediated by membrane-bound murein hydrolases specifically located adjacent to the attachment site of the replisome, which cleave the murein, thus allowing the two lipid bilayers to adhere to each other at the future site of division.

If the origin attachment protein were attached to both sides of the membrane throughout (not shown), rearrangements of the lipid bilayers to produce two separate inside-out-cells (g) might be relatively simple, perhaps induced simply by Brownian motion or convection currents causing a shearing force between the two halves. The evolution of some such replication and regular segregation mechanism would be a major step forward and have allowed a marked increase in genome size compared with that of the randomly fragmenting inside-out-cell with only RNA genes shown in Fig. 1.2.

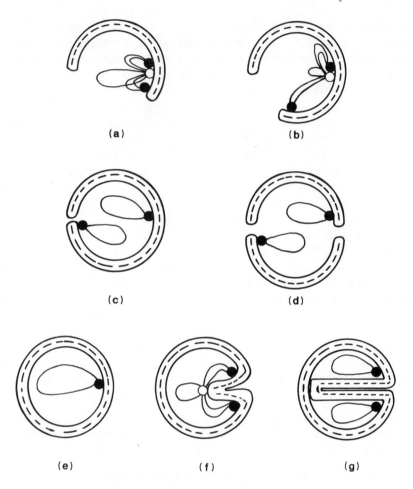

Fig. 1.4.   Origin of the first true cell division cycle from a curved inside-out-cell.

Curvature of the growing inside-out-cell would be advantageous if the DNA was on the inside of the curve since NTPs would have more chance of being used for replication before diffusing away (a). Growth of such a cell (b) and division as in Fig. 1.3 would produce two curved daughters. Still greater curvature would produce an almost closed cell (c) but division by the same mechanism would still produce two quite open half inside-out-cells (d). Formation of a true cell depends on three key things: (1) the complete fusion of the lips of the curved inside-out-cell (e); (2) a new transport mechanism across the now distinct inner (plasma) and outer membranes; and (3) a new cell division mechanism (f, g) by septum formation: septum formation could occur by a change in the growth direction of the murein growth rate adjacent to the old origin attachment site. Since a true cell cannot exist and evolve unless it can divide in a controlled manner, and since such a mechanism is crucial to the origin of the first cell, it is unlikely that it could evolve until after the evolution of DNA replication and a regular precellular segregation mechanism such as that of Fig. 1.3 had allowed a considerable increase in genome size.

accept the idea (Stackebrandt and Woese, 1981) that archaebacteria and eubacteria acquired their prokaryotic nature independently; the concept of the progenote (Woese and Fox, 1977) on which it is based, is too vague to be scientifically criticized. Bioenergetic considerations (Dickerson, 1980) suggest that the most primitive gram-negative bacteria, and therefore in my view the most primitive existing cells, are the anaerobic green photosynthetic bacteria (Chlorobiaceae) (but see Broda, 1978, for an alternative view).

Hypertrophy of the peptidoglycan to make a thick layer would prevent the transfer of lipid and protein molecules to the outer membrane and therefore inevitably cause its loss (Cavalier-Smith, 1980c), and I suggest that gram-positive bacteria originated in just that way. Mutational loss of muramic acid in a gram-positive bacterium, leading to the loss of murein or its conversion into pseudomurein, would yield mycoplasmas (which 16S rRNA sequences show are derived from gram-positive bacteria), and archaebacteria and eukaryotes (which the sharing of tRNA splicing (Kaine *et al.*, 1983) and similarities in other molecules show to have a common ancestor that is not shared by eubacteria). In the case of mycoplasmas and archaebacteria, which retained their basic prokaryotic features, this caused no major changes in genome size (Herdman, Chapter 2), but in the line that led to eukaryotes a major genetic revolution (Cavalier-Smith, 1975, 1980c, 1981) led eventually to vast increases in genome size.

The above adumbration of possible reasons for the origin of a chromosome consisting of a single DNA molecule and a single unit of replication, and of the first prokaryote cell, provides a preliminary answer to Dawkins (1982) question as to why genes have aggregated to make organisms. Before that happened it was indeed true that 'selfish genes' (Dawkins, 1976) were the units of selection. But the thesis that this is still true of organisms (Williams, 1966; Dawkins, 1982) is based on a confusion between population genetic accounting, where it is often simpler to consider genes in isolation, and the actual mode of action of conventional organismic selection, which is by the death and reproduction of individual organisms. The gene or replicator as defined by Williams and Dawkins is a unit not of selection but of inheritance, or of evolutionary 'hitchhiking' of neutral or quasineutral genes, and does not correspond at all with the units of function that molecular geneticists call by these names (Cavalier-Smith, 1985). In some instances, when discussing heterozygote advantage it is essential to consider genotypes rather than genes as units of selection. The important contrast between intragenomic selection where 'selfish' replicators do indeed compete directly with each other *within* the organism, and conventional organismic selection where it is the multiplication and death of organisms that *indirectly* changes allele frequencies, is totally obscured by thinking of 'the gene' or 'the replicator' as the universal 'unit of selection'.

The prokaryote–eukaryote transition

If prokaryotes are more primitive than eukaryotes (Cavalier-Smith, 1981) then the origin of eukaryotes marked the second major step in the evolution of genome size. It is sometimes said that prokaryotes are limited in complexity because they lack large genomes, and sometimes implied that larger genomes had to evolve before the eukaryote cell could evolve. I have argued that on the contrary it was the origin of the eukaryote cell and in particular of its new mode of replication and segregation that allowed the evolution of larger genomes (Cavalier-Smith, 1978, 1981). The greater complexity of a eukaryote cell does not necessarily require more genes than are present in some prokaryotes. Vegetative yeast cells probably have only 4000 active genes. Yet the genome size of some cyanobacteria is sufficient to contain several times this number of genes (Chapter 2). This emphasizes that it was not the lack of DNA *per se* that prevented the increased complexity of the prokaryote cell but the lack of specific genes, notably the absence of the actomyosin- and tubulin-based motility and cytoskeletal systems (Cavalier-Smith, 1975, 1980c, 1981). The origin of these, and of probably concomitant changes in the membranous organelles (Cavalier-Smith, 1980c, 1981), provided a wholly new capacity for evolutionary complexity. Possibly fewer new genes were involved in the evolution of the basic features of the eukaryote cell than in the evolution of cyanobacterial photosynthesis. It was probably the radical nature of the total cellular transformation that made the step so inherently improbable rather than the absence of 'spare' DNA.

NON-CODING DNA AND ORIGIN OF NEW GENES

In fact the whole idea that organisms need spare non-coding DNA for progressive evolution is implausible. It is far more probable that new genes evolve by modification of old ones following gene duplication than that they arise *de novo* from non-coding DNA; this is because gene duplicates will already have promoters and terminators for RNA synthesis, and start and stop signals for protein synthesis. The evidence for such a process of gene duplication and divergence is very strong (Ohno, 1970). It is possible that every new gene evolves from a pre-existing gene—*omnis gena e gena*. If so the amount of non-coding DNA may be irrelevant to the evolutionary accumulation of informational DNA. It also seems highly probable that illegitimate recombination between duplicated genes, perhaps involving transposable elements, has been of key importance in the origin of new genes. In some cases this might involve introns (Gilbert, 1978; Blake, 1978) but there is no fundamental necessity for this. One case where extensive DNA splicing to form new genes probably did occur is in the addition of signal sequences to proteins that are inserted into or secreted across cell membranes (Blobel *et al.*, 1979).

If new genes arise mainly or entirely from pre-existing genes then it is entirely inappropriate to speak of a lack of *sufficient* DNA as imposing any limit to evolutionary progress. However, in so far as transposable elements might cause important progressive mutations, especially the origin of new genes and major alterations of old ones (rather than relatively trivial tinkering by the odd point mutation), then the existence of some non-coding DNA to provide a suitable home for transposable elements (Cavalier-Smith, 1978), might be an important facilitating factor in macroevolution. But the existence of such elements even in highly economical prokaryotic genomes makes it improbable that more than a tiny amount of secondary DNA would be 'needed' for this. There is no reason to suppose that even the largest prokaryote genomes could not be further increased by gene duplication if the products of such duplication proved useful to the cell.

However, it is also possible that new genes could arise by DNA-splicing events involving a mixture of promoter and other controlling sequences derived from old genes and segments of secondary DNA. This could provide more radically different proteins than conventional gene duplication, and large amounts of secondary DNA might favour such a process. It should be emphasized that this is not being suggested as a function of secondary DNA. Not all consequences are functions. Saying that a structure has a particular function is to assert that it has evolved, and is now still there, because it increases the number of grandchildren of the organism bearing it compared with what they would be if it were absent. If it merely *might*, perhaps millions of years later, help to make a new gene that *might* have this effect, this is not a function. Some speculation on the 'function' of split genes (e.g. Gilbert, 1978) has fallen into this anthropomorphic trap of endowing natural selection with foresight that it just does not have. The forelimbs of mammals did not evolve so that they could later evolve into bats' wings or seals' flippers. Progressive evolution is an opportunistic making use of the consequences of all earlier evolution without thought or foresight. But we have to explain the present existence of all structures, including secondary DNA, in terms of what happened in the past, not what may happen in the future. Therefore, though it may well be true that many new genes are derived from duplicated old ones via intermediate stages that are transcriptionally silent (Ohno, 1970) this cannot be regarded as a reason for the existence of non-coding DNA or of any particular genome size.

EUKARYOTE GENOME SIZE AND PROGRESSIVE EVOLUTION

Kimura (1961) regarded eukaryote evolution since the Precambrian in terms of a steady increase in genetic information which he equated with the accumulation of DNA. If eukaryotes did evolve from prokaryotes, the first eukaryote probably had a low genome size, although perhaps not as low as its prokaryote ancestor since it is probable that a distinct increase in genome size occurred as a result of duplication and transposition during the transition itself (Cavalier-Smith, 1975, 1980c, 1981). Therefore genome size must have greatly increased in those lineages whose surviving descendants have large genomes. But the magnitude of the increase has been highly variable in different lineages derived from the first eukaryote: in some there can have been little or no increase over the past 1 500 000 000 years, whereas in others the increase is anything up to 80 000-fold. This means that there is no universal inexorable increase in DNA content, or accumulation of genetic information. The differences in the amount of DNA accumulated can hardly be explained in terms of differences in its rate of formation by mutation and transposition but must reflect differences in the selective forces acting on individuals in the different lineages.

The evidence reviewed in Chapter 3, suggesting that gene numbers vary by no more than 40-fold in different eukaryotes, also shows that one must not equate accumulation of DNA and accumulation of genetic information. There has undoubtedly been a 12–40-fold accumulation of genetic information, along *certain* lineages, during eukaryote evolution. But total genome size provides no measure of this, and is probably largely irrelevant to qualitative aspects of eukaryote evolution such as increases in complexity or qualitative diversification. On the other hand, it is probably highly relevant to such quantitative features of eukaryote evolution as cell and nuclear volumes and growth rates (Chapter 4). In prokaryotes, however, there is probably a much closer relationship between genome size and the number of genes (Chapter 2), since non-informational DNA seems rare.

It is also often said that progressive evolution may be limited by lack of DNA, the implication being that organisms with small genomes have a poor evolutionary potential. In eukaryotes there is no evidence for this. The large number of species (100 000) and great success of the fungi, which have the lowest genome size of all eukaryotes, does not support it; nor does the tremendous evolutionary burgeoning of beetles (Coleoptera), butterflies (Lepidoptera), ants, bees and wasps (Hymenoptera) or flies (Diptera), which also have small genomes compared with insects like grasshoppers and cockroaches, which have many fewer species despite their large genomes; among amphibians, frogs, with small genomes, are far more diverse and successful than urodeles with uniformly larger genomes.

In fact it could more reasonably be argued, and has been (Hinegardner, 1976), that large genomes are an evolutionary burden and somehow slow down or inhibit evolution. The exceptionally large genomes of lungfish, the paedomorphic salamanders, and the relatively large genomes of old groups such as pteridophytes and grasshoppers and cockroaches might be cited to support this. However, in the case of the lungfish it is more probable that large genome size is a result rather than a cause of evolutionary stasis. Because of the correlation between cell volume and genome size one can estimate the genome size of fossil lungfish by measuring cell volumes in their bones (see Chapter 4). This indicates that ancestral lungfish had a small genome size and the bulk of the present DNA accumulated *after* evolution had slowed down. This might appear to support the idea of an inexorable increase in 'selfish' DNA in the absence of counterselection. But such a view would leave it unexplained why selection against the extra DNA was relaxed in lungfish but not in other fish. Since we are fairly ignorant of the selective forces acting on lungfish cell and genome size, it is premature to rule out the possibility that it was a response to positive selection for larger cell size. Szarski (1983) argues that larger cell size and a corresponding lower metabolic rate could be a positive selective advantage in certain organisms: lungfish are particularly sluggish, and in Chapter 4 I argue that they provide one of the best examples in animals of positive selection for large cell size.

It has been suggested (Cavalier-Smith, 1978) that the tendency for rapidly evolving species-rich groups to have small genomes is because such rapid evolution often occurs in small rapidly multiplying organisms with large populations (Simpson, 1953), and because such organisms will have relatively small cells, since those with large cells have relatively slow development; therefore, whatever the reasons for the correlation between cell volume and genome size, such organisms will have small genomes. If this, perhaps over-simple, view is correct, then large cell volume, rather than the large genome size *per se*, might well be a factor tending to restrict the colonising ability and speciation rate of certain groups.

Another suggestion that has been made is that specialized species tend to have small genomes (Hinegardner, 1976). The difficulty with this is that assessment of 'degree of specialization' is highly subjective. This view is based mainly on teleost fish, where it is argued that the least typically 'fishy'-looking fish have the smallest genomes. I should want an objective measure of 'fishyness' before becoming convinced of its truth even for teleosts. It is not clear that this fishy argument can be applied at all to other groups. What could be more specialized than lungfish and paedogenetic salamanders? Yet their genomes are larger than in any other animals. Most angiosperm plants have lower genome size than gymnosperms but are they more specialized? Are frogs as a whole more specialized than salamanders? Are birds more specialized than mammals? In fungi, protozoa and algae, species with unusually *large* cells and genomes might conventionally be regarded as specialized. But unless a specific measure of specialization applicable to all eukaryotes can be suggested, the suggestion that genome size is inversely related to specialization will remain in the realm of metaphysics rather than science.

NOTES ADDED IN PROOF:

1. Since Chapter 11 was written introns have been commonly found in protein-coding as well as in tRNA and rRNA genes of chloroplasts. Like those in fungal mitochondrial DNA they probably evolved from transposable elements that entered the organelles from the nuclei (Cavalier-Smith, T. (1985). Selfish DNA and the origin of introns. *Nature*, in press). The fact that both organelle genomes are drastically reduced compared with those of their ancestral symbionts, despite the evident capacity for selfish DNA to spread into them strongly implies that selection against useless DNA is so strong that the vast amounts of non-coding DNA in the nucleus must be being positively selected for a nuclear function absent from the cytoplasmic organelles. As the time needed for replication of organelle DNA is over an order of magnitude shorter than the cell cycle, their small genomes cannot be maintained by intracellular competition, for this would favour mutations of the replicon origin increasing the frequency of initiation of replication, not ones making the molecule shorter.
2. The most probable locale for the inside-out phase of precellular evolution was *small* aqueous droplets trapped in rock crevices (Van Holde (1980). In H. O. Halvorson and K. E. Van Holde (Eds), *The Origin of Life and Evolution*, Liss, New York pp. 31–46). Only after true cells evolved, enabling the NTPs to concentrate inside, could life permanently colonize large bodies of water like the sea.
3. Dawes (1981. In Carlile, M. J., Collins, J. F. and Moseley, B. E. B. (Eds), *Molecular and Cellular Aspects of Microbial Evolution*, Cambridge University Press, pp. 85–130) suggested a conceivable, but implausible, mechanism.

REFERENCES

Avdulov, N. P. (1931). Karyo-systematische Untersuchung der Familie Gramineen. *Bull. Appl. Bot. Genet and Plant Breed.* Suppl., **44**, 1–428 (In Russian, 73 pp. German summary).

Bennett, M. D. (1971). The duration of meiosis. *Proc. R. Soc. Lond. B*, **178**, 259–275.

Bennett, M. D. (1972). Nuclear DNA content and minimum generation time in herbaceous plants. *Proc. R. Soc. Lond. B*, **181**, 109–135.

Bennett, M. D. (1973). Nuclear characters in plants. *Brookhaven Symposia in Biology*, **25**, 344–366.

Bennett, M. D. (1976). DNA amount, latitude, and crop plant distribution. *Environ. Exp. Bot.*, **16**, 93–108.

Bennett, M. D. and Smith, J. B. (1976). Nuclear DNA amounts in angiosperms. *Phil. Trans. Roy. Soc. B*, **274**, 227–274.

Blake, C. (1978). Do genes-in-pieces imply proteins-in-pieces? *Nature*, **273**, 267.

Blobel, G. (1980). Intracellular membrane topogenesis. *Proc. Natl. Acad. Sci. USA*, **77**, 1496–1500.

Blobel, G., Walter, P., Chang, C. N., Goldman, B. M., Erickson, A. H. and Lingappa, V. R. (1979). Translocation of proteins across membranes: the signal hypothesis and beyond. *Symp. Soc. Exp. Biol.*, **33**, 9–36.

Boivin, A., Vendrely, E. and Vendrely, C. (1948). L'acide desoxyribo-nucléique du noyau cellulaire dépositaire des caractères héréditaires; arguments d'ordre analytique. *C.R. Hebd. Seanc. Acad. Sci. Paris*, **226**, 1061–1063.

Bolton, E. T., Britten, R. J., Cowie, D. B., Roberts, R. B., Szafranski, P. and Waring, M. J. (1965). 'Renaturation' of the DNA of higher organisms. *Carnegie Institute of Washington Yearbook*, Vol. 64, 316–333.

Bouchard, R. A. (1982). Moderately repetitive DNA in evolution. *Int. Rev. Cytol.*, **76**, 113–193.

Britten, R. J. and Kohne, D. E. (1968). Repeated sequences in DNA. *Science N.Y.*, **161**, 529–540.

Broda, E. (1978). *The Evolution of the Bioenergetic Processes*, revised reprint, Pergamon Press, Oxford.

Callan, H. G. (1967). The organization of genetic units in chromosomes. *J. Cell Sci.*, **2**, 1–7.

Callan, H. G. and Lloyd, L. (1960). Lampbrush chromosomes of crested newts *Triturus cristatus* (Laurenti). *Phil. Trans. R. Soc. B*, **243**, 135–219.

Cavalier-Smith, T. (1975). The origin of nuclei and of eukaryotic cells. *Nature, Lond.*, **256**, 463–8.

Cavalier-Smith, T. (1978). Nuclear volume control by nucleoskeletal DNA, selection for cell volume and cell growth rate, and the solution of the DNA C-value paradox. *J. Cell Sci.*, **34**, 247–278.

Cavalier-Smith, T. (1980a). How selfish is DNA? *Nature, Lond.*, **285**, 617–618.

Cavalier-Smith, T. (1980b). *r*- and *K*-tactics in the evolution of protist developmental systems: cell and genome size, phenotype diversifying selection, and cell cycle patterns. *BioSystems*, **12**, 43–59.

Cavalier-Smith, T. (1980c). Cell compartmentation and the origin of eukaryote membranous organelles. In W. Schwemmler and H. E. A. Schenk (Eds), *Endocytobiology: Endosymbiosis and Cell Biology*, de Gruyter, Berlin, pp. 831–916.

Cavalier-Smith, T. (1980d). The nature and origin of life. *New Humanist*, **95**, 137–141.

Cavalier-Smith, T. (1981). The origin and early evolution of the eukaryotic cell. In M. J. Carlile, J. F. Collins and B. E. B. Moseley (Eds), *Molecular and Cellular Aspects of Microbial Evolution*, Society for General Microbiology Ltd, Symposium **32**, Cambridge University Press, Cambridge, pp. 33–84.

Cavalier-Smith, T. (1982a). Skeletal DNA and the evolution of genome size. *Ann. Rev. Biophys. Bioeng.* **11**, 273–302.

Cavalier-Smith, T. (1982b). The origins of plastids. *Biol. J. Linn. Soc.*, **17**, 289–306.

Cavalier-Smith, T. (1983a). Genetic symbionts and the origin of split genes and linear chromosomes. In H. E. A. Schenk and W. Schwemmler (Eds), *Endocytobiology II: Intracellular Space as Oligogenetic Ecosystem*, de Gruyter, Berlin, pp. 29–45.

Cavalier-Smith, T. (1983b). Cell volume and genome size. In P. E. Brandham and M. D. Bennett (Eds), *Kew Chromosome Conference II*, George Allen and Unwin, London, p. 332.

Cavalier-Smith, T. (1983c). A 6-kingdom classification and a unified phylogeny. In H. E. A. Schenk and W. Schwemmler (Eds), *Endocytobiology II: Intracellular Space as Oligogenetic Ecosystem*. de Gruyter, Berlin, pp. 1027–1034.

Cavalier-Smith, T. (1983d). Endosymbiotic origin of the mitochondrial envelope. In H. E. A. Schenk and W. Schwemmler (Eds), *Endocytobiology II: Intracellular Space as Oligogenetic Ecosystem*, de Gruyter, Berlin, pp. 265–279.

Cavalier-Smith, T. (1985). Review of Dawkins' *The Extended Phenotype*. *Biol. J. Linn. Soc.*, in press.

Commoner, B. (1964). Roles of deoxyribonucleic acid in inheritance. *Nature, Lond.*, **202**, 960–968.

Cosmides, L. M. and Tooby, J. (1981). Cytoplasmic inheritance and intragenomic conflict. *J. Theor. Biol.*, **89**, 83–129.

Crick, F. H. C. (1967). The genetic code—yesterday, today, and tomorrow. *Cold Spring Harbor Symp. Quant. Biol.*, **31**, 3–9.

Dale, J. W. and Smith, J. T. (1979). The effect of a plasmid on growth and survival of *E. coli*. *Antonie van Leeuwenhoek*, **45**, 103–111.

Darlington, C. D. (1937). *Recent Advances in Cytology*, 2nd edn, Churchill, London.

Darlington, C. D. (1939). *Evolution of Genetic Systems*, Cambridge University Press, Cambridge.

Darlington, C. D. (1955). The chromosome as a physico-chemical entity. *Nature, Lond.*, **176**, 1139–1144.

Darlington, C. D. (1956). *Chromosome Botany*, Allen and Unwin, London.

Darlington, C. D. and La Cour, L. (1941). The detection of inert genes. *J. Hered.*, **4**, 114–121.

Darwin, C. (1868). *The Variation of Animals and Plants under Domestication*, vol. 2, Murray, London.

Dawid, I. B. (1966). Evidence for the mitochondrial origin of frog egg cytoplasmic DNA. *Proc. Natl. Acad. Sci. USA*, **56**, 269–276.

Dawkins, R. (1976). *The Selfish Gene*, Oxford University Press, London.

Dawkins, R. (1982). *The Extended Phenotype: the Gene as the Unit of Selection*, Freeman, Oxford and San Francisco.

Dickerson, R. E. (1980). Cytochrome c and the evolution of energy metabolism. *Scientific American*, **242** (3), 98–110.

Doolittle, W. F. and Sapienza, C. (1980). Selfish genes, the phenotype paradigm and genome evolution. *Nature, Lond.*, **284**, 617–618.

Eigen, M. and Schuster, P. (1979). *The Hypercycle*, Springer-Verlag, Berlin.

Gall, J. G. (1963). Kinetics of DNAase action on chromosomes. *Nature, Lond.*, **198**, 36–38.

Gilbert, W. (1978). Why genes in pieces? *Nature, Lond.*, **271**, 501.

Goldschmidt, R. B. (1955). *Theoretical Genetics*, University of California Press, Berkeley and Los Angeles.

Gray, M. W. and Doolittle, W. F. (1982). Has the endosymbiont hypothesis been proven? *Microbiol. Rev.*, **46**, 1–42.

Greenberg, J. R. and Perry, R. P. (1971). Hybridization properties of DNA sequences directing the synthesis of mRNA and hn RNA. *J. Cell Biol.*, **50**, 774–787.

Grime, J. P. (1983). Prediction of weed and crop response to climate based upon measurements of nuclear DNA content. In *Aspects of Applied Biology*, Volume 4. *Influence of environmental factors on herbicide performance and crop and weed biology*. The Association of Applied Biologists, National Vegetable Research Station, Wellesbourne, Warwick.

Gruzder, A. D. and Reznik, N. A. (1981). Evidence for the uninemy of eukaryote chromatids. *Chromosoma*, **82**, 1–8.

Haldane, J. B. S. (1964). Data needed for a blueprint of the first organism. In Fox, S. W. (Ed.), *The Origins of Prebiological Systems and of their Molecular Matrices*, Academic Press, New York, pp. 11–15.

Heitz, E. (1926). Der Nachweis der Chromosomen: Vergleichende Studien über ihre Zahl, Grösse und Form in Pflanzenreich I. *Zeits. f. Bot.*, **18**, 625–681.

Heitz, E. (1927). Ueber multiple und aberrante Chromosomenzahlen. *Abh. d. Naturwiss. Ver. Zu Hamburg*, **21**,

Heitz, E. (1928). Das Heterochromatin der Moose T. *Jahrb. wiss. Bot.*, **69**, 762–818.

Hertwig, O. (1884). Das Problem der Befructung und Teilung des Tierischen Eies, eine Theorie der Vererbung. *Jenaische Zeitschr.*, **8**.

Hinegardner, R. (1976). Evolution of genome size. In F. J. Ayala (Ed.), *Molecular Evolution*, Sinauer, Sunderland, pp. 179–199.

Joklik, W. K. (1974). Evolution in viruses. In M. J. Carlile and Shekel, J. J. (Eds), *Evolution in the Microbial World: 24th Symp. Soc. Gen. Microbiol.*, Cambridge University Press, Cambridge, pp. 293–320.

Kaine, B. P., Gupta, R. and Woese, C. R. (1983). Putative introns in tRNA genes of prokaryotes. *Proc. Natl. Acad. Sci. USA*, **80**, 3309–3312.

Kavenoff, R. and Zimm, B. H. (1973). Chromosome sized DNA molecules from Drosophila. *Chromosoma*, **41**, 1–28.

Kemble, R. J. and Bedbrook, J. R. (1980). Low molecular weight circular and linear DNA in mitochondria from normal and male-sterile *Zea mays* cytoplasm. *Nature*, **284**, 565–6.

Kimura, M. (1961). Natural selection as the process of accumulating genetic information in adaptive evolution. *Genetical Research*, **2**, 127–140.

Laird, C. D. (1971). Chromatid structure: relationship between DNA content and nucleotide sequence diversity. *Chromosoma*, **32**, 378–406.

Larson, A. (1981). A reevaluation of the relationship between genome size and genetic variation. *Amer. Nat.*, **118**, 119–125.

Lauer, G. D. and Klotz, L. C. (1975). Determination of the molecular weight of *S. cerevisiae* nuclear DNA. *J. Mol. Biol.*, **95**, 309–326.

Levin, D. A. and Funderburg, S. W. (1979). Genome size in angiosperms: temperate versus tropical species. *Amer. Nat.*, **114**, 784–795.

Lewin, B. (1980). *Gene Expression*, vol. 2, *Eukaryotic Chromosomes*, 2nd edn, Wiley, New York.

Martin, P. G. and Shanks, R. (1966). Does *Vicia faba* have multi-stranded chromosomes? *Nature, Lond.*, **211**, 650–651.

Maynard Smith, J. and Haigh, J. (1974). The hitch-hiking effect of a favourable gene. *Genet. Res. Camb.*, **23**, 27–35.

Miescher, F. (1871). Ueber die chemische Zusammensetzung der Eiterzellen. In Hoppe-Seyler, F. (Ed.), *Medicinisch-chemische Untersuchungen*, August Hirschwald, Berlin, pp. 441–460.

Mirsky, A. E. and Ris, H. (1951). The DNA content of animal cells and its evolutionary significance. *J. gen. Physiol.*, **34**, 451–462.

Muller, H. J. and Painter, T. S. (1932). The differentiation of the sex chromosomes of *Drosophila* into genetically active and inert regions. *Z. indukt. Abstamm. u. Vererblehre.*, **62**, 316–365.

Nägeli, C. (1884). *Mechanische-physiologische Theorie der Abstammungslehre*, München, Leipzig.

Ohno, S. (1970). *Evolution by Gene Duplication*, Springer-Verlag, Berlin.

Ohno, S. (1972). So much 'junk' DNA in our genome. In H. H. Smith (Ed.), *Evolution of Genetic Systems, Brookhaven Symposium in Biology*, vol. 23, pp. 366–370.

Orgel, L. E. and Crick, F. H. C. (1980). Selfish DNA: the ultimate parasite. *Nature, Lond.*, **284**, 604–607.

Orgel, L. E., Crick, F. H. C. and Sapienza, C. (1980). Selfish DNA. *Nature, Lond.*, **288**, 645–646.

Östergren, G. (1945). Parasitic nature of extra fragment chromosomes. *Bot. Notiser*, **2**, 157–163.

Palmer, J. D. and Shields, C. R. (1984). Tripartite structure of the *Brassica campestris* mitochondrial genome. *Nature, Lond.*, **307**, 437–440.

Petes, T. D. and Fangman, W. L. (1972). Sedimentation properties of yeast chromosomal DNA. *Proc. Natl. Acad. Sci. USA*, **69**, 1188–1191.

Petes, T. D., Byers, B. and Fangman, W. L. (1973). Size and structure of yeast chromosomal DNA. *Proc. Natl. Acad. Sci. USA*, **70**, 3072–3076.

Pierce, B. and Mitton, J. B. (1980). The relationship between genome size and genetic variation. *Amer. Nat.*, **116**, 850–861.

Rees, H. (1972). DNA in higher plants. In H. H. Smith (Ed.), *Evolution of Genetic Systems, Brookhaven Symposium in Biology*, vol. 23, Gordon and Breach, New York, pp. 394–418.

Rees, H. and Jones, R. N. (1972). The origin of the wide species variation in nuclear DNA content. *Int. Rev. Cytol.*, **32**, 53–92.

Rosbash, M., Ford, P. J. and Bishop, J. O. (1974). Analysis of the C-value paradox by molecular hybridization. *Proc. Natl. Acad. Sci. USA*, **71**, 3746–3750.

Rothfels, K., Sexsmith, E., Heimburger, M. and Krause, M. O. (1966). Chromosome size and DNA content of species of *Anemone* L. and related genera (Ranunculaceae). *Chromosoma*, **20**, 54–74.

Roux, W. (1883). *Über die Bedeutung der Kernteilungsfiguren*, Leipzig.

Sachs, J. (1882). *Vorlesungen über Pflanzen-physiologie*, Leipzig.

Schuster, P. (1983). Coevolution of proteins and nucleic acids. In H. E. A. Schenk and W. Schwemmler (Eds), *Endocytobiology II: Intracellular Space as Oligogenetic Ecosystem*, de Gruyter, Berlin and New York, pp. 3–28.

Simpson, G. G. (1953). *The Major Features of Evolution*, Columbia University Press, New York.

Stackebrandt, E. and Woese, C. R. (1981). The evolution of prokaryotes. *Symp. Soc. Gen. Microbiol.*, **32**, 1–31.

Stebbins, G. L. (1966). Chromosomal variation and evolution. *Science, N.Y.*, **152**, 1462–1469.

Stebbins, G. L. (1971). *Chromosomal Evolution in Higher Plants*, Edward Arnold, London.

Stebbins, G. L. (1976). Chromosome, DNA and plant evolution. *Evolutionary Biology*, **9**, 1–34.

Stern, D. B. and Lonsdale, D. M. (1982). Mitochondrial and chloroplast genomes of maize have a 12-kilobase DNA sequence in common. *Nature, Lond.*, **299**, 698–702.

Stoeckenius, W. and Bogomolni, R. A. (1982). Bacteriorhodopsin and related pigments of Halobacteria. *Ann. Rev. Bioch.*, **51**, 587–616.

Strasburger, E. (1893). Über die Wirkungssphäre der Kerne und die Zellengrösse. *Histol. Beitr.*, **5**, 97–124.

Suyama, Y. and Bonner, W. D. (1966). DNA from plant mitochondria. *Plant Physiol.*, **41**, 383–388.

Swift, H. (1950). The constancy of desoxyribose nucleic acid in plant nuclei. *Proc. Natl. Acad. Sci. USA*, **36**, 643–654.

Szarski, H. (1970). Changes in the amount of DNA in cell nuclei during vertebrate evolution. *Nature, Lond.*, **226**, 651.

Szarski, H. (1983). Cell size and the concept of wasteful and frugal evolutionary strategies. *J. theor. Biol.*, **105**, 201–209.

Thomas, C. A. (1971). The genetic organization of chromosomes. *Ann. Rev. Genet.*, **5**, 237–256.

Vendrely, R. and Vendrely, C. (1948). La teneur de noyau cellulaire en ADN à travers les organes, les individus et les espèces animales. *Experientia*, **4**, 434–436.

de Vries, H. (1889). *Intracelluläre Pangenesis*, Fischer, Jena.

Williams, G. C. (1966). *Adaptation and Natural Selection*, Princeton University Press, Princeton, New Jersey.

Williamson, D. H. (1970). The effect of environmental and genetic factors on the replication of mitochondrial DNA in yeast. *Symp. Soc. Exp. Biol.*, **24**, 247–276.

Woese, C. R. and Fox, G. E. (1977). The concept of cellular evolution. *J. Mol. Evol.*, **10**, 1–6.

Zünd, P. and Lebek, G. (1980). Generation time-prolonging R plasmids: correlation between increases in the generation time of *Escherichia coli* caused by R plasmids and their molecular size. *Plasmid*, **3**, 65–69.

The Evolution of Genome Size
Edited by T. Cavalier-Smith
© 1985 John Wiley & Sons Ltd

CHAPTER 2

The Evolution of Bacterial Genomes

Michael Herdman
Unité de Physiologie Microbienne, Département de Biochimie et Génétique Microbienne, Institut Pasteur, 28 rue du Docteur Roux, 75724 Paris, France

SUMMARY

Bacteria exhibit a surprising variation in genetic complexity: some obligately intracellular parasitic or symbiotic forms contain genomes of size $0.1–0.3 \times 10^9$ daltons, while the genomes of morphologically and physiologically complex strains are as large as 8×10^9 daltons. Bacterial genome sizes appear to have changed during evolution by several different processes, the most major change involving fusions of, or duplications of, small ancestral genomes. The result was an abrupt doubling in genome size, possibly the first dramatic change since the evolution of the ancestral DNA genome. This process seems to have occurred independently in different groups of bacteria which had already diverged phylogenetically. It is proposed that such increases in genetic complexity occurred (or were maintained) only after the appearance of the aerobic atmosphere, possibly as a consequence of the efficient respiratory energy metabolism which then evolved.

INTRODUCTION

The aim of this chapter is to survey the genome sizes of bacteria and to suggest the mechanisms by which their genetic information content was modified during the course of evolution.

The literature survey produced genome sizes for more than 600 organisms. There is insufficient space to list them all and therefore the data are presented as mean values for the individual species, together with an indication of the number of strains of each species examined, in Tables 2.1–2.4. The tables also indicate the references to the original publications, which are rarely cited in the text. The bacterial genera are grouped into the four divisions of the kingdom Procaryotae proposed by Gibbons and Murray (1978): the

Mollicutes, Mendocutes, Firmacutes and Gracilicutes. The generic names originally published have, in many cases, been subsequently changed; they are given here in their latest accepted form and the changes are indicated in the footnotes of the tables. Justification for these changes can be found in *Bergey's Manual of Determinative Bacteriology* (1974) or in Starr *et al.* (1982).

Genome size is defined as the size (given here as the molecular weight in daltons) of a single copy of the bacterial chromosome. The chromosome of all bacteria studied is a circular, double-stranded DNA molecule; although several chromosomes may exist in the same cell, they are all identical, and in this respect the bacteria differ from eukaryotic organisms. Total cellular DNA content (genome size multiplied by number of genome copies) is often erroneously confused with genome size.

Bacterial genome sizes are conveniently measured from the kinetics of renaturation of denatured (single-stranded) DNA, followed spectrophotometrically as a decrease in absorbance. Providing that certain parameters (ionic strength, temperature, DNA fragment size and concentration) are carefully controlled, the rate of renaturation gives a direct measure of genetic complexity; at present the most commonly employed methods are those of Wetmur and Davidson (1968), Gillis *et al.* (1970) and Gillis and DeLey (1979). Most of the data presented in this chapter were obtained from renaturation kinetic studies; occasionally, other methods which are listed in the text were employed.

GENOME SIZES OF THE MAJOR GROUPS OF BACTERIA: A CORRELATION WITH THEIR MORPHOLOGICAL AND PHYSIOLOGICAL DIVERSITY?

The Gracilicutes

Members of the division Gracilicutes (possessing a gram-negative cell wall) are divided into two classes, the Scotobacteria (with a non-phototrophic metabolism) and the Photobacteria (Gibbons and Murray, 1978).

The Scotobacteria (Table 2.1) show a wide range of genome size, from 0.4×10^9 daltons (*Chlamydia* and some parasites of ciliated eukaryotes) to 4.8×10^9 daltons (e.g. *Alcaligenes*, *Chromobacterium*). The family Enterobacteriaceae is the most thoroughly studied group, represented by the first 11 genera of Table 2.1, and well illustrates the variation of size within single genera and species. The overall range of genome size, from 2.02×10^9 to 3.98×10^9 daltons, is relatively large, but members of different species within a single genus usually show rather similar genome sizes, e.g. *Shigella*. The exception to this rule is one strain of *Klebsiella pneumoniae* with a genome of 3.70×10^9 daltons, which contrasts to 13 *K. pneumoniae* strains and

Table 2.1. Genome sizes of bacteria of the division Gracilicutes: the class Scotobacteria.

Organism	No. of strains studied	Genome size (daltons × 10⁻⁹)			References
		Range	Mean	Mean of genus	
Erwinia herbicola	1		3.05		1
Erwinia uredovora	2	3.24, 3.28	3.26		1
				3.19	
Morganella morganii	1[a]		2.02		2
Salmonella pullorum	4	2.77–2.92	2.84		1
Salmonella typhimurium	4	2.74–3.15	2.92		1, 3
				2.88	
Proteus vulgaris	1		2.09		2
Shigella sonnei	1		2.09		2
Shigella flexneri	1		2.56		4
Shigella dysenteriae	2	2.69, 2.82	2.76		4
Shigella boydii	1		2.30		4
				2.49	
Serratia marcescens	4	see text	3.98		1, 2, 5, 6
Yersinia pseudotuberculosis	1		3.75		2
Enterobacter aerogenes	1		2.6		7
Enterobacter cloacae	1		2.6		7
				2.6	
Klebsiella ozaenae	1		2.36		2
Klebsiella rubiaceurum	1		2.88		1
Klebsiella pneumoniae	13	2.2–3.0	2.56		7
	1		3.70		1
				2.64	
Citrobacter freundii	6	2.37–2.99[b]	2.65		3
Citrobacter diversus	2	2.92, 2.87[c]	2.90		3
Citrobacter amalonaticus	1[d]		2.84		3
				2.73	
Escherichia coli	26	2.20–2.97	2.50		1, 2, 7–11
Pseudomonas fluorescens	2	see text	3.26		1, 5
Pseudomonas putida	1		2.99		1, 5
Pseudomonas aeruginosa	2	see text			2, 5
Pseudomonas stutzerii	1	4.21, 4.65	4.43		2, 5
Pseudomonas oleovorans	1	see text			2, 5
Pseudomonas trifolii	1		3.77		1
Pseudomonas palleronii	3	2.5–3.1	2.8		12
Pseudomonas pseudoflava	2	3.7, 4.4	4.05		12
Pseudomonas flava	1		3.1		12
Pseudomonas facilis	1		2.8		12
Pseudomonas saccharophila	1		3.5		12
Pseudomonas rubescens	2	3.04, 3.40	3.22		6
Pseudomonas putrefaciens	20	3.04–4.22	3.37		6

Table 2.1. Continued

| Organism | No. of strains studied | Genome size (daltons × 10⁻⁹) | | | References |
		Range	Mean	Mean of genus	
Pseudomonas piscicida	1		3.94		6
Pseudomonas solanacearum	12	2.52–3.26	3.05		13
				3.29	
Xanthomonas pelargonii	1		2.96		1
Agrobacterium (3 races)	not given		3.4		14
Rhizobium trifollii	1		3.14		15
Rhizobium meliloti	1		3.31		15
Lotus sp. *rhizobia*	2	3.14, 3.54	3.34		15
				3.28	
Azotobacter agilis	2	1.75, 1.75	1.75		1
Azotobacter vinelandii	1		2.4		16
				1.94	
Alcaligenes paradoxus	2	4.7, 4.8	4.75		12
Alcaligenes eutrophus	2	4.1, 5.1	4.6		12
Alcaligenes odorans	1		2.03		1
Alcaligenes sp. AB717	1		2.75		1
				2.39, 4.68	
Vibrio metschnicovii	1		2.26		2
Gluconobacter oxydans	11	1.35–1.78	1.57		13
Acetobacter aceti	8	2.04–2.87	2.51		13
Acetobacter rancens	1		1.81		1
				2.43	
Chromobacterium	4	2.67–3.03	2.86		5
violaceum	1		4.85		2
Janthinobacterium lividum	6	2.74–3.49	3.23		17
Flavobacterium balustinum	1		2.99		19
Flavobacterium odoratum type 1ᵉ	5		2.72		18
Flavobacterium odoratum type 2	3		2.59		18
Flavobacterium odoratum type 3	2		3.57		18
Flavobacterium breve	10	2.72–3.85	3.21		19
'*Flavobacterium breve*-like'	3	2.90–3.12	3.02		19
Flavobacterium meningosepticum	3	2.50–2.87	2.74		19, 20
Flavobacterium group IIb	4	2.85–3.64	3.27		19, 20
Flavobacterium group IIf	3	1.64–1.81	1.75		19, 20
Flavobacterium group IIk	1		2.51		19
				1.75, 3.04	
Nitrosococcus sp.	1		2.05		1

Table 2.1. Continued

Organism	No. of strains studied	Genome size (daltons × 10⁻⁹)			References
		Range	Mean	Mean of genus	
Nitrosomonas sp.	1		1.40		1
'Nitrifying organism 25'	1		2.57		1
Paracoccus denitrificans	3	2.5–3.0	2.7		12
Legionella pneumophila	1		2.5		21
Myxococcus xanthus	3	see text			
Stigmatella aurantiaca	1	see text			
Hyphomicrobium sp. B-522	1		2.5		22
Caulobacter crescentus	1		2.5		23
Campylobacter jejuni biotype 1	1		2.30		24
Campylobacter jejuni biotype 2	1		2.16		24
Campylobacter coli	1		2.01		24
Campylobacter fetus	3	1.54–2.31	1.81		24
				1.98	
Bdellovibrio bacteriovorans	6	1.30–1.41	1.35		9, 10
Bdellovibrio starrii	1		1.70		9, 10
Bdellovibrio stolpii	1		1.49		9, 10
Bdellovibrio W	1		1.32		10
Bdellovibrio (host-dependent)	1		1.33		10
				1.40	
Zymomonas mobilis	22	1.14–1.35	1.25		13
Pasteurella multocida	1		1.13		2
Haemophilus influenzae	4	1.0–1.52	1.30		1, 2, 5, 8, 25
Haemophilus aegyptius	1		1.17		2
				1.28	
Neisseria crassa	1		1.73		2
Neisseria subflava[f]	1		1.45		26
Neisseria gonorrhoeae	2	0.99, 1.28	1.14		2, 26
Neisseria meningitidis	1		1.12		26
Neisseria sicca	1		1.45		26
				1.34	
Moraxella (Branhamella) catarrhalis[g]	2	1.04, 1.52	1.28		2, 26
Moraxella osloensis	1		1.45		1
				1.34	
Acinetobacter calcoaceticus[h]	3	1.44–1.75	1.64		1, 2
Rickettsia prowazekii	4	1.06–1.14	1.10		27
Rickettsia typhi	3	1.07–1.09	1.08		27

Table 2.1. Continued

Organism	No. of strains studied	Genome size (daltons × 10⁻⁹)			References
		Range	Mean	Mean of genus	
Rickettsia rickettsia	1		1.01		26
				1.08	
Coxiella burnetii	1		1.04		28
Rochalimaea quintana[i]	3	1.01–1.05	1.02		26, 27
Chlamydia trachomatis	1		0.40		26
Chlamydia psittaci	1	0.57, 0.66	0.62		26, 29
				0.54	
Mu (*Pseudocaedobacter conjugatus*)	1		0.56		30
Pi (*Pseudocaedobacter falsus*)	1		0.59		30
Lambda (*Lyticum flagellatum*)	1		0.39		31
Omikron	1		0.5		32
Xenosome	1		0.34		33

[a] As *Proteus*. [b] Sizes given as % relative to *Escherichia coli* K-12. [c] As *Levinea malonatica* and [d] *L. amalonatica*; listed here as *Citrobacter* after Brenner and Fanner, 1982. [e] Colony types 1 and 3 shared 80–100% DNA homology; type 2 was unrelated. [f] As *N. flava*. [g] As *Neisseria catarrhalis* and [h] as *Mima polymorpha, Moraxella flucidolytica* and *Achromobacter anitratum*. [i] As *Rickettsia quintana*.

References:
1. Gillis and DeLey, 1975; 2. Bak *et al.*, 1970; 3. Crosa *et al.*, 1974; 4. Brenner *et al.*, 1973; 5. Gillis and DeLey, 1979; 6. Owen *et al.*, 1978; 7. Seidler *et al.*, 1975; 8. Bak *et al.*, 1969; 9. Seidler *et al.*, 1972; 10. Torella *et al.*, 1978; 11. Brenner *et al.*, 1972; 12. Auling *et al.*, 1980; 13. DeLey *et al.*, 1981; 14. DeLey, 1974; 15. Crow *et al.*, 1981; 16. Sadoff *et al.*, 1979; 17. DeLey *et al.*, 1978; 18. Owen and Holmes, 1978; 19. Owen and Holmes, 1980; 20. Owen and Snell, 1976; 21. Brenner *et al.*, 1978; 22. Moore and Hirsch, 1973; 23. Wood *et al.*, 1976; 24. Owen and Leaper, 1981; 25. MacHattie *et al.*, 1965; 26. Kingsbury, 1969; 27. Myers and Wissman, 1980; 28. Myers *et al.*, 1980; 29. Higashi, 1975; 30. Soldo and Godoy, 1974; 31. Soldo and Godoy, 1973; 32. Schmidt and Heckmann, 1980; 33. Soldo *et al.*, 1983.

members of two other species with a mean genome size of about 2.5 × 10⁹ daltons. The variation in genome size within strains of the same genus is little greater than that found within a single species. *Escherichia coli* strains showing identical phenotypes and isolated from the same geographical area varied at least 22% (equivalent to 600 genes) in genome size (Brenner *et al.*, 1972). In calculating the mean of *Serratia marcescens*, two reported genome sizes have been ignored: Bak *et al.* (1970) described *S. marcescens* ATCC

274 and MMCA 55 as 5.56×10^9 and 5.02×10^9 daltons, respectively; Gillis and DeLey (1979) found 3.98×10^9 for ATCC 274, which fits well with the reported size of CCEB 293 (Gillis and DeLey, 1975; 1979) and NCTC 1377 (Owen *et al.*, 1978). Within the family Pseudomonadaceae, *Pseudomonas* itself is the most thoroughly studied genus of all the bacteria. The data for three species require further comment. *P. fluorescens* ATCC 13525 was reported to contain a genome of size 4.83×10^9 (Bak *et al.*, 1970) or 3.48×10^9 (Gillis and DeLey, 1979); the latter value is probably correct since it agrees closely with the size of *P. fluorescens* CCEB 488 (Gillis and DeLey, 1975, 1979). The genome size of *P. aeruginosa* is difficult to establish, having been described as 6.96×10^9 (Bak *et al.*, 1970), 4.20 (Gillis and DeLey, 1979) and, from electron microscopy and velocity sedimentation studies, 2.1×10^9 daltons (Pemberton, 1974). Similarly, *P. oleovorons* ATCC 8062 was reported to contain a genome of 4.04×10^9 (Bak *et al.*, 1970) or 3.06×10^9 daltons (Gillis and DeLey, 1979). With these uncertain strains removed (and not considered in the calculations in the rest of this chapter), the genus shows a relatively large range of genome size (2.8×10^9 to 4.1×10^9 daltons), again with wide variations in strains of the same species (e.g. *P. solanacearum*).

Many of the remaining genera of Table 2.1 (*Xanthomonas, Agrobacterium, Rhizobium, Azotobacter, Vibrio, Chromobacterium, Janthinobacterium, Nitrosococcus, Nitrosomonas, Paracoccus* and *Legionella*) contain genomes which fall within the range of size of those of the enteric bacteria and pseudomonads. *Alcaligenes* divides into two subgroups (with means of 2.39×10^9 and 4.68×10^9 daltons) as does *Flavobacterium* (1.75×10^9, 3.04×10^9 daltons), sufficient to warrant further study of the taxonomy of these genera. *Gluconobacter* strains contain unusually small genomes like *Acetobacter rancens*, significantly smaller than those of the other acetic acid bacterium, *Acetobacter aceti*. With the exception of *Legionella pneumophila* (the causative agent of Legionnaires disease) and the enteric bacteria, most of the strains listed above are non-pathogenic, free-living, aerobic or facultatively anaerobic. They exhibit a relatively simple morphology and have no complex growth requirements.

The genome sizes of the more complex gliding bacteria *Myxococcus xanthus* and *Stigmatella aurantiaca* were recently estimated from both renaturation kinetics and the quantitation of restriction enzyme fragments to lie within the range 3.1×10^9 to 3.8×10^9 (Yee and Inouye, 1981), with the most likely estimate being 5.69×10^6 base pairs (Yee and Inouye, 1982), equivalent to 3.2×10^9 daltons. However, the genome size of *M. xanthus* was previously described as 8.2×10^9 daltons from viscoelastic measurements (unpublished results quoted by Zusman *et al.*, 1978) and 4.9×10^9 daltons from chemical measurement of DNA content of the microcyst (Zusman and Rosenberg, 1968). If the recent estimates are correct, then these bacteria, with a complex developmental cycle, contain no more genetic information than many other bacteria. Similarly, the budding bacteria (*Hyphomicrobium, Caulobacter*) contain relatively small genomes.

The remaining members of the Gracilicutes are unique in containing small genomes and, with the exception of *Zymomonas mobilis* and *Acinetobacter*, are all parasites or pathogens of other organisms. *Z. mobilis* is a free-living, microaerobic to anaerobic bacterium which ferments simple sugars to ethanol and carbon dioxide; its mean genome size of 1.25×10^9 daltons perhaps defines the minimum genetic information content required for a free-living existence. *Pasteurella multocida* and the two species of *Haemophilus* are facultative anaerobic animal pathogens with complex growth requirements; their genome sizes have means of 1.13×10^9 and 1.28×10^9 daltons, rather similar to the aerobic animal pathogens *Neisseria* and *Moraxella* which also have complex growth requirements. In contrast, members of the genus *Campylobacter* (microaerobic to anaerobic animal pathogens with complex growth requirements) have comparatively large genomes, with a mean of 1.98×10^9 daltons. The *Bdellovibrio* strains are unique parasites of other bacteria; even the 'host-independent' strains have complex growth requirements and small genomes identical in size to those of the host-dependent strains. The rickettsias may be divided into two groups: *Rickettsia* and *Coxiella*, obligate intracellular parasites, and *Rochalimaea*, extracellular parasites, all of which have small (1.01×10^9 to 1.14×10^9 daltons) genomes. The two species of *Chlamydia* are metabolically deficient even in comparison with *Rickettsia*, lacking an ATP-generating system, and are obligate intracellular parasites with a reduced genome size of mean 0.54×10^9 daltons. Rather similar genomes, the smallest of all the members of the Scotobacteria, are to be found among the intracellular parasites of *Paramecium* and other ciliates: *Mu, Pi, Lambda, Omikron* and xenosomes.

The class Photobacteria is divided into two subclasses, the Anoxyphotobacteriae (which do not utilize water as the electron donor for photosynthesis, and thus do not liberate oxygen) and the Oxyphotobacteriae. It is unfortunate that information for the genome sizes of members of the former group is at present limited to *Rhodomicrobium vannielii* (Table 2.2). Within the Oxyphotobacteriae, *Prochloron* is the only existing genus of the order Prochlorales, and has a genome of size (Table 2.2) not unlike that of many other bacteria. This organism is unique among the prokaryotes in containing both chlorophylls *a* and *b*, like the chloroplasts of eukaryotes.

The cyanobacteria (order Cyanobacteriales) exhibit a wide range of morphological and physiological diversity, accompanied by a wide variation in genome size from 1.57×10^9 to 8.58×10^9 daltons (Herdman *et al.*, 1979a) (see Table 2.2). Among the unicellular forms, *Synechococcus* and *Synechocystis* both contain subgroups whose properties are so different that these genera must be divided into further genera (Herdman *et al.*, 1979b; Rippka *et al.*, 1979); together with *Gloeobacter, Gloeocapsa* and *Chamaesi-*

Table 2.2. Genome sizes of bacteria of the division Gracilicutes: the class Photobacteria.

Organism	No. of strains studied	Genome size (daltons × 10⁻⁹) Range	Mean	References
Subclass Anoxyphotobacteriae				
Rhodomicrobium vannielii	1		2.1	1
Subclass Oxyphotobacteriae				
Order Prochlorales				
Prochloron sp.	1		3.59	2
Order Cyanobacteriales				
Synechococcus sub-group 1[a]	5	2.55–4.20	3.16	3
Synechococcus sub-group 2	8	1.57–2.52	1.90	
	2	3.09, 3.11	3.10	
Synechococcus sub-group 3	5	2.01–2.44	2.27	
Synechocystis sub-group 1	4	2.12–2.50	2.30	
Synechocystis sub-group 2	8	1.79–2.35	2.15	
	1		3.50	
Gloeobacter violaceus	1		2.69	
Gloeocapsa	4	2.90–3.47	3.20	
Gloeothece	2	5.02, 5.22	5.12	
Chamaesiphon	2	3.64, 3.75	3.70	
Dermocarpa	2	3.07, 4.39		
Xenococcus	1		3.99	
Dermocarpella	1		3.33	
Myxosarcina	1		3.53	
Chroococcidiopsis sub-group 1	1		3.31	
Chroococcidiopsis sub-group 2	3	4.06–4.75	4.48	
Pleurocapsa group	2	3.13, 3.43	3.28	
Oscillatoria	6	2.50–4.38	3.62	
Pseudanabaena	7	2.14–3.77	3.05	
Spirulina	1		2.53	
LPP group	2	2.58, 2.63	2.61	
	12	3.16–4.08	3.72	
	10	4.53–5.19	4.86	
Anabaena species 1	2	3.17, 3.56	3.37	
Anabaena species 2	4	3.74–3.89	3.81	
Nodularia	1		3.34	

Table 2.2. Continued

Organism	No. of strains studied	Genome size (daltons × 10⁻⁹)		References
		Range	Mean	
Cylindrospermum	2	5.71, 6.15	5.93	
Nostoc	11	4.00–6.42	5.09	
Calothrix	7	5.07–5.46	5.30	
	3	7.75–8.58	8.16	
Scytonema	1		7.40	
Chlorogloeopsis	2	4.2, 5.24	4.72	
Fischerella	4	3.62–4.75	4.24	
Cyanelle of *Cyanophora paradoxa*	1		0.12	4

[a] The sub-groups correspond to Genera, but have not yet been named.

References:
1. Potts *et al.*, 1980; 2. Herdman, 1981; 3. Herdman *et al.*, 1979a; 4. Herdman and Stanier, 1977.

phon, these unicellular organisms contain genomes in the range 1.57×10^9 to 3.70×10^9 daltons. Certain unicellular strains (*Gloeothece* and some members of *Synechococcus* subgroup I) contain much larger genomes, and together share the ability to fix atmospheric nitrogen (Rippka *et al.*, 1979); members of the genus *Gloeothece* are among the few unicellular prokaryotes capable of doing so under strictly *aerobic* conditions and possess a genome larger than that of any other unicellular bacterium.

The baeocyte-forming cyanobacteria (*Dermocarpa, Xenococcus, Dermocarpella, Myxosarcina* and *Chroococcidiopsis*) are characterized by a complex developmental cycle involving the production of small cells (baeocytes) by rapid successive binary fission (Waterbury and Stanier, 1978). They contain relatively large genomes (3.07×10^9 to 4.75×10^9 daltons), not unlike the filamentous non-heterocystous cyanobacteria (*Oscillatoria, Pseudanabaena, Spirulina* and the 'LPP-group'). The 'LPP-group' is a provisional assemblage of strains (Rippka *et al.*, 1979) showing a wide range of both DNA base composition (Herdman *et al.*, 1979b) and genome size (Table 2.2) and must be divided into further genera when more information is available.

The peak of morphological and physiological complexity is reached by the heterocystous filamentous cyanobacteria; these all produce heterocysts, specialized cells which are the sites of nitrogen fixation under aerobic conditions and which protect nitrogenase from oxygen produced by photosystem II activity of the vegetative cells. Many strains (e.g. *Nostoc*) show a complex developmental cycle involving the production of hormogonia (filaments composed of small cells, produced by rapid division, which act as dispersal mechanisms); other strains form akinetes ('spores'), and others show complex morphology with tapering trichomes (*Calothrix*) or even branching trichomes (e.g. *Fischerella*). In general, this increase in complexity is accompanied by an increase in genome size, from 3.4×10^9 in the 'simple' *Anabaena* strains to 4.0×10^9 to 6.4×10^9 daltons in *Nostoc*, reaching 8.6×10^9 daltons in a *Calothrix* strain; the latter is the largest known genome of any prokaryote, larger even than those of the complex gram-positive actinomycete genera (e.g. *Streptomyces*). The order Cyanobacteriales could also, perhaps, claim the smallest known bacterial genome. The flagellate protozoan *Cyanophora paradoxa* contains photosynthetic organelles identical with certain unicellular cyanobacteria in every known respect except two: the loss of the lipopolysaccharide layer of the cell wall, and a genome size of 0.12×10^9 daltons (Herdman and Stanier, 1977). This 'cyanelle' may be considered analogous to the gram-negative parasites (*Mu*, *Pi*, etc., see Table 2.1) of *Paramecium* and other ciliates, and is probably an intermediate stage in the evolution of a true chloroplast from an endosymbiotic cyanobacterium (Herdman and Stanier, 1977; see also Chapter 11).

The Firmacutes

Although comparatively few members of the division Firmacutes (bacteria possessing a Gram-positive cell wall) have been studied, their known range of genome size (0.9×10^9 to 7.2×10^9 daltons) is almost as great as that of the Gracilicutes. Small genomes of size 0.5×10^9 daltons have not been reported, perhaps because organisms corresponding to *Chlamydia* do not exist. Many of the Gram-positive bacteria which possess relatively small genomes (around 1×10^9 daltons) are pathogens: *Staphylococcus aureus* and several species of *Corynebacterium* (Table 2.3); non-pathogenic *Corynebacterium* species (*C. glutamicum*, *C. (Brevibacterium) ammoniagenes*) contain larger genomes. *Lactobacillus* strains (non-pathogenic) also have small genomes, perhaps associated with their complex growth requirements. The streptococci are an extremely diverse group in terms of genome size, no differences being evident between the pathogenic and lactic (e.g. *S. cremoris*, *S. lactis*) species. *Caryophanon* is a free-living bacterium producing trichomes and flagella, and having few growth requirements (thiamine, biotin); its small genome size (0.9×10^9 to 1.2×10^9 daltons) may define the minimum required for autonomous existence, analogous to the Gram-negative *Zymomonas mobilis*. The majority of the *Mycobacterium* spp. have large genomes and, indeed, exhibit a wide range of genome size; no differences exist between the pathogens (the first 12 spp. in Table 2.3) and the saprophytes (the next 5 spp.), except for the 'host-dependent' *M. leprae* which has a small genome of 1.3×10^9 daltons. Imaeda *et al.* (1982) have discussed

Table 2.3. Genome sizes of bacteria of the division Firmacutes.

Organism	No. of strains studied	Genome size (daltons × 10⁻⁹)			References
		Range	Mean	Mean of genus	
Bacillus anthracis	1		2.78		1
Bacillus cereus	1		2.60		1
Bacillus megaterium	1		2.78		2, 3
Bacillus polymyxa	1		2.75		1
Bacillus subtilis	2	2.18, 2.39	2.29		2, 3
				2.58	
Micrococcus luteus	3[a]	2.68–2.82	2.76		1
			1.86		4
Staphylococcus aureus	2[b]	1.12, 1.43	1.28		1
Deinococcus radiodurans[c]	1		2.0		5
Streptococcus bovis	3	3.0–3.89	3.48		6
Streptococcus uberis	1		2.1		6
Streptococcus dysgalactiae	1		2.4		6
Streptococcus pneumoniae	1[d]		1.45		1
Streptococcus agalactiae	2	1.20, 3.2			1, 6
Streptococcus faecalis	2	1.47, 3.1			1, 6
Streptococcus pyogenes	1		1.27		1
Streptococcus faecium	1		5.0		6
Streptococcus cremoris	4	1.9–4.8			7, 8
Streptococcus lactis	6	3.0–4.6	4.0		7
	2	1.7, 1.75	1.73		8
Streptococcus raffinolactis	5	2.2–3.1	2.55		6, 7
				2.85	
Lactobacillus plantarum	1		1.06		9
Lactobacillus casei	1		1.28		9
				1.17	
Caryophanon latum	10	1.1–1.2	1.15		10
Caryophanon tenue	3	0.9–1.0	0.97		10
				1.11	
Corynebacterium minutissimum	1		1.4		11
Corynebacterium diptheriae	1		1.2		12
Corynebacterium renale	2		1.2		12
Corynebacterium sp.	1		1.1		13
Corynebacterium glutamicum	1		1.8		2, 3
C. (Brevibacterium) ammoniagenes	1		1.98		1
C. (Brevibacterium) liquefaciens	1		1.8		12
C. (Brevibacterium) vitarumen	1		1.2		12
				1.22, 1.86	
Mycobacterium tuberculosis	5	2.0–3.13	2.53		11, 13, 14

Table 2.3. Continued

Organism	No. of strains studied	Genome size (daltons × 10⁻⁹)			References
		Range	Mean	Mean of genus	
Mycobacterium bovis	3	3.02–3.3	3.13		11, 15
Mycobacterium kansasii	3	3.0, 4.2, 4.29			13, 14, 15
Mycobacterium marinum	3	2.5, 3.8, 4.51			13, 14, 15
Mycobacterium intracellulare	3	2.0, 3.1, 4.29			13, 14, 15
Mycobacterium scrofulaceum	2	2.4, 4.57			13, 15
Mycobacterium xenopi	1		3.29		15
Mycobacterium avium serotype II	1		3.92		15
Mycobacterium avium serotype III	4	2.1–2.6	2.4		11, 13, 15
	1		4.0		15
Mycobacterium fortuitum	3	2.3, 2.8, 4.16			11, 14, 15
Mycobacterium farcinogenes	2	4.27, 4.40	4.34		15
Mycobacterium chelonei	1		2.5		13
Mycobacterium gordonae	2	, 4.57			13, 15
Mycobacterium gastri	1		4.20		15
Mycobacterium smegmatis	7	3.0–5.55	4.43		14, 15
Mycobacterium phlei	2	3.5, 4.3	3.9		14, 15
Mycobacterium vaccae	1		2.5		13
Mycobacterium stercoides	1		3.8		14
Mycobacterium lepraemurium	1		1.8		13
Mycobacterium leprae	2		1.3		13
				3.36	
Norcardia caviae	1[e]		3.2		13
Nocardia asteroides	1		2.8		11
Nocardia corynebacteroides	1		2.2		11
				2.7	
Streptomyces coelicolor	1	7.09–7.23	7.17		16, 17
Streptomyces rimosus	1	6.33–6.77	6.55		16
				6.86	

[a] As *M. flavus*, *M. lysodeikticus* and *Sarcina lutea*. [b] One strain as *S. albus*. [c] As *Micrococcus radiodurans* (see Brooks and Murray, 1981). [d] As *Diplococcus pneumoniae*. [e] As *N. otitidus-caviarum*.

References:
1. Bak *et al.*, 1970; 2. Gillis *et al.*, 1970; 3. Gillis and DeLey, 1975; 4. Gillis and DeLey, 1979; 5. Hansen, 1978; 6. Garvie and Farrow, 1981; 7. Garvie *et al.*, 1981; 8. Jarvis and Jarvis, 1981; 9. Srirangnathan, 1974; 10. Adcock *et al.*, 1976; 11. L. Barksdale, personal communication; 12. Lanéelle *et al.*, 1980; 13. Imaeda *et al.*, 1982; 14. Bradley, 1973; 15. Baess and Mansa, 1978; 16. Benigni *et al.*, 1975; 17. Antonov *et al.*, 1978.

the possible loss of DNA in this organism. The other genera of gram-positive bacteria studied (*Bacillus*, *Micrococcus*, *Deinococcus* and *Nocardia*) have medium-sized genomes around 2–3 × 10⁹ daltons; the two spp. of *Streptomyces*, which produce complex aerial mycelia and spores, contain the largest genomes of any bacterial strains except those cyanobacteria with an analogous degree of morphological complexity, described in the previous section.

The Mollicutes

The mycoplasmas, members of the division Mollicutes (Table 2.4), are the smallest organisms capable of self-reproduction, and are unique among the bacteria in being incapable of synthesizing a cell wall. The range of genome size within each genus is so small that the separate species are not listed in Table 2.4; only the ranges and means of the genera are given. *Mycoplasma* and *Ureaplasma*, extracellular animal parasites or pathogens requiring sterols for growth, have extremely small genomes (mean values 0.48 × 10⁹ and 0.45 × 10⁹ daltons, respectively). *Acholeplasma*, also an extracellular animal pathogen (but with no sterol requirement), and the plant pathogen *Spiroplasma* have larger genomes of 1.02 × 10⁹ and 1.01 × 10⁹ daltons, respectively.

Table 2.4. Genome sizes of bacteria of the division Mollicutes.

Organism	No. of strains studied	Genome size (daltons × 10⁻⁹)		References
		Range	Mean of genus	
Mycoplasma (15 species)[a]	26	0.40–0.68	0.48	1–7
Ureaplasma (2 species)[b]	8	0.41–0.48	0.45	1, 8
Acholeplasma (3 species)[a]	6	0.95–1.11	1.02	1, 7, 9
Spiroplasma citri	2	1.01	1.01	9

[a] Together with several strains not identified to species level. [b] Two clusters separated at 40–60% DNA homology level (Christiansen *et al.*, 1981).

References:
1. Bak *et al.*, 1969; 2. Morowitz *et al.*, 1967; 3. Bode and Morowitz, 1967; 4. Seidler *et al.*, 1972; 5. Allen, 1971; 6. Ryan and Morowitz, 1969; 7. Askaa *et al.*, 1973; 8. Black *et al.*, 1972; 9. Saglio *et al.*, 1973.

The Mendocutes

The members of the final major grouping of bacteria, the archaebacteria (division Mendocutes) share many unique traits which separate them from the eubacteria, from which they are thought to have diverged at a very early stage in evolution (see, for example, Fox *et al.*, 1980). Two strains of *Thermoplasma acidophilum* (an inhabitant of burning coal refuse tips and acid hot springs) contain genomes of 0.94×10^9 to 1.0×10^9 daltons (Christiansen *et al.*, 1975); this genus, previously classified with the Mollicutes, quite clearly belongs with the archaebacteria (Woese *et al.*, 1980). Two strains of *Methanobacterium thermoautotrophicum*, sharing 46% DNA relatedness, have genomes of size 1.1×10^9 to 1.22×10^9 daltons (Mitchell *et al.*, 1979; Brandis *et al.*, 1981). Four isolates of the extremely halophilic bacteria (*Halobacterium salinarium*, *H. cutirubrum* (= *salinarium*), *Halobacterium* sp. isolate III and 'moderate halophile isolate I') all contain genomes of 4.1×10^6 nucleotide pairs, about 2.3×10^9 daltons (Moore and McCarthy, 1969).

Within the bacteria as a whole, therefore, there is a general, but not universal, trend apparent: bacteria which are obligatory parasitic or pathogenic tend to contain small genomes, while free-living forms tend to contain larger genomes whose size is often further related to the morphological (and also, therefore, physiological?) complexity of the organism. This trend is examined in detail in the remainder of this chapter.

EVOLUTIONARY CHANGES IN BACTERIAL GENOME SIZE

The bacteria in their entirety contain genomes which show a remarkable range of complexity: from 0.12×10^9 daltons (if the cyanelle of *Cyanophora paradoxa* can really be considered to be a prokaryote) or 0.3×10^9 daltons (if the parasites of the ciliates can be taken as true bacteria) or 0.4×10^9 daltons (*Chlamydia trachomatis*) up to 8.6×10^9 daltons (complex cyanobacterial strains). The genome sizes of the 605 strains of bacteria so far published are presented as a frequency distribution in Fig. 2.1. It is striking that this distribution is discontinuous, showing major peaks with modal values (daltons $\times 10^{-9}$) of about (1) 0.5, (2) 1.0–1.25, (3) 2.5–2.75, (4) a rather long tail extending up to 4.75 and (5) a few very large genomes.

It was indeed this kind of unequal distribution of genome sizes that led Wallace and Morowitz (1973) to suggest, on the basis of the known genome sizes of 96 bacterial strains, that the larger genomes evolved from primitive small genomes by successive duplications. It was noted that the size difference between the two classes of mycoplasmal genomes (0.4×10^9 to 0.6×10^9 daltons and about 1×10^9 daltons) was consistent with the idea that the larger evolved from the smaller by a duplication event involving the entire genome. The alternative hypothesis, that the small genomes arose during 'degenerative' evolution of *Mycoplasma* strains from other bacteria (containing larger genomes) was discounted, since progressive deletions of genetic material should lead to all values of genome size from 1.0×10^9

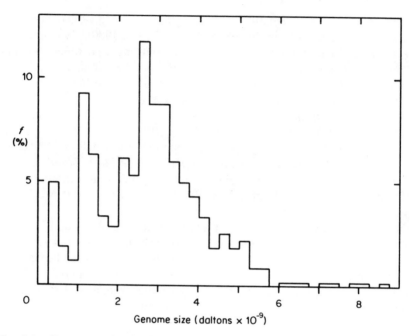

Fig. 2.1.   Frequency distribution of the genome sizes of the 605 strains of
bacteria discussed in the text.

daltons down to the minimum possible for a living cell, thereby giving a
continuous size distribution. The larger genomes of other bacteria, in the
range of 2 × 10⁹ daltons and greater, were presumed to have arisen by
further duplications of the largest mycoplasmal genomes. The newly created
DNA copies would, following mutagenic change and natural selection, create
the new phenotypes of the evolving bacteria. Zipkas and Riley (1975)
suggested that many pairs of genes whose products are functionally related
(and, therefore, possibly evolved from a common ancestral gene) lie either
at 90° or 180° apart on the chromosome map of *Escherichia coli*. It was
suggested that two sequential duplication events of an ancestral genome gave
rise to the larger *E. coli* genome, followed by mutation and divergence of
the genes which nevertheless remained sufficiently similar in function as to
be identified as related. However, Riley *et al.* (1978) re-examined this hypothesis on the basis of the recalibrated linkage map of *E. coli*: of the genes
analysed, only those coding for the enzymes of glucose metabolism showed
a significant departure from a random distribution on the genetic map: such
genes tended to lie at either 90° or 180° from one another, and most lay in
only four clusters on the genome. This distribution was consistent with the
hypothesis that the *E. coli* genome arose by two sequential doublings of an
ancestral genome one-quarter of the size (i.e. about 0.6 × 10⁹ daltons) of its

modern descendant; on the other hand, since very few genes are involved, such a distribution might easily arise by chance.

Other mechanisms can be proposed to account for changes in genome size, as thoroughly discussed by Riley and Anilionis (1978) and in Chapter 12 of this book. These include:

1. Gradual additions to a small ancestral genome, resulting from gene transfer, integration of plasmids and viruses, or duplications of small segments of the genome; the consequence of this type of change would be a normal (continuous) distribution of genome size, not a discontinuous one.
2. Gradual losses from the ancestral genome. The ancestral genome would have been large, which, as we shall see, is unlikely to be the case, and again this mechanism should create a random distribution of sizes. Such deletions may have occurred among the intracellular parasitic or symbiotic strains, removing genetic information that was no longer required in the new, protected, environment.
3. Genome duplications, accompanied by (1) and (2). This would create distinct groups of genome size, slightly deviating, perhaps, from true doublings as a result of the succeeding small changes. The data in Fig. 2.1 are consistent with this hypothesis. In addition, this combination of mechanisms would first create a pattern in which related genes lie at 90° or 180° intervals, according to the hypothesis of Zipkas and Riley (1975); subsequently, the pattern would be destroyed as additions and deletions took place (Riley and Anilionis, 1978).

Small duplications, deletions and transposition events do indeed occur in bacteria (see Riley and Anilionis, 1978; see also Chapter 12.) The occurrence of genome duplication events, involving the entire genome, has not yet been proven in any satisfactory way. In trying to assess the occurrence of this mechanism of genome evolution, it is useful to compare the genome sizes of bacteria which are known to be related. As suggested below, the data from all four divisions of the bacteria are consistent with the evolution of their members by genome duplications.

Within the division Mendocutes, although very few strains have been analysed, *Halobacterium* spp. possess genomes almost exactly double the size of those of *Methanobacterium* and *Thermoplasma*. The mycoplasmas, (division Mollicutes) show a doubling, as described above, but this is somewhat obscured by the disproportionate number of strains of *Mycoplasma* studied (Fig. 2.2(a)). Members of the Firmacutes (Fig. 2.2(b)) fall into no less than four distinct groups of mean genome size ($\times 10^9$ daltons), 1.2, 2.1, 2.9 and 4.3, together with a few larger genomes of varied sizes. The four distinct groups could have arisen as a result of fusions of an ancestral genome of size 0.7×10^9 daltons, producing larger genomes containing two ($1.4 \times$

10⁹ daltons), three (2.1 × 10⁹ daltons), four (2.8 × 10⁹ daltons) and six (4.2 × 10⁹ daltons) unit genomes; these hypothetical multiple genomes are extremely similar in size to the real ones. Within the division Gracilicutes, subclass Scotobacteria, the small genomes of the symbionts of the Ciliophora, together with *Chlamydia*, should possibly be excluded from the analysis: they cannot be considered as being equivalent to the ancestral bacterial genome since these organisms are obligate intracellular parasites and therefore must be regarded as extremely advanced in evolutionary terms. The genomes of the remaining bacteria of this subclass fall clearly into two groups (Fig. 2.2(c)) of mean sizes 1.3 × 10⁹ and 2.9 × 10⁹ daltons. These correspond remarkably to those which would be obtained by one duplication of a genome of 0.7 × 10⁹ daltons (creating a genome of 1.4 × 10⁹ daltons) and by further duplication of the latter to yield a genome of 2.8 × 10⁹ daltons. The size of the hypothetical ancestral genome is identical to that of the Gram-positive bacteria discussed above. The frequency distribution (Fig. 2.2(d)) of members of the Photobacteria, with data on *Rhodomicrobium* and *Prochloron* added, is little different to that first presented by Herdman *et al.*,

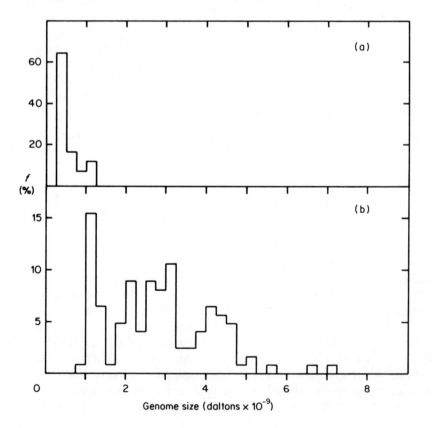

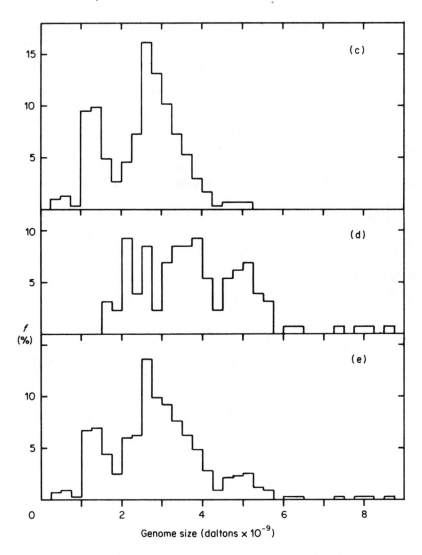

Fig. 2.2. Frequency distributions of the genome sizes of: (a) 42 members of the division Mollicutes, bacteria lacking a cell wall; (b) 123 members of the division Firmacutes, the gram-positive bacteria; (c) 305 strains of the division Gracilicutes, subclass Scotobacteria (non-photosynthetic gram-negative bacteria) and (d) 130 members of the subclass Photobacteria, the photosynthetic gram-negative bacteria. The data for all members of the Gracilicutes (435 strains) are shown in (e). The frequency distribution of genome sizes of the Archaebacteria (division Mendocutes) is not shown, because insufficient data are available (see text).

(1979a). The genome sizes fall into four distinct groups with mean values of 2.2, 3.6, 5.0 and 7.4 ($\times$ 10^9) daltons. These values correspond closely to theoretical ones involving a series of fusions of an ancestral genome of 1.2 $\times$ 10^9 daltons, creating genomes containing two (2.4 $\times$ 10^9), three (3.6 $\times$ 10^9), four (4.8 $\times$ 10^9) and six (7.2 $\times$ 10^9 daltons) copies of the original genome.

Strikingly similar trends are therefore evident in all of the major bacterial groups, except that the hypothetical ancestral genome of the cyanobacteria is already about twice the size of that of the Scotobacteria and of the Firmacutes. *Mycoplasma* and the intracellular parasitic bacteria may be special cases which evolved by loss of DNA, and will be discussed separately.

GENOME SIZE AND TRUE PHYLOGENETIC RELATIONSHIPS

The groupings of bacteria discussed above were created primarily for taxonomic convenience, and do not necessarily reflect true phylogenetic relationships. In order to further examine the mechanisms of genome evolution, it is appropriate to analyse the differences in genome size between bacteria that can be shown to be genetically related.

As a result of the pioneering work of Woese and colleagues, a reliable phylogenetic classification of the bacteria is emerging, based on the comparative analysis of the 16S rRNA oligonucleotide sequences (see Fox *et al.*, 1980; Stackebrandt and Woese, 1981 for a full discussion and further refer-

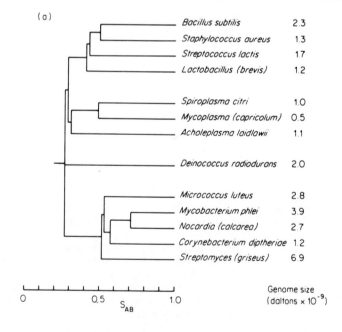

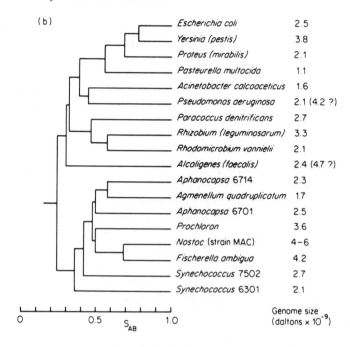

Fig. 2.3. Phylogenetic relationships (determined from 16S rRNA oligonucleotide sequence comparisons) and genome sizes of all those bacteria for which both sets of data are available: (a) the divisions Mollicutes and Firmacutes; (b) the division Gracilicutes. Members of the division Mendocutes (Archaebacteria) are not included, because insufficient data on genome size are available (see text). The inclusion of the specific name in parentheses indicates that the genome size was not determined for a member of that species, but on a different species of the genus; in this case, the mean value for the genus is given wherever possible. If two values of genome size are given, the published data do not permit an unambiguous determination (see text). According to a recent taxonomic revision of the cyanobacteria (Rippka *et al.*, 1979), *Aphanocapsa* 6701 and 6714 are members of subgroups I and II respectively of the genus *Synechocystis; Agmenellum quadruplicatum* (*Synechococcus* 7002) and *Synechococcus* 6301 are members of subgroup II of *Synechococcus*, markedly different from *Synechococcus* 7502 (subgroup I). These organisms are listed under their respective subgroups in Table 2.2.

ences). Fig. 2.3 describes the phylogenetic relationships among all those bacteria for which both 16S rRNA catalogues and genome size data are available. The phylogenetic data are taken mostly from Fox *et al.* (1980) and Stackebrandt and Woese (1981) who summarized the published data. More recent results on *Staphylococcus aureus* (Ludwig *et al.*, 1981) and *Prochloron* (Seewaldt and Stackebrandt, 1982; Stackebrandt *et al.*, 1982) have been added. The similarity coefficient, S_{AB}, describes the degree of relatedness of any pair of organisms and, assuming a constant rate of mutational change, is a measure of the evolutionary time scale. The point in evolution at which any two organisms diverged from each other can therefore be estimated. The genome sizes of the organisms are given next to the specific names; the original publications are listed in Tables 1–4. It is evident (Fig. 2.3) that within each of the major phylogenetic groups of bacteria sudden and marked changes in genome size occurred at specific times during evolution. These changes, and the organisms in which they took place, will now be described in some detail.

As discussed above, Wallace and Morowitz (1973) suggested that the genome of ancestral bacteria was similar to that of present-day *Mycoplasma* species, around 0.5×10^9 daltons in size, and genome duplication gave rise to the *Acholeplasma* type (and, as is now evident from Fig. 2.3, also to *Spiroplasma*). This hypothesis, however, implies that *Mycoplasma* spp. are descendants of a primitive organism that preceded all of the typical 'modern' bacteria. However, *Mycoplasma* may have evolved from more 'advanced' bacteria: as extracellular parasites, their specialized habitats could permit a reduction in genome size and loss of cell wall. From a comparison of their 16S rRNA oligonucleotide catalogues, Woese *et al.* (1980) concluded that the mycoplasmas (*Mycoplasma*, *Acholeplasma* and *Spiroplasma*) arose by degenerative evolution, as a deep branch of the subline of clostridial ancestry that led to the *Bacillus–Streptococcus–Lactobacillus* group (see Fig. 2.3), and did not necessarily share a common ancestor that was itself wall-less. In stronger terms, *Acholeplasma* and *Spiroplasma* may not have evolved from a *Mycoplasma*–like organism, this grouping being then an artificial one based on a rather unusual common characteristic, the absence of a cell wall. Certainly, however, the phylogenetic grouping of the mycoplasmas with this group of gram-positive bacteria is supported by a large number of other characteristics. *Streptococcus* spp. and *Lactobacillus* spp. are facultative anaerobes, oxygen-tolerant, lack cytochromes and have complex nutritional requirements, and in these respects resemble many mycoplasmas. *Acholeplasma* (but not *Mycoplasma*) strains contain a lactate dehydrogenase which is activated by fructose-1,6,-bisphosphate (Neimark and Lemcke, 1972), a control mechanism observed only in the lactic acid bacteria (the lactic streptococci and two species of *Lactobacillus*); *Acholeplasma* strains are related immunologically to the *Lactobacillus–Streptococcus* group (Neimark and London, 1982). The similarities in genome size of these organisms are thus not surprising: *Lactobacillus* 1.2×10^9 daltons, *Acholeplasma* 1.1×10^9 daltons, some *Streptococcus* strains also lying in the same range. *Staphylococcus aureus* was later shown to cluster within the same group, on the basis of 16S rRNA similarity (Ludwig *et al.*, 1981): note the genome size (1.3×10^9 daltons) of this species. The only reasonable conclusion from these data is that the *Bacillus–Staphylococcus–Streptococcus–Lactobacillus* group shared a common ancestor with the mycoplasmas. This ancestor may have been a

Mycoplasma-like organism (lacking cell wall and cytochromes, with a fermentative metabolism) with a genome size of 0.5×10^9 daltons. In this case, genome doublings gave rise to all of the other genera of the group. However, such doublings must have occurred many times, independently: once in the *Acholeplasma* line, once in the *Spiroplasma* line *after* the divergence of this genus from *Mycoplasma*, and at least once in the line leading to the gram-positive group of organisms. If, on the other hand, the ancestor had a larger genome (around 1×10^9 daltons), *Mycoplasma* must have evolved by loss of genetic information (although the ancestral strain must still have had a fermentative metabolism and lacked cytochromes). Is it too much of a coincidence that almost exactly 50% of the genome was lost, or is it more likely that a genome doubling occurred, with a *Mycoplasma*-like organism as the ancestor?

At the present time, it appears to be impossible to distinguish between the above two possibilities. However, an exciting conclusion can be drawn from this group of gram-positive bacteria. *Bacillus subtilis* (genome size 2.3×10^9 daltons), unlike its other relatives in the group, evolved a typical *aerobic* (respiratory) metabolism; the data (see Fig. 2.3) demonstrate convincingly that a genome duplication occurred after its divergence from its small-genomed relatives. Furthermore, it is possible to estimate the time in evolution at which this occurred, as described below.

It is generally accepted that the primordial atmosphere was anaerobic, becoming aerobic as the result of the activity of early prokaryotes which performed oxygenic photosynthesis—the cyanobacteria or their ancestors. The early bacteria would thus be anaerobic with a fermentative metabolism which, as shown by modern anaerobes, is relatively inefficient (2–3 moles of ATP produced by substrate-level phosphorylation per mole of glucose consumed). The clostridia and lactobacilli persist today as examples of this ancient group (see for example Hall, 1971; Horvath, 1974) in relatively stable microenvironments. The formation of an aerobic atmosphere resulted in the evolution of aerobic metabolism, producing up to 34 moles of ATP per mole of glucose; this required the evolution of haem electron carrier proteins, the cytochromes. As elegantly discussed by Fox *et al.* (1980), the phylogenetic arrangement of bacteria based on 16S rRNA oligonucleotide comparisons demonstrates clearly the change from an anaerobic, reducing environment to the modern aerobic atmosphere. The oldest groups of bacteria, for example the clostridia and relatives, are those which show the greatest range of S_{AB} values and are all basically anaerobic. All strictly aerobic groups form far shallower groupings, in other words, they evolved much later. *Bacillus* and its aerobic relatives diverged from their anaerobic ancestors quite late in bacterial evolution, showing minimal S_{AB} values of 0.50. It is reasonable to assume, therefore, that S_{AB} values of about 0.5 mark the age of the aerobic atmosphere. Thus the divergence of *Bacillus* (genome size 2.3×10^9 daltons) from *Lactobacillus* (1.2×10^9 daltons) and *Staphylococcus* (1.3×10^9 daltons) with S_{AB} values of about 0.5 indicates that a dramatic doubling of genome size occurred either at, or somewhat after, the appearance of O_2 in the atmosphere.

Aerobic respiratory metabolism appears to have arisen independently several times: all of the major phylogenetic groups of bacteria were already well separated in evolution before O_2 appeared (Fig. 2.3), yet they all contain examples of aerobic organisms. Although the available data are rather limited, a good example can be seen among the enteric, Gram negative, bacteria. The strains of *Escherichia*, *Yersinia*, *Proteus* and *Pasteurella* listed are all facultative anaerobes, but only *Pasteurella* has a fermentative metabolism. The genome size of this organism is 1.1×10^9 daltons, while its relatives contain genomes at least twice as large. The ancestor of this group therefore probably contained a small genome like that of *Pasteurella*, duplication of which gave rise to the genomes of the newly evolving genera which again diverged from *Pasteurella* near the time of formation of the aerobic atmosphere (S_{AB} values around 0.5). It would be extremely interesting to have 16S rRNA oligonucleotide sequence data for members of the genus *Haemophilus*, organisms very similar to *Pasteurella* in many phenotypic respects (Broom and Sneath, 1981), including genome size: a close relationship would be predicted. Similarly, within the *Bacillus* group of gram-positive bacteria, genome size data for members of the genus *Clostridium*, which clusters phylogenetically with this group (Fox *et al.*, 1980), are sadly lacking. In accordance with the hypothesis presented here, a genome size in the range 1.0×10^9 to 1.2×10^9 daltons is to be expected for many of the members of this (phylogenetically diverse) genus. A further example of the evolution of respiratory metabolism, together with an apparent genome duplication, is to be found among the archaebacteria. Of the very few strains examined, *Halobacterium* spp. contain genomes of 2.3×10^9 daltons, these organisms being strict aerobes with a respiratory metabolism; the strictly anaerobic *Methanobacterium* and members of the genus *Thermoplasma* (which, although obligate aerobes, have a fermentative energy metabolism) possess genomes half the size of those of *Halobacterium*.

Comparatively little phylogenetic data is available for the cyanobacteria. The strains so far examined form a closely related group (Fig. 2.3), indicating that they have diverged quite recently and that the ancestral forms no longer exist (or have not been studied). The absence of small genomes in the 128 strains examined, representative of all of the major taxonomic subdivisions (Herdman *et al.*, 1979a), is consistent with the view that as a result of their O_2-producing metabolism and photosynthetic ability, the cyanobacteria possessed an efficient energy-yielding metabolism long before any other bacteria, possibly even before the appearance of the aerobic atmosphere. Evidence for genome duplication events in this group is presented earlier in this chapter. One of the later duplication events may be evident in the limited phylogenetic relationship shown in Fig. 2.3, producing the large genomes in the cluster *Prochloron–Nostoc–Fisherella*. Associated with this event was a burst of phenotypic and metabolic diversity, creating organisms capable of producing several different differentiated cell types and able to fix atmospheric nitrogen under aerobic conditions. In addition, the cyanobacteria which contain large genomes are also characterized by their larger cells. Consistent with a demand for increased production of cell material in such organisms, Nichols *et al.* (1982) have shown that three strains with large genomes (*Anabaena cylindrica*, *Anabaena* CA and *Nostoc* MAC) contained a greater number of copies of the rRNA genes than did *Synechococcus* (*Anacystis*) PCC 6301, a unicellular organism with a smaller (2.12×10^9 daltons) genome. Such multiple copies of identical genes correlated rather well with the differences in genome size, supporting the concept of evolution by genome duplication.

SELECTIVE PRESSURES LEADING TO CHANGES IN GENOME SIZE

The organisms which persist today with a fermentative metabolism are modern representatives of those which existed in the anaerobic atmosphere. These presumably contained relatively small genomes, like those of their present-day relatives. The switch from anaerobic to respiratory metabolism, occurring separately in each of the major bacterial lines of descent, was accompanied by one or more doublings of genome size. The exceptions to this apparently general rule are the obligate parasitic or pathogenic forms, which retained small genomes irrespective of the energy metabolism (fermentative or respiratory) of the cell. Those with genomes of about 1×10^9 daltons may have retained the true ancestral type of genome, or may have evolved from bacteria with larger genomes by loss of DNA sequences. Unfortunately, the lack of phylogenetic data for these strains prevents a reasonable choice between these possibilities. Genomes of size 0.5×10^9 daltons (e.g. *Mycoplasma*, *Chlamydia*) may have evolved by further reduction, although it may be considered that the loss of exactly 50% of the genome is not in accordance with this possibility.

Irrespective of the size (0.5×10^9 or 1.0×10^9 daltons) of the ancestral genome, all of the available data are in agreement with my hypothesis that a first *verifiable* marked increase in genome size in the evolution of the aerobic bacteria (from about 1×10^9 to approximately 2×10^9 daltons) corresponds to an event which can only be interpreted as a genome duplication. This evidently occurred independently in the different groups of bacteria, but always at a rather specific time in evolution: after the appearance of the aerobic atmosphere. Further duplications of the genome may have given rise to larger types, although it would seem prudent to await further information on the phylogeny and genome size of these bacteria before regarding the latter duplication events as proven. However, the available data on the cyanobacteria would tend to support this conclusion.

Bacterial genomes, therefore, appear to have evolved in several ways which can be linked either to the habitat or metabolism of the organism: loss of DNA in many parasitic bacteria, and certainly one (and probably more) genome duplications which occurred during the evolution of bacteria possessing a respiratory metabolism. To these must be added small scale additions and deletions. The very small genomes found in intracellular 'bacteria' in ciliates (*Mu, Pi, Lambda, Omikron*, xenosomes) or in *Cyanophora paradoxa* may be considered to be a logical consequence of adaptation to the intracellular environment, where selective pressure no longer exists for the retention of genes normally required in free-living forms. It is striking that the few genes which remain in such organisms are present in multiple copies, because many copies of the small genome are present: 7 in *Omikron* (Schmidt and Heckmann, 1980), 8–9 in *Mu* and *Pi* (Soldo and Godoy, 1974) and xenosomes (Soldo *et al.*, 1983) and 60 in the photosynthetic cyanelles of *Cyanophora paradoxa* (Herdman and Stanier, 1977). A similar evolutionary process, but with genetic information content reduced to the minimum required for functional organelles, would give rise to the mitochondria and chloroplasts of the eukaryotic cell. The abrupt doubling of genome size which accompanied the divergence of organisms with an aerobic metabolism is suggested in this chapter to be the direct result of the increased efficiency of aerobic energy-yielding metabolism. The low yield of ATP from fermentative metabolism would impose severe limitations on the cell, a logical consequence of which would be that only the minimum quantity of DNA could be maintained in that cell. Efficient respiratory metabolism would remove this constraint, permitting more DNA to be maintained and replicated. For the first time in evolution, two identical genomes could exist permanently (rather than transiently following replication) in the same cell, a prerequisite for genome size doubling by fusion of two genomes. The duplicated halves of the new genome would be available to create new genetic information by mutation, thus accounting for the increased morphological, physiological and ecological diversity of the aerobic bacteria. It is clear that increased genome size is not an obligate consequence of, and certainly not a prerequisite of, aerobic metabolism, since many aerobic strains (e.g. *Caryophanon*) have nevertheless retained their small, ancestral, genomes.

A simple proof of the effect of energy availability on total cellular DNA *content* has been provided (Herdman *et al.*, 1985). Chemostat cultures of the cyanobacterium *Oscillatoria agardhii* were maintained at constant dilution (growth) rate and low light intensity; the cells contained only a low (1–2 copies of the genome) amount of DNA. Increasing the light intensity supplied to the culture, with the growth rate held constant by the fixed dilution rate of the chemostat, led to a marked increase in the cellular DNA content, consistent with the increased energy supply from photosynthesis. If such increased energy availability influences DNA content of the cell, it seems reasonable to assume that during the course of evolution it would permit the appearance (and persistence) of larger genomes which would arise from fusion of the multicopy small genomes.

CONCLUDING REMARKS

The bacteria in their entirety, in exhibiting at least a 30-fold variation in genome size, therefore contribute to the 'C-value paradox', one of the central themes of this book. As suggested above, the combination of new reliable techniques for the study of their phylogeny with improved estimates of their genome size has permitted a clarification of both the evolutionary stimuli for, and the mechanisms of, genome evolution.

It is unfortunate that data on the genome sizes of certain anaerobic bacteria which may be considered to be examples of 'primitive' organisms (e.g. *Clostridium* and the photosynthetic genera *Chlorobium* and *Chromatium*) are at present lacking. It was therefore not possible to discuss these organisms in this chapter. However, the general correlations made above would lead to the prediction that such bacteria contain relatively small genomes.

ACKNOWLEDGEMENTS

I thank Dr L. Barksdale for kindly providing copies of results prior to publication, and Dr N. G. Carr for suggesting certain improvements to the ideas outlined in this chapter.

I am indebted, for countless reasons, to the late Professor R. Y. Stanier, to whom this chapter is dedicated.

NOTE ADDED IN PROOF:

Since the completion of the literature survey (June, 1982) genome size data have been published for several additional strains, some of which are of interest because they support the hypothesis presented above. *Azotobacter chroococcum* strains CW8 and MCD-1 contain genomes of 1.18 $\times$ 10^9 and 1.29 $\times$ 10^9 daltons, respectively (Robson *et al.*, 1984. *J. Gen. Microbiol.*, **130**, 1603–1612). These strains are therefore markedly different from *A. agilis* and *A. vinelandii*, which contain genomes of 1.75 $\times$ 10^9 and 2.4 $\times$ 10^9 daltons (Table 2.1), perhaps indicating a genome duplication within a single genus. The genome size of *Desulfovibrio gigas* and *D. vulgaris* is 1.09 $\times$ 10^9 and 1.14 $\times$ 10^9 daltons, respectively (Postgate *et al.*, 1984. *J. Gen. Microbiol.*, **130**, 1597–1601). *Desulfovibrio* is a Gram negative obligate anaerobe which has probably changed little during the course of evolution: 16S rRNA cataloguing data indicates that this genus diverged from the members of the upper group of Fig. 2.3 (*E. coli* to *P. aeruginosa*) with a S$_{AB}$ value of 0.32 (Fox *et al.*, 1980) and can be considered, with the clostridia, to be one of the most ancient bacterial groups. The small genome size of *Desulfovibrio* further supports the hypothesis that the ancestral bacteria contained a genome of about 1 $\times$ 10^9 daltons.

REFERENCES

Adcock, K. A., Seidler, R. J. and Trentini, W. C. (1976). Deoxyribonucleic acid studies in the genus *Caryophanon*. *Can. J. Microbiol.*, **22**, 1320–1327.

Allen, T. C. (1971). Base composition and genome size of *Mycoplasma meleagridis* deoxyribonucleic acid. *J. Gen. Microbiol.*, **69**, 285–286.

Antonov, P. P., Ivanov, I. G., Markov, G. G. and Benigni, R. (1978). Reassociation analysis of DNA in studying the genome size of *Streptomyces*. *Studia Biophys.*, **69**, 67–74.

Askaa, G., Christiansen, C. and Ernø, H. (1973). Bovine mycoplasmas: genome size and base composition of DNA. *J. Gen. Microbiol.*, **75**, 283–286.

Auling, G., Dittbrenner, M., Maarzahl, M., Nokhal, T. and Reh, M. (1980). Deoxyribonucleic acid relationships among hydrogen-oxidizing strains of the genera *Pseudomonas*, *Alcaligenes* and *Paracoccus*. *Int. J. Syst. Bacteriol.*, **30**, 123–128.

Baess, I. and Mansa, B. (1978). Determination of genome size and base ratio on deoxyribonucleic acid from mycobacteria. *Acta Pathol. Microbiol. Scand. (Series B)*, **86**, 309–312.

Bak, A. L., Black, F. T., Christiansen, C. and Freundt, E. A. (1969). Genome size of mycoplasmal DNA. *Nature, Lond.*, **224**, 1209–1210.

Bak, A. L., Christiansen, C. and Stenderup, A. (1970). Bacterial genome sizes determined by DNA renaturation studies. *J. Gen. Microbiol.*, **64**, 377–380.

Benigni, R., Petrov, P. A. and Carere, A. (1975). Estimate of the genome size by renaturation studies in *Streptomyces*. *Appl. Microbiol.*, **30**, 324–326.

Black, F. T., Christiansen, C. and Askaa, G. (1972). Genome size and base composition of deoxyribonucleic acid from eight human T-mycoplasmas. *Int. J. Syst. Bacteriol.*, **22**, 241–242.

Bode, H. R. and Morowitz, H. J. (1967). Size and structure of the *Mycoplasma hominis* H39 chromosome. *J. Molec. Biol.*, **23**, 191–199.

Bradley, S. G. (1973). Relationships among mycobacteria and nocardiae based upon deoxyribonucleic acid reassociation. *J. Bacteriol.*, **113**, 645–651.

Brandis, A., Thauer, R. K. and Stetter, K. O. (1981). Relatedness of strains △H and Marburg of *Methanobacterium thermoautotrophicum*. *Zbl. Bakt. Mikrobiol. Hyg., I Abt. Orig. C*, **2**, 311–317.

Brenner, D. J., Fanning, G. R., Miklos, G. V. and Steigerwalt, A. G. (1973). Polynucleotide sequence relatedness among *Shigella* species. *Int. J. Syst. Bacteriol.*, **23**, 1–7.

Brenner, D. J., Fanning, G. R., Skerman, F. J. and Falkow, S. (1972). Polynucleotide sequence divergence among strains of *Escherichia coli* and closely related organisms. *J. Bacteriol.*, **109**, 953–965.

Brenner, D. J., Steigerwalt, A. G., Weaver, R. E., McDade, J. E., Feeley, J. C. and Mandel, M. (1978). Classification of the Legionnaires' disease bacterium: an interim report. *Current Microbiol.*, **1**, 71–75.

Brooks, B. W. and Murray, R. G. E. (1981). Nomenclature for '*Micrococcus radiodurans*' and other radiation-resistant cocci: *Deinococcaceae* fam. nov. and *Deinococcus* gen. nov., including five species. *Int. J. Syst. Bacteriol.*, **31**, 353–360.

Broom, A. K. and Sneath, P. H. A. (1981). Numerical taxonomy of *Haemophilus*. *J. Gen. Microbiol.*, **126**, 123–149.

Buchanan, R. E. and Gibbons, N. E. (1974). *Bergey's Manual of Determinative Bacteriology, 8th Edition*. Williams and Wilkins, Baltimore.

Christiansen, C., Black, F. T. and Freundt, E. A. (1981). Hybridization experiments with deoxyribonucleic acid from *Ureaplasma urealyticum* serovars I to VIII. *Int. J. Syst. Bacteriol.*, **31**, 259–262.

Christiansen, C., Freundt, E. A. and Black, F. T. (1975). Genome size and deoxyribonucleic acid base composition of *Thermoplasma acidophilum*. *Int. J. Syst. Bacteriol.*, **25**, 99–101.

Crosa, J. H., Steigerwalt, A. G., Fanning, G. R. and Brenner, D. J. (1974). Polynucleotide sequence divergence in the genus *Citrobacter*. *J. Gen. Microbiol.*, **83**, 271–282.

Crow, V. L., Jarvis, B. D. W. and Greenwood, R. M. (1981). Deoxyribonucleic acid homologies among acid-producing strains of *Rhizobium*. *Int. J. Syst. Bacteriol.*, **31**, 152–172.

DeLey, J. (1974). Phylogeny of Prokaryotes. *Taxon*, **23**, 291–300.

DeLey, J., Gillis, M. and De Vos, P. (1981). Range of the molecular complexities of bacterial genomes within some well established bacterial species. *Zbl. Bakt. Mikrobiol. Hyg., 1 Abt. Orig. C*, **2**, 263–270.

DeLey, J., Segers, P. and Gillis, M. (1978). Intra- and intergeneric similarities of *Chromobacterium* and *Janthinobacterium* ribosomal ribonucleic acid cistrons. *Int. J. Syst. Bacteriol.*, **28**, 154–168.

Fox, G. E., Stackebrandt, E., Hespell, R. B., Gibson, J., Maniloff, J., Dyer, T. A., Wolfe, R. S., Balch, W. E., Tanner, R. S., Magrum, L. J., Zablen, L. B., Blakemore, R., Gupta, R., Bonen, L., Lewis, B. J., Stahl, D. A., Luehrsen, K. R., Chen, K. N. and Woese, C. R. (1980). The phylogeny of Prokaryotes. *Science*, **209**, 457–463.

Garvie, E. I. and Farrow, J. A. E. (1981). Sub-divisions within the genus *Streptococcus* using deoxyribonucleic acid/ribosomal ribonucleic acid hybridization. *Zbl. Bakt. Mikrobiol. Hyg., 1 Abt. Orig. C*, **2**, 299–310.

Garvie, E. I., Farrow, J. A. E. and Phillips, B. A. (1981). A taxonomic study of some strains of streptococci which grow at 10°C but not at 45°C, including *Streptococcus lactis* and *Streptococcus cremoris*. *Zbl. Bakt. Mikrobiol. Hyg., 1 Abt. Orig. C*, **2**, 151–165.

Gibbons, N. E. and Murray, R. G. E. (1978). Proposals concerning the higher taxa of bacteria. *Int. J. Syst. Bacteriol.*, **28**, 1–6.

Gillis, M. and DeLey, J. (1975). Determination of the molecular complexity of double-stranded phage genome DNA from initial renaturation rates. The effect of DNA base composition. *J. Molec. Biol.*, **98**, 447–464.

Gillis, M. and DeLey, J. (1979). Molecular complexities of bacterial genomes determined by the initial optical renaturation rate method. *FEMS Microbiol. Letters*, **5**, 169–171.

Gillis, M., DeLey, J. and De Cleene, M. (1970). The determination of molecular weight of bacterial genome DNA from renaturation rates. *Eur. J. Biochem.*, **12**, 143–153.

Hall, J. B. (1971). Evolution of the prokaryotes. *J. Theoret. Biol.*, **30**, 429–454.

Hansen, M. T. (1978). Multiplicity of genome equivalents in the radiation-resistant bacterium *Micrococcus radiodurans*. *J. Bacteriol.*, **134**, 71–75.

Herdman, M. (1981). Deoxyribonucleic acid base composition and genome size of *Prochloron*. *Arch. Microbiol.*, **129**, 314–316.

Herdman, M., Janvier, M., Rippka, R. and Stanier, R. Y. (1979a). Genome size of cyanobacteria. *J. Gen. Microbiol.*, **111**, 73–85.

Herdman, M., Janvier, M., Waterbury, J. B., Rippka, R., Stanier, R. Y. and Mandel, M. (1979b). Deoxyribonucleic acid base composition of cyanobacteria. *J. Gen. Microbiol.*, **111**, 63–71.

Herdman, M. and Stanier, R. Y. (1977). The cyanelle: chloroplast or endosymbiotic prokaryote? *FEMS Microbiol. Letters*, **1**, 7–12.

Herdman, M., Van Liere, L. and Mur, L. R. (1985). Cellular macromolecular composition in continuous cultures of the cyanobacterium *Oscillatoria agardhii* under light energy- or inorganic nutrient limitation. *Arch. Microbiol.*, in press.

Higashi, N. (1975). Studies on chlamydia and togaviruses. *Ann. Rep. Inst. Virus Res. Kyoto Univ.*, **18**, 3–39 (cited by Becker, Y. (1978). *Microbiol. Reviews*, **42**, 274–306).

Horvath, R. S. (1974). Evolution of anaerobic energy-yielding metabolic pathways of the procaryotes. *J. Theoret. Biol.*, **47**, 361–371.

Imaeda, T., Kirchheimer, W. F. and Barksdale, L. (1982). DNA isolated from *Mycobacterium leprae*: genome size, base ratio, and homology with other related bacteria as determined by optical DNA-DNA reassociation. *J. Bacteriol.*, **150**, 414–417.

Jarvis, A. W. and Jarvis, B. D. W. (1981). Deoxyribonucleic acid homology among lactic streptococci. *Appl. Environ. Microbiol.*, **41**, 77–83.

Kingsbury, D. T. (1969). Estimate of the genome size of various microorganisms. *J. Bacteriol.*, **98**, 1400–1401.

Lanéelle, M.-A., Asselineau, J., Welby, M., Norgard, M. V., Imaeda, T., Pollice, M. C. and Barksdale, L. (1980). Biological and chemical bases for the reclassification of *Brevibacterium vitarumen* (Bechdel *et al.*) Breed (Approved lists, 1980) as *Corynebacterium vitarumen* (Bechdel *et al.*) comb. nov. and *Brevibacterium liquefaciens* Okabayashi and Masuo (Approved lists, 1980) as *Corynebacterium liquefaciens* (Okabayashi and Masuo) comb. nov. *Int. J. Syst. Bacteriol.*, **30**, 539–546.

Ludwig, W., Schleifer, K.-H., Fox, G. E., Seewaldt, E. and Stackebrandt, E. (1981). A phylogenetic analysis of staphylococci, *Peptococcus saccharolyticus* and *Micrococcus mucilaginosus*. *J. Gen. Microbiol.*, **125**, 357–366.

MacHattie, L. A., Berns, K. I. and Thomas, C. A. (1965). Electron microscopy of DNA from *Haemophilus influenzae*. *J. Molec. Biol.*, **11**, 648–649.

Mitchell, R. M., Loeblich, L. A., Klotz, L. C. and Loeblich, A. R. (1979). DNA organization of *Methanobacterium thermoautotrophicum*. *Science*, **204**, 1082–1084.

Moore, R. L. and Hirsch, P. (1973). Nuclear apparatus of *Hyphomicrobium*. *J. Bacteriol.*, **116**, 1447–1455.

Moore, R. L. and McCarthy, B. J. (1969). Base sequence homology and renaturation studies of the deoxyribonucleic acid of extremely halophilic bacteria. *J. Bacteriol.*, **99**, 255–262.

Morowitz, H. J., Bode, H. R. and Kirk, R. G. (1967). The nucleic acids of *Mycoplasma*. *Ann. N.Y. Acad. Sci.*, **143**, 110–114.

Myers, W. F., Baca, O. G. and Wisseman, C. L. (1980). Genome size of the rickettsia *Coxiella burnetii*. *J. Bacteriol.*, **144**, 460–461.

Myers, W. F. and Wisseman, C. L. (1980). Genetic relatedness among the typhus group of *Rickettsiae*. *Int. J. Syst. Bacteriol.*, **30**, 143–150.

Neimark, H. and Lemcke, R. M. (1972). Occurrence and properties of lactic dehydrogenases of fermentative mycoplasmas. *J. Bacteriol.*, **111**, 633–640.

Neimark, H. and London, J. (1982). Origin of the mycoplasmas: sterol nonrequiring mycoplasmas evolved from streptococci. *J. Bacteriol.*, **150**, 1259–1265.

Nichols, J. M., Foulds, I. J., Crouch, D. H. and Carr, N. G. (1982). The diversity of cyanobacterial genomes with respect to ribosomal RNA cistrons. *J. Gen. Microbiol.*, **128**, 2739–2746.

Owen, R. J. and Holmes, B. (1978). Heterogeneity in the characteristics of deoxyribonucleic acid from *Flavobacterium odoratum*. *FEMS Microbiol. Letters*, **4**, 41–46.

Owen, R. J. and Holmes, B. (1980). Differentiation between strains of *Flavobacterium breve* and allied bacteria by comparison of deoxyribonucleic acids. *Current Microbiol.*, **4**, 7–11.

Owen, R. J. and Leaper, S. (1981). Base composition, size and nucleotide sequence similarities of genome deoxyribonucleic acids from species of the genus *Campylobacter*. *FEMS Microbiol. Letters*, **12**, 395–400.

Owen, R. J., Legros, R. M. and Lapage, S. P. (1978). Base composition, size and sequence similarities of genome deoxyribonucleic acids from clinical isolates of *Pseudomonas putrefaciens*. *J. Gen. Microbiol.*, **104**, 127–138.

Owen, R. J. and Snell, J. J. S. (1976). Deoxyribonucleic acid reassociation in the classification of flavobacteria. *J. Gen. Microbiol.*, **93**, 89–102.

Pemberton, J. M. (1974). Size of the chromosome of *Pseudomonas aeruginosa* PAO. *J. Bacteriol.*, **119**, 748–752.

Potts, L. E., Dow, C. S. and Avery, R. J. (1980). The genome of *Rhodomicrobium vannielii*, a polymorphic prosthecate bacterium. *J. Gen. Microbiol.*, **117**, 501–507.

Riley, M. and Anilionis, A. (1978). Evolution of the bacterial genome. *Ann. Rev. Microbiol.*, **32**, 519–560.

Riley, M., Solomon, L. and Zipkas, D. (1978). Relationship between gene function and gene location in *Escherichia coli*. *J. Molec. Evol.*, **11**, 47–56.

Rippka, R., Deruelles, J., Waterbury, J. B., Herdman, M. and Stanier, R. Y. (1979). Generic assignments, strain histories and properties of pure cultures of cyanobacteria. *J. Gen. Microbiol.*, **111**, 1–61.

Ryan, J. L. and Morowitz, H. J. (1969). Partial purification of native rRNA and tRNA cistrons from *Mycoplasma* sp. (Kid). *Proc. Natl. Acad. Sci. (USA)*, **63**, 1282–1289.

Sadoff, H. L., Shimei, B. and Ellis, S. (1979). Characterization of *Azotobacter vinelandii* deoxyribonucleic acid and folded chromosomes. *J. Bacteriol.*, **138**, 871–877.

Saglio, P., Lhospital, M., Laflèche, D., Dupont, G., Bové, J. M., Tully, J. G. and Freundt, E. A. (1973). *Spiroplasma citri* gen. and sp. n.: a mycoplasma-like organism associated with 'stubborn' disease of citrus. *Int. J. Syst. Bacteriol.*, **23**, 191–204.

Schmidt, H. J. and Heckmann, K. (1980). DNA of *Omikron*. In W. Schwemmler and H. Schenk (Eds), *Proceedings of the International Colloquium on Endosymbiosis and Cell Research* (Tubingen), de Gruyter, Berlin and New York, pp. 108–112.

Seewaldt, E. and Stackebrandt, E. (1982). Partial sequence of 16S ribosomal RNA and the phylogeny of *Prochloron*. *Nature, Lond.*, **295**, 618–620.

Seidler, R. J., Mandel, M. and Baptist, J. N. (1972). Molecular heterogeneity of the bdellovibrios: evidence of two new species. *J. Bacteriol.*, **109**, 209–217.

Seidler, R. J., Knittel, M. D. and Brown, C. (1975). Potential pathogens in the environment: cultural reactions and nucleic acid studies on *Klebsiella pneumoniae* from clinical and environmental sources. *Appl. Microbiol.*, **29**, 819–825.

Soldo, A. T., Brickson, S. A. and Larin, F. (1983). The size and structure of the DNA genome of symbiont xenosome particles in the ciliate *Parauronema acutum*. *J. Gen. Microbiol.*, **129**, 1317–1325.

Soldo, A. T. and Godoy, G. A. (1973). Molecular complexity of *Paramecium* symbiont Lambda deoxyribonucleic acid: evidence for the presence of a multicopy genome. *J. Molec. Biol.*, **73**, 93–108.

Soldo, A. T. and Godoy, G. A. (1974). The molecular complexity of Mu and Pi symbiont DNA of *Paramecium aurelia*. *Nucl. Acids Res.*, **1**, 387–396.

Sparrow, A. H. and Newman, A. F. (1978). Evolution of genome size by DNA doublings. *Science*, **192**, 524–529.

Srirangnathan, N. (1974). Deoxyribonucleic acid hybridization and plasmid studies in the genus *Lactobacillus*. *Ph.D. Thesis, Oregon State University, Corvallis* (cited by Adcock *et al.*, 1976).

Stackebrandt, E., Seewaldt, E., Fowler, V. J. and Schleifer, K. H. (1982). The relatedness of *Prochloron* sp. isolated from different didnemnid ascidian hosts. *Arch. Microbiol.*, **132**, 216–217.

Stackebrandt, E. and Woese, C. R. (1981). The evolution of prokaryotes. In M. J. Carlile, J. F. Collins and B. E. B. Moseley (Eds), *Molecular and Cellular Aspects of Microbial Evolution*, Symposium 32 of the Society for General Microbiology, Cambridge University Press, Cambridge, pp. 1–31.

Starr, M. P., Stolp, H., Trüper, H. G., Balows, A. and Schlegel, H. G. (Eds) (1982). *The Prokaryotes*, Springer-Verlag, Berlin, Heidelberg, N.Y.

Torrella, F., Guerrero, R. and Seidler, R. J. (1978). Further taxonomic characterization of the genus *Bdellovibrio*. *Can. J. Microbiol.*, **24**, 1387–1394.

Wallace, D. C. and Morowitz, H. J. (1973). Genome size and evolution. *Chromosoma*, **40**, 121–126.

Waterbury, J. B. and Stanier, R. Y. (1978). Patterns of growth and development in pleurocapsalean cyanobacteria. *Bacteriol. Reviews*, **42**, 2–44.

Wetmur, J. C. and Davidson, N. (1968). Kinetics of renaturation of DNA. *J. Molec. Biol.*, **31**, 349–370.

Woese, C. R., Maniloff, J. and Zablen, L. B. (1980). Phylogenetic analysis of the mycoplasmas. *Proc. Natl. Acad. Sci. (USA)*, **77**, 494–498.

Wood, N. B., Rake, A. V. and Shapiro, L. (1976). Structure of *Caulobacter* deoxyribonucleic acid. *J. Bacteriol.*, **126**, 1305–1315.

Yee, T. and Inouye, M. (1981). Reexamination of the genome size of myxobacteria, including the use of a new method for genome size analysis. *J. Bacteriol.*, **145**, 1257–1265.

Yee, T. and Inouye, M. (1982). Two-dimensional DNA electrophoresis applied to the study of DNA methylation and the analysis of genome size in *Myxococcus xanthus*. *J. Molec. Biol.*, **154**, 181–196.

Zipkas, D. and Riley, M. (1975). Proposal concerning the mechanism of evolution of the genome of *Escherichia coli*. *Proc. Natl. Acad. Sci. (USA)*, **72**, 1354–1358.

Zusman, D. R., Krotoski, D. M. and Cumsky, M. (1978). Chromosome replication in *Myxococcus xanthus*. *J. Bacteriol.*, **133**, 122–129.

Zusman, D. and Rosenberg, E. (1968). Deoxyribonucleic acid synthesis during microcyst germination in *Myxococcus xanthus*. *J. Bacteriol.*, **96**, 981–986.

CHAPTER 3

Eukaryote Gene Numbers, Non-coding DNA and Genome Size

T. Cavalier-Smith

SUMMARY

The 40-fold variation in the number of protein-coding genes is insufficient to account for the 80 000-fold variation in eukaryote genome size; neither protein-coding genes nor recombinator genes increase in numbers in proportion to genome size. The number of ribosomal RNA genes, replicator genes and centromere genes does increase in direct proportion to genome size, but together they constitute far too small a fraction of total DNA for this variation to account for the C-value paradox. The several-fold variation in the size of protein and RNA genes caused by the presence of introns is also far too small to be able to explain the C-value paradox. Comparison of gene numbers and size with the genome size of different eukaryotes shows that the amount of non-genic DNA per genome must vary from about 0.003 pg to over 300 pg (a 10^5-fold range) and constitute anything from less than 30% to about 99.998% of the genome.

This chapter reviews the evidence for these conclusions, and argues that taken in conjunction with the much greater sensitivity to mutagens and radiation of large genomes, and the cost in energy and nutrients of such large amounts of non-genic DNA, they show that there must be a major positive evolutionary force that favours larger genomes in larger cells. I also present a model for the microtubule-binding function of centromere genes in relation to their variation with genome size, to their tendency, like that of telomere genes, to generate highly repetitive DNA sequences in eukaryotes with larger genomes, and to the phasing of nucleosomes with respect to DNA repeats.

INTRODUCTION

Estimating the number of genes in different species is important for the theme of this book because of the need to establish what fraction of the total genome is genic DNA; without knowing this one cannot say how much non-genic DNA there is to be explained in terms other than the established functions of genes.

The amount of DNA in eukaryotes is often said to be 'an order of magnitude in excess of that required for the known gene-coding capacities'; but such oversimplified statements are misleading. In this chapter I show that the relationship between eukaryote gene numbers and genome size is more complex and that the ratio of non-coding to coding DNA must vary greatly from species to species—even in closely related ones: it may be as high as 50 000 or as low as 0.5. This tremendously wide range in the genic fraction of total DNA is the crux of the C-value paradox, and the key to its solution.

THE DIFFERENT CLASSES OF GENE

The first problem in trying to count genes is to agree on the definition of a gene. Classically, a gene was a hypothetical unit that influenced the phenotype in a specific way and which could undergo mutation and recombination and be mapped at a specific chromosomal locus. In classical molecular biology (Watson, 1976) the gene came to be a segment of DNA coding for a polypeptide with a specific function or specifying a transfer RNA (tRNA) or ribosomal RNA (rRNA) molecule. (Sometimes the latter function is loosely referred to as 'coding' for tRNA or rRNA, but since the genetic code is not involved in the formation of such non-translated RNA molecules, I prefer to use the term 'specification' rather than 'coding' to describe the genetic control of non-translated RNA sequences.) However, there is no reason to restrict the term gene to sequences with these two functions. If a specific DNA segment can be shown, for example by deletion, to be essential for a specific function, it deserves to be called a gene even if it plays no role in coding for proteins. To say otherwise would be to redefine 'gene' in terms of the best established function of genes and to ignore the possibility that genes with other functions may exist. Three universal genetic functions for specific DNA segments are so far known in addition to coding for polypeptides and the specification of non-translated RNA molecules:

1. In prokaryotes, plasmids, viruses, and probably also eukaryotes, the initiation of DNA replication depends on specific *replicator genes* at the origin of each replicon which may serve as binding sites for specific initiator or repressor molecules as well as for the transcription of transient oligonucleotide primers (Chapter 7).
2. Site-specific recombination involves the breakage and rejoining of DNA molecules at specific sites. It occurs during eukaryote meiotic recombi-

nation (Whitehouse, 1982) as is shown, for example, by the fact that chiasmata never occur on lampbrush loops, and its molecular basis has been studied in detail in prokaryotes, for example in the integration of phage lambda. It is also involved in transposition, the mechanisms of which are discussed by Dyson and Sherratt (Chapter 12). One can call such specific sites *recombinator genes* (Holliday, 1964).

3. Segregation is even less well understood than replication and recombination, but there is little doubt that the membrane-attachment of the origin and terminus of the bacterial chromosome is mediated by specific sequences (Hendrikson *et al.*, 1981) and is important for segregation. A role for specific replicon origin sequences in segregation is clear in bacterial plasmids (Meacock and Cohen, 1980). In eukaryotes the specific location of centromeres on each chromosome strongly indicates at least some kind of genetic specificity for centromere DNA, and some authors have referred to *centromere genes* (Darlington, 1969).

Therefore at present we can distinguish five types of genes:

1. protein-coding genes that code for polypeptides, comprising sequences that control the transcription (and splicing) of messenger RNA molecules;
2. RNA genes specifying functional RNA molecules, notably tRNA, rRNA, signal recognition particle 7SL RNA, and (in eukaryotes) small nuclear RNA molecules (snRNA) which may be important in RNA splicing;
3. replicator genes that specify sites for replication initiation or termination without themselves coding for proteins;
4. recombinator genes that provide specific recognition sites for recombination enzymes;
5. *segregator genes* that provide specific sites for attachment to the segration machinery.

In eukaryotes segregator genes include *centromere genes* for attachment to the spindle and *telomere genes* for attachment to the nuclear envelope during meiotic prophase; interstitial segregator genes for attachment to the synaptonemal complex and mitotic chromosome 'scaffold' have also been postulated but not experimentally demonstrated.

The DNA constituting these five categories of genes will be referred to as genic DNA: other DNA will be referred to as non-genic or secondary DNA. Note that non-coding DNA is not a synonym for non-genic DNA but a broader category. This is because four of the five genic functions do not involve coding. Moreover, even in protein-coding genes some of the genic DNA consists of non-coding sequences which may be transcribed (as 5′ leader or 3′ trailer sequences (Baralle, 1983), introns and spacer sequences in multigenic operons), or non-transcribed (as promoters or terminators):

non-coding DNA sequences may be highly conserved (e.g. promoter sequences) or evolve rapidly.

Since the relationship between gene numbers and overall genome size differs for the different classes of gene, they will be considered in turn.

PROTEIN-CODING GENES

A reasonable estimate of the maximum possible number of protein-coding genes in any organism can be derived from the size of its genome and the average size of its genes, if one assumes that only one copy of each gene is present per genome and that all the DNA codes for protein. The average size of the coding part of a gene can readily be calculated from the average size of polypeptide chains (about 500 amino acids) by multiplying by the coding ratio (3) to give the mean number of nucleotides, i.e. 1500.

Only in bacteria does such a simple calculation give a good estimate of the varying numbers of protein-coding genes in different species, as Herdman explained in Chapter 2. The length of non-coding sequences per bacterial gene is small compared with that of the coding sequences, so on average a bacterial gene can be taken as 1600 bp. Given the range in genome sizes of bacteria (0.56–13.6 fg a 24-fold range (Chapter 2)) the possible range in gene numbers in different species is from 400–8000. These figures are only approximate because even in prokaryotes the assumption that all the DNA consists of single-copy protein-coding genes is not strictly justified.

In eukaryotes such calculations are much more difficult because many genes are split by non-coding introns and the average size of eukaryote genes depends on the length of their introns as well as on their coding sequences (exons). Since some eukaryote genes have no introns and others have introns whose total length may be ten or more times that of the coding sequences, the size of protein-coding genes can be much more variable in eukaryotes than in prokaryotes. Variation in the size of eukaryote genes resulting from such variation in the number and lengths of introns might therefore, in principle, in conjunction with a variation in the number of genes, account for the great interspecific variation in the genome size of different species, i.e. for the C-value paradox; Raff and Kaufman (1983) even asserted that 'the existence of introns provides much of the answer to the frustrations of the C-value paradox'. In order to test this possibility one would need to measure gene numbers and mean gene sizes in eukaryotes of greatly differing genome size.

Counting protein-coding genes in eukaryotes

A variety of measurements bearing on the problem have been made, which together strongly suggest that variation in gene size and numbers is quantitatively insufficient to explain the C-value paradox, and that Raff and Kaufman's assertion is incorrect. This conclusion, however, rests partly on indirect arguments because the measurements so far have been mainly on multicellular organisms which are less well suited than unicellular ones for such tests.

The number of protein-coding genes in an organism could in principle be estimated by extracting the messenger RNA molecules from every stage in the life history and determining how many different mRNA molecules are present. Such determinations can be done in two ways. One is to use reverse transcriptase to make radiolabelled DNA complementary to the purified mRNA fraction, which is therefore called cDNA, and then to measure the rate at which this hybridizes with a large excess of unlabelled mRNA. The rate of hybridization depends on the number of different mRNA sequences present; the number of different mRNA molecules present can be calculated from the rate if their mean length is known. The second method measures not the rate of hybridization but the extent of hybridization between labelled genomic DNA and a large excess of messenger RNA; so long as the reaction proceeds to completion and is done with the unique DNA fraction (i.e. total genomic DNA from which repetitive DNA has first been removed) the fraction of the DNA that forms hybrids gives a direct measure of the fraction of unique DNA for which transcripts are present in the mRNA. The number of different mRNA molecules can be calculated from this and from their mean length if the total genomic content of unique DNA is known.

These methods are described in detail by Paul (1979), Lewin (1980) and Van Ness and Hahn (1982). As both methods when applied to the same cell type give similar, though seldom identical, numerical results, they give a reasonable estimate of the number of protein-coding genes being transcribed. However, if one is to count not merely the protein-coding genes active in a particular cell type or tissue but all those in the whole genome it would be necessary to obtain mRNA from every cell type at every stage in the life cycle. Because this goal is still far from being achieved for multicellular organisms, and because determination of gene size has proved more straightforward in unicellular eukaryotes (protists), these are discussed first.

Gene size and number in protists

The average size of eukaryote protein-coding genes can be estimated from the size of the primary transcripts in nuclei. In animals these are often several times the size of mRNA and very heterogeneous in size and so are referred to as heterogeneous nuclear RNA (hnRNA). Although the same term is used for non-ribosomal nuclear RNA in protists, it is in this case a misnomer, for the limited results so far indicate that nuclear RNA in protists is usually very little larger and no more heterogeneous than mRNA. Table 3.1 shows

Table 3.1. Gene size and gene numbers in protists

Organism	Genome size (pg)	% of genome transcribed into total poly A+ RNA or Hn RNA	Mean size of RNA (nucleotides) nuclear	mRNA	Number of different mRNAs
FUNGI					
Saccharomyces cerevisiae	0.009[a]	40[b]		1500[b]	3000[b] 4000[b]
CHROMISTA†					
Achlya ambisexualis	0.05[c]	9.5[d]	1150[e]	1150[e]	2600[d]
PROTOZOA					
Dictyostelium discoideum	0.036[f]	56–80	1500[g]	1200[g]	~6500
Physarum polycephalum	0.57[h]		1533[h]	1339[h]	<20 000
Oxytricha nova	0.4[i]		<2200[i]*		24 000[i]*
Euplotes aediculatus	0.3[i]		<1836[i]*		40 000[i]*
PLANTAE					
Chlamydomonas reinhardii	0.12[k]	12.0[k]			12 000**

[a] Galeotti *et al.*, 1981; [b] Hereford and Rosbash, 1977; [c] Hudspeth *et al.*, 1977; [d] Rozek *et al.*, 1978; [e] Timberlake *et al.*, 1977; [f] Sussman and Rayner, 1971; [g] Firtel and Lodish, 1973; [h] Melera, 1980; [i] Chapter 14; [j] Swanton *et al.*, 1982; [k] Howell and Walker, 1977; * Refers to macronuclear DNA molecules, not RNA; ** Assuming size of mRNA to be 1000 nucleotides. † A new eukaryotic kingdom characterised by chloroplast endoplasmic reticulum and/or tubular mastigonemes, and comprising the heterokont algae and 'pseudofungi', the Cryptophyta, and Haptophyta. See Cavalier-Smith, T. (1983c; reference on p. 33) for the classification followed here.

that in the oomycete 'pseudofungus' *Achlya* nuclear RNA and mRNA are the same size while in the protozoa *Physarum* and *Dictyostelium* nuclear RNA is, respectively, only 15% and 25% larger than mRNA. There are similar observations in *Paramecium* and *Amoeba*.

Evidence that the smaller size of nuclear transcripts is not the result of degradation during isolation comes from the finding in yeast and *Dictyostelium* that the sequence complexity of total cellular RNA is essentially the same as that of mRNA. Therefore, RNA processing in the nucleus of premessenger RNAs can at most remove only a small fraction of the nucleotides. The similarity in the mean size of genes in *Physarum* and *Dictyostelium* (1500 base pairs), even though *Physarum* has twelve times as much DNA per genome, makes it highly improbable that variations in genome size have anything to do with variations in gene size. Table 3.2 shows that genome size in protists varies by a factor of well over 10^4, in marked contrast to mean gene size which varies between species by less than a factor of 2 (Table 3.1). One corollary of this is that although split genes do occur in protists (e.g. Woudt *et al.*, 1983), most protist genes must be unsplit or else their intron lengths must be short in comparison with their coding sequences. In any case it seems clear that the presence of introns can do nothing to explain the $<10^4$-fold variation in protist genome size.

The same is true of variation in gene numbers. However, Table 3.1, which shows only about a 10-fold variation in gene numbers in different species, does not include data on gene numbers from protists having really large genomes, which is clearly needed before one can be entirely confident that variation in gene number is irrelevant to the C-value paradox. However, the observed 10-fold variation in numbers of different proteins between the morphologically simple fungi and the highly complex hypotrich heterokaryote protozoa is as big a variation as one might expect from the varying organismic complexity of protists. The nuclear DNA content of *Amoeba dubia* and *A. proteus* is 10^4 times that of *Dictyostelium* even though the two amoebae are less structurally and developmentally complex than *Dictyostelium*. Though the ploidy of the two *Amoeba* species is not known, so the nuclear DNA content may somewhat overestimate their genome size, one can hardly escape the conclusion that the variation in genome size in amoeboid protozoa has nothing to do with structural complexity.

Consideration of unicellular algae leads to the same conclusion. Their nuclear DNA content varies over a 5000-fold range. Yet they are structurally and developmentally similar in complexity; one would not expect the number of their protein-coding genes to vary even by as much as a factor of 5, which is very small compared with the several thousand-fold variation in genome size.

The above arguments strongly indicate that the vast majority of the extra DNA sequences in high C-value protists do not code for proteins. This

Table 3.2. Genome size of protists

Organism	Genome size (pg)	References
FUNGI	0.009–1.5	(167-fold range)
Ascomycota		
Saccharomyces cerevisiae	0.009	Galeotti *et al.* 1981
Schizosaccharomyces pombe	0.012	Thuriaux, 1977
Neurospora crassa	0.017	Minegawa *et al.*, 1959
Aspergillus nidulans	0.0216	Thuriaux, 1977
Erysiphe graminis	0.63	Manners and Myers, 1975
E. cichoracearum	1.5	Manners and Myers, 1975
Basidiomycota		
Coprinus lagopus	0.047	Thuriaux, 1977
CHROMISTA[†]	0.036–25	(694-fold range)
Oomycetes		
Achlya ambisexualis	0.05	Hudspeth *et al.*, 1977
Diatomophyceae (diatoms)	0.036–25	
Thallassiosira fluviatilis	1.65	Shuter *et al.*, 1983
Navicula pelliculosa	0.036	Shuter *et al.*, 1983
Ditylum brightwellii	6.5	Shuter *et al.*, 1983
Skeletonema costatum	0.17	Shuter *et al.*, 1983
Coscinodiscus asteromphalus	25	Shuter *et al.*, 1983
Chrysophyceae	0.071–2.36	
Monochrysis lutheri	0.71	Shuter *et al.*, 1983
Syracosphaera elongata	2.36	Shuter *et al.*, 1983
PROTOZOA	0.024–700	(29 167-fold range)
Sarcodina	0.036–700	(19 444-fold range)
Dictyostelium discoideum	0.036	Sussman and Rayner, 1971
Physarum polycephalum	0.3	Mohberg and Rusch, 1971
	0.57	Melera, 1980
		Fouquet *et al*, 1974
Entamoeba histolytica	0.4–0.6	Gelderman *et al.*, 1971
E. histolytica (like)	0.08–0.09	Gelderman *et al.*, 1971
E. moshkovski	1	Gelderman *et al.*, 1971
E. invadens	0.32–0.4	Gelderman *et al.*, 1971
Amoeba dubia	700[a]	Friz, 1968
A. proteus	300[a]	Friz, 1968
Heterokaryota (Ciliophora)	0.024–8.8	(367-fold range)
Hypotrichs	0.024–5	(208-fold range)
Paraurostyla cristata	2.4	Table 1 Chapter 14
Oxytricha sp.	0.4	Table 1 Chapter 14
Euplotes minuta	0.024–0.04	Table 1 Chapter 14
E. aediculatus	0.3	Table 1 Chapter 14
Stylonychia mytilus	5.0	Table 1 Chapter 14

Table 3.2. Continued

Organism	Genome size (pg)	References
Non-hypotrich heterokaryota	0.06–8.8	(147-fold range)
Paramecium aurelia	0.18–0.21	(Chapter 14, p. 433)
P. bursaria	3.8	Shuter *et al.*, 1983
	2.3	
P. caudatum	8.8	Shuter *et al.*, 1983
P. trichium	1.6	Shuter *et al.*, 1983
Cyclidium glaucoma	0.06	Shuter *et al.*, 1983
Tetrahymena pyriformis	0.21	Shuter *et al.*, 1983
Didynium nasutum	1.1	Shuter *et al.*, 1983
Glaucoma scintillans	0.14	Shuter *et al.*, 1983
G. frontata	1.1	Shuter *et al.*, 1983
Colpidium campylum	0.27	Shuter *et al.*, 1983
C. colpoda	0.58	Shuter *et al.*, 1983
Halteria grandinella	0.46	Shuter *et al.*, 1983
Dinozoa (Dinophyta; dinoflagellates)	1.4–100	(71-fold range)
Crypthecodinium cohnii	3.9	Allen *et al.*, 1975
Amphidinium carteri	1.4	Shuter *et al.*, 1983
Gonyaulax polyedra	100	Shuter *et al.*, 1983
Gymnodinium nelsoni	71.5	Shuter *et al.*, 1983
G. breve	50.8	Shuter *et al.*, 1983
Prorocentrum micans	21	Shuter *et al.*, 1983
Cochonina niei	5	Shuter *et al.*, 1983
Gyrodinium cohnii	3.5	Shuter *et al.*, 1983
Peridinium trochoideum	17	Shuter *et al.*, 1983
Euglenozoa		
Leishmania donovani	0.1	Leon *et al.*, 1978
Euglena gracilis	2.4	Shuter *et al.*, 1983
Parabasalia		
Trichomonas gallinae	0.7	Shuter *et al.*, 1983
T. vaginalis	1.0	Shuter *et al.*, 1983
Sporozoa		
Plasmodium	0.059	Whitfield, 1952

[a] Whole cell value including mitochondrial DNA. The ploidy of *Amoeba* and *Entamoeba* is not known. † See footnote table 3.1.

strong indication needs to be made conclusive by studies of messenger RNA complexity in a few of the protists with really large genome sizes; it is likely that such studies will show that even protists with the largest genomes have no more than 40 000 protein-coding genes. Assuming this to be the case and the mean gene size to be 1500 nucleotides, gives an upper limit of 6×10^7 nucleotides for the protein-coding part of protist genomes. This is approximately six times the size of the whole genome of the yeast *Saccharomyces*, which has the lowest genome size of any protist: therefore, as far as protein coding is concerned, no protists would need a genome even 10 times larger than yeast! Protists with the largest genomes therefore have about 10^4 times as much DNA as they need for protein coding. In such protists 99.99% of the DNA does not code for proteins.

In yeast, however, at least 35% of the genome codes for proteins. This figure is a minimum estimate because studies of messenger RNA complexity have been confined to vegetative cells. One does not know how many extra proteins are needed in sexual cells and the spore stages. Though it is conceivable that all the extra DNA in yeast is taken up with such functions, it seems unlikely that these extra functions would require more different proteins than vegetative functions. If so, probably not more than 70% and perhaps no more than 50% of the *Saccharomyces* yeast genome codes for proteins. The other figures for gene numbers in Table 3.1 also need to be increased somewhat to allow for extra sexual and/or sporulation functions. Even so it is likely that all protist gene numbers will fall within the 16-fold range, 2500–40 000 genes, and that the fraction of non-coding DNA will vary from less than 30% to about 99.99%.

One therefore has to explain why the amount of non-coding DNA varies from about 0.003 pg to over 300 pg. This contrast between the 16-fold range in coding DNA and the 10^5 range in non-coding DNA amounts is the central problem of the C-value paradox. Does the conclusion that the number of protein-coding genes varies about 10^4 times less than genome size apply also to multicellular organisms?

Protein-coding gene numbers in animals and flowering plants

Table 3.3 shows that the range in protein gene numbers in cells and tissues of animals and plants is similar to that in protists. What is less easy to determine is the total number of protein genes in the genome, because only selected tissues and/or developmental stages have been examined. Some estimate is needed of what fraction of the genes active in a particular tissue or stage are uniquely active there and what fractions are also active in other tissues. The various methods for studying this question are evaluated by Lewin (1980). The main conclusion is that there is a great deal of overlap between the genes active in different tissues: a small fraction may be unique to that tissue but a larger fraction is shared with other tissues.

This overlap is particularly striking in sea urchins, where all the messengers in the gastrula stage are also active in the oocyte and most of these are active in the ovary and blastula stages also. The oocytes and blastula also have as many genes again that are not expressed in the gastrula, but adult tissues have several times fewer genes active than the embryonic stages: however, the majority of these are also expressed in embryonic stages. It seems therefore that the majority of genes are expressed in early embryonic stages but that the majority are switched off in particular tissues. Without studying every stage and other adult tissues in much more detail one could not be sure to detect all the active genes, and this is hardly feasible. Nonetheless Lewin's (1980) estimate from the observed overlap of a maximum of 25 000 protein genes in sea urchins seems reasonable.

Less work has been done on embryonic stages of vertebrates, but there is clear evidence for considerable overlap between mRNA populations in different adult tissues. Moreover, the number of mRNA sequences in oocytes of the frog *Xenopus* is similar to that in sea urchin oocytes, as is mRNA complexity in mouse embryos and sea urchin gastrulae. The main difference from sea urchins is the several-fold larger number of mRNA sequences in adult tissues (and tissue culture cells). However, this may simply reflect differences in the tissues chosen rather than a fundamental developmental difference. The sea urchin tissues studied were tube-feet, coelomocytes and intestine. It might be expected that such non-growing tissues would be making fewer different proteins than actively growing tissue culture cells, or mammalian brain and liver which make a great variety of proteins. The biggest uncertainty at present concerns mammalian brain; some results suggest an mRNA complexity similar to liver, and others a figure several times greater (Bernstein *et al.*, 1983). A reasonable estimate of the number of mammalian genes would be about 30 000, taking the lower figure, or about 100 000 if the higher figure is correct.

Mutational load arguments (Muller, 1950) have been used to suggest that the number of animal genes cannot exceed about 40 000 (Ohno, 1972). However, more recent estimates of mean and median mutation rates are respectively about 10 times and 100 times lower than those assumed in Muller's calculations (Cavalli-Svorza and Bodmer, 1971) so the mutation load limit on gene numbers may be correspondingly greater. The potential limit to gene numbers set by mutational load is therefore probably very much greater than the actual numbers of protein-coding genes suggested by the mRNA complexity data, and therefore neither of evolutionary significance nor any help in estimating the numbers of protein-coding genes or the amounts of non-coding DNA. The mRNA complexity data, however, strongly indicate that well over 90% of vertebrate DNA must consist of non-coding sequences.

The evidence from flowering plants is much less extensive but the number of genes active in a plant organ (leaf) appears to be similar to that in liver or kidney. Since a plant has fewer different organs than an animal, it is unlikely that flowering plants have any more genes than do animals, even though their genome size is frequently much greater.

The lower limit for gene numbers in animals is equally poorly known, and would probably be found in the simplest ones such as sponges and coelenterates. Since even cultured *Drosophila* cells have nearly 7000 genes

Table 3.3. Gene size and gene numbers in animals and flowering plants

Organism and tissue	Genome size (pg)	% of genome transcribed into total poly A+ RNA or Hn RNA	% of genome expressed as mRNA	Mean size of RNA nuclear	Mean size of RNA mRNA	Number of different mRNAs
ANIMALIA						
Insects						
Drosophila melanogaster						
Schneider cultured cells[b]	0.18			4 200[b]	2 100[b]	6 679[a]
Aedes[b]						
tissue culture	0.7–0.96			8 400	2 100	
Echinoderms						
Strongylocentrotus purpuratus[c]	0.89					
intestine		17.9	0.75			3 000
gastrula		14	2			8 000
blastula		3.1		8 800	2 100	13 000
oocyte			4.5			18 500
T. gratilla[d]						
blastula		16.6				
pluteus		17.3				
Vertebrates						
Xenopus laevis[e]	3.15					
oocyte		1				15 000
Gallus domesticus[f]	1.25					
oviduct			2.5			1 500
						12 508
liver			2.8			17 000
						11 100

Homo sapiens[g]						
Hela cells	3.5			10 000	2 100	9 400
Rattus norvegicus	3				1 500	6 500
brain[h]		15.6	5			
cerebellum[i]		12.7				
liver[h]		10.9				
kidney[h]		5.3				
Mus musculus[g]	3					
embryo						10 420
L-cells						6 232
liver		1.6–4.8				1 070
kidney		1.8–3.9				1 229
Friend cell		5.4–5.9				12 315
embryonal carcinoma		3				
brain		7.8–21.2				
PLANTAE						
Nicotiana tabaccum[j]	3.9					16 580
leaf						18 000

[a] Levy and McCarthy, 1975; [b] Lengyel and Penman, 1975; [c] Hough *et al.*, 1975; Wold *et al.*, 1978; Galau *et al.*, 1974, 1976; Nemer *et al.*, 1974, 1975; [d] Dubroff and Nemer, 1976; [e] Kleene and Humphreys, 1977; [e] Davidson and Hough, 1971; [f] Axel *et al.*, 1976; [g] Lewin, 1980; [h] Chikaraishi *et al.*, 1978; [i] Bernstein *et al.*, 1983; [j] Goldberg *et al.*, 1978.

it is unlikely that any animals have as few genes as some protists. Genetic estimates suggest the number of genes that can mutate so as to be lethal is about 5000 in *Drosphila* and about 2000 in the nematode *Caenorhabditis*. However, these estimates have a fairly wide margin of error, since it is not known what fraction of protein-coding genes can mutate to lethality (O'Brien, 1973). As there are at least some protein-coding genes that can be completely dispensed with, it is unlikely that any animal or higher plant has less than 5000 protein-coding genes.

The most probable range for gene numbers in animals and higher plants is therefore 5000 to 100 000. This 20-fold range greatly overlaps the 16-fold range for protists of 2500–40 000 genes, so that the overall range in gene numbers in eukaryotes is probably no more than about 40-fold. Although there are uncertainties in these figures it is improbable that any of them are in error by more than a factor of two or three.

Split-genes and gene size in multicellular organisms

Table 3.3 shows that the size of mRNA is only slightly, if at all, larger than in protists (1400–2200 nucleotides rather than 1200–1500), whereas nuclear RNA (hn RNA) is considerably larger. In invertebrates hn RNA is on average from 2–4 times the length of mRNA and in vertebrates it is about 5 times the length of mRNA. This means that on average the genes of animals are from about 3–7 times longer than those of protists. This difference is probably mainly attributable to the presence of introns and extensive RNA splicing. Although data are so far available for only a few split genes, mainly from vertebrates, which may not be representative of the whole genome, the ratio between the lengths of their primary transcripts and the final processed mRNA is usually in the range 1.3–6.3 though occasionally it may be vastly in excess of this. No explanation other than the presence of introns is therefore needed for the larger size of genes in multicells compared with protists.

Even though genes of multicells are on average larger than those of protists, this does nothing to explain the C-value paradox, for the following reasons:

1. There is no evidence that C-values are on average greater in multicells than in protists; in fact the spread of C-values and therefore the seriousness of the C-value paradox is much greater in protists.
2. The extra size is small compared with the amount of DNA whose function is unaccounted for.

Even allowing for the greater gene size there are vast amounts of non-coding DNA in multicells. Table 3.4 and 3.5 show the overall ranges in C-value of the major groups and Table 3.6 shows the percentage of the genome that probably codes for proteins in selected organisms, based on reasonable estimates

Table 3.4. Genome size ranges of animals

Organism	Genome size range	Ratio $\frac{highest}{lowest}$	References
ANIMALIA	0.055–142	2581.8	
Sponges	0.055		White, 1973
Cnidaria	0.33–0.73	2.2	White, 1973
Annelida	0.9–5.3	5.9	Hinegardner, 1976
Crustacea	0.7–22.6	32	Hinegardner, 1976
Mollusca	0.43–5.4	13	Hinegardner, 1976
Aschelminthes	0.088–2.5	28	Sulston and Brenner, 1974; White, 1973
Insecta	0.1–7.5	75	Hinegardner, 1976
Echinodermata	0.54–3.3	6.1	Hinegardner, 1976
Chordata	0.16–142	887.5	
Tunicata	0.16–0.205		White, 1973
Cephalochordata	0.61		White, 1973
Vertebrata	0.39–142	346	
Agnatha	0.65–2.75	4	Olmo, 1983
Chondrichthyes	1.5–16.15	11	Stingo *et al.*, 1980
Actinopterygii	0.39–4.4	11	Hinegardner and Rosen, 1972; Olmo, 1983
Coelacanth	3.1		
Choanata	0.95–142	150	
(a) Anamniote choanata	0.95–142	150	
Lungfish	80–142	1.8	Olmo, 1983
Amphibia	0.95–82.5	87	Olmo, 1983
Anura	0.95–10.55	11	Olmo, 1983
Urodela	15.1–82.5	5	Olmo, 1983
(b) Amniote Choanata	0.85–5.8	6.8	
Reptiles	1.25–5.45	4.4	Olmo, 1983
Birds	1.7–2.3	2.2	Olmo, 1983; Hinegardner, 1976
Mammals	1.45–5.8	4	Olmo, 1983; Hinegardner, 1976
Bats	1.45–1.95	1.3	Olmo, 1983

Table 3.5. Genome size ranges of vascular plants.

Organism	Genome size range	Ratio $\dfrac{\text{highest}}{\text{lowest}}$	References
Pteridophytes	0.1–131	1310	Bouchard, 1976; Price *et al.*, 1972
Gymnosperms	18.0–69.3 (5.3–78.5	3.85 14.8	Miksche, 1967 Price *et al.*, 1973)*
Angiosperms	0.2–127.4	637	Bennett and Smith, 1976; Bennett *et al.*, 1982

* Estimates from nuclear volume, not direct DNA measurements.

Table 3.6. Estimates of coding and non-coding genomic fractions in selected eukaryotes.

Organism	Genome size (pg)	Estimated number of genes	Percentage of genome coding for proteins	not coding for proteins
Protists				
Saccharomyces cerevisiae	0.009	4 000	69	31
Chlamydomonas reinhardii	0.12	8 000	10	90
Oxytricha nova	0.4	25 000	13	87
Gonyaulax polyedra	100	15 000	0.002	99.998
Distyostelium discoideum	0.036	6 500	19	81
Physarum polycephalum	0.57	15 000	3.6	96.4
Animals				
Caenorhabditis elegans	0.088	5 000	25	75
Drosophila melanogaster	0.18	10 000	33	67
Strongylocentrotus purpuratus	0.89	25 000	25	75
Homo sapiens	3.5	(a)30 000 (b)90 000	9 27	91 73
Triturus cristatus	19	(a)30 000 (b)90 000	1.5 4.5	98.5 95.5
Protopterus aethiopicus	142	(a)30 000 (b)90 000	0.4 1.2	99.6 98.8
Flowering Plants				
Arabidopsis thaliana	0.2	30 000	31	69
Nicotiana tabaccum	3.9	30 000	0.7	99.3
Fritillaria assyriaca	127.4	30 000	0.02	99.98

The gene number estimates and gene sizes used for calculating the figures in the third column are based on those from tables 3.1 and 3.3. For the vertebrates conservative figures (a) are given based on the lower estimates for gene numbers in the nervous system, as well as perhaps more realistic higher figures (b) based in the higher estimates.

for the size and number of protein genes. Even for animals this fraction is rather variable, ranging from about 1% in *Protopterus* to about 33% in *Drosophila*. The variability is less than in protists primarily because the range in genome size is lower. In all cases most of the DNA seems not to code for protein. It is important to stress that even in animals, where genes are larger than in protists, most of the non-coding DNA lies between rather than within genes. Therefore, although it is important to establish why animals should have larger genes than protists, answering this question in itself would not solve the C-value paradox because it is the total amount of non-coding DNA that is at issue here and not simply the minor fraction present in introns.

GENES SPECIFYING NON-INFORMATIONAL RNA

In protists the number of different genes specifying non-informational RNA must be much lower than the number of different protein-coding genes, since as discussed above the kinetic complexity of total cell RNA is scarcely distinguishable from that of mRNA. But such genes could still make up a large fraction of the total genome and, therefore, be relevant to the C-value paradox, if they were present in vast numbers of copies in the genome. The evidence from renaturation and hybridization studies does indicate some repetition for the known major classes of such 'RNA genes', i.e. rRNA, 5S-RNA, tRNA and snRNA. But the degree of repetition (from about 10-fold to several thousand-fold) is far too small to contribute significantly to the vast amounts of DNA found in eukaryote genomes.

In the case of rRNA there is a positive correlation between the degree of repetition of the rRNA genes and genome size (Birnstiel *et al.*, 1971; Fig. 3.1). But since rRNA on average constitutes only about 1%, or at maximum (for eukaryote cells) only 3.7%, of total cellular genome size, and forms a relatively small fraction of total DNA in larger genomes, such variation cannot be regarded as a significant cause of the increased genome size. A simple explanation for the correlation is that it is the indirect consequence of the coevolution of genome size and cell volume as discussed in detail in Chapter 4. Gall (1968) suggested that the temporary amplification of rRNA genes in certain animal oocytes was an adaptation to the large volume of these cells and the associated need to synthesize a vast amount of rRNA in a much shorter time than would be possible with the limited number of template gene copies found in somatic cells. If increased cell size is not a temporary phenomenon of only one cell type but a permanent feature of the organism, then permanent increases in the gene dosage of rRNA genes would be advantageous. Evolutionary increases in cell volume and therefore in ribosome numbers per cell would simultaneously lower the concentration of the rRNA genes and increase the number of transcripts that would have to be made per cell cycle; this would cause rDNA duplications to undergo

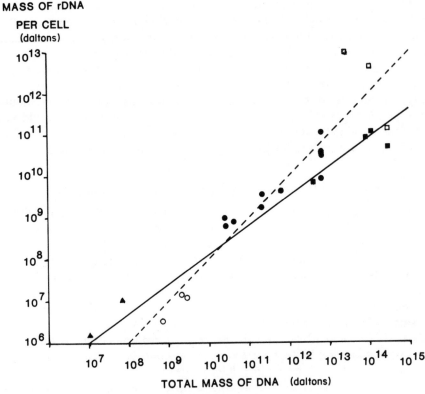

Fig. 3.1. The proportionality between the mass of ribosomal DNA (rDNA) and genome size. Amplification of rDNA in amphibian oocytes may cause some of their rDNA values (□) to fall well off the regression line (solid line) calculated by least squares from all the non-amplified examples (■ amphibian somatic cells; ● other eukaryotes; ○ bacteria; ▲ mitochondria). The slope of the regression line (0.70) shows that although the amount of ribosomal DNA increases greatly with overall genome size it forms a steadily decreasing fraction of total DNA as genome size increases (from nearly 25% in animal mitochondria to as little as 0.016% in the amphibian *Necturus* with the largest genome size). For eukaryotes the values plotted on both ordinate and abscissa are masses per nucleus (i.e. diploid values, except for the amphibian oocytes) while for bacteria and mitochondria they are the masses per DNA molecule. The dashed line is based on the assumption that the mass of rDNA is isometrically related to genome size and forms a constant 1% of total cellular DNA irrespective of genome size; this appears to be most nearly true for nuclei of organisms that have no ribosomal gene amplification (circles), but even they have relatively less rDNA in larger genomes. Data from Birnsteil *et al.* (1971).

strong positive selection until the number of copies of rRNA genes was sufficient not to limit the rate of growth. The fact that species without ribosomal gene amplification do not deviate much from the isometric (dashed) line in Fig. 3.1, taken together with the isometric relationship between genome size and cell volume discussed in the next chapter, implies that the concentration of ribosomal genes is held roughly constant during cellular evolution over a 10^4-fold range in genome size and cell volume, and is approximately the same in prokaryotes and eukaryotes. Decreases in cell volume would be accompanied by selection, probably weaker, for rDNA deletions, since unnecessary DNA would prove a metabolic burden and occupy valuable space in the nucleus. Since both duplications and deletions of rDNA readily occur (Cullis, 1982), such selective forces would be able rapidly to adjust the number of copies to the 'needs' of the cell.

Similar considerations apply to 5S and tRNA and to the signal recognition 7S RNA (Walter and Blobel, 1982) which larger cells will also require in larger numbers. Since larger cells might also require larger numbers of at least some messengers there may also be an increased requirement for small nuclear RNAs involved in RNA splicing and other nuclear functions. However, the total sequence complexity of known species of non-informational RNA and the number of copies per genome of each gene is so low that together they comprise usually well under 2% of the total genome and therefore do not contribute significantly to overall genome size.

NUMBER OF REPLICATOR GENES

The constancy in replicon size in eukaryotes implies, as discussed in more detail in Chapter 7, that the number of replicator genes increase in proportion to genome size. But since the replicator regions occupy only a very small fraction of each replicon (probably well under 1%) they do not directly contribute much to overall genome size. It can reasonably be argued that larger genomes need proportionately more replicators in order to replicate their extra DNA rapidly enough (see Chapter 7), in which case their extra numbers can be regarded as a necessary *consequence* rather than a *cause* of increased genome size. However, it is also likely that duplication of whole replicons (i.e. replicator + the other sequences) is one of the most important modes of DNA increase in genome evolution; so that in practice replicators may increase in numbers at the same time as other sequences. But the key question is why such changes in replicon number are associated with proportional changes in cell volume. (See Chapters 4 and 7 for a detailed discussion.)

NUMBER OF RECOMBINATOR GENES

It is a common misconception that making a eukaryote chromosome longer would correspondingly increase the chances of recombination. However, Thuriaux (1977) has shown that the overall recombination frequency and the frequency of recombination within a gene are both independent of genome size (Fig. 3.2). Therefore the average probability of crossing-over per unit length of DNA must decline markedly with and in about direct proportion to increased genome size. There are three possible explanations for this:

1. The probability of recombination at any one site declines in direct proportion to the increase in the number of potential recombination sites.
2. The number of potential recombination sites stays constant as genome size increases.
3. There is a less than proportional increase in recombination sites as genome size increases coupled with a less than proportional decline in the frequency of recombination at each.

There is independent genetic evidence (Whitehouse, 1982) that crossing-over is not initiated at random but only at a limited number of specific sites: the recombinator genes. The universal absence of chiasmata from lampbrush loops suggests that recombinator sites do not overlap with units of transcription in eukaryotes with large genomes. In eukaryotes with small genomes, such as fungi, *Drosophila* and *Chlamydomonas*, however, there is abundant evidence for intragenic recombination. This does not mean that the recombinator site need lie within the transcribed region; all that is necessary is that the region of hybrid DNA associated with crossing-over (Whitehouse, 1982) extends into the transcribed region. These observations and those of Thuriaux are compatible with the idea that the number of recombinator sites

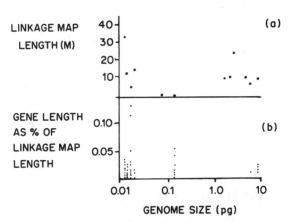

Fig. 3.2. The independence of recombination frequencies and genome size; (a) Total genetic map length is about 10 morgans (M) over a nearly 1000 times range in genome size. (b) The length of the genetic map of individual genes, measured by intragenic crossing over, is about 0.1 cM irrespective of genome size. Data from Thuriaux (1977).

stays fairly constant in eukaryotes despite vast increases in genome size. The idea that each replicon contains a recombinator site (Whitehouse, 1963) might be true for low C-value eukaryotes, but Thuriaux's data argues against it for those with large genomes, in which it appears that most replicons neither contain coding sequences nor recombinator sequences.

It is clear that the synaptonemal complexes that mediate meiotic chromosome pairing are longer in organisms with larger genomes than in those with small genomes; this implies that such organisms need more sites for interaction with the complex. However, it is not known whether such sites have specific sequences or are randomly selected; nor is there any reason to equate them with recombinator sequences or to see the variations in length of synaptonemal complexes of different species as anything other than a necessary consequence of variations in genome size.

NUMBERS OF SEGREGATOR GENES

Centromeric sequences

The major DNA sequences directly involved in chromosome segregation are those in the centromere regions. The classical comparative evidence (Darlington, 1937; Swanson, 1958) as well as more recent data from recombinant DNA studies of centromeres (Fitzgerald-Hayes *et al.*, 1982) shows that they have a definite specificity. Electron microscopy has shown that microtubules bind to specific centromeric regions, the kinetochores, and that the number of microtubules per centromere is highly variable from species to species. In the lowest C-value species with minute chromosomes, e.g. yeast and some protozoa, there is only one microtubule per kinetochore, whereas in high C-value species with large chromosomes, e.g. *Haemanthus*, there are many hundreds. This suggests that the number of microtubules per centromere is evolutionarily adapted to the size of the chromosome, presumably for mechanical reasons.

This means that the number of microtubule-binding sites at the centromere must also vary in proportion to genome size. The similarity in diameter of the 25 nm microtubules and the basic 30 nm chromatin fibre suggests that microtubules may bind directly to such threads. Since the binding sites must be genetically determined, and since the 30 nm chromatin threads have DNA on their outer surface, such binding almost certainly involves specific DNA sequences. The number of such DNA sequences should therefore vary in proportion to genome size. Bennett *et al.* (1981) have shown that in the interphase nuclei of flowering plants centromeric chromatin is ultrastructurally distinguishable from bulk chromatin; the total centromeric volume can therefore be measured from electron micrographs of serial sections, and shown to be linearly related to genome size (Fig. 3.3). Though there is no direct evidence that such chromatin consists of a special category of DNA one can argue *a priori* that centromeric DNA as a whole must have some specificity. Firstly, is the need to ensure that microtubule-binding sites appear on the surface and with the correct, and opposite, orientation in both daughter chromatids; secondly, there must be some specificity to account for the later (anaphase) splitting of centromeres compared with the rest of the chromosome. These are functional reasons why centromeric DNA should contain some specific sequences and why it should vary in amount proportionally with genome size. The need for a large number of correctly oriented microtubule-binding sites in species with larger genomes means that their centromeric DNA must contain repetitive sequences. It is well known that in many organisms highly repetitive satellite DNA is found in centromeric regions, often, but by no means always, exclusively so (John and Miklos, 1979). The fact that some organisms lack highly repetitive satellite DNAs altogether does not contradict my thesis since there is no necessary functional requirement for the repeated sequences in centromeric DNA to be contiguous, nor for their base composition to differ markedly from that of bulk DNA, both of which are necessary for the detection of highly repetitive satellites. The functional requirements are for identical microtubule-binding sites to be interspersed in other DNA sequences that are constrained to fold

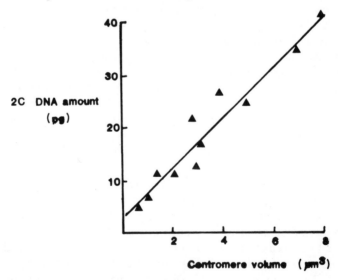

Fig. 3.3. Proportionality between the total volume of centromeric chromatin and DNA content per nucleus in 11 species of flowering plants of differing genome size. From Bennet *et al.* (1981); reprinted by permission of the Company of Biologists Ltd.

so as to orient them correctly. The exact sequence of most centromeric DNA might be irrelevant to its function: what seems likely to be important is the *length* of the interspersing sequences that lie between the microtubule binding sites. Such interspersing sequences might also be repetitive but they need not be: if their sequence is not under selective constraints, their detectability as highly repetitive DNA, and the degree of conservation of any repeats, would simply depend on how long ago they originated, whether by unequal crossing-over (Smith 1976) or by some other duplication mechanism.

Fitzgerald-Hayes *et al.* (1982) cloned the centromeric DNA from chromosomes 3 and 11 of the yeast *Saccharomyces*, and showed that each contains three different sequence elements of 14, 11, and 11 base pairs (bp) that are identical for the two centromeres and separated from each other by longer non-conserved sequences of 87 and 248 bp in CEN 3 and 88 and 251 bp in CEN 11. The first two conserved sequences (called elements I and III) are identical at 12 and 9 sites, respectively, to similar sequences in *Drosophila* satellite DM 359 that are also separated by 90 bp; since deletion mutants of plasmids lacking elements I and III, but retaining element IV, lose their mitotic stability (Bloom *et al.*, 1983) these sequences are probably essential for the attachment of spindle microtubules, which may be mediated by the five sequence-specific DNA-binding polypeptides isolated by Bloom *et al.* (1983). Sequences closely similar to elements I and II, and separated by 89 base bases are found in the cloned centromere of yeast chromosome 6 (Panzeri and Philippsen, 1982); there are element IV sequences in chromosome 6, however, each of 10 bp, which are also more closely spaced than in the other 2 chromosomes.

The conservation of the centromeric sequence elements I and III and their separation by an AT-rich segment of defined length in both yeast and *Drosophila* must be reconciled with the fact that the third yeast-conserved element (IV) was not detected in the *Drosophila* satellite. This implies that the arrangement of the centromeric DNA-binding sites may be different in yeast from other eukaryotes. Since yeast chromatin threads appear to be heterogeneous in thickness, with some regions of 25–30 nm, others of 10 nm and others of about 4 nm (Rattner *et al.*, 1982), whereas other eukaryotes have more uniform 25–30 nm solenoidal chromatin threads, I suggest that the microtubule binding site(s) of yeast centromeres are in a 10 nm thick region consisting of a single row of nucleosomes. This would allow a different mode of microtubule attachment for yeast (Fig. 3.4(a)) compared with organisms with typical solenoids (Fig. 3.4(b)).

A model for microtubule/chromatin interactions in yeast (Fig. 3.4(a)) and other eukaryotes (Fig. 3.4(b)) is suggested by the conserved spacing of the AT-rich region between sequence elements I and III. This spacing is such that when wrapped around the histone core of a nucleosome the two sequences would be adjacent: therefore it is suggested that they together

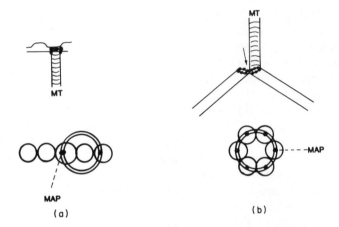

Fig. 3.4.   (a) Model for microtubule/chromatin interaction
at the yeast centromere. Each microtubule (MT) is bound to
a 10 nm region of the chromatin fibril by two microtubule-
associated-proteins (MAPs) shown in black: one forms a
strong primary attachment to the element I/III binding site
on the kinetochore nucleosome, while the other forms a
weaker secondary attachment to element IV. (b) Model for
microtubule/chromatin interaction in eukaryotes with more
regular 30 nm chromatin fibres. The chromatin is assumed
to take the form of a solenoid that can be kinked (arrow) so
as to allow the end of a microtubule to bind end-on to 1 gyre
of the solenoid. This enables several adjacent kinetochore
nucleosomes to bind via their element I/III binding sites to
different MAPs attached to the same microtubule so as to
form a stronger multiple attachment to the microtubule:
therefore no element IV is present.

form a single binding site for a microtubule that is conserved throughout all
eukaryotes. This nucleosome probably has a unique structure since nuclease
digestion studies show that it contains 220 bp rather than the 165 bp typically
found in yeast nucleosomes (Bloom *et al.*, 1983): I refer to it as the kinetochore nucleosome. Since tubulin does not bind to DNA, but microtubule-
associated-proteins (MAPs) do (Villasante *et al.*, 1981), I suggest that the
element I/III site binds specifically to a MAP. I further suggest that in yeast
the 11 base pair element IV forms a separate MAP-binding site on another
nucleosome for a separate MAP molecule on the opposite side of the microtubule; this second site would stabilize but be unable to initiate the binding
of a microtubule. The fact that in chromosomes 3 and 11 five bases (TTTAG)
are identical in elements III and IV, but inverted with respect to each other,
is in keeping with this hypothesis since the MAPs on opposite sides of the
microtubule must also be inverted with respect to each other. Electron
micrographs suggest that the microtubule/chromatin junction at the yeast

centromere may in fact be very slender (smaller in diameter than the micro-tubule itself—Fig. 1 of Peterson and Ris, 1976).

In eukaryotes with more regular 25–30 nm chromatin solenoids a stronger mode of attachment is possible if the solenoid is kinked so as to allow a complete gyre to be bound directly to the end of a microtubule (Fig. 3.4(b)). In this case no element IV binding site is needed: if there are 6 nucleosomes in a gyre the I/III compound binding site on each can bind to 6 MAPs attached to the complementary gyre at the end of the spindle microtubule. Not only will this provide 6 compound binding sites per microtubule, and therefore much stronger attachment, but it also means that the structure of the centrometic binding sites are inherently repetitive, with the repeats phased with the nucleosomes. In African green monkeys periodicity of the α-satellite is in phase with the nucleosomes (Maio *et al.*, 1977); this satellite is located at the centromere, as well as in the nucleolus; other primates have similar satellites and these have short regions of similarity to satellites in many other mammals (Singer, 1982). However, this phasing may have nothing to do with the phasing of the kinetochore nucleosomes themselves: because of the large amount of the satellite it more probably functions to fold the centromeric chromatin surrounding the kinetrochore nucleosomes in a regular paracrystalline fashion so as to ensure that all their microtubule-binding sites are oriented in the same direction. Even in yeast several nucleo-somes on either side of the kinetochore nucleosome are strictly phased, this phasing being controlled by the DNA sequence of their own DNA, and not merely by the presence of the kinetochore nucleosome (Bloom *et al.*, 1983); I suggest that this strict phasing is necessary for the packing of the perikineto-chore chromatin in such a way as to orient the two sister kinetochores in opposite directions.

It is interesting to consider whether the yeast pattern of variable thickness chromatin fibres, and centromeres with, probably, only two dissimilar MAP-binding sites, is primitive or derived from that with several tandemly arranged identical MAP-binding sites postulated for eukaryotes with regular 25 nm solenoidal chromatin threads. Though I once postulated that yeasts might be the most primitive eukaryotes (Cavalier-Smith, 1980, 1981) a better under-standing of protozoan classification and phylogeny (Cavalier-Smith, 1983a and in preparation) and stronger arguments for the symbiotic origin of mito-chondria (Cavalier-Smith, 1983b) have caused me to abandon this view. I now think that the numerous protozoa in the subkingdom Archezoa (Cavalier-Smith, 1983a) which all lack mitochondria, are the most primitive protozoa, and that the apparent simplicity of yeasts is the result of simplific-ation from the more complex ciliated chytridiomycote fungi which in turn probably evolved from the choanociliate protozoa or close relatives. Among the Archezoa, the Microspora which, like bacteria, have 70s ribosomes (Ishi-hara and Hayashi, 1968), may be the most primitive surviving eukaryotes, even though they are all now intracellular parasites. Like yeasts they have only a single microtubule per centromere (Raikov, 1982), which is probably the primitive condition in eukaryotes (Cavalier-Smith, 1981). It will be important to determine whether they have yeast type or more typical 30 nm chromatin fibres, and to clone these centromeres for sequence comparison with yeast and with animals; I predict that the Microspora will have 30 nm threads and that the yeast condition will eventually be shown to be second-arily derived. One also may speculate that the basic packing of the 30 nm solenoid initially evolved in a common ancestor of all eukaryotes as a direct consequence of the attachment of 10 nm threads in a gyre at the end of a microtubule, the original function of which has been proposed to be chromo-some segregation (Pickett-Heaps, 1974; Cavalier-Smith, 1975).

If kinetochore DNA in all eukaryotes except yeast is necessarily a tandem hexamer, as the model suggests, it will be very prone to unequal crossing-over, which could generate repeats much longer than are functionally necessary. The spacer sequences of centromeric DNA would therefore also tend to become highly repeated and the number of microtubule-binding sites per centromere to increase. The importance of multiple binding sites and of many microtubules per centromere in species with larger genomes is unlikely to be simply because of the greater force needed to pull the chromosome, since this is so small ($\sim 10^{-13}$ N/chromosome, Taylor, 1965) and of the same order of magnitude as the force generated by a *single* myosin ATPase in striated muscle (Bagshaw, 1982) that a single microtubule ought easily to be able to pull a mammalian chromosome at the rate observed. A key factor may be the greater cell volume (Chapter 4) of high C-value species, coupled with the fact that spindle microtubules grow from the poles and only attach secondarily to the centromere. If there is only one chromosomal spindle microtubule and one primary microtubule-binding site per chromosome, the time taken to 'find' each other by random movement increases greatly as the cell volume increases; such drastic increases in the time needed for mitotic premetaphase could be prevented by having numerous spindle microtubules and microtubule-binding sites per centromere, which I suggest may be the selective advantage of having more spindle microtubule in larger cells. Alter-natively if spindle microtubules, as many now think, function not to generate the force but to *resist* in a controlled way the real motive force generated by *another* spindle component so as to slow down the rate of chromosome movement (King, 1983) then an even simpler explanation is obvious: larger spindles (needed to segregate larger chromosomes in larger cells) might generate more force to be resisted, so would require more microtubules.

Transposition of centromeric DNA to numerous sites along the chromo-somes can account for the evolution of 'diffuse' or polycentromeric centro-meres as in the plant *Luzula*. Such transpositions would be viable only if they were interspersed with the correct periodicity with respect to the chromosomal folding pattern to ensure the correct orientation of the micro-tubule-binding sites in the metaphase chromosomes. Moreover, they would probably be strongly selected against in high C-value organisms since large monocentric chromosomes can be packed much more easily than large poly-centric chromosomes into the metaphase plate, which inevitably has a limited area; chromosome size increases in direct proportion to genome size (Chapter 4), whereas the area of the metaphase plate only increases with the two-thirds power of genome size (this is because cell volume increases in exact proportion to genome size (Chapter 4) and the metaphase plate increases in area with the two-thirds power of cell volume). Therefore cells with large genomes will be under stronger selection than those with small genomes against the origin of polycentric chromosomes. Not surprisingly, species with diffuse centromeres have relatively small genomes, usually 1 pg or less; the largest recorded value, for *Luzula purpurea* of 4.3 pg, is 20 times less than the maximum for species with monocentric chromosomes.

Telomere DNA sequences

Telomeres have three functions or properties that must depend on specific DNA sequences.

1. They confer a stability on natural chromosome ends much greater than that of broken chromosomes or DNA molecules (Muller, 1938).
2. They bind specifically to the nuclear envelope during meiotic prophase; and also during interphase, at least in dinozoan protozoa but perhaps in all eukaryotes (this is discussed further in Chapter 4).
3. There is a fundamental problem in the replication of the ends of linear DNA molecules which have a basic tendency to get shorter with each replication (Watson, 1972); the various possible ways of avoiding this problem require special sequences and enzymes able to recognize them (Cavalier-Smith, 1974, 1983c).

There is much evidence that in high C-value eukaryotes telomeres commonly contain heterochromatin and/or highly repeated DNA. Recently telomeres or probable telomeres have been cloned from protozoa (*Oxytricha* minichro-mosomes; *Tetrahymena* rDNA plasmid) and from the low C-value yeast, *Saccharomyces*; in all cases they contain highly-repeated CA-rich DNA sequences (Cavalier-Smith, 1983c). Since the protozoan telomeres confer replicative stability on artificial plasmids in yeast it is clear that functions 1 and 3 at least are highly conserved in eukaryotes and depend on specific highly-repetitive DNA sequences. It is not known if the same sequences are responsible for chromosome binding to the nuclear envelope; it seems unlikely that all the telomeres of *Oxytricha* minichromosomes can be bound to the nuclear envelope. But in high C-value species, at least, each telomere will probably have several binding sites for the nuclear envelope, just as the centromeres do for microtubules. It has been suggested (Cavalier-Smith, 1982) that DNA-binding to the nuclear envelope is not direct but via the lamin proteins of the nuclear lamina. Telomeres of large chromosomes should therefore contain highly repeated lamin-binding sites as well as spacer sequences that need not necessarily be highly repeated; but for the same reasons as in the centromere, highly repeated DNA would be very likely to evolve as a result of mutational accidents. It is possible, though not *a priori* essential, that the postulated telomeric lamin-binding sites are the same as the sequences that confer replicative stability on the telomeres; the stability of telomeres might even result from their binding to lamins. It is clear that *Tetrahymena* cells have a mechanism able to multiply the number of repeats at the ends of a DNA molecule, since this clearly occurs during the formation of rDNA by gene amplification during formation of the macronucleus. Such a mechanism could provide the basis for correcting the chromosome short-ening that would otherwise tend to occur during replication (Cavalier-Smith, 1983c). If all eukaryotes have such a mechanism, then occasional mistakes in it could provide a saltatory increase in the amount of highly repetitive DNA, as has been frequently postulated.

CONSTITUTIVE HETEROCHROMATIN AND TELOMERIC AND CENTROMERIC DNA

If, as argued previously, telomeric and centromeric functions both require some highly repetitive sequences, they can provide the fundamental biological explanation for the occurrence of such sequences in eukaryotes and their absence in prokaryotes. But these functions do not *necessitate* the occurrence of large enough blocks of highly-repeated DNA of sufficiently different base sequence to be detectable on density gradients as satellites, or as constitutive heterochromatin by chromosome banding techniques. Indeed, the variations in the *amounts* of such sequences in different eukaryotes bears no clear relationship either to the amounts needed for established centromeric or telomeric functions or to genome size. The simplest expla-nation for this is that the overall amount of highly-repetitive DNA is unre-lated to the functional needs of the organism, but results from what I shall call *intragenomic drift*, caused by accidental amplifications and deletions, that makes its amount fluctuate within a total DNA amount determined by the selective forces that act on the overall genome size, to be discussed in Chapter 4. The late-replicating properties of constitutive heterochromatin may result from a relative paucity of replicons and/or its special folding properties (Chapter 7).

Though constitutive heterochromatin and highly-repetitive DNA are usually concentrated at centromeres or telomeres they are also found interstitially; interstitial localization could arise by transposition of centromeric or telomeric sequences. However, if any interstitial sequences (e.g. replicon origins, replicator sequences) have inherently a small-scale repetitive structure they also might be prone secondarily to evolve into highly repetitive DNA. Each replicon is probably attached to the proteinaceous nuclear matrix (Maul, 1982); if specific matrix-binding sites are found at intervals along the DNA they would be dispersed throughout the genome like middle-repetitive DNA, though the actual binding sites would be 10-fold shorter than the middle-repetitive sequences. Such sites also could evolve into highly-repetitive DNA if they were internally repetitive: however the relative rarity of interstitial heterochromatin perhaps suggests that specific matrix-binding sites are usually not internally repetitive.

ABSENCE OF CORRELATION OF GENE NUMBERS WITH GENOME SIZE

The various categories of genes discussed above can be divided into two classes:

1. Those whose numbers and size do not correlate with genome size, i.e. protein-coding genes, recombinator genes;
2. Those whose numbers but not size, do increase with genome size, i.e. rRNA and tRNA genes, replicon origins, centromere genes.

As the latter class makes up only a small fraction of total DNA its contribution to total genome size is relatively small; the same is true of telomeric DNA, for which the relationship with genome size is unknown, and for the hypothetical matrix-binding sequences. Moreover, for those genic sequences that do increase with genome size, the increase is explicable as the functionally necessary *consequence* either of significant increases in genome size itself (in the case of replicon origins) or for the universally correlated increases in cell volume (in the case of rRNA, tRNA and centromere genes): there is no reason why an increase in any or all of such sequences should necessarily cause a proportional increase in genome size as a whole.

Thus all available evidence shows that it is wrong to equate genome size with the number of genes. The accumulation of large amounts of DNA in higher C-value eukaryotes therefore has nothing to do with 'accumulating genetic information in adaptive evolution' (Kimura, 1961). The extra DNA must either have a non-genic function or be useless.

One non-genic function sometimes proposed is that of 'protection' of the genic DNA, presumably against mutagens. There is in fact extensive experimental evidence for a universal relationship between genome size and sensitivity to mutagens and the lethal effects of radiation. However, as Fig. 3.5 shows, large genomes are proportionally *more* sensitive to radiation and

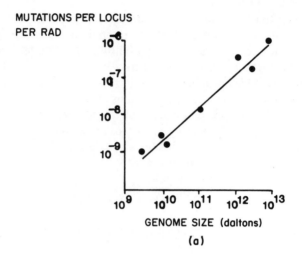

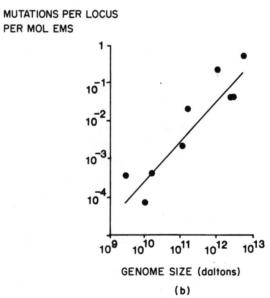

Fig. 3.5. Proportionality between induced mutation rate and genome size. (a) X-ray- induced mutations. (b) Mutations induced by ethyl methane sulphonate. Data from Heddle and Athanasiou (1975).

chemical mutagens. This means that non-genic DNA, far from having a protective effect, makes cells possessing it more likely to be killed or mutated.

The evolution of elaborate DNA repair mechanisms indicates that cells are under continuous powerful selection to minimize such damage (Maynard Smith, 1978). Extra non-functional, non-genic DNA will therefore always be harmful, and selected against, and such counter-selection will be stronger in cells with larger genomes. Evolutionary increases in genome size can therefore occur only if there is a positive evolutionary force sufficient to counterbalance not only this disadvantage, but also the extra demands for energy and nutrients such as phosphate that useless DNA would place on the cell. The next chapter presents the evidence for my theory that this force is positive selection for extra DNA that functions as a nuclear skeleton so as to increase the volume of the nucleus in larger cells, which I argue occurs because nuclear volume has to coevolve with cell volume for basic functional reasons.

REFERENCES

Allen, J. R., Roberts, T. M., Loeblich, A. R. III and Klotz, L. C. (1975). Characterization of the DNA from the dinoflagellate, *Crypthecodinium cohnii*, and the implications for nuclear organisation. *Cell*, **6**, 161–169.

Axel, R., Feigelson, P. and Schutz, G. (1976). Analysis of the complexity and diversity of mRNA from chicken oviduct and liver. *Cell*, **7**, 247–254.

Bagshaw, C. R. (1982). *Muscle Contraction*. Chapman and Hall, London.

Baralle, F. E. (1983). The functional significance of leader and trailer sequences in eukaryotic mRNAs. *Int. Rev. Cytol.*, **81**, 71–106.

Bennett, M. D. and Smith, J. B. (1976). Nuclear DNA amounts in angiosperms. *Phil. Trans. Roy. Soc. B*, **274**, 227–274.

Bennett, M. D., Smith, J. B. and Heslop-Harrison, J. S. (1982). Nuclear DNA amounts in angiosperms. *Proc. R. Soc. B*, **216**, 179–199.

Bennett, M. D., Smith, J. B., Ward, J. and Jenkins, G. (1981). The relationship between nuclear DNA content and centromere volume in higher plants. *J. Cell Sci.*, **47**, 91–115.

Bernstein, S. L., Gioio, A. E. and Kaplan, B. B. (1983). Changes in gene expression during postnatal development of the rat cerebellum. *J. Neurogen.*, **1**, 71–86.

Birnstiel, M. L., Chipchase, M. and Spiers, J. (1971). The ribosomal RNA cistrons. *Progr. Nucleic Acid Res. Molec. Biol.*, **11**, 351–389.

Bloom, K. S., Fitzgerald-Hayes, M. and Carbon, J. (1983). Structural analogues and sequence organisation of yeast centromeres. *Cold Spring Harbor Symp. Quant. Biol.*, **47**, 1175–1185.

Boswell, R. E., Klobutcher, L. A. and Prescott, D. M. (1983). Inverted terminal repeats are added to genes during macronuclear development in *Oxytricha nova*. *Proc. Nat. Acad. Sci. U.S.A.*, **79**, 3255–3259.

Bouchard, R. A. (1976). DNA amount and organization in some lower vascular plants. PhD thesis. University of Chicago.

Cavalier-Smith, T. (1974). Palindromic base sequences and replication of eukaryote chromosome ends. *Nature*, **250**, 467–470.

Cavalier-Smith, T. (1975). The origin of nuclei and of eukaryotic cells. *Nature*, **256**, 463–468.

Cavalier-Smith, T. (1980). How selfish is DNA? *Nature*, **285**, 617–618.

Cavalier-Smith, T. (1981). The origin and early evolution of the eukaryotic cell. In

Carlile, M. J., Collins, J. F. and Moseley, B. E. B. (Eds), *Molecular and Cellular Aspects of Microbial Evolution*, Society for General Microbiology Symposium *32*, Cambridge University Press, Cambridge, pp. 33–84.

Cavalier-Smith, T. (1982). Evolution of the nuclear matrix and envelope. In G. G. Maul (Ed.), *The Nuclear Envelope and the Nuclear Matrix*, Liss, New York, pp. 307–318.

Cavalier-Smith, T. (1983a). A 6-kingdom classification and a unified phylogeny. In H. E. A. Schenk and W. Schwemmler (Eds), *Endocytobiology II. Intracellular Space as Oligogenetic Ecosystem*, de Gruyter, Berlin, pp. 1027–1034.

Cavalier-Smith, T. (1983b). Endosymbiotic origin of the mitochondrial envelope. In H. E. A. Schenk and W. Schwemmler (Eds), *Endocytobiology II. Intracellular Space as Oligogenetic Ecosystem*, de Gruyter, Berlin, pp. 265–279.

Cavalier-Smith, T. (1983c). Cloning chromosome ends. *Nature*, **301**, 112–113.

Cavalli-Sforza, L. L. and Bodmer, W. F. (1971). *The Genetics of Human Populations*, Freeman, San Francisco.

Chikaraishi, D. M., Deeb, S. S. and Sueoka, N. (1978). Sequence complexity of nuclear RNAs in adult rat tissues. *Cell*, **13**, 111–120.

Cullis, C. A. (1982). Quantitative variation of the ribosomal RNA genes. In E. G. Jordan and C. A. Cullis (Eds), *The Nucleolus*, Cambridge University Press, Cambridge, pp. 102–112.

Darlington, C. D. (1937). *Recent Advances in Cytology* (2nd ed.) Churchill, London.

Darlington, C. D. (1969). What we do *not* know about chromosomes. In C. D. Darlington and K. R. Lewis (Eds), *Chromosomes Today*, Oliver and Boyd, Edinburgh.

Davidson, E. and Hough, B. R. (1971). Genetic information in oocyte RNA. *J. Mol. Biol.*, **56**, 491–506.

Dubroff, L. M. and Nemer, M. (1976). Developmental shifts in the synthesis of hn RNA classes in the sea urchin embryo. *Nature*, **260**, 120–124.

Firtel, R. A. and Lodish, H. F. (1973). A small nuclear precursor of messenger RNA in the cellular slime mould *Dictyostelium discoideum*. *J. Mol. Biol.*, **79**, 295–314.

Fitzgerald-Hayes, M., Clarke, L. and Carbon, J. (1982). Nucleotide sequence comparisons and functional analysis of yeast centromere DNAs. *Cell*, **29**, 235–244.

Fouquet, H., Bierweiler, B. and Sauer, H. W. (1974). Reassociation kinetics of nuclear DNA from *Physarum polycephalum*. *Eur. J. Biochem.*, **44**, 407–410.

Friz, C. T. (1968). The biochemical composition of the free-living amoebae *Chaos chaos*, *Amoeba dubia* and *Amoeba proteus*. *Comp. Biochem. Physiol.*, **26**, 81–90.

Galau, G. A., Britten, R. J. and Davidson, E. H. (1974). A measurement of the sequence complexity of polysomal mRNA in sea urchin embryos. *Cell*, **2**, 9–22.

Galau, G. A., Klein, W. H., Davis, M. M., Wold, B. J., Britten, R. J. and Davidson, E. H. (1976). Structural gene sets active in embryos and adult tissues of the sea urchin. *Cell*, **7**, 487–506.

Galeotti, C. L., Sriprakash, K. S., Batum, C. M. and Clark-Walker, G. D. (1981). An unexpected response of *Torulopsis glabrata* fusion products to X-irradiation. *Mutation Research*, **81**, 155–164.

Gall, J. G. (1968). Differential synthesis of the genes for ribosomal RNA during amphibian oogenesis. *Proc. Natn. Acad. Sci. USA*, **60**, 553–560.

Gelderman, A. H., Bartgis, I. L., Keister, D. B. and Diamond, L. S. (1971). A comparison of genome sizes and thermal denaturation-derived base composition of DNAs from several members of *Entamoeba* (histolytica group). *J. Parasitol.*, **57**, 912–916.

Goldberg, R. B., Hoschek, G., Kamalay, J. C. and Timberlake, W. E. (1978).

Sequence complexity of nuclear and polysomal RNA in leaves of the tobacco plant. *Cell*, **14**, 123–132.

Heddle, J. A. and Athanasiou, K. (1975). Mutation rate, genome size and their relation to the rec. concept. *Nature*, **258**, 359–361.

Hendrickson, W., Yamaki, H., Murchie, J., King, M., Boyd, D., Schaechter, M. (1981). Specific binding of *Escherichia coli* replicative origin DNA to membrane preparations. *ICN-UCLA Symposium on Molecular and Cellular Biology* Vol. 22, pp. 79–90.

Hereford, L. M. and Rosbash, M. (1977). Number and distribution of polyadenylated RNA sequences in yeast. *Cell*, **10**, 453–462.

Hinegardner, R. (1976). Evolution of genome size. In F. J. Ayala (Ed.), *Molecular Evolution*, Sinauer, Sunderland, pp. 179–199.

Hinegardner, R. and Rosen, D. E. (1972). Cellular DNA content and the evolution of teleostean fishes. *Am. Nat.*, **106**, 621–644.

Holliday, R. (1968). Genetic recombination in fungi. In W. J. Peacock and R. D. Brock (Eds), *Replication and Recombination of Genetic Material*, Australian Academy of Sciences, Canberra, pp. 157–174.

Hough, B. R., Smith, M. J., Britten, R. J. and Davidson, E. H. (1975). Sequence complexity of hn RNA in sea urchin embryos. *Cell*, **5**, 291–300.

Howell, S. H. and Walker, L. L. (1977). Transcription of the nuclear and chloroplast genomes during the vegetative cell cycle in *Chlamydomonas reinhardi*. *Devel. Biol.*, **56**, 11–23.

Hudspeth, M. E. S., Timberlake, W. E. and Goldberg, R. B. (1977). DNA sequence organization in the water mold *Achlya*. *Proc. Natl. Acad. Sci. USA*, **74**, 4332–4336.

Ishihara, R. and Hayashi, Y. (1968). Some properties of ribosomes from the sporoplasm of *Nosema bombycis*. *J. Invert. Pathol.*, **11**, 377–385.

John, B. and Miklos, G. L. G. (1979). Functional aspects of satellite DNA and heterochromatin. *Int. Rev. Cytol.*, **58**, 1–114.

Kimura, M. (1961). Natural selection as the process of accumulating genetic information in adaptive evolution. *Genet. Res.*, **2**, 127–140.

King, S. M. (1983). A regulatory model for spindle function during mitosis. *J. Theor. Biol.*, **102**, 501–510.

Kleene, K. C. and Humphreys, T. (1977). Similarity of hn RNA sequences in blastula and pluteus stage sea urchin embryos, *Cell*, **12**, 143–155.

Lengyel, J. and Penman, S. (1975). hn RNA size and processing as related to different DNA content in two dipterans: *Drosophila* and *Aedes*, *Cell*, **5**, 281–290.

Leon, W., Fouts, D. L. and Manning, J. (1978). Sequence arrangement of the 16S and 26S rRNA genes in the pathogenic haemoflagellate *Leishmania donovani*. *Nucl. Acids Res.*, **5**, 491–504.

Lewin, B. (1980). Gene expression 2nd ed. Vol. 2. Eukaryotic chromosomes. Wiley, New York.

Levy, W. B. and McCarthy, B. J. (1975). Messenger RNA complexity in *Drosophila melanogaster*. *Biochemistry*, **14**, 2440–2446.

Maio, J. J., Brown, F. L. and Musich, P. R. (1977). Subunit structure of chromatin and the organisation of eukaryotic highly repetitive DNA: recurrent periodicities and models for the evolutionary origins of repetitive DNA. *J. Mol. Biol.*, **117**, 637–656.

Manners, J. G. and Myers, A. (1975). The effect of fungi (particularly obligate pathogens) on the physiology of higher plants. *Symp. Soc. Exp. Biol.*, **29**, 279–296.

Maul, G. G. (Ed.). (1982). *The Nuclear Envelope and the Nuclear Matrix*, Liss, New York.

Maynard Smith, J. (1978). *The Evolution of Sex.* Cambridge University Press, Cambridge.

Meacock, P. A. and Cohen, S. N. (1980). Partitioning of bacterial plasmids during cell division: a cis-acting locus that accomplishes stable plasmid inheritance. *Cell*, **20**, 529–542.

Melera, P. W. (1980). Transcription in the myxomycete *Physarum polycephalum.* In W. F. Dove and H. P. Rusch (Eds), *Growth and Differentiation in* Physarum polycephalum. Princeton University Press, Princeton, New Jersey, pp. 64–97.

Miksche, J. P. (1967). Variation in DNA content of several gymnosperms, *Can. J. Genet. Cytol*, **9**, 717–722.

Minagawa, T., Wagner, B. and Strauss, B. (1959). The nucleic acid content of *Neurospora crassa. Arch. Biochem. Biophys.*, **80**, 442–445.

Mohberg, J. and Rusch, H. P. (1971). Isolation and DNA content of nuclei of *Physaram polycephalum. Exp. Cell. Res.*, **66**, 305–316.

Muller, H. J. (1938). The remaking of chromosomes. *Collect. Net.*, **13**, 181–195, 198.

Muller, H. J. (1950). Our load of mutations. *Amer. J. Hum. Genet.*, **2**, 111–176.

Nemer, M., Dubroff, L. M. and Graham, M. (1975). Properties of sea urchin embryo mRNA containing and lacking poly (A). *Cell*, **6**, 171–178.

Nemer, M., Graham, M. and Dubroff, L. M. (1974). Coexistence of non-histone mRNA species lacking and containing poly (A) in sea urchin embryos. *J. Mol. Biol.*, **89**, 435–454.

O'Brien, S. J. (1973). On estimating functional gene numbers in eukaryotes. *Nature New Biol.*, **242**, 52–54.

Ohno, S. (1972). Simplicity of mammalian regulatory systems. *Devel. Biol.*, **27**, 131–136.

Olmo, E. (1983). Nucleotype and cell size in vertebrates: a review. *Bas. Appl. Histochem.*, **27**, 227–256.

Panzeri, L. and Philippsen, P. (1982). Centromeric DNA from chromosome VI in *Saccharomyces cerevisiae* strains. *The Embo Journal*, **1**, 1605–1611.

Peterson, J. B. and Ris, H. (1976). Electron microscopic study of the spindle and chromosome movement of the yeast *Saccharomyces cerevisiae. J. Cell Sci.*, **22**, 219–242.

Pickett-Heaps, J. (1974). Evolution of mitosis and the eukaryotic condition. *BioSystems*, **6**, 37–45.

Price, H. J., Levis, R. W., Coggins. L. W. and Sparrow, A. H. (1972). High DNA-content of *Sprekelia formosissima* Herbert (Amaryllidaceae) and *Ophioglossum petiolatum* Hook. (Ophioglossaceae). *Exptl. Cell. Res.*, **73**, 187–191.

Price, H. J., Sparrow, A. H. and Naumann, A. F. (1973). Evolutionary and development considerations of the variability of nuclear parameters in higher plants. I Genome volume, interphase chromosome volume and estimated DNA content of 236 gymnosperms. *Brookhaven Symp. Biol.*, **25**, 390–421.

Raff, R. A. and Kaufman, T. C. (1983). *Embryos, Genes and Evolution: the Developmental-Genetic Basis of Evolutionary Change*, Macmillan, New York.

Raikov, I. B. (1982). *The Protozoan Nucleus: Morphology and Evolution*, Springer-Verlag, Wien and New York.

Rattner, J. B., Saunders, C., Davie, J. R. and Hamkalo, B. A. (1982). Ultrastructural organization of yeast chromatin. *J. Cell Biol.*, **93**, 217–222.

Rozek, C. E., Orr, W. C. and Timberlake, W. E. (1978). Diversity and abundance of polyadenylated RNA from *Achlya ambisexualis. Biochemistry*, **17**, 716–722.

Shuter, B. J., Thomas, J. E., Taylor, W. D. and Zimmerman, A. M. (1983). Phenotypic correlates of genomic DNA content in unicellular eukaryotes and other cells. *Amer. Nat.*, **122**, 26–44.

Singer, M. F. (1982). Highly repeated sequences in mammalian genomes. *Int. Rev. Cytol.*, **76**, 67–112.

Smith, G. P. (1976). Evolution of repeated DNA sequences by unequal crossovers. *Science*, **191**, 528–535.

Stingo, V., du Buit, M.-H. and Odierna, G. (1980). The genome size of some selachian fishes. *Bollettino di Zoologia*, **47**, 129–137.

Sulston, J. E. and Brenner, S. (1974). The DNA of *Caenorhabditis elegans*. *Genetics*, **77**, 95–104.

Sussman, R. and Rayner, E. P. (1971). Physical characterization of DNAs in *Dictyostelium discoideum*. *Arch. Biochem. Biophys.*, **144**, 127–137.

Swanson, C. P. (1958). *Cytology and Cytogenetics*, Macmillan, London.

Swanton, M. T., Heumann, J. M. and Prescott, D. M. (1980). Gene-sized DNA molecules of the macronuclei in three species of hypotrichs: size distributions and absence of nicks. DNA of ciliated protozoa VIII. *Chromosoma*, **77**, 217–227.

Taylor, E. W. (1965). *Proc. Int. Cong. Rheol.* 4th. pp. 222–260.

Timberlake, W. E., Shumard, D. S. and Goldberg, R. B. (1977). Relationship between nuclear and polysomal RNA populations of *Achlya*: a simple eukaryotic system. *Cell*, **10**, 623–632.

Thuriaux, P. (1977). Is recombination confined to structural genes on the eukaryotic genome? *Nature, Lond.*, **268**, 460–462.

Van Ness, J. and Hahn, W. E. (1982). Physical parameters affecting the rate and completion of RNA driven hybridization of DNA: new measurements relevant to quantitation based on kinetics. *Nucl. Acids. Res.*, **10**, 8061–8077.

Villasante, A., Corces, V. G., Manso-Martinez, R. and Avila, J. (1981). Binding of microtubule protein to DNA and chromatin: possibility of simultaneous linkage of microtubule to nucleic acid and assembly of the microtubule structure. *Nucl. Acids Res.*, **9**, 895–908.

Walter, P. and Blobel, G. (1982). Signal recognition particle contains a 7S RNA essential for protein translocation across the endoplasmic reticulum. *Nature, Lond.*, **299**, 691–698.

Watson, J. D. (1972). Origin of concatemeric T7 DNA. *Nature New Biol.*, **239**, 197–200.

Watson, J. D. (1976). *Molecular Biology of the Gene*, 3rd edn, Benjamin, London.

White, M. J. D. (1973). *Animal Cytology and Evolution*, 3rd edn, Cambridge University Press, Cambridge.

Whitehouse, H. L. K. (1963). *Towards an Understanding of the Mechanism of Heredity*. Edward Arnold, London.

Whitehouse, H. L. K. (1982). *Genetic Recombination: Understanding the Mechanisms*, Wiley, Chichester.

Whitfield, P. R. (1952). Nucleic acids in erythrocytic stages of a malaria parasite. *Nature, Lond.*, **169**, 751–752.

Wold, B. J., Klein, W. H., Hough-Evans, B. R., Britten, R. J. and Davidson, E. H. (1978). Sea urchin embryo mRNA sequences expressed in the nuclear RNA of adult tissues. *Cell*, **14**, 941–950.

Woudt, L. P., Pastink, A., Kempers-Veenstra, A. E., Jansen, A. E. M., Meyer, W., Planta, R. J. (1983). The genes coding for histone H3 and H4 in *Neurospora crassa* are unique and contain intervening sequences. *Nucl. Acids. Res.*, **11**, 5347–5360.

The Evolution of Genome Size
Edited by T. Cavalier-Smith
© 1985 John Wiley & Sons Ltd

CHAPTER 4

Cell Volume and the Evolution of Eukaryotic Genome Size

T. CAVALIER–SMITH

SUMMARY

Eukaryotic genome size varies over several orders of magnitude in direct proportion to both nuclear volume and cell volume. The scaling between the three variables is fundamentally similar in unicellular and multicellular eukaryotes; it is usually isometric, but may be slightly allometric in some cell types. The DNA C-value paradox can be solved only by giving satisfactory explanations for these correlations, and for the vast range in eukaryotic cell volumes. A simple explanation for the correlation would be that nuclear DNA contents causally determine both cell volume and nuclear volume; the volume of non-dividing cells, which can vary greatly in volume independently of DNA content, clearly cannot be thus determined, but it is a real possibility for actively proliferating cells. The evidence for and against such nucleotypic control of proliferating cell volume is at present equivocal. Nuclear volume, on the other hand, probably is directly determined by the nuclear DNA content by means of attachment of the telomeres and centromeres to the nuclear envelope and by control over the degree of folding or unfolding of the DNA: it is not necessary to postulate the direct causal determination of cell volume by genome size: if DNA contents determine nuclear volume but not cell volume, the correlation between genome size and cell volume could simply be the indirect evolutionary result of selection for an optimal, and approximately constant, ratio between cell and nuclear volumes so as to optimize cell growth in cells of differing volume.

The selective forces that favour large or small cells are complex and various and are discussed separately for unicellular and multicellular eukaryotes. One factor probably of widespread importance is cell growth rates, which show a weak, non-isometric, inverse relationship to genome size and cell volume; I argue that this is an indirect evolutionary consequence of increased cell size rather than evidence for a causal determination of growth rates by genome size. In animals the lower respiratory rates of larger cells, and in plants the greater shoot elongation rates allowed by large cells, may be key factors favouring the evolution of larger cells and correspondingly larger genomes. The diversity in eukaryote cell volumes probably results from a varying balance, in species occupying different niches, between such selective forces favouring larger cells and those, such as selection for rapid cell multiplication rates, that favour smaller cells.

In contrast to my thesis that genome size is functionally important because of its skeletal role in determining nuclear volume is the thesis that mutation pressure and/or intragenomic selection, coupled with organismic selection against excessive amounts of 'junk' or 'selfish' DNA, can provide sufficient explanation for the existence of the correlation between genome size and cell volume. These alternatives will be discussed in some detail; the balance of the evidence, especially that from heterokaryote protozoa, clearly favours the skeletal DNA hypothesis.

The need to pack DNA economically into nuclei of a finite and functionally significant size, and to vary nuclear volume at different stages of development, imposes constraints on genome evolution that I discuss in relation to the evolution of heterochromatin, chromosome size and arm lengths, and chromatin diminution.

Finally I shall discuss the problem of elucidating the direction of changes in genome size during phylogeny. Though eukaryote genome size has often increased in evolutionary history, this is simply because the first cells were small and their descendants underwent adaptive radiation to produce cells of all sizes, and therefore corresponding DNA contents.

INTRODUCTION: THE UNIVERSAL CORRELATIONS BETWEEN GENOME SIZE, CELL VOLUME, AND NUCLEAR VOLUME

When Mirsky and Ris (1951) first measured cellular DNA contents of a variety of animal species they came to two fundamental conclusions. Firstly, there was no correlation between C-value and organismic complexity (or the number of genes one might expect an organism to have, as discussed in the preceding chapter). Secondly, there was a strong positive correlation between C-value and the cell size of different species, so long as one compared cells from the same tissue. Both conclusions were later confirmed for flowering plants (McLeish and Sunderland, 1961; Price *et al.*, 1973) and for unicellular eukaryotes such as algae (Holm-Hansen, 1969) and protozoa (Shuter *et al.*, 1983), and have repeatedly been reconfirmed for many groups of animals (e.g. reptiles, Olmo and Odierna, 1982; amphibia, Horner and Macgregor, 1983; vertebrates as a whole (357 species), Olmo, 1983).

Neither conclusion would have surprised classical cytologists. Strasburger

(1893) showed that when different species were compared the volume of the nucleus was closely related to the volume of the cell; for several decades the approximate constancy of the ratio between cell and nuclear volume (the cytonuclear ratio) was intensively studied and widely discussed (see review by Trombetta, 1942). Likewise, long before DNA was shown to be the genetic material, the size of chromosomes and the overall volume of a haploid chromosome set were known to vary tremendously from species to species in a way that bore no relationship to organismic complexity. Certain protozoa, for example, had chromosomes larger than those of mammals (Darlington, 1937; Goldschmidt, 1955). As early as 1931 Navashin showed that in different species of *Crepis* chromatin mass (as indicated by the overall length of the chromosome set) was directly proportional to their differing cell volumes (Fig. 4.1). Since it had long been known that DNA was the main chromosomal constituent that stained with classical chromosome dyes, and that the amount of stainable material was greater in species with larger nuclei and a larger total chromosome volume, an early cytologist familiar with the above facts ought to have been able to predict on purely logical grounds that DNA C-values must be unrelated to organismic complexity but would correlate with cell volume.

However, cellular DNA contents appeared to have no great significance until actual measurements showed that diploid cells had a 2C and haploid cells a 1C amount of DNA, which helped to re-establish the temporarily discounted idea that DNA was the genetic material or germ plasm. As the

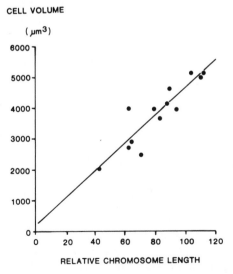

Fig. 4.1. Correlation between the total length of the chromosome set and cell size in 13 species of the flowering plant *Crepis*. Data from Navashin (1931). Regression line calculated by least squares.

primary function of DNA in coding for proteins became firmly established, the divergence between the probable number of genes and DNA C-values discussed in Chapter 3 became a real problem. Yet even then discussion of this C-value paradox concentrated on Mirsky and Ris's negative findings of a lack of correlation between C-value and presumed genic complexity, and usually totally ignored their positive findings of a correlation with cell volume. This repeated neglect of the correlation of cell volume and C-values over a thirty-year period highlights the fragmented nature of biology and the inefficient communication between biologists with different backgrounds and primary concerns; it may intrigue future historians of science.

Though most attempts to explain the C-value paradox have ignored the correlation with cell volume, many could be modified so as to be compatible with it. Callan's master-slave theory, which postulated that high C-value species had more copies of each gene than low C-value ones, could, for example, be supplemented by the postulate that larger cells require more copies of each gene than do smaller cells. Likewise, to the idea that extra DNA in high C-value species is useless junk (Ohno, 1972) one could add the postulate that larger cells could accommodate more useless DNA before it became enough of a burden to be a selective disadvantage. Some other theories, however, seem hardly to lend themselves to such modification, for example the broad class of theories that suggest that the extra DNA is needed for gene-regulation (e.g. Stebbins, 1966; Zuckerkandl, 1976). Such theories must expect DNA content to correlate with organismic complexity, and not with cell volume. Only if a theory can explain *both* the lack of correlation with organismic complexity *and* the positive correlation with cell volume can it be considered a serious contender for an explanation of the C-value paradox. Szarski (1968, 1970, 1976, 1983) has been the most consistent advocate of the importance of the connection between genome size and cell volume in the case of animals. But even he tended to sidestep the key issue as to the explanation for this correlation. As I previously emphasized (Cavalier-Smith, 1978, 1982a) and Fig. 4.2 makes clear, any explanation has to be equally applicable to unicellular and to multicellular organisms and, therefore, to reflect universal and basic properties of eukaryote cells. It must also take account of the close correlation between genome size and nuclear size, first clearly established by Vialli (1957, see Fig. 4.3).

Two extreme kinds of explanation are possible for the variation in genome size and cell volume shown in Fig. 4.2:

1. Functionalist or adaptationist interpretations in which genome size and cell volume are both positively selected so as to be optimal for survival and/or reproductive success, in which case their diversity reflects a diversity of niches with different requirements.
2. Non-functionalist or neutralist explanations according to which genome size and cell volume are irrelevant to survival and reproductive success

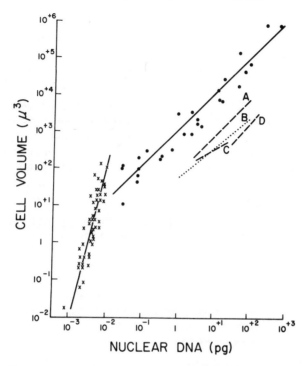

Fig. 4.2. Correlation between cell volume and haploid genome size for 50 species of prokaryotes (X) and between cell volume and G_1 nuclear DNA content for 27 species of unicellular eukaryotes (●: algae, protozoa and yeasts). The two solid lines are least squares fits with slopes of 0.97 (eukaryotes) and 3.52 (prokaryotes). The other lines are least squares regression lines for multicellular eukaryotes: A: flowering plant root meristem cells. B. fish red blood cells. C. frog red blood cells. D. salamander red blood cells. Though the two amphibian regression lines (C, D) appear to differ in slope from each other, recent more precise measurements (Horner and MacGregor, 1983) show a single amphibian regression line that closely parallels the flowering plant line (A). From Shuter *et al.* (1983). Reprinted by permission of the University of Chicago.

and their diversity is simply the result of mutational processes and genetic drift. Of course, the functionalist interpretation might be correct for cell volume and the non-functionalist one for genome size, or vice versa, making four possibilities in all:

1. Genome size and cell volume are both optimally adapted.
2. Genome size and cell volume are both neutral characters.
3. Cell volume is adaptive but genome size is neutral.
4. Genome size is adaptive but cell volume is neutral.

The situation is similar to the debate between selectionist and neutralist interpretations of protein polymorphisms and evolutionary substitutions of amino acids. As in that debate, it may be difficult to find evidence that convincingly discriminates between the extreme alternatives. Nonetheless, present evidence argues strongly against the view that genome size and cell volume are completely neutral characters, and favours the view that cell volume at least is functionally important. Whether genome size is also functionally important depends on the cause of the correlation between cell and volume and genome size, which I shall consider in detail.

Similar considerations apply to the relationship between genome size and nuclear volume, which I argue is also of basic functional importance to the cell, and will discuss later in the chapter.

POSSIBLE CAUSES OF THE CORRELATION BETWEEN C-VALUE AND CELL VOLUME

Two very different explanations are possible: on the one hand, the correlation could be the *indirect* result of a balance of *evolutionary* forces causing genome size and cell volume to coevolve; on the other, it could be caused by a *direct causal* relationship between genome size and cell volume. Since genome size is automatically inherited by DNA replication, advocates of a direct causal

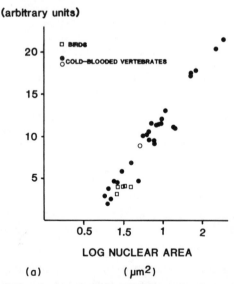

(a)   (μm²)

Fig. 4.3.(a)   Correlation between nuclear DNA content and nuclear size in the red blood cells of 34 species of vertebrate. Note that the homeothermic birds (□) have uniformly low genome and nuclear size, compared with the highly variable values for the poikilothermic species: anamniotes (fish and amphibians) (●); reptile (○)). After Vialli (1957).

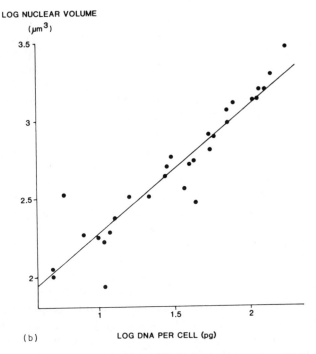

(b)

Fig. 4.3.(b)   Correlation between nuclear volume and nuclear DNA content in apical meristems of 30 species of herbaceous flowering plants. Data from Baetke *et al*. (1967). Since the mean nuclear volume was slightly different in root and shoot meristems (root nuclear volumes averaged only about 90% of shoot nuclei, though in some species they were larger) each point is the mean of the values given for shoot and root nuclei. Regression line (slope = 0.826) fitted by least squares.

relationship have suggested that the amount of DNA per cell determines the volume of the cell in a direct way that depends not on specific genes or particular DNA sequences but simply on the total mass of DNA present (Commoner, 1964; Bennett, 1972). Since such a mechanism of hereditary control over phenotypic characters would be distinct from classical genic control, Bennett (1971, 1972) called it *nucleotypic*, and the overall amount of DNA the *nucleotype*. Nucleotypic control has been proposed not only for cell volume (Commoner, 1964; Szarski, 1970 (by implication)), but also for cell cycle length (Van't Hof and Sparrow, 1963; Bennett, 1971, 1972), nuclear volume (Bennett, 1972; Cavalier-Smith, 1978), and a variety of other quantitative characters (Bennett, 1972) that correlate with DNA content.

It is crucially important to determine whether or not there is nucleotypic control of phenotypic characters. If the nucleotype causally controls cell volume, and if differences in nucleotype are the main cause of the large differences in cell volume between different eukaryotes, then one might only need to understand the mechanism of such control in order to have a complete understanding of the correlation between cell volume and genome size: the explanation would ultimately be in terms of cell and molecular biology. If, however, cell volume is not nucleotypically controlled, or if the large differences in cell volume in different eukaryotes are mainly caused by ordinary genic differences, then cell and molecular biology alone are insufficient to explain the correlation. This is because the genes controlling cell volume could mutate independently of changes in genome size and because genome size could undergo large mutational changes without affecting cell volume; to explain the correlation we would have to suppose either that the net effect of mutation and selection on genome size is a function of cell volume or that the net effect of mutation and selection on cell volume is a function of genome size: in either case the explanation involves population genetic and ecological arguments as well as cell biology.

This contrast may not be as strong as I have implied. This is because the correlation between cell volume and genome size is not absolute. Fig. 4.2 shows that for a given C-value the cell volume of unicellular eukaryotes can vary by five to ten-fold; this variation is unlikely to be totally explicable by experimental errors. Moreover, studies of mutants, reviewed by Nurse in Chapter 5, show that single gene mutations in the absence of any change in genome size can cause up to a two-fold variation in cell volume. Gene mutations therefore clearly can change cell volume. The nucleotypic theory would have to explain why such mutations are relatively unimportant compared with nucleotypic differences in producing evolutionary changes in cell volume. Unless the mechanism of the genetic control of cell volume, jointly by the genes and the nucleotype were such that genic mutations causing large changes (more than about ten-fold) in cell volume were inherently impossible, the nucleotypic theory would be insufficient on its own to explain the observed correlation. One would have to add on extra conditions, to explain why such mutations were much rarer than, or selected against compared with nucleotypic mutations having the same effect. Only a very special type of nucleotypic control could therefore allow one to dispense with a population genetic explanation.

EVIDENCE AND ARGUMENTS CONCERNING THE NUCLEOTYPIC CONTROL OF CELL VOLUME

Nucleotypic control has been postulated primarily for eukaryotes. However, in prokaryotes also there is a clear positive correlation between cell volume and genome size (Fig. 4.2). If nucleotypic control is the explanation of this correlation in eukaryotes why should it not also be so in prokaryotes? But to postulate this leads directly to the difficulty that the quantitative relationship is fundamentally different in the two superkingdoms. In unicellular eukaryotes cell volume increases in direct proportion to and isometrically with genome size, whereas in prokaryotes it increases much more steeply (with the 3.5 power of genome size). This seems irreconcilable with a universal mechanism for control of cell volume by sheer DNA amount. Moreover, in prokaryotes, although the mean correlation between genome size and volume (over a 10^4-fold range in cell volume and 10-fold range in genome size) is highly significant one can readily find two prokaryote species with the same genome size differing 100-fold in cell volume; genome size therefore cannot directly determine prokaryote cell volume irrespective of other factors.

Likewise, in eukaryotes, the cell volume/genome size ratio is not the same in multicellular organisms as in unicells, even though in both groups the slope of the correlation is unity. For the examples shown in Fig. 4.2 the cell volume is 15–60 times greater in unicells of a given genome size than in multicells. However, this number is heavily dependent on the multicell tissue chosen. Cell size can vary by many orders of magnitude in different tissues: if an unusually large cell type, e.g. certain nerve cells in animals, or microspore mother cells or megasporocytes in plants, had been studied instead, one could have found the reverse—a greater (perhaps 100-fold or more) volume for multicells than for unicells of the same genome size. Clearly, therefore, cell size in multicells and the ratio of cell volume to cellular DNA content can vary by 10^4 times or more in one individual. It makes no sense in multicells to speak of genome size as determining cell size in general. Genome size clearly does not itself determine the cell volume of the different kinds of differentiated cells, which is adapted to their specific functions in the life cycle. But it could reasonably be argued that it is the major factor determining the size of undifferentiated embryonic or meristematic cells. In most cases the cell volume of proliferating cells is controlled primarily by the setting of the sizer mechanisms that control the initiation of DNA replication (Cavalier-Smith, 1980c; Craigie and Cavalier-Smith, 1982; Cavalier-Smith, 1985) as discussed in more detail in Chapters 5 and 7; these mechanisms could either be purely genic or partly genic and partly nucleotypic. A purely nucleotypic mechanism can be argued against on several grounds. One is that plant apical meristem cells have about a 15 times lower cell volume than do unicellular eukaryotes of the same genome size. Moreover this fact, coupled with the nearly 10-fold variation in volume of unicellular species having the same genome size (not all of which can be attributed to experimental errors) indicates that even in proliferating eukaryote cells genes can mutate so as to vary cell volume at least a 100 times without any corresponding variation in genome size; this rules out a universal mechanism in eukaryotes for the determination of cell volume *solely* by genome size.

Cellular DNA content can be experimentally altered and cell volume studied. Miklos (1982) increased the overall DNA content of *Drosophila* by adding extra heterochromatic DNA. But he failed to observe any corresponding increase in the size of the facets of the *Drosophila* eye, though this might have been expected if purely nucleotypic control over cell volume is the fundamental explanation of the unitary slope of Fig. 4.2. However, this observation is not decisive since he was studying the surface area of only one face of the cells, which might be controlled independently of volume in such specialized differentiated cells. Studies of artificial autopolyploidy have often been cited as possibly indicating a nucleotypic control over cell volume, since cell volume usually increases in direct proportion to ploidy and therefore overall DNA content (Sinnott, 1960). However, the increased volume could instead be caused by the increased copy number of particular genes which occurs during polyploidy. Studies of mutants in yeast clearly show that changes in gene dosage can change cell volumes (see Chapter 5). Two classic findings suggest that increased gene dosage rather than total DNA content may be the cause of the volume increase in artificial autopolyploids (and the decrease in haploids):

1. An artificial autopolyploid moss strain after 11 years spontaneously reduced its cell volume without reducing its chromosomal content: a gene mutation could have caused this (Wettstein, 1937; reference p. 196).
2. The degree of increase in cell volume in polyploid mosses is much lower in hybrids between two species than in either of the two parents (Wettstein, 1937).

Though these observations suggest that genic effects are involved in volume increases resulting from polyploidy, a mixture of genic and nucleotypic effects would be consistent with available evidence (Chapters 5 and 7).

If nucleotypic effects were the major cause of differences in cell volume in organisms of different C-value, one would expect cell volume and genome size to be linked and not to segregate independently in crosses between such species. Unfortunately in the few such crosses that have been made (see Chapter 9) the segregation of cell volume has not been studied.

Only Commoner (1964) postulated a mechanism for nucleotypic control of cell volume, i.e. that DNA replication used up nucleotides, thus reducing the cellular nucleotide pool, influencing protein synthesis and cell size. He asserted that if extra DNA had to be synthesized it would reduce the nucleotide pool still more. He suggested that such reduction would increase cell size, but did not explain why: one might expect a reduction in the nucleotide pool to *reduce* protein synthesis and cell volume. Moreover, since growing cells make far more RNA than DNA the effect of DNA synthesis on the pool would be relatively insignificant; in fact protein synthesis and cell growth both continue uninterruptedly through the cell cycle, and the occurrence of DNA replication does not alter their rate (Lloyd *et al.*, 1982), so Commoner's hypothesis is incorrect. Let us consider a more plausible nucleotypic model for the control of cell volume in proliferating cells (see also pp. 191, 228–232).

THE CONTROL OF CELL VOLUME IN PROLIFERATING CELLS

Proliferating cells normally grow continuously, with only a temporary slowing of growth during mitosis; their size therefore fluctuates two-fold during the cell cycle from a maximum just before division to a minimum just after. Mean cell volume depends partly on the growth rate, but mainly on the control of the timing of cell division. Since growth is essentially continous delaying division increases cell size and advancing it reduces cell size.

Nucleotypic control over cell volume would therefore have to act by controlling the frequency of division. In formal terms one can imagine a mechanism for this involving the stoichiometric binding of division-initiation molecules to DNA irrespective of its nucleotide sequence (Fig. 4.4). If the initiator molecules are synthesized continuously as a constant proportion of the cell mass and have to saturate all the DNA in order to trigger division, then evolutionary changes in DNA content would lead to proportional changes in cell volume so long as the amount of initiator synthesized per unit cell mass remained constant. This follows from the two assumptions of the model, which may be expressed algebraically:

$$I/M = \alpha \tag{1}$$

$$I_i = \beta D \tag{2}$$

where I is the number of initiator molecules per cell, D the amount of DNA per cell, I_i is the number of initiators needed to saturate the DNA, M the mass of the cell, and α and β are constants. Substituting $M = V\rho$ where V is cell volume and ρ cell density, into equation (1) gives $I/V\rho = \alpha$, or rearranging

$$V = I/\alpha\rho.$$

At the initiation of replication the volume, V_i, will be given by

$$V_i = I_i/\alpha\rho.$$

Substituting for I_i from equation (2) gives $V_i = \beta D/\alpha\rho$.

Clearly, therefore, if α, β and ρ are constant, then evolutionary changes in cellular DNA content, D, will cause proportional changes in the cell volume at initiation of replication. ρ is likely to vary only very slightly during evolution; β will be determined by the stoichiometric binding of the initiator to DNA and therefore the structure of the initiator's binding site. A possible mechanism for maintaining the constant initiator/mass ratio α is the autorepressor model of Sompayrac and Maaløe (1973). Note that the postulated control mechanism is not *purely* nucleotypic, but is partly genic since cell volume depends on α, β and ρ which are all genically controlled, as well as on the amount of DNA. However, nucleotypic control by such a mechanism would be compatible with the data of Fig. 4.2 only if $\beta/\alpha\rho$ had undergone about a 15-fold change between unicellular eukaryotes and flowering plant

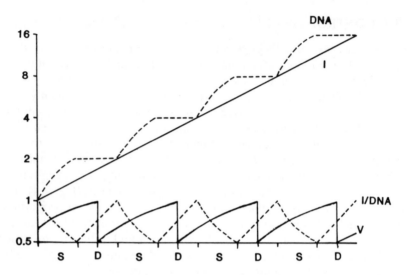

Fig. 4.4.   Model for the nucleotypic control of cell volume in an exponentially multiplying cell population. An initiator of DNA replication (I) is
synthesized in direct proportion to total cell mass and therefore increases
exponentially. DNA increases at a linear rate (see Chapter 7) during S-
phase but remains constant throughout G_2, M and G_1. The ratio I/DNA
(i.e. of the number of initiator molecules to the total number of binding
sites for them on the DNA) therefore fluctuates about a mean value as
shown, and causes initiation of an S-phase whenever it reaches the value of
unity. Mean cell volume (*V*) is halved at every cell division (D) and increases
linearly (for the evidence for linear volume increases see Cavalier-Smith,
1985) between divisions, and therefore fluctuates over a 2-fold range as
shown.
   Since initiation of replication leads on automatically after a fairly constant
time interval to division (D), which halves the cell volume, the volume of
pre-division cells and daughter cells (and therefore the mean cell volume in
the steady state) depends primarily on the cell volume at the time of initiation
of S-phase (which depends on the total mass of nuclear DNA and the ratio
I_i/D = ß), and only to a small extent on the rate of growth in the period
between initiation and division. The model would have to be modified for
some cell cycles (e.g. *Physarum*) to make the initiator control division
directly, rather than indirectly by initiating replication as seems to occur in
most cell cycles (Cavalier-Smith, 1985); the modified model would be a
special case of the initiator accumulation model of Sachsenmaier (1981).

meristems but was essentially constant within these two groups. There is
no reason to doubt that mutation in the components of the autorepressor
mechanism could allow a 15-fold change in α; but if this is so one would
have to ask why do such changes not occur *within* a major group: one could
reasonably argue that mutational changes in genome size were commoner
and/or less likely to be disadvantageous than mutations of the autorepressor
mechanism affecting α.

Another problem with the model is how the saturation of the DNA by a controlling molecule could actually trigger cell division: such a mechanism in effect postulates the cooperative summing up of the effect of something all over a number of different chromosomes. One possibility is that the concentration of free initiator, unbound to the DNA, is the key factor that initiates division; this concentration would rise sharply as soon as all the DNA became saturated with the initiator (i.e. when the I/DNA ratio of Fig. 4.4 reached 1. It might, for example, activate an enzyme that modified DNA or histones in such a way as to initiate the processes leading to division. However, such a nucleotypic mechanism appears rather elaborate compared with the genic control mechanisms discussed in Chapter 7 in connection with the control of the initiation DNA replication. Because of this, and because there is no compelling evidence requiring the postulation of nucleotypic control, it is simpler to assume that direct nucleotypic control of cell volume is absent in both eukaryotes and prokaryotes and to adopt the working hypothesis that the correlations shown in Fig. 4.2 are the indirect result of the coevolution of genome size and cell volume rather than the direct developmental control of volume by genome size. It might be thought that the problem of cooperative triggering of division could be simply avoided if the molecule removed by binding to the DNA is not an activator of division but a repressor synthesized in a burst at division and later removed from solution by binding to the newly synthesized DNA at S-phase; the target site of the repressor could then be a single cellular site or a single gene per genome. But such a model would be counter to the nucleotypic theory since mutational increase in DNA content would accelerate division and produce smaller, not larger cells. (The unstable activator theory of Wheals and Silverman (1982) referred to in Chapter 7 (p. 230) would, if the rate of degradation of the activator depended on total DNA rather than the dosage of specific genes be a nucleotypic theory: but it would predict that cell volume should increase with the square root of genome size.)

The positive nucleotypic sizer mechanism outlined above could be directly tested by injecting DNA into cells not yet committed to division: subsequent divisions should be delayed, the more so as the amount of DNA injected is increased. For a true nucleotypic effect the delay should occur irrespective of the source of the DNA and should also occur with random synthetic polydeoxyribonucleotides. If such delay, with a consequent increase in cell volume, does not occur the theory of nucleotypic control of cell volume by total DNA amount as the basic explanation of the C-value paradox could be more confidently rejected.

DNA injection experiments in *Xenopus* oocytes (Newport and Kirschner, 1982) provide positive evidence for nucleotypic control of a developmental transition—the midblastula transition, where transcription is activated and longer cell cycles begin. DNA from the bacterial plasmid pBR 322 was quantitatively as effective as *Xenopus* DNA, showing the control to be nucleotypic rather than genic. Since the transition was advanced rather than delayed by the extra DNA it is probable that the transition is triggered by the titration of suppressor molecules by the total DNA content. Therefore this particular nucleotypic control mechanism does not have the right properties to account for the evolutionary correlation between cell DNA contents and cell volumes. However, such a clear demonstration of an example of nucleotypic control should make one open to the possibility that others may exist.

CELL VOLUME CONTROL IN DIFFERENTIATED NON-DIVIDING CELLS OF MULTICELLULAR ORGANISMS

When such cells begin to differentiate their size is very similar, and is determined primarily by the sizer mechanism of the cell cycle during the preceding proliferating phase; their differences in cell volume arise subsequently by different amounts of extra growth in the absence of cell division (in very rare instances, notably animal sperm, there may be instead a reduction in volume during differentiation). The sizes of each cell type are characteristic of each species, so there must be mechanisms causing the cessation of growth at a genetically predetermined volume. This implies the existence of volume-dependent blocks to cell growth and therefore of specific sizer mechanisms that in effect measure the cell volume and switch off growth: virtually nothing is known about such sizers.

There are two basically different developmental mechanisms that produce the tremendous differences observed in the volumes of non-dividing differentiated cells within a multicellular organism. In one, common in invertebrates and flowering plants, but very rare among vertebrates, DNA replication continues in step with cell growth and cells therefore become highly polyploid, the degree of polyploidy being proportional to cell volume (d'Amato, 1977; Nagl, 1978); most such endopolyploid cells have essentially the same DNA/cell volume ratio as diploid dividing cells, so neither a simple gene-titration nor a nucleotypic control would appear to provide a suitable mechanism for sizer control over the cessation of their growth. In the second mechanism, characteristic of vertebrates and many other organisms, not just division but also DNA replication is blocked. Therefore, as the cell grows its DNA/cell volume ratio becomes progressively smaller; in this case either a genic or a nucleotypic titration mechanism could provide a basis for size control over growth cessation. But if the mechanism were nucleotypic it would have to be set very differently in different cell types, to account for their vastly different volumes; if it were capable of such flexibility its mere existence would be unable to explain the approximate constancy of the DNA/

cell volume ratio when the same cell type is compared in species of vastly differing genome sizes.

The fundamental differences in the genetic control mechanism over cell volume in proliferating cells (primarily by a sizer control over initiation of DNA replication, and thereby over division) and in differentiated non-dividing cells (primarily by a sizer control over cessation of growth) means that it is essential to consider them separately when discussing the relationship between cell volume and genome size. In particular the studies by Miklos on *Drosophila* eye cells and Wettstein on moss leaf cells referred to above argue against nucleotypic control over the final size of differentiated cells. But they do not provide evidence against nucleotypic sizer control over replication in proliferating cells; for this, comparable studies of cell volume in embryonic or meristematic cells would be needed.

COEVOLUTION OF GENOME SIZE AND CELL VOLUME IN THE ABSENCE OF NUCLEOTYPIC CONTROL

Though the arguments against nucleotypic control over cell volume of proliferating eukaryote cells are much weaker then for non-dividing cells, there is no direct evidence for it. Since genic control over DNA replication appears to offer a slightly simpler explanation for the control of cell volume in proliferating cells (see Chapter 7), and since it would allow a basic continuity in mechanism between DNA replication control in bacteria and eukaryotes, I adopted it as a provisional working hypothesis (Cavalier-Smith, 1978, 1980c, 1982a) in preference to the nucleotypic theory of cell volume control.

But if cell volume is not largely or exclusively controlled nucleotypically, then C-value and cell volume could in principle evolve independently of each other. The scatter around the regression lines suggests that this occurs to some extent. But as the scatter for unicellular eukaryotes is about 10^4-fold smaller than the overall range in cell volume and genome size, the most striking feature of Fig. 4.2 is the very strong overall correlation between cell volume and genome size. This shows that the two variables coevolve; small changes in genome size occur without corresponding changes in cell volume (and vice versa) but large changes do not.

The fundamental question is whether cell volume or genome size is the primary factor in this coevolution. If cell volume is not determined nucleotypically, it is hard to see why large evolutionary changes in genome size would necessarily be followed or accompanied by corresponding changes in cell volume. On the other hand, there are several reasons why large changes in cell volume might be have followed by corresponding changes in genome size. I have therefore argued that large changes in genome size are the secondary evolutionary result of primary changes in cell volume. If this is correct it raises two fundamental questions which I shall discuss in turn:

1. What are the evolutionary forces governing evolution of differing cell volumes.
2. Given this diversity of cell volumes, what are the secondary evolutionary forces that ensure that on average large cells evolve larger genomes and small cells evolve smaller genomes?

THE EVOLUTION OF CELL VOLUME

Understanding the nature of mutations that change cell volume, such as the *wee* genes of yeast that have now been cloned (Beach *et al.*, 1982), will be essential for elucidating the actual mechanisms for the genetic control of cell volume, which are discussed elsewhere (Cavalier-Smith, 1985; Nurse, Chapter 5). But to clarify the C-value paradox it is equally essential to understand the nature of the selective forces acting on mutants having different volumes. This section therefore discusses why cell volumes should vary so greatly in different species. Cell volume might conceivably be adaptively neutral and evolve purely by mutation and random drift; alternatively it may by adaptively important and universally optimized by selection. It is also possible that cell volume is selectively important for some species, but not for others, or that large changes are selectively important but small changes are more or less neutral. The comparative evidence seems to me to rule out universal neutrality. Though it does not rule out universal optimality, differences in cell volume do appear to be of much greater selective importance in some groups than others; this means that a thorough discussion of the problem is impossible in a short space. No example can be representative and no generalization applies to all kinds of organisms. It is especially important to discuss unicellular and multicellular organisms separately, since changes in cell volume have very different consequences in the two groups. But even within these broad categories cell volume changes have profoundly different consequences for different types of organism. The best I can do is to choose examples that illustrate these differences so as to substantiate my thesis that the selective importance of cell volume differs from group to group, and that in most groups there is a trade off between the advantages and disadvantages of a given cell size.

Unicellular organisms

Differences in body size of multicellular organisms are of great adaptive significance; if body size changes in evolution numerous characters undergo correlated changes (Banse, 1982a; Peters, 1983). Such correlated changes usually obey allometric scaling laws, since small organisms are never simply isometrically scaled-down versions of larger ones. In unicellular organisms body size and cell size are the same thing, so scaling considerations apply directly to cell size.

The classical arguments about the effect of changes in cell volume relate to the decreasing surface/volume ratio as cell size increases. Maynard Smith (1969) emphasized also the longer diffusion path-lengths within larger cells which will also tend to slow down growth. Both factors mean that selection for rapid reproduction will tend to favour small cells with a high density of active transport molecules in their plasma membrane. But the impact of such selection on cell volume will differ greatly according to the type of nutrition of the cell and the transport mechanisms it has. Microbes that are photosynthetic (i.e. algae) or phagotrophic (i.e. protozoa feeding by phagocytosis or micropinocytosis) will be less dependent on high rates of active transport through the plasma membrane than are non-photosynthetic and non-phagotrophic saprophytes (i.e. fungi and most bacteria); only in the latter will the transport of soluble nutrients into the cell commonly be the rate-determining factor for growth. Therefore selection for small cell size should be much more stringent in saprotrophs than in phototrophs or phagotrophs: the variation in cell size in both prokaryotes and eukaryotes of different nutritional types supports this.

In prokaryotes the saprotrophic bacteria, whether unicellular, or filamentous like actinomycetes, are with few exceptions very thin, usually 0.25–1.0 μm in diameter; phototrophic cyanobacteria are often much larger than this. Efficient light trapping is favoured by a greater area and more layers of photosynthetic membranes. Since the light-harvesting thylakoids are dispersed throughout the cyanobacterial cell a thicker cell can absorb a greater fraction of the incident light. Large cell size might particularly benefit cells growing in brighter light; perhaps this might be why the photosynthetic but anaerobic purple non-sulphur bacteria, which usually grow at some depth and in dimmer light than cyanobacteria, have smaller cells than most cyanobacteria, though their cells are larger than in typical saprophytic bacteria. The purple non-sulphur photosynthetic bacteria however have cells the same size as saprotrophs; I suggest this may be because they are facultative saproptrophs—under aerobic conditions they respire using external organic substrates similarly to *E. coli*—and are selected for small size in this phase of their life cycle. The green photosynthetic bacteria are also small, like saprotrophic bacteria; I suggest that this is because their photophosphorylation systems are restricted to the plasma membrane—in the absence of internal photosynthetic membranes there can be no advantage in a greater cell thickness.

Eukaryotes show similar differences between saprotophs and phototrophs. Fungi have small cells, hyphae usually being 5–10 μm in diameter, whereas unicellular algae have a tremendous size range (from 1 μm–10 000 μm). It is interesting to consider why the diameter of a fungal cell is about 10 times that of a prokaryotic saprotroph with very similar habits, such as an actinomycete. One possibility is that the plasma membrane has a greater transport capacity for metabolites per unit area in fungi than in bacteria because the respiratory electron transport chain and oxidative phosphorylation enzymes, which must occupy much of the bacterial plasma membrane, are segregated into mitochondria (the existence of infolded mitochondrial cristae also means that a eukaryote does not need to maintain a high plasma membrane/cell volume ratio to support a high metabolic rate). A second possibility is that the existence of active intracytoplasmic transport, e.g. by cytoplasmic streaming, in fungi but not bacteria, may lessen the importance of diffusion path lengths in limiting cell size. As Maynard-Smith (1969) pointed out, if a molecule has to diffuse from one specific site to another in a cell the time taken depends on the cube of the distance. A 10-fold increase in cell diameter will therefore multiply such diffusion times 1000-fold, whereas times for directed transport such as cytoplasmic streaming increase only 10-fold. Increase in size from 1 μm to 1 mm would increase transport times by cytoplasmic streaming only 1000 $\times$ but will increase diffusion times by 10^9 times.

I suggest that the vastly increased size of eukaryote cells compared with bacteria depended on two key innovations: (1) the evolution of directed intracellular transport involving actomyosin and tubulin-based microtubules, both unknown in prokaryotes, and (2) the evolution of exocytosis, endocytosis and the related origin of the rough endoplasmic reticulum and Golgi apparatus, all unknown in bacteria (Cavalier-Smith, 1981). Cavalier-Smith (1980b, 1981) suggested that exocytosis of chitosomes in fungi accounts for their much more rapid hyphal elongation compared with the cell-surface based peptidoglycan synthesis of bacterial cell walls. The same argument applies to cell wall synthesis in unicellular algae and multicellular plants (and the synthesis of intercellular matrix and other secretions in multi-cellular animals): because the internal endoplasmic reticulum/Golgi membranes have assumed the biosynthetic role, this frees more of the plasma membrane for transport rather than biosynthetic functions, and means that the biosynthetic rate of envelope and extracellular materials can be increased indefinitely in proportion to cell volume however large the cell. In prokaryotes, by contrast, synthesis of wall materials will be limited by their surface area, so that as cell size increases this will tend to limit severely their potential growth rate.

Endocytosis (phagocytosis, pinocytosis and micropinocytosis) is especially important in protozoa since it allowed the evolution of predation by engulfing food, either macromolecules or other cells. If the food is macromolecular, micropinocytosis suffices and large size is not advantageous, as in the many intracellular parasites of the protozoan phyla Sporozoa (e.g. malarial parasites, *Plasmodium*) and Microspora. If the food is cellular (or even multicellular), as in most free-living protozoa as well as symbionts in animal guts, then large size is often advantageous. A motile protozoan cell can swim faster if it is longer and has more or larger cilia and can ingest a larger number of bacteria per unit time and also take prey that are too large to be captured by a smaller cell. Large size can be favoured in non-motile phago-trophs that catch more prey by greatly extended pseudopodia in a net-like fashion, as in foraminifera (up to ~10 cm) and radiolaria; not all have giant cells, any more than do all plants become giant sequoias or all animals whales or giant squids; this is because within each broad adaptive zone a diversity of niches exists suitable for organisms of greatly different size.

The protozoan phylum Sporozoa well illustrates the fundamental differ-ences in the selective forces acting on cell size in intracellular and extracellular parasites, and also the differences in selective advantage of different cell size at different stages of the life cycle of the same organism (phenotype diver-sifying selection, discussed elsewhere in detail for protists with special refer-ence to cell volume: Cavalier-Smith, 1980c). In the Sporozoa the infective stage (sporozoite) is a small cell, but the feeding stage (trophozoite) grows continuously manyfold in size without division; eventually it undergoes multiple fission to produce numerous small infective cells. It seems clear that large size and/or the absence of division is favoured during the feeding stage; for the infective stage it is probably not small size *per se* that is of selective advantage, but the production of as many infective units as possible from a given mass of protoplasm, which necessarily means small size. The size of the mature trophozoite cell depends on whether or not it is intracellular or extracellular. In the classes Coccidea and Hematozoa (Vivier, 1982), such as the malaria parasite *Plasmodium*, trophozoites are intracellular and the maximum size is set by and depends on the size of the host cell. In *Plasmo-dium* which infects first liver cells and then red blood cells which have 100 times lower volume, the size of the mature trophozoite is much larger in the liver parenchyma cells (enough to produce 10 000–30 000 small intermediate infective cells (merozoites)) than in the red cells (only enough to yield 6–24 merozoites). In the class Gregarinea the mature trophozoite cell is extracellular and can therefore grow much larger than in Coccidea or Hematozoa—even up to 16 mm in length (Hyman, 1940) and can be up to about 10^6-fold the volume of a human red blood cell, or a *Plasmodium* trophozoite inhabiting a red cell!

Protozoa symbiotic in animal guts have a marked tendency to increase immensely in size, as do tapeworms that feed in a similar fashion; groups as disparate as gregarines, opalinids, polymastigotes, hypermastigotes and entodiniomorph heterokaryotes show this phenomenon. If, and only if, they are uninucleate (i.e. gregarines and hypermastigotes) they also have corre-spondingly large genomes. The fact that gregarines and hypermastigotes had chromosomes of the same size, shape, number and detailed structure as those of the 'highest' animals and plants long ago rightly made Goldschmidt (1955) sceptical of the notion that changes in the amount of the genetic material was responsible for increases in the complexity of organisms. The fact that the genome size of malarial parasites is among the lowest measured for eukaryotes (0.059 pg; Table 3.2, p. 77), whereas that for the related gregarines must be very large judging from their chromosome size, is simply an example of the general rule illustrated in Fig. 2 that for uninucleated eukaryote cells genome size is directly proportional to cell volume. The large cell size of the hypermastigotes is particularly noteworthy since they are anaerobic and entirely lack mitochondria. This shows that the capacity of eukaryote cells to evolve larger cells did not depend on the evolution of mitochondria.

The preceding discussion mentioned some of the selective forces that may be involved in influencing cell size and some of the ways they may differ in different kinds of unicells. There is, however, a general selective force that probably applies in some degree to all cell types; this is selection for rapid cell proliferation. Natural selection involves both selection for greater individual survival, or viability, and for greater reproductive success, and there has been much recent discussion of the relative trade-off between selection for viability and for a high reproductive rate, particularly using the concepts of r-selection and K-selection (Pianka, 1970; Southwood, 1976).

This distinction is based on the idea that the relative importance of selection for maximal reproductive rate (r-selection, or more properly r_{max} selection where r_{max} is the intrinsic rate of increase under optimal conditions where survival and growth are both the maximum possible for that organism) and selection for maximum survival may differ from species to species. Selection under conditions much less than optimal because of crowding or nutrient or energy shortage is referred to as K-selection. For the present discussion what is important is the effect of cell size on these two aspects of Darwinian fitness. Much more study has been devoted to the effect of cell size on maximal reproductive rates than on cell survival. Both for heterotrophic unicells (Fenchel, 1974; Taylor and Shuter, 1981) and photoautotrophic unicells (Malone, 1980; Banse, 1982b) it has frequently been demonstrated that r_{max} is lower for larger cells. This is inversely related to the length of the cell cycle in the absence of cell death by the equation $r_{max} = (\ln 2^n)/T_c$ where T_c is the length of the cell cycle and n is a positive integer ($n = 1$ for binary fission; $n > 1$ for multiple fission) (Cavalier-Smith, 1980c, 1982a). However, as discussed later (p. 144ff.) in connection with the length of the cell cycle, the decrease in r_{max} (or increase in T_c) with cell size is much less than isometric. Empirical studies originally suggested that $r_{max} = kW^{-0.29}$, where k is a constant and W is cell mass (Fenchel, 1974), but recent work shows that the situation is more complex (Banse, 1982b) and that in some groups r_{max} is more strongly dependent on cell volume than in others. Whatever the complexities the fact that small cells do on average have significantly greater

r_{max} than large ones means that algae and protozoa cannot all be purely r-selected—if they were the large-celled species would soon become extinct and be replaced by smaller ones. The fact that species of vastly different cell volumes coexist in the biosphere means that there must be compensating advantages of large cell size sufficient to balance the disadvantage of a lower r_{max}. This advantage must be an increase in growth rate or viability in suboptimal conditions.

The advantages of large cell size seem less clear for algae than for predatory protozoa. The problem has been most thoroughly discussed for phytoplankton (see especially Smayda, 1970; Malone, 1980; Banse, 1982b; Shuter, 1978). Even though many important cell properties are known (e.g. resistance to grazing, sinking rates, phosphorus and nitrogen subsistence quotas) or suspected (e.g. respiration rate, survival and growth rates in suboptimal environments) to be related to cell volume, there is no concensus concerning the factors that favour large cells. The fact that large algal cells occur with high frequency in certain specific environments, indicates that the balance of advantage is a subtle one that can shift from place to place (e.g. in freshwater large-celled desmids abound in oligotrophic conditions, but are relatively rare in eutrophic ones, whereas at sea larger algal cells are commoner in coastal waters and rarer in the oligotrophic epipelagic zone of the open ocean). It can also shift markedly from season to season, and much effort has been devoted to the study of the seasonal succession of algal species. Small-celled r-selected species predominate early in the season when nutrient levels are high and grazing low, but as nutrients run out and grazing pressures increase the balance of advantage tends to shift to larger-celled species (Reynolds, 1982, 1984). Grazing by filter feeding zooplankton appears to be a major factor in causing this shift since cells over 60 μm are virtually immune to ingestion. Slower growing larger cells often seem better able to continue growing when nutrient levels are low. One also needs to consider the dormant resting phases of the life cycle: many of the smaller rapidly growing cells lack such a stage, but they are common in larger species. Since in small-celled species that do have resting stages the spores are commonly markedly larger than are vegetative cells, as in *Chlamydomonas*, it is likely that large size is advantageous for dormant spores as it would enable them to store more food and energy resources.

Some molecular biologists have said that the great diversity in cell volume of different organisms must be unimportant and selectively neutral as it seems to make no difference to their success as shown by the simple fact of their coexistence in the biosphere. But this fallacious argument could also be used to 'prove' that *all* differences between organisms are neutral and nonadaptive—it does not matter whether you have leaves or jaws, a brain or not, you can survive or reproduce equally well! In one sense all organisms with a stable population are equally successful and the *overall* differences between them are overall neutral. But this *overall* neutrality conceals the non-neutrality of component characters. It is a waste of time arguing whether the possession of eyes or teeth or legs is of selective advantage or not. Eyes, teeth and legs have all been gained and lost on numerous occasions. They have been selected for in some organisms and against in others at different points in history. It is better not to speak of a character as being selectively advantageous or not, but of a *change* in a character. Since organisms do not live *in vacuo*, but in a specific habitat, and characters do not exist in isolation but together with other specific characters with which they are coadapted, what *changes* are selectively advantageous, disadvantageous or neutral, will depend (a) on that habitat and (b) on the other characters of the organism. The examples given indicate that in unicells this is as true for cell size as for other characters (see Simpson (ref. p. 35) for a more detailed discussion).

Multicellular organisms

The situation is even more complex in multicellular organisms where it is essential to distinguish between two types of organism:

1. Organisms where adult cell number is constant, as in most members of the animal phylum Aschelminthes (i.e. rotifers, nematodes, gastrotrichs, kinorhynchs and acanthocephalans). Since these organisms contain a fixed number of cells (usually about 1000) irrespective of body size, those with larger bodies have proportionally larger cells. Mean cell size therefore directly determines body size and will be directly affected by all the selective forces that influence body size. Mutations that affect cell size necessarily affect body size and vice versa.

2. Organisms that do not show cell constancy but where the number of cells per individual can vary freely. In such organisms, that include most animals and plants, cell size does not determine body size. Cell number per individual and cell size does not determine body size. Cell number per individual and cell size can both be changed by mutations and can both affect body size. Selection for changes in body size can therefore change either cell number or cell size, or both. There is no doubt that selection for different body sizes has been profoundly important; its effect on cell volume depends firstly on whether or not cell volume is a neutral character and secondly on the relative frequency of mutations that change cell number or cell volume by a comparable factor. If cell volume is neutral and of no adaptive significance for multicells, and if mutations affecting cell volume are as frequent as those affecting cell number, then selection for large size would on average result in both more and larger cells; selection for small size would result in both fewer and smaller cells. There would be a good overall positive correlation between cell volume

and body size, but with scatter about the curve and with a slope of about 0.5; the rarer the mutations affecting cell volume the smaller would be the slope.

Since organisms with a vast range in body size such as vertebrates, molluscs and flowering plants show no overall correlation between body size and cell volume it follows either that mutations changing cell size are very rare or that cell volume is not neutral but is an adaptive character that is optimized by selection and not allowed to drift or be selected for secondarily because of its effect on body size.

The fact that cell volume varies so greatly in unicells (even within a genus, e.g. the desmid *Cosmarium*, it can vary 10 000 fold) implies that mutations affecting cell volume are not particularly rare. But what is at issue is not their absolute frequency but their frequency compared with those altering cell numbers. Even so, demonstration that they occur with high frequency would, taken together with the lack of correlation between body size and cell volume, be *prima facie* evidence against the assumption that they are neutral. Perhaps the strongest argument comes from considering polyploidy, which invariably causes a proportional increase in cell volume. In humans triploidy is the commonest single cause of spontaneous abortions, and is estimated to occur in about 1% of conceptions. This frequency is several orders of magnitude higher than that of *any* gene mutations, which argues strongly against the neutrality of cell volume in humans. In fact human triploids, and the rarer tetraploids, are absolutely inviable. those exceptional few that do not die in embryonic or foetal life succumb within hours of birth with multiple defects including respiratory failure. I have suggested that death is caused ultimately by the greater cell size of triploids (Cavalier-Smith, 1978). One has to suppose that at least one aspect of human development or physiology is so sensitive to a 50% increase in cell volume that it becomes upset and eventually kills the organism. Triploidy *per se* does not cause inviability: although it does cause aberrant meiosis and so greatly reduces fertility, and can therefore seldom spread in natural sexual populations. Natural triploids have been found in many groups of plants, invertebrates and lower vertebrates such as fish and amphibians and can be produced artificially with no obvious differences in viability from diploids. The lethality of triploids in humans strongly suggests that mammals are exceptionally sensitive to changes in cell size. Since viable human diploid/triploid mosaics are known, the exact cell volume cannot be equally important for all cell types. If lethality is caused by hypersensitivity of one or more cell types to changes in volume one should be able to determine which these are by determining which cell types in such mosaics are invariably diploid. It has been suggested that red blood cell size may be of key importance (Cavalier-Smith, 1978): this hypothesis predicts that viable human diploid/triploid mosaics will never be found with purely triploid erythroblasts; it is well known that human red blood cell volume is remarkably constant.

Red blood cell size has been measured in nearly 1000 mammals and shows very little variation (2-fold) compared with 100-fold variation in amphibians. One reason may be that mammals are warm-blooded and have a higher and more uniform rate of metabolism; relatively slight variations in the respiratory and circulating capacity of their red cells could therefore be much more harmful than in the cold blooded amphibia. Birds also are warm-blooded and have a high and uniform metabolic rate; they also have rather uniform red blood cell volumes that differ only slightly from those of mammals (Dabrowski, 1968). In both birds and mammals red blood cells are the smallest somatic cells (ignoring platelets which are cell fragments, not whole cells); this small size is attained during differentiation by successive cell divisions in the absence of growth. Since this mechanism of reduction in volume is unique to red cells and must have been specifically selected for, it follows that in mammals and birds red cells must be under stronger selection for small size than are other cells. Coupled with the greater uniformity of red cell volume compared with other mammalian and avian cells, this supports the idea that strong stabilizing selection for optimal red cell volume is a major selective force that maintains a relatively uniform cell volume in mammals and birds (and secondarily causes the uniformity in C-values).

Not all mammalian and bird cells are uniform in size from species to species: muscle and nerve cells both tend to be larger in larger animals, which implies that their size is adapted to that of the whole animal. This characteristic of nerve and muscle cells seems true of all animals. Egg cells also differ greatly in volume* from species to species; these differences are probably adaptive and have been related to a variety of ecological variables, especially in amphibia (Kaplan and Salthe, 1979), (*but see note 5 p. 178.)

Although it can be seen why nerve, muscle, egg, sperm and red cells should have an optimal size, it is harder to decide whether or not this is true of most other cell types, at least in medium to large animals. In very small animals, however, especially microscopic ones such as copepods or parasitoid hymenoptera, the sheer size of the animal imposes a strong evolutionary constraint on cell volume: since the whole animal is smaller than the cells of large protists, they can make complex multicellular organs only if their cells are very small. If behavioural complexity depends in part on the number of nerve cells, as is generally supposed, then complex behaviour will only be possible for a very small animal if it has unusually small cells. If diploid *Drosophila melanogaster* cells were as big as those of mammals, a *Drosophila* brain would have 20 times fewer cells than it does. For medium and large animals this would not pose any constraint. Nonetheless, all groups of multicellular organisms show much less variation in the volume of typical or proliferating somatic cells than do unicellular algae and protozoa. This strongly suggests that their cell volume is under stronger evolutionary constraints than in many unicellular groups and that each group has an optimum range in cell volume. I argue that this narrower range in cell volume arises simply because body size and cell size can be substantially dissociated in multicells: body size can therefore be adapted to highly diverse niches without corresponding changes in cell volume, which can therefore be optimized according to functional criteria that may be hard to discern if they are merely quantitative and general, and not obviously related to particular environmental needs or specialized functions. Two such general cell properties that might be important have been suggested. One is the tendency for small cells to have shorter cell cycles. As discussed later in the chapter (pp. 144–155) several authors have argued that this could affect developmental rates of multicellular plants and animals in an adaptively significant way; in both amphibians and flowering plants there is an inverse relationship between developmental rates and cell volume. The second idea is that the altered surface/volume ratio of larger cells has numerous physiological consequences (Szarski, 1976). In particular, solute leakage across the plasma membrane would be expected to be relatively less, with correspondingly lower needs for active transport which consumes a high fraction of the basal metabolic energy in animals (about 50% in humans according to Keynes, 1975). More systematic study is needed of the physiological correlates of cell size, since present data is rather sparse. Smith (1925) and subsequent authors cited by Szarski (1983) have shown that in amphibia the basal respiratory rate per unit body mass does indeed decline as cell size increases (Fig. 4.5). For an active species basal metabolism will constitute only a small fraction of its total energy requirements and high rates of metabolism associated with smaller cells may often be of net benefit by allowing relatively higher rates of food capture. But for sluggish animals that spend much of their time under starvation conditions the more economical energy metabolism made possible by larger cells may be sufficiently important to constitute a net selective force in favour of larger cells. Many authors (e.g. Szarski, 1983; Olmo, 1983) have noted that those amphibia with relatively large C-values, and therefore large cell volumes, have just such sluggish life styles: those with the highest C-values are all paedogenetic, that is they never undergo metamorphosis to form air-breathing adults but become sexually mature as large sluggish tadpoles that breathe by gills rather than lungs. Since the plethodontid salamanders, many of which also have unusually large genome size, are also lungless and breath through their moist skins, one is tempted to suggest that dispensing with lungs allows amphibia to evolve larger cells, perhaps because of selection for smaller red blood cells in the pulmonary circulation. But such a suggestion meets the objection that the fish with the largest C-values and cell volumes are the lungfish; therefore air-breathing by lungs cannot *per se* be an obstacle to the evolution of large cells. Probably therefore there

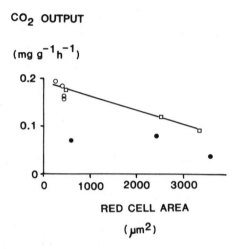

Fig. 4.5. Relationship between cell size and basal metabolic rate of seven species of amphibia (circles). For the species shown by closed circles, which had a very much greater body size than the others, separate observations were made on individual animals closely matched for body weight (□). The data show that larger cells and larger body size both tend to reduce the basal metabolic rate. Data replotted from Smith (1925).

is no direct connection between the absence of lungs and large cell volumes; instead it seems likely that retention of external gills in lieu of developing lungs and large cell sizes are independent adaptations to optimize the uptake and minimize consumption of oxygen in giant aquatic salamanders.

In fact the case of the lungfish gives cogent support to the thesis that selection for low levels of basal metabolism can favour larger cells. All species of lungfish are able to survive levels of oxygen so low that other fish die; there can be little doubt that a low rate of basal metabolism is essential for this adaptation. Though modern lungfish all have very large genomes, there is a clear difference between the Lepidosirenidae (*Lepidosiren* with 121 pg of DNA per haploid genome: *Protopterus* with 142 pg per genome) on the one hand, which have the largest genomes of any animals, and the Ceratodontidae (*Neoceratodus* with 80 pg per genome) on the other (Thomson, 1972). *Neoceratodus* also has markedly smaller cells than the lepidosirenids (Fig. 4.6). I attribute this major difference to the fact that the lepidosirenids alone undergo aestivation; in the dry season they dig into the mud, leaving a small opening for breathing, and can remain in this torpid state for six months. Clearly this habit will subject them to much stronger selection for low basal metabolism and therefore larger cells than in the non-aestivating *Neoceratodus*, which has a genome size similar to the largest found in amphibians (i.e. *Necturus* which also inhabits muddy freshwater that may often be prone to deoxygenation leading to hypertrophy of its

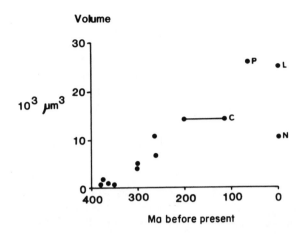

Fig. 4.6. Changes in osteocyte volume during the evolutionary history of lungfish. N = *Neoceratodus*, P = *Protopterus*, L = *Lepidosiren*. Data from Thomson (1972). Note that the cell size of *Ceratodus* (C), a mesozoic relative of *Neoceratodus*, remained constant for nearly 100 million years.

external gills). Fossil remains of such burrows associated with lungfish bones show that aestivation dates back to the Permian (Young, 1962). Fig. 4.6 shows that lepidosirenid cell size more than doubled since the mid Permian (250 Ma ago); I suggest that this doubling was the direct result of positive selection for larger cells caused by the evolution of aestivation in the Permian. *Neoceratodus* which did not evolve aestivation shows no increase over the past 250 Ma; if anything, its cells are slightly smaller than those of its close Mesozoic relative *Ceratodus*.

Possibly the marked rise in lungfish cell volume during the Carboniferous (280–350 Ma ago) was because they lived in tropical freshwater prone to stagnation which favoured a low basal metabolic rate and larger cells. By contrast their crossopterygian relatives the coelacanths, which diverged from them in the Devonian when lungfish cells were still quite small, colonized the well-oxygenated sea in the Triassic where selection will not have favoured exceptionally large cells. The fact that coelacanths have a typical vertebrate C-value of 3.1 pg (Thomson, 1972) and a correspondingly small cell volume (Thomson and Murazko, 1978) fits this adaptive interpretation of the two major rises in cell volume in lungfish history; since coelacanths have not changed significantly over the past 200 million years it also contradicts the idea that the rise in lungfish cell volume is non-adaptive and an inevitable consequence of evolutionary stasis.

Though cell volume may have greatly increased only twice during lungfish history, evidence from the volume of osteocyte spaces in amphibian fossils implies that several different groups of amphibia independently evolved larger cell volumes (Thomson and Muraszko, 1978). The evolutionary trade-off in these freshwater anamniotes between larger cell size, slow development and greater energy economy on the one hand, against the mutually incompatible smaller cell size, higher metabolic rate and more rapid development on the other, may be compared to the more familiar trade off between homeothermy, which allows a more active behaviour at the expense of greater energy costs, and poikilothermy, which gives greater energy savings at the expense of lower activity under cooler conditions. In both cases niches are available where either a more frugal or a more wasteful use of energy in optimal (Szarski, 1983).

Though amphibians of differing genome size can 'inhabit the same ponds' and 'each is successful' (Horner and Macgregor, 1983), this does not mean that there are no ecologically significant correlates of their differing cell volumes. Wherever the ecology of coexisting animal species has been properly studied, which has been done most frequently for birds, important ecological differences between them and their niches have invariably been found (Hutchinson, 1978); even in amphibia a large number of differences have been found. Quantitative studies of amphibian ecological energetics in relation to genome sizes are needed to test the present thesis.

The diversity of plant reproductive tactics and the considerable degree of understanding of them (Grime, 1979) makes it easier than for animals to attempt to correlate such features with cell volume and genome size. Since earlier data on genome sizes mostly referred to species for which little ecological data exist, Grime followed up my suggestion to study genome sizes in the species that he has investigated (Grime and Mowforth, 1982; Grime, 1983), and found a remarkable correlation between genome size and the rate of leaf extension in the critical spring growing period (Fig. 4.7). Grime (1983) showed that leaf cell size in these species is positively correlated with genome size, as previously established for apical meristem cells, and suggested (Grime and Mowforth, 1982) that larger cells allow more rapid rates of shoot expansion under cool conditions. He has shown that species where expansion occurs earlier in the spring season (and therefore at a lower temperature) have larger genome sizes (Fig. 4.8) and therefore larger cell volumes; selection for rapid expansion in early spring may be the most important selective force favouring larger cells in such herbaceous flowering plants of temperate regions.

The more rapid expansion of leaves with larger cells is an interesting contrast to the longer cell cycles and therefore slower growth of meristematic cells. I suggest that it arises because leaf expansion is brought about primarily by cell expansion rather than by cell multiplication, and that large size increases the rate of cell expansion, despite reducing the rate of cell multiplication. Cell expansion depends on two things:

Mean leaf extension

(mm day^{-1})

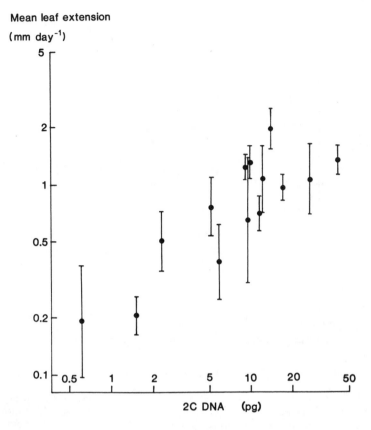

Fig. 4.7. Relationship between genome size and leaf extension rate from 7 March to 1 July 1983 for grassland plants in Derbyshire, England. From Grime (1983). Reprinted by permission of The Association of Applied Biologists.

1. vacuolization and the redistribution of the existing cytoplasm around the cell periphery;
2. additional synthesis and assembly of new cytoplasmic components and, especially, membrane and cell wall materials.

It is reasonable to suppose that the rate of these processes will depend on the total amount of the biosynthetic and secretory machinery on which they depend, and that the initial amount of such machinery prior to cell expansion will be greater in larger cells, which therefore ought to be able to increase their volume at a greater rate. A similar explanation was suggested for the greater rate of cell volume increase that we have observed in larger cells of the unicellular plant, *Chlamydomonas* (Cavalier-Smith, 1985). In the *Chlamydomonas* cell cycle there is a contrast between the limiting factors for cell

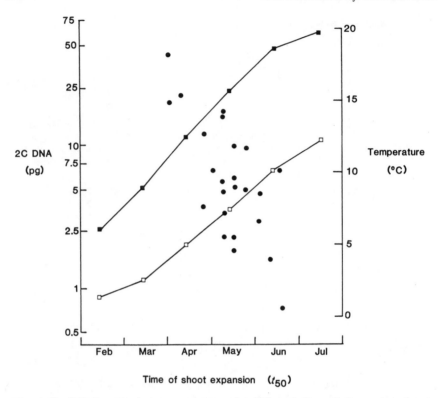

Fig. 4.8.   Relationship between genome size (●) and time of the spring shoot expansion in 24 flowering plants growing near Sheffield, England. Also shown are average daily minimum (□) and maximum (■) temperatures at Sheffield. From Grime and Mowforth (1982). Reprinted from *Nature, London*, **299**, 152, 1982, by permission of MacMillan Journals Ltd.

volume increase and for cell or protein mass increase, as shown by the linear growth of the former and the exponential growth of the latter (Cavalier-Smith, 1985); this is analogous to the contrast in higher plants between the factors limiting cell expansion rates and cell cycle lengths. Later in this chapter I argue that cell cycles take longer in larger cells because net exponential growth rates tend to be limited by gene concentrations, which for unique genes are inversely proportional to cell volume. If, however, cell expansion rates depend on the amount of gene products initially present at the beginning of the expansion phase, and which have been accumulated earlier 'at leisure' (e.g. during the previous year's growing period, as in spring bulbs such as the bluebell *Endymion non-scriptus* (C-value 21 pg)) then the lower gene concentrations of larger cells will not limit cell and leaf expansion rates, and the net advantage will be for larger cells if the need for rapid expansion outweighs that for rapid cell multiplication. This is the case for early spring bulbs, which need to trap as much solar energy as possible before being shaded out by the regrowth of tree leaves and/or taller herbs later in the spring; such bulbs usually have large C-values and cell volumes. By contrast strongly *r*-adapted species, such as ephemeral weeds like *Arabidopsis thaliana* (C = 0.2 pg) that can go through several generations per year, need the highest possible overall exponential growth rate so as to exploit as rapidly as possible newly cleared areas, and will have a premium instead on short cell cycles and therefore small cells. Clearly many species will be in between and the balance of advantage between short cell cycles and rapid shoot expansion will differ according to their exact niche. One such example, discussed later, is the annual grass *Poa annua* (C = 2.9 pg); in this case the C-value shows considerable intraspecific variation (Grime, 1983), perhaps indicative of a variety of ecotypes or a fluctuating balance of advantage between the two factors. The broad spectrum of cell sizes and genome sizes in flowering plants may therefore simply reflect this differing balance in different species between these two major forces favouring small and large cells.

The idea that cell volume is generally optimized in multicells is also supported by the fact that in Hymenoptera, where the males are haploid and the females diploid, cell volume is the same in both sexes; since, in all studied organisms, artificially made haploid cells have half the volume of diploid ones it follows that the equality in cell volume in the two sexes must be positively selected, and subject to genic control.

WHY IS GENOME SIZE CORRELATED WITH CELL VOLUME?

If cell volume is not exclusively determined nucleotypically by genome size, but can in principle evolve independently of it, we have to explain why in fact cell volume and genome size do broadly speaking evolve together; this is so irrespective of whether cell volume is a selectively significant or a neutral character. In either case we have to suppose that an evolutionary increase in cell volume will tend to be followed by a corresponding increase in genome size, and a decrease in cell volume by a decrease in genome size. The genome size of any species depends on the balance between the mutational forces that change genome size and selection for or against particular genome sizes; therefore cell size must influence either the mutational or the selective forces acting on genome size, or both. As it is hard to see how altered cell size would systematically affect which kinds of mutation occur, I conclude that the selective forces on genome size probably depend on cell size.

The systematic correlation between cell and genome size implies a balance between two opposing evolutionary forces, at least one of which depends on cell size. If mutation is random with respect to genome size and if mutational increases and decreases in genome size are equally probable, then the two opposing forces must both be selective—decreases in genome size must have been selected for in cells that have recently become smaller and increases in those that have become larger. This implies that larger cells need more DNA, as my skeletal DNA hypothesis argues (Cavalier-Smith, 1978, 1982a). However, if mutation is universally and consistently biased towards increasing genome size, cells could evolve large genomes even if they were of no selective advantage to them; mutation pressure towards larger genomes could in principle be balanced by an equally general selective pressure favouring smaller genomes (Cavalier-Smith, 1980a, 1982a): if the strength of the selective disadvantage of larger genomes declined as cell size increased, then larger cells would evolve larger genomes. But this would produce the observed correlation between genome size and cell volume only if the selective disadvantage of larger genomes was a function of genome size as well as of cell volume; such counter-selection would have to be a step function of the genome size/cell volume ratio as explained in Fig. 4.9. The assumption by Orgel and Crick (1980) that selection against extra DNA is a function of the ratio of coding to non-coding DNA cannot explain the observed correlation.

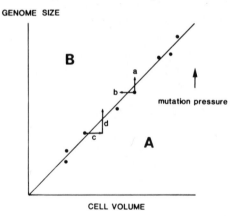

Fig. 4.9.   Genome size/cell volume correlation as the product of a hypothetical equilibrium between mutation pressure and selection. The line separating regions A and B represents an evolutionarily stable state for genome size and cell volume if (1) there is a universal upward mutation pressure always tending to increase genome size *and* (2) selection is always against such increases, but for species in region A its strength is always too weak to counteract mutational increases whereas as soon as mutational increase in genome size (a), *or* decrease in cell volume (b), moves a species across the line into region B the strength of counter-selection rises so sharply as to prevent any further increases in genome size. A mutational increase in cell volume (c) would sooner or later be followed by further increases in genome size (d), which would only be subject to effective counter-selection when the species crosses the equilibrium line. (For a powerful argument against this model see note 3 p. 178).

Such a complex dependence on both cell and genome size of selection against extra DNA is not impossible, but it would remain a purely *ad hoc* hypothesis with no empirical or theoretical basis unless one could (a) provide convincing evidence that mutation pressure *is* consistently towards larger genomes and (b) explain *why* the selective disadvantage of extra DNA should vary universally with cell volume in just such a way as to produce a constant ratio of DNA mass to cell volume; since no one has provided such an explanation* I prefer my skeletal DNA hypothesis, which provides a simple, functional explanation for the constant DNA/cell volume ratio. But before discussing this I shall consider the possibility of a mutational bias towards larger genomes. [*See also the important note (3) on p. 178.]

Mutation pressure and genome size

Possible sources of extra DNA are tandem duplication, duplicative transposition, retroposition, trisomy, polyploidy and integration of DNA viruses or plasmids, which I shall consider in turn in order to assess whether one can expect them to be balanced by deletion. Their physical mechanisms are discussed in Chapter 12 by Dyson and Sherratt.

Tandem duplications arise by unequal crossing-over, either between sister chromatids at mitosis or homologous chromosomes at meiosis, or by replication errors. Consider first the number of repeats (n) in already tandemly repeated DNA. Unequal crossing-over can both increase and decrease the number of tandem repeats in a chromosome, and there is no reason to suppose that increases will be more frequent than decreases. However, deletion of tandem repeats by internal crossing-over within a single chromosome would appear to be much more likely the larger the number of repeats. Such deletions are common in *Drosophila* ribosomal genes and yield bobbed mutants; in general they will tend to put a downward pressure on the overall number (n) of tandem repeats in the absence of selection to increase n. Now consider the case where $n = 1$; unequal crossing-over will then be by illegitimate rather than homologous recombination, and will probably be rarer than changes in n where $n \geq 2$ and will create one chromosome with $n = 2$ and one with $n = 0$. If non-coding DNA only is involved both recombinants will be viable. But it must sometimes involve coding DNA, where $n = 0$ may be inviable; on average therefore recombinants with $n = 0$ will less often be viable than those with $n = 2$. This *selective* force will somewhat bias genome increases in the upward direction, but if such changes are less frequent than intrachromosome deletions where $n = 2$ there may still be an overall tendency to reduce the amount of tandemly repeated DNA in the absence of positive selection for it. Though some animals have appreciable fractions of tandemly-repeated DNA, the above analysis is supported by the fact that very many animals and plants lack detectable tandemly highly-repeated non-genic DNA (John and Miklos, 1979). If there were a universal mutational bias in favour of accumulation of tandem repeats by unequal crossing-over, and this were the sole reason for its accumulation, all eukaryotes with large genomes would be expected to have large amounts of such DNA.

Now consider replication errors which can lead to the doubling of whole replicons, as appears to have happened to many replicons in *Chironomus thummi thummi* (Keyl and Pelling, 1963). Though the exact mechanism of such increases is unclear (for one possibility see Maclean, 1973), it appears to be a mutational process with an inbuilt bias towards increasing the overall amount of DNA. Although the repeated DNA would be expected to be especially prone to deletion by internal crossing-over, the fact that most bands in *C. thummi thummi* have 2^n times as much DNA as those of *C. thummi piger* suggests in *C. thummi* such deletions are rare compared with the original duplications, or else strongly selected against. The fact that so many bands of *thummi* have increased, but none appear to have increased in *piger* compared with *thummi*, implies either that numerous mutations causing replicon doublings occurred in a concerted fashion in *thummi*—with none in *piger*—or that they occurred equally in both subspecies but that selection was so strong in *piger* as to prevent any duplications becoming established in the population. Since the latter is hard to accept, it seems more probable that what happened was a rare but concerted saltatory event. Such events, even if rare compared with deletions at individual loci, might, by causing replicon doublings, quadruplings or octuplings at hundreds of separate loci, place an upward mutation pressure on genome size. But such tandem duplications would be highly prone to deletion (pp. 364–368).

Duplicative transposition also has a potential to impose an upward pressure on genome size (Cavalier-Smith, 1978; Doolittle and Sapienza, 1980; Orgel and Crick, 1980). The possibility that at least some eukaryote inverted repeats (Cavalier-Smith, 1977, 1978) and middle repetitive DNA (Doolittle and Sapienza, 1980) may either be, or be derived from, 'selfish' transposons is becoming increasingly accepted. Nonetheless, the idea that duplicative transposition is the major cause of increased C-values in eukaryotes (Doolittle and Sapienza, 1980) does not stand up to close examination. As emphasized previously (Cavalier-Smith, 1978) selection will act strongly to prevent an indefinite increase in transposon numbers per cell, and to limit the damage done by the insertion of transposons into genic DNA. It is now clear that transposons are subject to self-control of their transposition rates (Dyson and Sherratt in Chapter 12) and of their population size in a cell. But they also stimulate deletions, and lack the rampant drive towards self-multiplication often conveyed by the term 'selfish'. Mutations derepressing transposition could, by leading to the death of the cell or the selective disadvantage of the animal or plant host, be almost as harmful to the 'parasite' as the host. I agree with Dawkins (1982) that in the evolutionary battle between such 'outlaw' genes and their host there will tend always to be a bias of advantage to the outlaw, and that this will tend to put an upward pressure on genome size. However, the key question is how important is this upward 'transposition pressure' (Cavalier-Smith, 1982a) compared with that from other sources, notably duplication pressure from replicon doubling?

To assess the relative strength of the two forces it is necessary to know (a) the relative frequency of increases in the copy number of transposons compared with replicon doublings and (b) the relative selective disadvantages of such increases.

First consider the disadvantages: it seems evident that duplicative transposition will more often be harmful than replicon doubling; transposons will frequently insert into and inactivate protein-coding genes, which will often be lethal, whereas gene duplication will very rarely be lethal. In the absence of data on the relative frequency of transposition rate mutations and replicon duplications, we can only speculate on their likely frequency. Since transposition is mediated and controlled by specific proteins, mutations affecting it might have the usual frequency for gene mutations: many such mutations would probably abolish transposition or lower its frequency rather than increase it. The very high frequency with which duplications can occur in cultured cells (see Chapter 6) makes it likely that such duplications occur at least as frequently as those affecting transposition rates. Though hard data would be preferable to such general arguments, they suggest that replicon duplication may be quantitatively more important than duplicative transposition in causing genome size increase: the case of *Chironomus thummi* strongly suggests this, the 27% increase in genome size has been produced entirely by replicon duplication (Keyl, 1965).

It might be argued that the above analysis ignores the fact that duplicative transposition can potentially increase numbers in the absence of gene mutations. But such increases, which may be thought of as a form of selection—'intragenomic selection' (Cavalier-Smith, 1980a, 1982a) or 'nonphenotypic selection' (Doolittle and Sapienza, 1980)—cannot occur indefinitely. In practice the cellular 'population size' of a particular tranposon is remarkably constant from generation to generation, implying rather close regulation. Clearly, if hybridization occurs between a strain lacking a particular transposon and one having it, the transposon may transiently increase by intragenomic selection (Charlesworth, Chapter 16). But apart from this, intragenomic selection will tend to increase transposon copy numbers only after the occurrence of a mutation in one of the genes that control copy number. It might also be argued that some kind of accident to the cell might, even in the absence of mutation, transiently derepress transposition, causing a sudden increase in copy numbers of transposons that might be inherited stably thereafter. However, exactly the same argument applies to replicon duplication for it seems likely that the ancestral *Chiron-*

omus thummi thummi had just such an accident, allowing a saltatory burst
of replicon doubling. It must not be forgotten that transposons and ordinary
replicons are both subject to regulatory control of their replication and
probably to the same basic kinds of mutation.

Despite recognizing the probable importance of duplicative transposition
and intragenomic selection in the evolution of certain dispersed repetitive
sequences is recognized, (Cavalier-Smith, 1977, 1978, 1980a; see Chapter 15)
I am sceptical of the view (Doolittle and Sapienza, 1980) that duplicative
transposition is the major force that has led to C-value increases. As argued
above, replicon duplication is likely to be more frequent and less harmful
than mutations affecting transposition mechanisms, and replicon duplication
was the actual mechanism in *Chironomus thummi*. The evidence that replicon
size is largely independent of C-value but that replicon number increases
in proportion to C-value, discussed in Chapter 7, is also fully consistent
with my view (Cavalier-Smith, 1978, 1980a) that replicon duplication has
been the major cause of C-value increase. However, these features of
replicon size and numbers could also result from duplicative transposition,
provided the transposons contained the same density of replicon origins as
the rest of the genome, (and from such mechanisms as polyploidy and
trisomy). But the uniformity of replicon size in organisms of different
C-values could not result directly by virtue of the mode of increase if it were
predominantly by duplications several-fold shorter than the mean length of
replicons.

It now appears that much dispersed repetitive DNA in animals may have
originated by the insertion into DNA of complementary DNA (cDNA)
copies of RNA molecules, presumably made by a reverse transcriptase. Such
sequences have been called retroposons (Rogers, 1983); I suggest that this
mode of origin be called retroposition. If retroposons lack the capacity to
transpose, once inserted, they could not spread by intragenomic selection
and would pose less mutagenic threat to the cell than transposons, and
perhaps therefore be less strongly selected against. But if as I suspect they
can be deleted, their abundance may be a sign of rapid sequence turnover
rather than of upward mutation pressure on genome size.

A source of slight upward mutation pressure is trisomy: trisomics are
commonly selectively disadvantageous, especially in mammals, but if the
extra chromosome consists largely of non-coding rather than genic DNA it
may become established in the population as a B-chromosome. Mutations
causing B-chromosomes preferentially to enter gametes can lead to their
increase in the population even if they are slightly harmful, (Jones, Chapter
13), but the very rarity of B-chromosomes is one of several arguments
showing the great strength of organismic selection against selfish DNA
(Chapter 8 and note 3 p. 178).

A final source of upward mutation pressure on cellular DNA content is polyploidy. Polyploidy is especially important in certain taxa, e.g. flowering plants, and may often be selected for because it reduces or eliminates the infertility of hybrids. Since polyploidization is far commoner than depolyploidization it will tend to increase DNA contents. In a recently formed polyploid one would not speak of an increase in genome size or C-value, but as the two formerly separate genomes undergo gene mutations, translocations and other rearrangements, and changes in chromosome number, there will eventually come a time when they are in effect merged into a single new genome. A sufficiently ancient polyploid will not be distinguishable from a diploid. Therefore in the long run polyploidy may reasonably be considered as a force tending to lead to genome size increase.

Virus and plasmid integration can both increase genome size, but both are reversible.

Natural selection on genome size as a function of cell volume

If there is a general upward mutation pressure on genome size then there would be no necessity to postulate positive selection for the extra DNA or a positive nucleotypic function for DNA, only if the strength of selection against extra non-coding DNA is inversely related to cell volume, and as explained in Fig. 4.9 is such as to exactly balance the upward mutation pressure and give a constant DNA/cell volume ratio in unicells; in multicells it would also have to give a constant DNA/cell volume ratio, but one which differed not only from that of unicells but also from tissue to tissue. But the strong selection against extra DNA implied by the arguments in Chapters 3 and 8 and note 3 p. 178 argue strongly against such an interpretation. The strong correlation between C-value and nuclear volume, and the fact that virtually all secondary DNA accumulates in the nucleus rather than in mitochondria and chloroplasts favours positive selection for nuclear volume as the key factor.

THE IMPORTANCE OF NUCLEAR VOLUME

Nuclear volume is directly proportional to C-value in both animals and plants as was shown in Fig. 4.3. As in the case of the correlation between cell volume and genome size this could reflect either an indirect evolutionary correlation of the two variables or a directly causal (nucleotypic) determination of nuclear volume by the DNA content. The skeletal DNA theory argued that DNA acts directly as a nucleoskeleton that is the primary determinant of interphase nuclear volume, and that the changes in nuclear volume that occur in different diploid somatic cells of multicells are brought about primarily by controlling the degree of folding or unfolding of the DNA (Cavalier-Smith, 1978); in endopolyploid somatic cells which are common in invertebrates and plants (d'Amato, 1977; Nagl, 1978) nuclear volume increases in direct proportion to DNA content. I also argued that nuclear volume is a strongly adaptive character and that larger cells need to have larger nuclei (Cavalier-Smith, 1978). The extensive evidence favouring both these hypotheses has been reviewed elsewhere (Cavalier-Smith, 1982a). Here I wish to stress that these two proposals, though initially presented together, are logically independent and need not stand or fall together. This is implicitly recognized by Orgel *et al.* (1980) who seem to accept the proposal that DNA contents may control nuclear volume (and even cell volume, though they do not specify a mechanism) while being sceptical of the idea that nuclear volume is functionally important; their view could be made more explicit by postulating that:

1. nuclear volume is a non-functional neutral character;
2. DNA contents evolve through a changing mutation-pressure-selection balance as shown in Fig. 4.9;
3. nuclear volume is nucleotypically determined by the nuclear DNA content: nuclear volume would on that hypothesis be determined purely by the position of the mutation-selection equilibrium.

This model will be referred to as the neutralist hypothesis for nuclear volume evolution since it asserts that neither large C-values nor large nuclei are positively selected for. Its flaw is that the idea that nuclear volume can be *purely* nucleotypically determined in untenable, because of the great variability in nuclear volume in different tissues of the same multicellular organisms. A purely nucleotypic theory of nuclear volume control is compatible with those nuclear volume variations caused by endopolyploidy, which is common in some cell types in invertebrates and flowering plants, but it cannot explain the extreme variation in nuclear volume found in non-polyploid germ line cells or of diploid somatic cells such as vertebrate nerve cells. In these latter cases only partial nucleotypic control is possible: genic control must also be postulated. This is also shown by the fact that nuclear volume is the same for comparable cell types in haploid male and diploid female Hymenoptera (Darlington, 1939).

The skeletal DNA theory is not purely nucleotypic, but involves a dual nucleotypic/genic control—specific genes must be involved in controlling the unfolding of the DNA in larger nuclei and in controlling the synthesis and assembly of extra nuclear matrix and envelope. Since mutations in such genes could cause large changes in nuclear volume it follows that nuclear volume cannot depend simply on the amount of DNA it contains: such genes must be subject to selection in order to maintain the characteristic nuclear volumes of different cells of the same organism.

In my view the evidence that nuclear volume is adaptive is stronger than the arguments for nucleotypic control of nuclear volume. So much so that one must seriously consider the possibility that nuclear volumes are purely genically controlled, and evolve by conventional mutation and selection to a nuclear volume optimally adapted to each cell type. Once a particular nuclear volume was set by the functional requirements of the cell it might be expected to exert a powerful constraint on the upward mutational drift of genome size: all nuclei of a particular size could accommodate just so much and no more extra non-coding DNA before its presence would begin to interfere with nuclear function and be selected against. This model agrees with the neutralist model in that it denies that large C-values are positively selected for, but agrees with the skeletal DNA hypothesis in assuming that nuclear volume is an adaptive character; I would therefore prefer it to the neutralist model, though it seems inferior to the skeletal DNA hypothesis for the following reason.

Consider again the wide range of nuclear volumes in multicellular organisms, and also compare the DNA/nuclear volume ratio in these with that in unicellular organisms. Since the cytonuclear ratio (the cell volume/nuclear volume ratio) is similar in the multicellular and unicellular eukaryotes represented in Fig. 4.2, it follows from the data of Fig. 4.2 that the DNA/ nuclear volume ratio must be about 20–80 times greater in the multicellular cell types shown here than in the unicellular ones. How can we explain the fact that the evolutionarily stable value for the DNA/nuclear volume ratio is about 20 × greater for plant apical meristem cells and about 80 × greater for amphibian and fish red blood cells than for unicellular organisms if the extra DNA is not selected for? The skeletal DNA hypothesis, however, readily explains why this is so (Cavalier-Smith, 1978, 1980c, 1982a). Some cell types, notably certain nerve cells of vertebrates and the haploid vegetative cell of plant pollen, have interphase nuclear volumes far greater than those of red cells or meristems: in these large cells the chromatin is decondensed as it is in unicellular eukaryotes, and their DNA/cell volume ratio is comparable to that of unicells. This is just what is expected if DNA determines nuclear volume, and if the differing nuclear volumes in the various cell types of a multicellular organism are primarily controlled by differential folding of the DNA. Selection for economy in the amount of skeletal DNA will keep the DNA/nuclear volume ratio as low as possible *at that stage of the life history that needs the largest nuclei*: such stages will always have fully decondensed chromatin (Cavalier-Smith, 1978, 1980c, 1982a) and a similar DNA/nuclear volume ratio to unicells. But if the organism also has other stages that require a lower nuclear volume (in animals, sperm and red blood cells are the most extreme examples) they will evolve mechanisms to fold their DNA more tightly (such mechanisms will be absent in those unicells in which nuclear volume changes little during the life cycle, but present in those where large changes in nuclear volume occur, which seems to be the case (Cavalier-Smith, 1980c)), as constitutive heterochromatin.

The hypothesis that there is no selective advantage for larger genomes, on the contrary, does not explain why the DNA/nuclear volume is much greater in apical meristem cells of plants compared with, say, unicellular algae. One would have to introduce the *ad hoc* hypothesis that selection against extra non-coding DNA is higher in unicellular organisms than in multicellular ones; but unless we could explain why this should be so this additional hypothesis would have no explanatory value. Unlike the skeletal DNA hypothesis, neutralist theories of genome size evolution do not adequately explain why the mechanism for the much greater unfolding of the chromatin in very large nuclei should have evolved: if the non-coding DNA is totally functionless, why bother?

GENOME SIZE, CELL VOLUME AND CELL GROWTH RATES

In plants and protists cell growth rates have been shown to be inversely related to genome size (Shuter *et al.*, 1983). Genome size also tends to be inversely related to developmental rates in amphibia (Oeldorf *et al.*, 1978) and flowering plants (Bennett, 1972); very rapidly developing species invariably have small genomes while those with large genomes always develop slowly. Several authors have therefore suggested that cell growth rates and the developmental rates of multicells may be nucleotypically determined (Commoner, 1964; Bennett, 1972). However, the quantitative relationship between genome size and growth rate is very different from that with cell and nuclear volumes. When cell or nuclear volume are plotted against genome size on a log/log plot a slope of 1, or slightly more, is observed. But if minimum doubling times of eukaryote cells (inversely related to maximum growth rates) are plotted against genome size on a log-log plot (Fig. 4.10(a)) the slope is only 0.2.

A satisfactory theory should do the following:

1. explain why the slope should have this particular value;
2. provide a physical mechanism for the relationship;
3. make a minimum of novel or *ad hoc* assumptions;
4. be compatible with and relatable to knowledge in related areas.

The idea of nucleotypic determination falls down on all four counts. It provides no plausible mechanism or quantitative explanation, and makes a major *ad hoc* assumption for which there is no independent evidence. Perhaps more importantly it ignores the fact that if genome size correlates strongly with cell volume it will also show correlations with any other organismic property that varies systematically with cell volume. Physiologists have long been interested in the systematic variation in numerous biologically important quantitative parameters with body size, i.e. the problem of scaling or allometry. A very large number of characters show such a correlation, which can usually be described by the allometric equation $y = ax^b$ or its equivalent $\log y = \log a + b \log x$, where x is body mass or volume and y the physiological or anatomical variable. The values of the constants a and b depend on the character in question (and sometimes also on the type of organism being studied, e.g. whether cold-blooded or warm-blooded). For unicells body size is the same as cell size and, as mentioned earlier, is correlated with a variety of variables, such as maximum exponential growth rate r_m, respiration rates, phosphorus and nitrogen subsistence quotas, sinking rates of phytoplankton, and susceptibility to predation, all of which are of ecological and adaptive importance. Since genome size is related to cell volume it will be related to all of these, but this does not mean that it need determine any of them. The simpler hypothesis is that almost every quantitative aspect of an organism is related either by simple physical scaling laws or in an adaptive manner to its size, and in unicells to its cell size, and that their correlation with genome size is the purely statistical consequence of this and of the correlation between genome size and cell volume, the possible causes of which have been discussed above. This view could be fully substantiated only if the causes of the correlations between cell volume and these other parameters are understood in detail. At present this is not the case, largely because cell and molecular biologists have almost totally neglected the problem; many are probably even unaware of it since it has mainly been the concern of ecologically oriented phycologists and marine and freshwater biologists.

Fig. 4.10(a) implies that minimum doubling time correlates with cell volume as it does with genome size. This is shown more directly in Figs. 4.10(b) and 4.11; Fig. 4.11 also shows that the quantitative relationship differs slightly from one taxon to another. However, as the slope of the curve

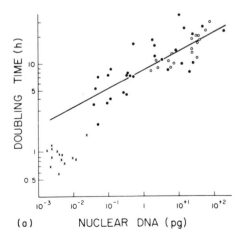

(a)   NUCLEAR DNA (pg)

DOUBLING TIME

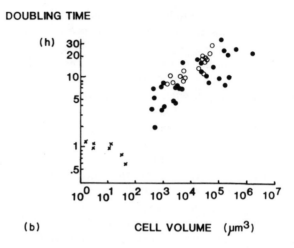

(b)   CELL VOLUME (μm³)

Fig. 4.10.   (a) Relationship between minimum cell doubling time and DNA content for bacteria (x) and unicellular eukaryotes (●) and multicellular eukaryotes (○: root meristem cells of flowering plants). The DNA content is that of one genome for bacteria and of one G_1 nucleus for eukaryotes. From Shuter *et al.* (1983) by permission of the University of Chicago. Note that for two-thirds of the unicellular eukaryotes the DNA values are not direct measurements but were calculated indirectly from their cell volume using the regression line of Fig. 4.2. It is therefore more meaningful to plot minimum doubling time directly against cell volume as shown in (b). Here the bacterial points have been plotted using the data in Table 2 of Shuter *et al.* and the flowering plant points have been moved as much to the left relative to the unicellular eukaryote points as necessary to compensate for their greater DNA/cell volume ratio (cf. Fig. 4.2).

is much less than for Fig. 4.2 and the overall scatter is very great, it is clear that other variables can influence maximum growth rates about as strongly as does cell volume: the situation is quite different from that shown for cell volume in Fig. 4.2 where the total range of the mean curve is nearly 10^4 times greater than the scatter for any one cell volume. Nonetheless, the fact that the quantitative relationships for flowering plant root meristems are very similar to those for unicellular algae, as shown in Fig. 4.10(b), suggests that there is a universal cause for the decline in the maximum growth rate of eukaryotic cells with size that is fundamentally the same for unicells and multicells.

Why are exponential growth rates lower for larger cells?

In multicellular organisms the minimization of cell cycle lengths (and therefore of cell volume and genome size) will be most important in species where cell multiplication throughout the life cycle is the major limit to the overall multiplication rate of the organism, as is probably the case for ephemeral

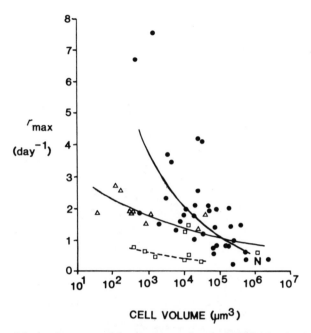

Fig. 4.11. Maximal exponential growth rate (r_{max}) at 20°C of unicellular euka-ryotes as a function of cell volume. Note that for cells of the same volume the polyploid heterokaryote protozoa (●) often have a greater growth rate than the diploid diatoms (△; continuous regression lines derived from least squares fit to log.log (Banse 1982b) plot) which in turn grow faster than the haploid dinozoa (□; dashed regression line (Banse 1982b)); yet in all three groups, as in the freshwater amoebae (Baldock *et al.*, 1980), growth rate is on average higher in smaller cells. N refers to the aberrant dinozoan *Noctiluca* referred to in the text. Data from Banse (1982b).

weeds. Grime (1983) has shown that the grass *Poa annua* varies in genome size, and that on average clones with smaller genomes do in fact have higher net rates of accumulation of dry matter (Fig. 4.12); I suggest this is because they have smaller cells and shorter cell cycles. Except for cases involving B chromosomes, only a few other reliable examples of intraspecific variations in genome size are yet known (Evans, 1968; Rees *et al.*, 1978; Price *et al.*, 1980; Price *et al.*, 1981; Kenton, 1983; Ainsworth *et al.*, 1983), in contrast to the large number of cases of autopolyploidy; earlier reports of intraspecific variation in genome size in conifers are now thought to have arisen from errors in measurement (Teoh and Rees, 1976). The intraspecific variation in genome size in *Poa annua* is especially important as it has been clearly correlated with ecologically and adaptively important characters that may reasonably be attributed to a variation in cell volume (see note 2 p. 178).

Biologists have long emphasized the importance of surface/volume ratios,

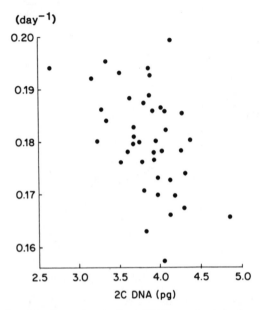

Fig. 4.12. Relationship between nuclear DNA content and mean relative growth rate (i.e. rate of production of dry matter) of seedlings of *Poa annua* grown for 6 weeks in a productive controlled environment (30°C day; 20°C night). From Grime (1983), reprinted by permission of The Association of Applied Biologists.

which decline as cells get larger. If surface/volume ratios were the basic determinant of growth rates doubling times should vary wth the one-third power of volume (i.e. $b = 0.33$). Though some early reports for heterokaryote protozoa were of this order, more extensive data for 35 species over a 5000-fold range in cell volume (Taylor and Shuter, 1981) suggest that $b = 0.247$, while it is lower still for non-polyploid photosynthetic protists (Banse 1982b; 0.11 for diatoms and 0.17 for dinozoa; when diatoms, dinozoa and green algae are lumped together as in Fig. 4.10, $b = 0.20$). This suggests that growth is not simply limited by surface/volume ratios.

Studies of the cell cycle have so focused on the mechanisms of cell division and its control that little attention has been given to the mechanisms of cell growth and even less to how these might be quantitatively influenced by changes in cell size. Cell growth depends on four processes:

1. net transport of molecules into the cell;
2. transport of molecules within the cell;
3. net synthesis of molecules;

4. assembly of molecules into subcellular structures such as membranes, microtubules and ribosomes.

In principle any of these four processes might be rate-limiting to growth (Cavalier-Smith, 1985). It is clear that a restricted supply of nutrients or energy reduces growth rates as a result of their lower than optimal rate of transport into the cell, but it is not obvious that transport rates into the cell would also be the basic determinant of maximum growth rates under optimum conditions. What about the other three processes?

Scaling considerations help to clarify the situation. It seems unlikely that the rate of assembly could be the rate-limiting step that is affected by changes in cell volume, since it ought to depend only on the direct interactions and local concentrations of the molecules in question. The same is true for the synthesis rates of the majority of molecules, which would depend on the local concentration of the enzymes, substrates, cofactors and (for informational molecules) of any necessary template, and on the fraction of the cell occupied by the compartment containing them. The fraction of the eukaryotic cell occupied by each compartment could easily be kept constant as cell size varies; this clearly is the case for the nuclear and cytoplasmic compartments. There are strong indications that it is true for chloroplasts (Butterfass, 1979) and one may expect the same to be true for mitochondria. If transport into or through the cell were not rate limiting then the concentration of each cell component could be maintained, as cell volume varied, by increasing or decreasing the number of copies per cell of each of the molecules that was involved in its synthesis in direct proportion to cell volume. For most molecules this increase or decrease would be automatic since their concentration is regulated by feedback controls that act directly via their concentration. The concentration of such molecules would automatically be maintained if the cell volume were to be altered whether temporarily during development or more permanently during evolution. The only molecules for which this is not true are the genes themselves: in the absence of endopolyploidy their synthesis, as Chapter 7 explains, is regulated so as to keep their numbers per *cell* constant and *not* their concentration (i.e. numbers per unit *volume*). I shall therefore argue that the basic reason why larger cells grow more slowly is because of their lower gene concentrations.

A gene mutation that doubled the cell volume, which for proliferating cells it could do only by altering the volume-dependent controls over replication, would halve the concentration of each gene. Therefore, some genes whose concentration was previously not rate-limiting to growth could become rate-limiting, and the larger cell therefore grow more slowly. But the genes in question could either undergo mutation of their promoters in order to increase their transcription rate so that their concentration was no longer rate-limiting to growth, or, if they had already evolved a maximal transcription rate, undergo gene duplication so as to increase their copy number and therefore concentration. For ribosomal genes it is clear from the fact that they are highly repeated in all eukaryotes, and that their copy number is positively correlated with C-value (see Chapter 3) and therefore with cell volume, that such duplication has occurred many times, as it has also for 5s RNA, 7s RNA and tRNA genes. But all these genes have RNA as their final product. For protein-coding genes translation provides a further amplification step, so their concentration is less likely to become rate-limiting for growth. The existence of numerous copies of histone genes in many animals is almost certainly attributable to the necessity for rapid histone synthesis to allow rapid DNA replication and division during embryonic cleavage when growth is not occurring, and is therefore not an example of a protein-coding gene limiting *growth*. However, the fact that histone-genes are repeated shows that protein-coding genes can become duplicated: the fact that most protein genes are unique suggests either that their concentration does not limit growth rates, or, as I shall argue, that there is insufficient selective advantage for higher growth rates to counterbalance any disadvantages that might be associated with repeated protein-coding genes (e.g. problems of maintaining homogeneity between the several copies, or proneness to unequal sister chromatid exchange or crossing-over). That low gene concentration can be a limiting factor is strongly suggested by the widespread occurrence of somatic polyploidy and polyteny, which both allow an increase in cell volume without decreasing gene concentrations. The fact that the maximum multiplication rate of heterokaryote protozoa (= Ciliophora) with polyploid* macronuclei is commonly several times faster than non-polyploid protozoa (e.g. dinozoa) of the same cell volume (cf. Fig. 4.11 and Shuter *et al.*, 1983) is consistent with the thesis that gene dosage may often be rate limiting to growth. Even the greater multiplication rate of diatoms compared with dinozoa of the same cell volume shown in Fig. 4.11 might partly result from their greater concentration of unique genes; as they are diploid rather than haploid they will have twice the gene concentration. Since diatoms are much more highly vacuolated than most dinozoa their effective relative gene concentration is even greater; their greater vacuolation also means that less organic material need be synthesized per cell cycle, which will further boost their exponential multiplication rate. Vacuolation therefore, like polyploidy, is a way of shifting the position of the trade-off between increased cell volume and reduced growth rates so as to allow a simultaneous increase in volume and growth rate. The aberrant giant dinozoan *Noctiluca* is also highly vacuolated and as Fig. 4.11 indicates has a growth rate more in keeping with that of diatoms than for more typical dinozoa. The unique characteristics of the heterokaryote protozoa make them a very important test case for theories of genome evolution, so they will be discussed in some detail. *(See notes 1 and 3 p. 178.)

When heterokaryote cell volumes are plotted against total nuclear DNA content they fall on the same curve as for other protists (Shuter *et al.*, 1983). But since most of their DNA is macronuclear, and their genome size is manyfold smaller than their total nuclear DNA content, this shows that it is *not* genome size but total DNA amount that coevolves with cell volume. This refutes the suggestion (Orgel and Crick, 1980) that it is the ratio of coding to non-coding DNA that evolutionarily determines genome size, since the ratio is very different for heterokaryotes and other eukaryotes. What is clearly important is the total amount of DNA irrespective of its composition. The relationship between doubling times and DNA content of heterokaryotes is equally revealing. Shuter *et al.* (1983) found that if heterokaryote doubling times are plotted against genome size (*not* total DNA content) they fall on the same curve as for other eukaryotes. Though this might be held to support the hypothesis that genome size determines doubling times, no mechanism is known for this and the simplest hypothesis is that doubling times are related to gene concentrations. If it is assumed that macronuclear polyploidy increases DNA contents and cell volume in direct proportion, as in other organisms, then it follows that the gene concentrations will be independent of the degree of polyploidy and that a polyploid heterokaryote will have the same gene concentration as a diploid protist of the same genome size. Therefore, if gene concentrations determine maximum growth rates, it should grow at the same rate; growth rates would show qualitatively the same correlation with genome size as in other eukaryotes, and this is observed.

Although the regression line for heterokaryote's growth rates against genome size is indistinguishable from that for other organisms the data points are more widely scattered about it than for other organisms such as flowering plants. The same is true for the relationships between cell volumes and growth rates where the heterokaryote scatter is distinctly greater than it is for diatoms or dinozoa (Banse, 1982). This greater scatter can also be explained by the present gene concentration hypothesis, since Taylor and Shuter (1981) showed that some heterokaryotes have a higher and others a lower degree of macronuclear polyploidy than would be expected from their size; this means that those with a higher degree of ploidy for their size will have an above average gene concentration, and those with lower ploidy below average concentrations, so species with an above average degree of ploidy for their size should grow faster and those with a lower than average ploidy more slowly than predicted by the regression line. Taylor and Shuter (1981) showed that this is the case, and also that there is a positive correlation between the degree of deviation from the regression line of r_m against volume and the relative DNA content of the micro- and macronuclei (i.e. the degree of polyploidy of the macronucleus). Taylor and Shuter (1981) found by multiple regression analysis that 88% of the variation in maximum growth rates in heterokaryotes can be predicted just from their cell volume and degree of macronuclear polyploidy, both of which will affect gene concentration in the way I have suggested. These observations strongly support the hypothesis that gene concentrations actually limit growth rates. A possible counter-example to this thesis is the observation that freshwater amoebae show exactly the same quantitative relationship between cell volume and growth rates as do heterokaryotes (Baldock *et al.*, 1980); unless these amoebae are polyploids, which is possible, this would suggest that the small to medium heterokaryotes grow faster than the photosynthetic diatoms and dinozoa not because they are polyploid but because they are heterotrophic: in bacteria also phototrophic species tend to grow more slowly than heterotrophic ones. Though gene concentrations may often be an important factor in influencing growth rates, other factors must also be important.

The gene concentration hypothesis does not predict exactly how growth rates should scale with genome size, but only that the allometric coefficient should lie between 0 and 1. If gene concentrations are actually rate-limiting in a particular species then evolutionary increases in cell volume unaccompanied by any evolutionary compensation, e.g. though gene doubling, would lead to a proportionate slowing in growth, so the allometric coefficient would be 1. Perfect compensation (by promoter mutations and/or gene duplication) would give an allometric constant of 0; partial compensation would yield intermediate ones, as are observed. The fact that the allometric coefficient may differ somewhat in different groups may reflect the differing selective advantages of different degrees of compensation in different groups. A possible instance of a total lack of compensation may be the rate of male meiosis in different flowering plants, where meiotic time scales directly with genome size ($b = 1$) (Bennett, 1973); the contrast with the plant mitotic cycle, where $b = 0.26$, could be explained if the limiting factor for pollen growth was one or more protein(s) needed in vastly greater amounts at that stage than for vegetative growth (synaptonemal complex proteins?). If the duration of meiosis were unimportant compared with the length of the mitotic cell cycles in determining the overall length of the life cycle it would not be expected to evolve any special compensation mechanism and therefore would show the full uncompensated gene concentration scaling effect ($b = 1$). The fact that tetraploids have the same meiotic times as diploids with half their cellular DNA content (and therefore half their cell volume) (Bennett, 1973) is exactly what the present gene concentration hypothesis would expect (since as tetraploids they would have the same gene concentrations) but directly contradicts the nucleotypic hypothesis that DNA amounts determine developmental rates. The fact that meiosis can take far longer and be less clearly related to genome size in animals is compatible with the present hypothesis which, unlike the nucleotypic hypothesis, postulates no direct connection between genome-size and developmental rates: many other factors could vary and prevent the simple relationship seen in plants from manifesting itself.

The preceding arguments assumed that transport within the cell did not limit the rate of growth. Is this assumption justified? Maynard Smith (1969) suggested that only when a substance is made in one part of the cell but has to diffuse to a specific target, which is the same size whether the cell is large or small, would one expect diffusion rates to limit growth. Since the time taken to diffuse in a specific direction will increase with the cube of the distance this hypothesis expects doubling times to scale directly with cell volumes. Since this is clearly not the case for ordinary mitotic cell cycles (though, as mentioned above it might be so for growth during meiosis) this cannot be the basic explanation of slower growth in larger cells. If instead diffusion to numerous targets (i.e. in any direction) were rate limiting, then as volume (V) increases, transport rates should scale with $b = 1/3$, transport needs directly with V ($b = 1$), and therefore growth rates with $b = 1/3$. Since observed values are markedly lower than this, this argues against generalized intracellular transport rates being rate limiting to growth. Another argument for the same view is that cyclosis is absent from the majority of growing cells; this is true for many orders of magnitude of cell size. Since cytoplasmic streaming could help to reduce intracellular transport times for macromolecules or macromolecular assemblies such as ribosomes compared with simple diffusion, its general absence from medium-sized and even fairly large cells suggests that transport only becomes limiting in exceptionally large or giant cells, e.g. slime mould plasmodia, foraminifera, or differentiated non-growing cells of higher plants and algae such as *Chara* that have enlarged greatly by vacuolization after divsion ceased.

It was suggested previously that transport rates of RNA could potentially limit growth rates (Cavalier-Smith, 1978), but if this were the major limiting factor one would also expect $b = 1/3$, which is probably significantly higher than observed values (Shuter *et al.*, 1983). Furthermore if, as suggested (Cavalier-Smith, 1978), nuclear surface area were the limiting factor one would expect nuclei to be aspherical or to have convoluted surfaces so as to provide a relatively larger surface area. That nuclei are usually non-spherical only in very large cells, e.g. some heterokaryotes, and show convolutions only in a few specially large cells, e.g. very large nerve cells, argues against the *general* limitation of growth rates by nucleocytoplasmic transport rates, though supports the importance of this limitation in specific instances. The data on unicellular algae and amphibia that suggested that nuclear volume increased more than proportionally with cell volume have been previously cited (Cavalier-Smith, 1978, 1982a) in support of my thesis that nuclear surface area was commonly a limiting factor in controlling growth rates, but there are two problems with this. First Holm-Hansen's algal data; these showed that when C-value was plotted against cellular carbon content $b = 1.22$. Since for flowering plants C-value and nuclear volume were directly proportional, with $b = 1$ (Baetke *et al.*, 1967) this was taken as evidence for an allometric relationship between nuclear volume and carbon-content. I further assumed that carbon-content was directly proportional to cell volume; but this assumption is invalid since carbon content is relatively lower in larger cells. Shuter *et al.* (1983) showed that when Holm-Hansen's algal data are corrected to allow for this (and compared with additional data) there is no evidence for an allometric relationship between nuclear volume and cell volume: for DNA content plotted against cell volume $b = 1.03$ which is not statistically distinguishable from the isometric condition where $b = 1$. Price *et al.* (1973) showed that for shoot meristem cells of 14 species of flowering plants $b = 0.99$, again not statistically distinguishable from 1. That these two independent studies, of unicellular eukaryotes on the one hand and multicellular plants on the other, should agree so closely suggests that it is a general rule that in *growing* eukaryote cells nuclear volume scales isometrically with cell volume. Data for growing animal cells is badly needed to test the universal applicability of the rule. The existing data for amphibians (Fig. 4.13) unfortunately does not do this since it is not for growing cells but for non-growing mature red blood cells. When their nuclear volume is plotted against cell volume, $b = 1.38$ (Fig. 4.13) which implies that nuclear volume does increase faster than cell volume. However if the selective forces acting on the cell and nuclear volumes of red blood cells differ from those on growing animals cells, the latter could have a quite different value for b. Since red blood cells are commonly smaller than tissue cells (6–8 times smaller for the two amphibians for which Horner and Macgregor give data), they probably are subject to selection to restrict their size: such selection would probably be relatively stronger in species with large tissue cells than in those with small ones, as is also implied by the fragmentation of the large red cells of certain amphibia into smaller units. Nuclear volume also is probably minimized in red cells so as to allow the maximum cytoplasmic space for haemoglobin, but selection could not reduce nuclear volume below the minimum packing space of the DNA and its associated histones, so nuclear volume will depend on the C-value and not the cell volume. Cell volume may therefore be easier to reduce during erythropoiesis than nuclear volume. Given those selective forces an allometric relationship between red blood cell volume and nuclear volume of $b = 1.38$ would be possible even if the relationship for growing cells is isometric as for plants and unicells. Though it may be premature totally to reject the idea that nuclear surface area is generally rate limiting to growth, the balance of evidence is against the idea and in favour of the gene concentration hypothesis. The fact that the relationship between nuclear volume and cell volume is actively proliferating cells appears to be isometric rather than allometric suggests that what is important to the cell is the *volume* of the nucleus rather than its surface area.

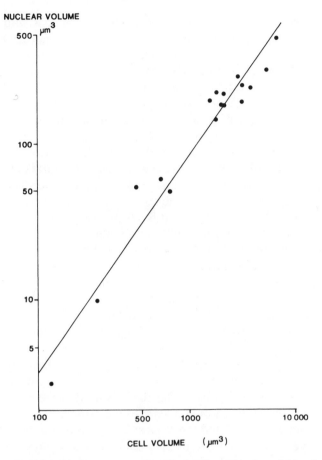

Fig. 4.13. Relationship between nuclear and cell volume in red blood cells of 17 species of amphibia. Data from Horner and Macgregor (1983). Regression line (slope = 1.38) fitted by least squares.

THE SIGNIFICANCE OF NUCLEAR VOLUMES: MODIFICATION OF THE SKELETAL DNA HYPOTHESIS

The isometric relationship means that the cytonuclear ratio (ratio of cell volume to nuclear volume) is fundamentally constant irrespective of cell size, as first suggested by Sachs and Strasburger. Two major exceptions to the constancy of the cytonuclear ratio led earlier workers to doubt its fundamental biological importance: (1) certain non-proliferating cells e.g. muscle where the cytonuclear ratio was unusually high, and (2) animal embryonic cleavage where the cytonuclear ratio steadily declines (e.g. from 550 at the beginning to 6 at the end of cleavage in a sea urchin (Brachet, 1950)). But neither case is counter to my present thesis that what is physiologically important is the cytonuclear ratio in actively growing cells with high levels of transcription and protein synthesis. Although cleavage cells are dividing they are not growing; by the end of sea urchin cleavage when the cytonuclear ratio is restored to its 'normal' level and growth recommences, its value is 6, the same as the mean calculated from the data of Price *et al.* for 14 flowering plants. This general constancy of the cytonuclear ratio in growing and proliferating cells, coupled with the great deviations in differentiated cells that are not both growing and dividing, suggests that a constant cytonuclear ratio is important for balanced growth. It is obvious that balanced growth is possible only if there is a balance between the rates of transcription, RNA processing, nucleocytoplasmic RNA transport, and translation. I suggest that this balance necessitates a constant ratio between the nuclear volume devoted to transcription and RNA processing on the one hand and the cytoplasmic volume devoted to protein synthesis on the other. In rapidly proliferating cells the bulk of the cytoplasm will usually be filled with ribosomes and the major bulk of the nucleus with the RNA synthesizing and processing machinery, so that this condition will imply a constant cytonuclear ratio irrespective of widely varying cell size; if the cytoplasm is greatly vacuolated or otherwise occupied by materials not involved in protein synthesis one would expect a correspondingly higher cytonuclear ratio.

Since, other things being equal, selection will always tend to maximize growth rates, it may be assumed that the structure of the basic transcriptional and translating machinery was optimized for maximum synthetic rates long before the origin of eukaryotes. After the evolution of the nuclear envelope selection for rapid growth would maximize the cytoplasmic concentration of the translation machinery and the nuclear concentration of the transcription machinery and adjust the cytonuclear ratio to that optimal for balanced steady state growth. In all subsequent eukaryote evolution, however much cell volume might change, stabilizing selection would maintain all these properties, which would provide the most fundamental quantitative constraints on cellular evolution and cause nuclear volume to scale isometrically with cell volume in non-vacuolated proliferating cells. Because of its prediction of isometric scaling this hypothesis provides a more satisfactory reason than nucleocytoplasmic transport why larger cells should have correspondingly larger nuclei.

If the above arguments are accepted we can explain the C-value paradox by the additional assumption that genome sizes always evolve in proportion to nuclear volume. According to my skeletal DNA hypothesis (Cavalier-Smith, 1978, 1980c, 1982a) this is because interphase nuclear volume is actually determined by the amount of DNA, its attachment to the nuclear envelope and its degree and manner of folding, i.e. the extra non-coding DNA in high C-value species has a skeletal role. Evidence for this idea, which is quite strong, has been recently reviewed (Cavalier-Smith, 1982) and will not be repeated here. (Newer evidence is discussed on pp. 161–164.)

NON-FUNCTIONALIST ALTERNATIVES TO THE SKELETAL DNA HYPOTHESIS

The skeletal DNA hypothesis assumes that overall genome size is of functional significance in controlling nuclear volume and also that nuclear volume is functionally important. These assumptions could be questioned in two contrasting but mutually incompatible ways:

1. We could accept that nuclear volume is nucleotypically determined by genome size but question the idea that nuclear volume is adaptive. This would explain the correlation between nuclear volume and genome size but not that between nuclear volume and cell volume. Because of this defect, and because the balanced-growth requirement for a constant balance between the overall rates of transcription and RNA processing and of protein synthesis seems to demand a (functionally significant) roughly constant cytonuclear ratio this hypothesis may be rejected.
2. The alternative hypothesis accepts this functional significance of nuclear volume but rejects the idea that genome size determines it; though this explains the correlation between cell and nuclear volume, additional assumptions would be needed to explain (a) the correlation between genome size and nuclear volume and (b) how nuclear volume *is* determined (problems exist with non-nucleotypic mechanisms (see Cavalier-Smith, 1982a)).

Here I consider only the correlation between genome size and nuclear volume. The correlations discussed earlier mean that within a group of species having widely varying cell volumes and genome sizes there is an evolutionary equilibrium that maintains the nuclear DNA concentration approximately constant irrespective of cell or nuclear volume. This constancy might be explained in a non-functionalist manner by supposing, first, that there is a universal upward pressure on genome size, whether by mutation pressure or duplicative transposition pressure and, second, that whenever nuclear DNA concentrations are below the equilibrium concentration, C_e, selection against such increases is too weak to prevent them but whenever C_e is reached selection is so strong as to prevent any further increase in genome size. Postulation of such 'threshold selection' is essential, since otherwise the existence of an upward mutational/transpositional pressure on genome size, (which is itself quite plausible and fully compatible with the skeletal DNA hypothesis) would be insufficient to explain the constancy of C_e.

Though the second hypothesis is compatible with the facts it is rather empty, as it gives no mechanistic explanation for the existence of a *constant* threshold C_e. The idea that the strength of selection against extra DNA depends on the ratio of non-coding to coding DNA (Orgel and Crick, 1980) is clearly wrong since this ratio must be many orders of magnitude greater in the largest cells compared with the smallest ones (see Table 3.6). It would be more reasonable to suppose that the energetic burden of non-functional DNA is relatively less in larger cells, compared with the overall energy needs of the cell (Doolittle and Sapienza, 1980) and that it therefore might be selected against less strongly. But this hypothesis is quantitatively defective: though energy consumption rates do increase with cell size, they do so less than proportionally (scaling with the 0.75 power of cell volume) so that for proliferating cells, where DNA amounts are directly proportional to cell volume, the fraction of energy expended on DNA replication ought to be greater in larger cells. The relative energy expenditure on non-coding DNA, the amount of which increases more than proportionally with cell volume, must increase even more in larger cells.

Another problem is the more than 10-fold higher value for C_e in proliferating cells of multicells than in unicells. This follows from their similar cytonuclear ratios coupled with the larger amount of DNA per unit cell volume of multicells (see Fig. 4.2). But there is no obvious reason why plant and animals cells should be able to tolerate 10 or more times as much non-coding non-functional DNA per unit volume of the nuclei as can unicells. The skeletal DNA hypothesis simply explains this higher DNA concentration as a result of the need for different cell and nuclear volumes at different stages of the life cycle. Certain differentiated cells, e.g. some animal nerve cells or plant pollen mother cells, are usually much larger than ordinary somatic proliferating cells and have correspondingly large nuclei and high rates of macromolecular synthesis; if DNA acts as a nuclear-volume-controlling skeleton, and selection adjusts the DNA content to the minimum necessary to achieve a given nuclear volume, then it is the nuclear volume of these exceptional cells that will be the determining factor in evolution of the genome size of multicells, and which would be expected to resemble unicells in their C_e. The smaller meristematic or embryonic cell nuclei will inevitably have a higher DNA concentration, simply because the extra skeletal DNA needed only in the cell types with much larger nuclei is not eliminated from those with small nuclei. If however, there is much of this extra skeletal DNA one would expect it to be folded as compactly as possible in the small nucleated cells; I suggested that this is the basic explanation for the existence of highly condensed chromatin or constitutive heterochromatin in so many animals and plants, especially in those with larger genomes. Its absence in most unicells, even in those with higher C-values than those of mammals, is expected if selection for economy keeps the amount of skeletal DNA to a minimum: clearly if relatively unfolded DNA can be used as a nuclear skeleton this is more economical than to use a larger amount of more compact DNA. The very few unicells that do have condensed chromatin appear to be, as the skeletal DNA theory predicts, those with unusual life cycles where nuclear volume does vary manifold at different stages (Cavalier-Smith, 1980c).

It is the capacity of the skeletal DNA theory to provide a quantitative explanation of the variations in genome size, and the connections between genome size, cell and nuclear volume, the presence or absence of hetero-chromatin, and the special case of the heterokeryote protozoa (note 3 p. 178), that makes it more attractive than the non-functionalist explanations.

The phenomena of chromosome elimination and chromatin diminution even more powerfully argues against the neutrality or 'junk' theory of non-coding DNA, since if the DNA eliminated from the somatic cells were retained by the germ line then it must be being selected for in the germ line (and against in the soma). I originally assumed that the positive selection in the germ line for the eliminatable DNA (E DNA) must be at the organismal level. However Doolittle (personal communication) has suggested that this DNA could be 'selfish DNA' and be being selected intragenomically. On this hypothesis the key feature that must be selected is the capacity to resist elimination in the germline: this could be through modification of the protected DNA, or by coding for an inhibitor or repressor of the excision nucleases. It is certainly feasible to suppose that these capacities first evolved in some kind of genetic symbiont. There are three possible scenarios for the origin of the difference in somatic and germline DNA contents.

1. The ancestral DNA amount was as in somatic cells, and the germ line amounts increased only after the origin of an elimination mechanism: on the skeletal DNA hypothesis this resulted from selection for larger and larger eggs and therefore for larger and larger nuclei; on the selfish-DNA hypothesis the increase of the DNA would not be caused by organismic selection for more skeletal DNA but either by the great multiplication in the germ line of a genetic symbiont with the capacity to excise itself in somatic cells or by the insertion into such a genetic symbiont of large amounts of other genetic symbionts, or simply by useless 'junk DNA' accumulating by mutation pressure; the upper level of such accumulation would be set by organismic selection (see note 4 p. 178).
2. The ancestral DNA amount was as in the germline, and the reduction in the soma occurred only after the origin of the elimination mechanism. On this scenario, there is no real difference between the skeletal and 'selfish DNA' hypothesis. In both cases it is *organismic* selection for smaller cells and nuclei that must have reduced the DNA content, which would have been equally beneficial to both host and genetic symbiont.
3. The original DNA amount was intermediate between the present somatic and germline level and the origin of the elimination mechanism allowed

the former to decrease and the latter to increase as in the two preceding scenarios.

It might be possible to distinguish between these scenarios by comparing cell and nuclear volume and DNA contents of organisms with chromatin elimination with their nearest relatives that lack it. It might be thought that the skeletal DNA and selfish DNA explanations could be distinguished by determining whether the genetic elements that code for the excision mechanism and those that prevent excision in the germline have any of the properties expected for transposable elements (as the selfish DNA idea would predict) or whether they appear to be normal cellular genes as the skeletal DNA hypothesis might expect. But this is not so, since the skeletal DNA hypothesis does not deny the possibility that these properties originated in genes coded by genetic symbionts. Indeed this seems to be the most likely explanation of their origin. The *de novo* origin of precise DNA excision mechanisms seems improbable; since they are widespread properties of genetic symbionts it is more likely that the cell took them over from genetic symbionts. If so, then chromatin elimination may be added to the telomere replication mechanisms and split genes, as cellular properties that may have originally evolved in 'selfish' genetic symbionts and later been incorporated as regular cellular mechanisms (Cavalier-Smith, 1983c).

The key distinction between the selfish DNA and skeletal DNA theories of chromatin diminution therefore lies not in the mode of origin of elimination but in the mechanism of its maintenance. Even if both the elimination mechanisms and their control originated from genetic symbionts they could only be maintained by *continual* selection to *maintain* the germ line inhibition mechanism. The skeletal DNA hypothesis supposes that mutations that would cause the total elimination of the E DNA from the germ line would reduce nuclear size in the germ line and therefore be selected against at the organismal level; the selfish DNA hypothesis must deny such organismic selection and postulate instead that the wild-type genes have a greater tendency to spread *intragenomically* (in the germ line) then their defective mutant alleles, and are tending continually to increase their nuclear dosage. Even if these genes coding for excision and its control formed part of the E DNA rather than non-diminuted DNA (for which there is no evidence) their ability to inhibit their own excision would not be sufficient to prevent their eventual loss by degrading mutations and random deletions if they were not being positively selected either intragenomically or organismically. On the grounds of its simplicity, and consistency with the evidence concerning cell and nuclear volumes and DNA amounts in other organisms, the skeletal DNA explanation of chromatin diminution is preferable; the indispensability of cecidomyid E chromosomes for gametogenesis supports the skeletal DNA interpretation, but selfish DNA supporters might argue that only part of the DNA is essential.

CHROMOSOME SIZE IN RELATION TO NUCLEAR AND GENOME SIZE

This section discusses the arrangement of non-coding DNA within chromosomes and the arrangement of chromosomes within the interphase nucleus, both of which must impose constraints on the addition or deletion of non-coding DNA during the evolution of different genome sizes. The absence of a clear correlation between chromosome number and genome size means that on average chromosomes size is directly proportional to genome size. How is the extra non-coding DNA of larger chromosomes distributed with respect to genic DNA? There are several reasons why it cannot be totally at random. Since definite sites are needed for replication origins (Chapter 7) non-genic DNA must be interspersed with some regularity between them: very long sequences without replicon origins could not be replicated in the time available during S-phase, so this places an upper limit in the length of totally meaningless sequences. For segregation the length of the lateral loops (Marsden and Laemmli, 1979) of the chromosomes must be reasonably uniform and sufficiently short that the chromosomes are compact enough not to suffer shearing damage to long projections during mitotic chromosome movements and cytokinesis. Since the volume of each metaphase chromosome of really high C-value organisms is much larger than the whole nucleus of the lowest C-value organisms these evolutionary constraints on the location of non-coding DNA are not trivial.

In many organisms non-coding DNA must be fairly evenly dispersed among coding DNA, as is shown by the structure of lampbrush chromosomes, polytene chromosomes, and the frequent absence of large blocks of constitutive heterochromatin. In some organisms, however, there are large blocks of constitutive heterochromatin. The erratic occurrence of such blocks in different species and the difficulty of correlating it with any phenotypic or ecological properties of the organisms in question (John and Miklos, 1979) strongly suggests that it mainly reflects the varying modes of origin of non-coding DNA (see Chapters 7 and 12 and pp. 96–7) and lacks functional significance. Some mechanisms of increase (e.g. unequal sister chromatid exchange) could readily produce large blocks of non-coding DNA, while others, e.g. duplicative transposition or random replicon doubling, would tend to produce more even dispersion. But the apparent absence of a specific functional role for such heterochromatin blocks does not necessitate a totally neutralist interpretation of their evolution. They will be subject, like euchromatin, to constraints on the distribution of replicon origins and attachment sites for the lateral loops to the chromosome core, and there will be selection against the evolution of sequences that mimic promoters or other biologically important signals that would cause waste or disrupt development. Their, often rather weak, effects on recombination (John and Miklos, 1979) will also influence the selective forces acting on them, but seem unlikely to be the major factor controlling their occurrence and amount. Organismic selection will limit the accumulation of heterochromatin by preventing the total DNA content from rising above a certain level; on the skeletal DNA hypothesis this will be close to the optimal nuclear volume, while on non-functionalist hypotheses it will be the level that imposes enough of a selective burden to balance the mutational pressures generating it.

DNA attachment to the nuclear envelope in interphase

The interphase nucleus is not a simple bag containing a random solution of DNA, RNA, and protein, but is structurally ordered. For example, in the *Drosophila* blastoderm the chromocentre (constitutive heterochromatin) is oriented towards the outer surface in every cell; in the salivary gland the polytene chromosomes are folded in the same pattern in each nucleus. These and similar regularities must depend on the folding pattern of each chromosome, and on its specific attachment to other structures like the nuclear matrix and the nuclear lamina that underlies the nuclear envelope.

These attachments are most obvious in the peridinean protozoa, since in most of these 'dinoflagellates' the chromosomes remain visible in interphase, as does the nuclear envelope during mitosis; in such nuclei the chromosomes are attached to the nuclear envelope by their ends (telomeres) during both interphase and mitosis. During mitosis they are attached by their centromeres, as are the centromeres of hypermastigote protozoa; evidence from fluorescent antibody studies of mammalian centromeres for attachment to the nuclear envelope throughout interphase, suggests that this may be true for all eukaryotes. Since in early prophase of meiosis in fungi, plants, and animals chromosomes are attached to the nuclear envelope by both telomeres (this attachment ceases later in meiosis), I suggest that telomeres and centromeres are both universally attached to the nuclear envelope during interphase. If this hypothesis is correct then each chromosome arm is anchored to the nuclear envelope at its two ends. If the chromosomes are condensed, as in the Peridinea, each arm is like a strut and its length must be the same as the direct separation of its attachment sites on the nuclear envelope. The longest arm would tend to span the whole diameter of the nucleus whereas shorter arms would span successively shorter chords. This geometrical constraint alone would tend to cause chromosome arms of similar length to lie adjacent to each other in the interphase nucleus even in the absence of any positive interaction between them. This is probably also true for chromosomes that decondense and become invisible in interphase, as in most other eukaryotes. The fact that they reappear in prophase at the same positions they occupied in the previous telophase is simply explained if their centromeres and telomeres attach to the nuclear lamina when it reassembles at telophase and if these sites are unable to move relatively to each other during interphase.

Such geometric constraints on the arrangement of chromosome arms, arising from their attachment to the nuclear envelope at both ends, could be sufficient explanation for the strong statistical regularities in the spatial arrangement of chromosome arms at metaphase in cereals (Bennett, 1982, 1983): non-homologous chromosome arms of similar size lie next to each other much more frequently than expected by chance, the order of chromosomes on the metaphase plate being predictable from the assumption that arms of similar length should be adjacent. This is comprehensible if arms are already arranged in this way during interphase because of centromeric and telomeric attachments to the nuclear lamina and if their relative arrangement does not alter during mitosis; such positional constancy from interphase to mitosis is to be expected in nearly all eukaryotes; only in the Peridinea is there evidence for 'nuclear cyclosis' that moves chromosomes around the nucleus prior to mitosis (Grassé, 1951).

The classic studies of Boveri (cited by Wilson, 1925, p. 890) on *Ascaris*, however, show that although telomeres appear always to be located on the nuclear envelope the rest of the chromosome is not folded in exactly the same way in all cells: variations in the detailed disposition of the chromosomes at telophase caused corresponding variations in the shape of the nucleus at subsequent interphase. This supports the basic thesis of the skeletal DNA hypothesis that it is the size and shape of the chromatin that determines the size and shape of nuclei. Forbes *et al.* (1983) have recently provided further experimental support for the skeletal DNA hypothesis by injecting *E. coli* viral or plasmid DNA into the cytoplasm of mature *Xenopus* eggs. They find that nuclear envelopes with typical pores and a nuclear lamina stainable by anti-lamin antibody assemble themselves around the injected DNA to form spherical 'nuclei'. Their experiments clearly show that there is a vast excess of nuclear envelope and lamina material in the unfertilized egg, sufficient to package 1000 times as much DNA as that found in the egg nucleus into structures with a typical nuclear morphology. The normal function of this store (probably mainly in the form of annulate lamellae) like that of the huge stores of histone (enough to package 20 000 nuclei) and snRNPs, is obviously to enable the rapid assembly of cleavage nuclei after the DNA is replicated during the rapid cleavage cell cycles. These observations show that the size of the egg nucleus is limited not by the availability of nuclear envelope or lamina material but by its DNA content.

Quantitative studies of the volume and DNA content of these artificial nuclei will provide the most powerful test of the skeletal DNA hypothesis. Such studies are necessary since the published observations so far only confirm that DNA is necessary to nucleate the assembly of the nuclear envelope (which was also indicated by the inhibition of assembly by psoralen cross-linking the DNA, Peterson and Berns, 1978), but do not prove that it controls the final volume as the skeletal DNA hypothesis requires.

Forbes *et al.* suggest that the nucleation step of nuclear assembly and the attachment of DNA to the nuclear envelope observed in their cycloheximide treated cells does not require specific DNA sequences longer than 6 bp. However if, as proposed (Cavalier-Smith, 1982b), the lamina proteins evolved from DNA-binding proteins attached to the bacterial plasma membrane, it is possible that their DNA-binding sites are sufficiently strongly conserved between prokaryotes and eukaryotes for specific lamin binding sites to be present in prokaryotic DNA. Similar studies using truly random polynucleotides are needed to rule out this possibility. It should be stressed that the nucleoskeletal DNA hypothesis does not specify whether the nucleation step requires specific sequences; it simply assumes that specific sequences are unimportant for the main bulk of the skeletal DNA. It seems likely that telophase assembly of the nuclear envelope can begin anywhere on the surface of the telophase chromosome, and therefore that interphase attachments of DNA to the nuclear envelope are mostly non-specific; which DNA segments are attached to the nuclear envelope might simply depend on which lie at the surface of the folded chromatin mass. However, the regular binding of telomeres and centromeres implies that their attachment must be more specific. These interphase attachments must generally be broken during nuclear division (Cavalier-Smith, 1982a, b); except in the protozoa with extranuclear spindles mentioned above, the chromosomes detach from the nuclear envelope, whether division is open (nuclear envelope breaks down) or closed (nuclear envelope remains intact). Since oocyte germinal vesicles are greatly enlarged prophase nuclei, in which the chromosomes are no longer attached to the nuclear envelope, a comparison of their lamina with that of interphase somatic cells might reveal the mechanisms of this change; the *Xenopus* oocyte lamina has only one lamin protein, not two or more different ones as in vertebrate somatic cells. Does the oocyte lamin provide the basic framework of the lamina, and the extra somatic lamin(s) provide DNA binding sites or sites of attachment for DNA binding proteins?

Since the chromosomes are not attached to the nuclear envelope of the germinal vesicle, and its volume is so much greater than that of somatic nuclei, its volume cannot be determined by genome size. The suggestion that oocyte nuclear volume is instead determined by skeletal RNA (Cavalier-Smith, 1978) has certain weaknesses (Macgregor, 1980; Cavalier-Smith, 1982a); Macgregor pointed out that the volume of amphibian germinal vesicles appears to be more closely related to this amount of amplified nuclear DNA rather than to the amount of RNA. Since this nucleolar DNA is attached to the nuclear envelope, possibly via a 148 kilodalton matrix protein (Scheer *et al.*, 1982) distinct from lamins, it is possible that nucleolar DNA plays a skeletal role in the oocyte comparable to chromosomal DNA in interphase. However, the nucleolar attachment to the envelope may simply reflect a need to have the site of assembly of ribosomal subunits as close as possible to the sites of transport through the envelope; it is possible that once the nuclear lamina is reorganized so as to lose its affinity for the chromosomes in meiotic prophase the nuclear envelope is no longer under the constraints imposed by the genome size and is free to grow to whatever extent the availability of raw materials allows. The fact that the germinal vesicle can be relatively large even in organisms like the cockroach *Periplaneta* (Macgregor, 1982) that have no nucleolar DNA amplification, and even fairly large in the meroistic *Drosophila* oocyte that makes no RNA, tends to support this latter view, according to which the volume-determining skeletal role of DNA is restricted to interphase cells. That the DNA-independent type of control over nuclear volume postulated here is restricted to meiotic prophase is implied by the sharp reduction in nuclear volume that accompanies the meiotic divisions to produce the small ovum nucleus comparable in volume to that of many somatic cells. Since Forbes *et al.* (1983) showed that ova contain a vast excess of nuclear envelope material, only requiring extra DNA to initiate its assembly, the changeover from the primary oocyte to the somatic type of nuclear lamina must take place before fertilization, i.e. during the later phases of meiosis; I suggest that it occurs after the disassembly of the envelope at the end of prophase I, and prior to meiotic telophase I so as to ensure that the *new* envelope attaches to and surrounds the surface of the telophase chromatin.

According to the preceding argument the large size of the germinal vesicle and the numerous nuclear pores in yolky eggs are adaptations to store nuclear material ready for the rapid assembly of daughter nuclei during cleavage rather than for accelerating RNA transport during oogenesis, which is maximal before the maximum nuclear surface area is attained. The nucleoskeletal RNA hypothesis (Cavalier-Smith, 1982a) should therefore be rejected. This rejection of nucleotypic control of nuclear volume in large growing oocytes does not conflict with the skeletal DNA hypothesis itself, since it appears to be universally true that DNA is essential for the telophase assembly of the nuclear envelope and is always attached to it during interphase: this fact, when contrasted with the evidence from the mature oocyte that it need not necessarily be so, strongly implies that there is a functional role for DNA attachment to the nuclear envelope and that cells cannot undergo a succession of cell cycles in its absence. This is a basic tenet of the skeletal DNA hypothesis.

Evolution of chromosome arm lengths and folding patterns

The argument that chromosome arms are the basic units of packing to form interphase nuclei has several evolutionary implications. Since the relative position of chromosomes persists from telophase to the following prophase, the interphase arrangement will largely reflect that at telophase. During anaphase the centromeres move ahead of the telomeres; therefore interphase chromosomes will tend to be arranged parallel to each other with their centromeres at one pole and telomeres at the other (Fig. 4.14(a), (b)). Selection for efficient packing into a spherical nucleus will favour not a uniform arm length but a spectrum of different arm lengths (Fig. 4.14(c)) such as will fit most economically into the spherical space; such a graded series of arm lengths is the commonest karyotype pattern. The classical hypothesis that chromosome length is adapted to the length of the mitotic spindle, and therefore to cell size (Stebbins, 1938; White, 1973, p. 407), would predict uniform arm lengths to be the general rule, which is not the case.

The *interphase packing hypothesis* also rationalizes the regular way in which chromosome size increases or decreases in related species with the same chromosome number and qualitatively similar karyotype (Narayan, 1983). If extra DNA were added purely randomly to existing DNA the amount of extra DNA in individual chromosomes within a karyotype would

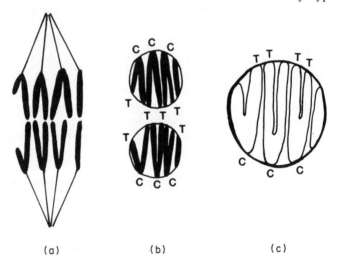

(a)   (b)   (c)

Fig. 4.14.   Interphase disposition of chromosomes as related to their anaphase and telophase orientations and arrangements. (a) At anaphase chromosomes are oriented with centromeres facing the poles and telomeres pointing to the equator. (b) In early telophase the nuclear envelope reforms around the chromosomes and they will retain this orientation. (c) The late telophase unfolding of the chromatin will increase the volume occupied by each chromosome, and therefore of the whole nucleus, but will not change their relative position or their points of attachment to the nuclear envelope: if each chromosome arm is attached to the nuclear envelope by its centromere (C) and telomeres (T), the length of each arm will influence how it is most efficiently packed in the nucleus and arms of similar length will tend to lie side by side.

be expected to be proportional to their size. Fig. 4.15 shows that this is not the case; instead the same total amount of DNA is added to each chromosome independently of its size, i.e. the percentage increase is greater for smaller chromosomes. For the *Lathyrus* data shown in Fig. 4.15 the chromosomes are ranked simply according to size, and there is no direct evidence that the corresponding chromosomes are actually homeologues. However for *Allium* (Jones and Rees, 1968), and *Festuca* and *Lolium* (Seal, 1982), study of meiotic pairing in interspecific hybrids differing in genome size shows which chromosomes are homeologues and that all homeologous pairs differ by a roughly similar amount of DNA (in these cases volume was measured, but it is well known that volume is proportional to DNA content (Fig. 4.16)). This non-random addition (or deletion) of DNA is expected on the interphase

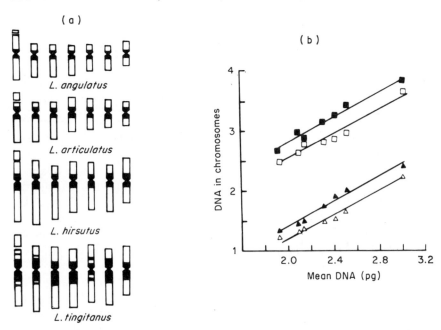

Fig. 4.15. Comparison of the DNA content of individual chromosomes in four species of *Lathyrus* with differing genome size. (a) the four karyotypes, C bands shown in black. (b) The chromosomes of each species were ranked in order of size and DNA content; the DNA content of each is plotted against the mean DNA amounts for the 7 size classes (i.e. the values for the smallest chromosomes of each of the four species is plotted against the interspecies mean of the four smallest chromosomes, and so on for the next smallest, the next-but-one and so on). If the 4 chromosomes in each size class are homeologues, the parallelism of the four regression lines would imply that the same amount of DNA is added to or subtracted from each member of the karyotype when the genome size changes. $\triangle$ *L. angulatus*, $\blacktriangle$ *L. articulatus*, $\square$ *L. hirsutus*, $\blacksquare$ *L. tingitanus*. From Narayan, (1983), reprinted by permission of Allen & Unwin.

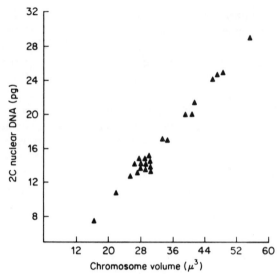

Fig. 4.16. The close correlation between total chromosome volume and nuclear DNA content for 24 species of *Lathyrus*. From Narayan (1983), reprinted by permission of Allen and Unwin.

packing hypothesis since efficient packing requires a similar absolute increase or decrease in the lengths of all the chromosome arms.

The above packing constraint will tend to make the chromosomes relatively more equal in size in high C-value than in low C-value species. Though this is the case in most animals and plants some animals, e.g. birds, reptiles, some Lepidoptera, show a marked disparity in size within the karyotype instead of the usual graded series of arm lengths (White, 1973); whether this reflects a special feature of interphase chromosome packing in the groups in question or has quite other causes is unclear.

It was assumed above that extra DNA increases the length but not the width of chromosomes. This appears to be generally true in flowering plants with medium to high C-value, though these are occasional exceptions; it appears also to be true in certain animal groups that do not have a high range of C-values. This, coupled with the uniformity of widths between the different members of a karyotype indicates a strong conservation in local folding patterns of a chromatid despite several-fold variations in chromosome lengths. It is possible that this is because in these groups the most common mode of increase (or decrease) is the addition (or deletion) of whole units of folding or lateral loop domains. This would be the case if, as discussed in Chapter 7, these loop domains correspond with replicons, and if DNA increases by the duplication of whole replicons rather than by parts of replicons. An indefinite increase in loop size, and especially a marked variation in loop size in different parts of the karyotype, would lead to problems during segregation, so selection would probably prevent such increases; this could be an example of what White (1973) called karyotypic orthoselection. But whether such selection would be sufficient or even strictly necessary to explain the observed regularities is unclear: a greater frequency of replicon duplication than of small scale deletions, coupled with nuclear-volume-dependent selection for overall arm length, could maintain uniform chromosome widths and loop sizes. Data on the relative frequency of occurrence of the different modes of DNA increase, as distinct from the frequency of their fixation in the population, is needed to decide whether the observed patterns depend mainly on such differential mutation pressure rather than selection. In the absence of such data it must not be assumed that the observed regularities in karyotype evolution in particular groups (the 'principle of homologous change' White, 1973) is caused by regularities in selection rather than regularities in the mode of increase itself.

Several authors, most recently Narayan (1983), have suggested that increases in genome size are quantized rather than continuous. This is based on a comparison of genome size in different species of plants of the same genus (Fig. 4.17). Much of the apparent periodicity in genome sizes and

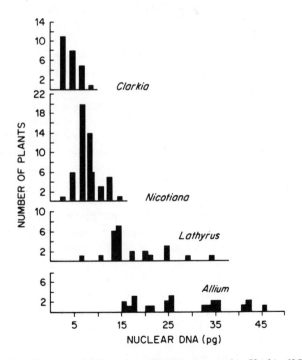

Fig. 4.17. Distribution of 2C nuclear DNA amounts in *Clarkia* (25 species), *Nicotiana* (51 species), *Lathyrus* (25 species) and *Allium* (25 species). From Narayan (1983), reprinted by permission of Allen and Unwin.

absence of intermediate values in earlier papers probably simply reflected a limited data base. This may still be true for the data for *Allium* where only 5% of the species have been sampled, and perhaps also for *Lathyrus* (though the clustering in Narayan's data is stated to be highly statistically significant), but for *Nicotiana* where 51 out of the 62 recognized species and for *Clarkia* where 25 out of 35 species have been sampled the data are more convincing. If we assume that the periodicity of the data is not a statistical artifact, then it is necessary to suppose either that the mode of increase is quantized or that intermediate values are weeded out by selection. Quantized increase could result from a saltatory multiplication of an episome-like element and its reinsertion into the chromosome. But it is hard to see why the overall amount of its multiplication should be the same in related species of different C-value (and therefore different cell volume) or why the amount inserted into each chromosome should be approximately the same. Moreover, the observations of regular pairing at meiosis in F_1 hybrids between plants differing in C-values is against the idea of extensive transposition or translocation or insertion of numerous identical sequences on different chromosomes. If, as this evidence and Keyl's studies on the banding pattern in *Chironomus thummi* suggest, the increase results instead from the local longitudinal duplications of DNA, there is a similar problem in explaining why the amount of such duplication should be more or less the same in all chromosomes even in species of different C-value. The idea of a quantized mode of increase by 2 or 4 pg per genome is also unattractive because there are numerous organisms with a total genome size varying widely in the range 0.05–1 pg, showing that much smaller increments in genome size than the postulated quantum step can occur (the data for *Poa annua* in Fig. 4.12 also clearly showed this); this argument would also tend to militate against the selection hypothesis unless the selective forces differed significantly for low and high C-value organisms; this might be the case, if the selective forces have to do with chromatin packing. I referred above to the evidence that in plants with C-values over about 2 pg increases in C-value largely involve increases in chromosome length, with no increase in width. One interpretation of this and of the quantized nature of their C-value would be that there exist subchromosomal units of packing of constant diameter and that increases in C-value arise mainly by the multiplication of such units. In *Lathyrus* where $2n = 14$ each unit would contain 3.95/14 = 0.28 pg DNA and in *Allium* with $2n = 18$, 16 or 14 it would have a 4.25/16 = 0.27 pg DNA, if we assume that each increment represents the duplication of 1 subchromosomal folding unit per chromosome; but for *Nicotiana* with $2n = 24$ it would be 1.95/24 = 0.081 pg.

However, in low C-value species, e.g. yeast or *Drosophila*, where chromosome widths are less than in these medium/high C-value plants, such packing units must be smaller, which would therefore allow much smaller increments in genome size; a continuous spectrum of sizes might even be possible below a certain genome size if evolutionary increases above a certain size were accompanied by a new level of chromatin folding that imposed new constraints on further increases that were absent below that level. The lack of such quantization pattern in animals, which show a continuous spectrum of genome size (Bachmann *et al.*, 1972), might be a sign that they evolved such a higher level of folding independently of plants, perhaps in a manner lacking such constraints. Such independent evolution would not be unreasonable since the most primitive animals, sponges and cnidaria, have low genome sizes (~0.05 and ~0.5 pg respectively); as this is unlikely to be a reduction from higher ancestral values (see next section) the protozoan common ancestor of plants and animals probably also had a small genome size.

However, even in higher C-value plants quantized C-values are probably not universal; for example in *Gibasis* (Fig. 4.18) they increase more or less continuously. Moreover the idea that duplication of a basic packing unit common to all higher C-value flowering plants is the basis of their C-value variation, is not supported by the difference in the increment size in *Lathyrus/ Allium* and *Clarkia/Nicotiana* or by the evidence from pairing in inter-specific hybrids of *Aloe* (Brandham, 1982), *Festuca* and *Lolium* (Seal, 1983) that the extra DNA is widely scattered along the length of the chromosomes. Perhaps the genera studied by Narayan do have a special quantized mode of DNA increase, even if the mechanism is unclear. It is desirable to ascertain whether their nuclear and cell volumes are equally well quantized, though it might be hard to obtain data with sufficient precision to test this, and also for independent workers to measure C-values in a large number of species in genera that apparently show quantization.

Though the existence of such quantization is not yet certain, other considerations discussed above suggest that interphase chromatin packing may pose important constraints on the relative sizes of chromosome arms and on the quantitative patterns during evolutionary changes in genome size and that these cannot be totally random. Interphase chromosome arrangements may also affect the patterns of transposition and translocation of DNA sequences during genome evolution since such events are obviously more probable between sequences that usually lie next to each other than between widely separated ones: the demonstration by Greilhuber and Loidl (1983) that chromosome G and Q bands lie in corresponding positions on chromosome arms of similar lengths (which should be adjacent during interphase according to Bennett's (1982, 1983) rule) suggests that transposition or mutual interchanges must have occurred preferentially between adjacent non-homologous chromosome arms at some stage in their history. Similar events presumably explain the spreading of rRNA genes to different chromosomes, since in species like man with nucleolar organizers on several chromosomes they tend to be located at similar positions with respect to the

GENOME SIZE
(pg)

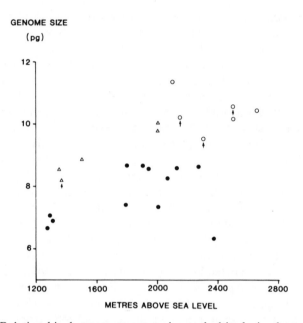

METRES ABOVE SEA LEVEL

Fig. 4.18.   Relationship between genome size and altitude in the tuberous perennial herb *Gibasis*. Both for the *G. linearis* alliance of species and subspecies (● broad-leaved species, ○ dwarf species) and for *G. karwinskyana* (△) there is a tendency for genomes to be larger at high altitudes: it would be valuable to study the phenology and leaf expansion rates of the various isolates since it seems possible that the colder shorter seasons at higher altitudes might favour more rapid expansion allowed by larger cells at the expense of slower growth later (cf. p 132–5). Data from Kenton (1983). Genome size was calculated by dividing nuclear DNA content by the ploidy; this would give a good indication of their relative cell volumes, except in the case of the 4 polyploid species (arrowed) which would probably have a relatively greater cell volume than implied by their genome size.

telomeres on arms of similar size (Stahl, 1982). The position of rRNA genes on the arms may also have functional significance: in a very wide variety of species they are close to a telomere (Lima-de-Faria, 1973, 1976, 1980). If, as postulated, telomeres remain attached to the nuclear envelope during interphase this means that nucleoli will also be located close to the nuclear envelope, which is usually observed (Bouteille *et al.*, 1982), and which would facilitate rapid transport of ribosomal subunits into the cytoplasm; the different nucleolar organizers will also be located near to each other and able to cooperate to make a single nucleolus.

Lima-de-Faria (1980) provided extensive evidence that a variety of other DNA sequences also tend to be located regularly with respect to centromeres and telomeres. He coined the phrase 'chromosome field' (Lima-de-Faria, 1954) to refer to this regularity in location within chromosome arms. Despite the great interest of these regularities his work has been neglected, perhaps because no mechanistic basis was apparent and his discussion was in very vague terms of the need for genes 'to find their optimum territory within the field' or of 'interactions' or 'communication' between genes, without specifying what these might be or why they should occur. Since many, if not all, the types of DNA sequences discussed by Lima-de-Faria are repeated sequences, their ordered position along the length of the chromosome can be explained without invoking such vague concepts, in terms of three principles:

1. Chromosome arms are attached to the nuclear envelope by their centromere and telomere, as postulated above, and tend to pack in interphase in such a way that arms of similar length are adjacent (Bennett, 1982, 1983).
2. Spread of repeated sequences by transposition occurs mainly between those chromatin regions that are adjacent during interphase (perhaps especially during S phase).
3. Selection against gross chromosomal rearrangements, arising from reduced fertility caused by meiotic mispairing, tends to conserve the relative position of such sequences within the arm (Cavalier-Smith, 1981); but in the long run it would allow many exceptions to Lima-de-Faria's rules to evolve, as has clearly happened.

It is not necessary to suppose that most genes have an optimal position within the chromosome arm. However, certain sequences may, like nucleolar organizers, have preferred locations if their position in the interphase nucleus is important. Thus the tendency for constitutive heterochromatin to be concentrated near the centromeres or telomeres may be so that at interphase it is concentrated at the periphery of the nucleus in the region between nuclear pores, as is often observed; this might be partly because in that location it will not obstruct free movement of molecules through and into and out of the nucleus, but a more important reason may be that it can then attach directly to the nuclear lamina and control its surface area by the degree to which it is compacted or extended, as postulated above.

The needs for telomere and centromere attachment to the envelope, and of chromosomal packing in the nucleus, also could be a major factor influencing haploid chromosome numbers, which mostly lie between 7 and 35; it is well known that arm numbers are more conservative than chromosome numbers owing to centric fissions and fusions. How much of the variability in chromosome numbers reflects variations in packing patterns and requirements in different groups and how much is the incidental result of the structural changes that happen to occur is not clear. As Charlesworth argues in Chapter 16 random drift in chromosome number is probably at least as important as Darlington's thesis of selection to optimize recombination rates.

PHYLOGENY AND LONG TERM TRENDS IN GENOME SIZE

Many authors have postulated a long-term orthogenetic tendency for genome size to increase. We have seen reasons why mutational and transpositional processes might produce an upward pressure on genome size. But this would lead to phylogenetic increases in genome size only if selection against them is relatively weak. Nonetheless, such upward pressure could make marked evolutionary reductions in genome size more difficult than corresponding increases, since stronger selection would be necessary to produce a downward shift. Evidence for and against such ideas can come only by correlating genome size with phylogeny.

Since genome size is so well correlated with cell volume, the cell volume of fossils is a guide to past genome size. In lungfish the size of the bony cavities occupied by osteocytes depends on their volume, and has been used, as discussed earlier (see Fig. 4.6), to show that their cell volume, and probably therefore their genome size, increased greatly for millions of years and then levelled off and has remained steady for the past 100 Ma or more, but at a different level in the two main lines of lungfish evolution. Similar studies on other vertebrate fossils and on vascular plants, where the presence of cell walls allows cell size to be determined, would be valuable.

The small genome size and cell size of prokaryotes compared with eukaryotes, together with the marked rise in the size of fossil cells about 1450 Ma ago (Schopf and Oehler, 1976) supports the idea that eukaryotes evolved from prokaryotes about 1500 Ma ago and that the associated changes in replication pattern allowed the subsequent increases in genome size so common in eukaryotes. It has been suggested (Cavalier-Smith, 1983a) that the second major rise in cell volume to modern levels, and a correlated major increase in protist complexity that occurred around 750 Ma ago (Brasier, 1979), were caused by the simultaneous symbiotic origin of chloroplasts and mitochondria and that in the period 1500–750 Ma ago the only eukaryotes were mitochondrionless protozoa of the subkingdom Archezoa (Cavalier-Smith, 1983b).

As present day protozoa and unicellular algae include species not only with the largest known genome sizes but also with the smallest known eukaryote genome sizes, there can be no inexorable evolutionary increase in genome size. The present diversity can be explained simply by supposing that the ancestral protozoa and algae diversified to fill a great variety of niches in which different cell volumes were optimal: those with the smallest cells retained their ancestral small genomes as a result of stabilizing selection while those that evolved larger cells evolved correspondingly larger genomes.

The fungi alone have almost uniformly low genome sizes (Chapter 3), presumably because they evolved from a small-celled low-genome protozoan ancestor such as a choanociliate (= Choanoflagellida), and because their hyphal mode of growth, and/or adaptive zone as saprophytes, places a premium on narrow or small cells and small nuclei. In none of the protozoa, algae and fungi is there good evidence for marked evolutionary reduction in genome size: it would be possible to explain the existing variation in terms of inheritance by some species of an ancestral low genome size plus increases in many different lines. Is this also true of animals and higher plants?

In animals the lowest genome sizes are found in sponges, which are almost certainly the most primitive animals and which probably evolved from choanociliate protozoa, which have small cells (C-value unknown). It is therefore highly probable that animals, like fungi, originated from low C-value protozoa. The fact that sponges which have been around for longer than lungfish have 1–2000 times less DNA per cell should make us sceptical of non-adaptive, neutralist or selfish DNA, interpretations of these and other increases that clearly occurred in many lines of animal evolution. I have recently critically re-examined the phylogeny of the animal kingdom (Cavalier-Smith, unpublished), and can fit the data on genome size onto their phylogeny without postulating any sizeable reductions (i.e. 2–3 times or more) in genome size at any stage of evolution. Formerly (Cavalier-Smith, 1978) the presence in 'primitive' panoistic insect groups, like the Orthoptera, of genomes up to 20–40 times that of more 'advanced' meroistic insects such as Diptera was interpreted as indicative of major reductions in genome size. But it is perfectly possible that the large genomes of the Orthoptera evolved *after* their divergence from the meroistic groups; in the lungfish such an increase clearly occurred *after* their divergence from other bony fish. In my view the ancestral insect probably had rather small cells.

Likewise in vascular plants I accepted the common notion that angiosperms evolved from gymnosperms, and therefore suggested that the much smaller genome size of most angiosperms compared with gymnosperms implied a marked evolutionary reduction in genome size resulting from the origin of vessels. But a critical examination of vascular plant phylogeny makes it improbable that angiosperms evolved from gymnosperms; instead it now appears more likely to me that they evolved from heterosporous water ferns (that already have vessels and a closed carpel-like sporangium) via aquatic monocytoledons of the Alismatidae. Since the heterosporous ferns have uniformly low genome sizes (but homosporous ones uniformly high ones) no marked reduction in genome size would be necessary during such an origin of the angiosperms.

Other authors have proposed reductions in genome size within the angiosperms on the basis of intrageneric comparisons (e.g. Stebbings, 1976; Rees and Jones, 1972). In several cases species regarded as 'advanced' by taxonomists have lower genome sizes than those regarded as 'primitive'. However, the judgement of what is advanced and primitive within an angiosperm genus is highly contentious and open to error; even where they are correct the 'lungfish phenomenon' cannot be ruled out of secondary increase in the more primitive group *after* the morphologically advanced species have diverged from it. Strictly speaking one requires fossil data to resolve the issue; but at the very least detailed critical phylogenetic analysis of the relationships within a group and between it and its nearest ancestor, coupled with C-value determinations is essential to clarify the phylogenetic direction of increases or decreases in genome size, and this has not yet been done for any group owing to the divorce between students of phylogeny and those interested in C-values. However, the classical phylogenetic studies on the grass family (Gramineae) by Avdulov (1931) indicated that it originated in the tropics where grass chromosomes and nuclei are small and that various evolutionary lines radiated into temperate zones accompanied by increases in nuclear volume and chromosome size. As Stebbins (1979) points out, the strong correlation between these parameters and nuclear DNA contents in grasses (Rees and Jones, 1972) means that Avdulov's data imply corresponding increases in DNA content.

This is not to deny the possibility of significant reductions in chromosomal DNA content, but simply to argue that they may be relatively rare and would probably require exceptionally strong selection for small nuclear size. Possible examples of such reduction are the exceptionally small genome sizes of yeasts and the very small chromosomes of the duckweeds *Lemna* (Blackburn, 1932). But in general the angiosperm data appear consistent with the view that evolutionary increases in genome size are much more frequent than decreases and that major decreases seldom occur.

It has been customary to regard increases in DNA content by polyploidy and increases in genome size in the absence of polyploidy as quite separate phenomena. Clearly they are separate as to the mechanism of increase, but in green plants especially they appear to be subject to very similar selective forces. In pteridophytes, homosporous ferns show both high levels of polyploidy and large genome sizes, while heterosporous ferns have low genome size and little or no polyploidy. In green algae and in angiosperms also the same families are prone to undergo evolution of large genomes and evolutionary polyploidy (e.g. Ranunculaceae, Liliaceae). Moreover the ecological correlates of the two phenomena in angiosperms are similar, both tending to be most marked in herbaceous perennials in temperate zones and least marked in woody plants, annuals and the tropics (Stebbins, 1971; Bennett, 1972; Grime, 1983). These similarities are to be expected since strong selection against increases in cell volume will prevent the evolution of both polyploids and larger genomes, whereas weak selection against, or even for, larger cells would favour both.

What are the reasons for the unusually high frequency of large genomes in perennial herbaceous temperate plants with fleshy subterranean storage organs such as many Ranunculaceae and monocotyledons such as lilies and crocuses (Darlington, 1939; Grime and Mowforth, 1982)? One possibility is net selection for large cells because of their greater efficiency as 'containers' of the food store; a related phenomenon occurs in seeds where the food storage organs often become endopolyploid and in succulents where water storage tissue is often endopolyploid; in all cases increasing the cell volume increases the ratio of enclosed space to cell wall materials. A second possibility is the seasonal need for rapid shoot elongation in cool conditions, which may be favoured by large cell volumes as discussed earlier in the chapter (Grime, 1983; Grime and Mowforth, 1982); the much discussed example of *Ranunculus ficaria* (Nicholson, 1983) well fits the pattern, that is, the larger-celled tetraploids do best in shaded sites where their rapid expansion in early spring is a real advantage while the smaller-celled diploids dominate the open habitats where sustained production throughout the season is more important. The evolutionary importance of autopolyploidy has been neglected in recent years but the many examples in plants reviewed by Muntzing (1936) show consistent ecological differences between diploid and tetraploid races that fit the theory that selection for different optimum cell sizes in slightly different niches has played a major role—perhaps the major role—in their differentiation. Even in allopolyploids their greater cell volume may be more important than usually realized; obviously the qualitatively intermediate character of an allopolyploid, fitting it for a niche that contains features of both of its different parents, must often be of special importance, but the significance of this hybridity of allopolyploids may often have been overemphasized in comparison with the less obvious but quantitively important contribution of their greater cell size to their ecological success. Just as some genera such as *Lathyrus* vary their nuclear volume predominantly by changing chromosome and genome size, others e.g. *Rumex* (Ichikawa *et al.*, 1971) do so predominantly by changing their ploidy.

Some authors have suggested that high amounts of DNA hinder evolution (Bier and Müller, 1969) or cause evolutionary senescence (Fredga, 1977), whereas others suggest the opposite, i.e. that low amounts hinder evolution and that high amounts confer evolutionary plasticity, or that large increases are necessary for quantum evolution. These contradictory views cannot both be correct; the comparative evidence reviewed in this chapter suggests that both are false, and that genome size is unrelated to qualitative aspects of evolution. The tendency for genome size to increase in evolution is undeniable, but can be regarded as the inevitable consequence of three things:

1. The coevolution of genome size and cell volume.
2. The small size of the first cells.
3. Subsequent diversification to produce a huge spectrum of cell volumes.

It appears unnecessary to postulate that genome size has any effect on phenotypes other than through its influence on nuclear volume (and possibly also on cell volume, though this is much more dubious). If this is so then the variations in genome size in different species ought eventually to be explicable in terms of variations in the selective forces that affect cell and nuclear volume in different kinds of organisms occupying different ecological niches.

This chapter has been able only to point to some of the considerations that will have to be evaluated more thoroughly in the future by critical and systematic study of the physiological, developmental and ecological significance of differing cell and nuclear volumes, and by cell and molecular biological studies of the causal basis for the genetic control of cell and nuclear volume. The importance of such quantitative interdisciplinary and comparative studies will grow as the exciting exploratory phase of cell and molecular biology draws to a close, for they will provide a vast field for testing our basic ideas and the opportunity to create a more unified biology.

NOTES ADDED IN PROOF:

1. Ammermann (p. 434) points out that macronuclei are not simply polyploid versions of the micronuclei, because some chromatin diminution and/or differential replication precedes genome multiplication. I propose the new term *macroploidy* to describe the multiplied state of DNA in the heterokaryote macronucleus.

2. Intraspecific variation in genome size in *Poa annua* and *Microseris douglasii* (p. 273) may be so readily detectable because they are predominantly self-fertilizing. In most species cross fertilization will homogenize genome size (and any nucleotypic character depending on it) within the species and prevent the evolution of local ecotypes.

3. Heterokaryote protozoa have between 50 and 500 times smaller genomes than other eukaryotes of the same size (chapter 14). A heterokaryote cell of a given size therefore has roughly 1000 × less non-coding DNA per genome than other eukaryotes. This shows that mutation pressures cannot be universally sufficiently strong to drive genomes up to the level of the line in fig. 4.9, p. 136. Thus in heterkaryotes mutation pressure and transcription are several orders of magnitude too weak to account for the vast amount of non-coding DNA in eukaryotes. Nonetheless the heterokaryote macronucleus shows exactly the same quantitative ratio between its DNA content and cell volume as in non-macroploid protists (Shuter *et al.*, 1983); moreover heterokaryotes of greatly differing size have a fairly constant cytonuclear ratio quantitatively similar to other protist cells (Taylor and Shuter., 1981) these observations are predictable from my theories of the functional significance of the cytonuclear ratio and the nucleotypic determination of nuclear volume. Because of macronuclear macroploidy the extra copies of the genic DNA can fulfil the skeletal role, and very much less non-genic skeletal DNA is needed than in other eukaryotes. These observations, and those on correlated variation of macronuclear DNA content and cell volume within a single heterokaryote species, are also consistent with the nucleotypic control of the cell volume by the macronucleus as discussed on p. 115–8.

4. The non-coding eliminatable DNA found in heterokaryotes with very large cells might possibly be such junk or selfish DNA. But even micronuclear volumes are loosely correlated with cell volume (Taylor and Shuter, 1981, who studied a much wider range in cell size than shown in fig. 14.1); if this is functionally significant much of the extra micronuclear DNA may be skeletal (Cavalier-Smith, 1978).

5. Egg cells increase less strongly with genome size than sonetic cells (Horner and Macgregor, 1983), so their size is probably less important for genome size evolution.

REFERENCES

D'Amato, F. (1977). *Nuclear Cytology in Relation to Development*. Cambridge University Press, Cambridge, London and New York.

Ainsworth, C. C., Parker, J. S. and Horton, D. M. (1983). Chromosome variation and evolution in *Scilla autumnalis*. In P. E. Brandham and M. D. Bennett (Eds), *Kew Chromosome Conference II*, Allen and Unwin, London, pp. 261–268.

Avdulov, N. P. (1931). Karyo-systematische Untersuchung der Familie Gramineen. *Bull Appl. Bot. Genet. Plant Breed.*, Suppl., **44**, 1–428 (in Russian, 73pp. German summary).

Bachmann, K., Goin, O. B. and Goin, C. J. (1972). Nuclear DNA amounts in vertebrates. *Brookhaven Symp. Biol.*, **23**, 419–447.

Baetke, K. P., Sparrow, A. H., Naumann, C. H. and Schwemmer, S. S. (1967). The relationship of DNA content to nuclear and chromosome volumes and to radiosensitivity (LD50). *Proc. Natl. Acad. Sci. USA*, **58**, 533–540.

Baldock, B. M., Baker, J. H. and Sleigh, M. A. (1980). Laboratory growth rates of six species of freshwater gymnamoebia. *Oecologia*, **47**, 156–159.

Banse, K. (1982a). Mass-scaled rates of respiration and intrinsic growth in very small invertebrates. *Mar. Ecol. Prog. Ser.*, **9**, 281–297.

Banse, K. (1982b). Cell volumes, maximal growth rates of unicellular algae and ciHates, and the role of ciliates in the marine pelagial. *Limnol. Oceanogr.*, **27**, 1059–1071.

Beach, D., Durkacz, B. and Nurse, P. (1982). Functionally homologous cell cycle control genes in budding and fission yeast. *Nature, Lond.*, **300**, 706–709.

Bennett, M. D. (1971). The duration of meiosis. *Proc. R. Soc. Lond. B*, **178**, 259–275.

Bennett, M. D. (1972). Nuclear DNA content and minimum generation time in herbaceous plants. *Proc. R. Soc. Lond. B*, **181**, 109–135.

Bennett, M. D. (1973). The duration of meiosis. In M. Balls and F. S. Billett (Eds), *The Cell Cycle in Development and Differentiation*, Cambridge University Press, Cambridge, pp. 111–131.

Bennett, M. D. (1982). Nucleotypic basis of the spatial ordering of chromosomes in eukaryotes and implications of the order for genome evolution and phenotypic variation. In G. A. Dover and R. B. Flavell (Eds), *Genome Evolution*, Academic Press, London and New York.

Bennett, M. D. (1983). The spatial distribution of chromosomes. In P. E. Brandham and M. D. Bennett (Eds), *Kew Chromosome Conference II*, Allen and Unwin, London, pp. 71–79.

Bier, K. and Müller, W. (1969). DNA – Messungen bei Inskekten und eine Hypothese uber retardierte Evolution und besonderen DNS–Reichtum im Tierreich. *Biol. Zbl.*, **88**, 425–449.

Blackburn, K. B. (1932). Notes on the chromosomes of the duckweeds (Lemnaceae) introducing the question of chromosome size. *Proc. Univ. Durham Philos. Soc.*, **9**, 84–90.

Bouteille, M., Hernandez-Verdun, D., Dupuy-Coin, A. M. and Bourgeois C. A. (1982). Nucleoli and nucleolar-related structures in normal, infected and drug-treated cells. In E. G. Jordan and C. A. Cullis (Eds), *The Nucleolus*, Cambridge University Press, Cambridge, pp. 179–211.

Brachet, J. (1950). *Chemical Embryology*, Interscience Publishers, New York.

Brandham, P. E. (1983). Evolution in a stable chromosome system. In P. E. Brandham and M. D. Bennett (Eds), *Kew Chromosome Conference II*. Allen and Unwin, London, pp. 251–260.

Brasier, M. D. (1979). The Cambrian Radiation Event. In M. R. House (Ed.), *The Origin of Major Invertebrate Groups*, Academic Press, London, pp. 103–159.

Butterfass, T. (1979). *Patterns of Chloroplast Reproduction: a Developmental Approach to Protoplasmic Plant Anatomy*, Springer, Wien.

Cavalier-Smith, T. (1977). Visualising jumping genes. *Nature, Lond.*, **270**, 10–12.

Cavalier-Smith, T. (1978). Nuclear volume control by nucleoskeletal DNA, selection for cell volume and cell growth rate, and the solution of the DNA C-value paradox. *J. Cell Sci.*, **34**, 247–278.

Cavalier-Smith, T. (1980a). How selfish is DNA? *Nature, Lond.*, **285**, 617–618.

Cavalier-Smith, T. (1980b). Cell compartmentation and the origin of eukaryote membranous organelles. In W. Schwemmler and H. E. A. Schenk (Eds), *Endocytobiology: Endosymbiosis and Cell Biology*, de Gruyter, Berlin, pp. 831–916.

Cavalier-Smith, T. (1980c). r- and K-tactics in the evolution of protist developmental systems: cell and genome size, phenotype diversifying selection, and cell cycle patterns. *BioSystems*, **12**, 43–59.

Cavalier-Smith, T. (1981). The origin and early evolution of the eukaryotic cell. In M. J. Carlile, J. F. Collins and B. E. B. Moseley (Eds), *Molecular and Cellular Aspects of Microbial Evolution, Society for General Microbiology, Symposium 32*, Cambride University Press, Cambridge, pp. 33–84.

Cavalier-Smith, T. (1982a). Skeletal DNA and the evolution of genome size. *Ann. Rev. Biophys. Biogen.*, **11**, 273–302.

Cavalier-Smith, T. (1982b) Evolution of the nuclear matrix and envelope. In G. G. Maul (Ed.) *The Nuclear Envelope and the Nuclear Matrix*. Liss, New York. pp. 307–318.

Cavalier-Smith, T. (1983a). Endosymbiotic origin of the mitochondrial envelope. In H. E. A. Schenk and W. Schwemmler (Eds), *Endoytobiology II. Intracellular Space as Oligogenetic Ecosystem*, de Guyter, Berlin, pp. 265–279.

Cavalier-Smith, T. (1983b). A 6-kingdom classification and a unified phylogeny. In *Endocytobiology II. Intracellular Space as Oligogenetic Ecosystem*, de Guyter, Berlin, pp. 1027–1034.

Cavalier-Smith, T. (1983c). Genetic symbionts and the origin of split genes and linear chromosomes. In H. E. A. Schenk and W. Schwemmler (Eds), *Endocytobiology II. Intracellular Space as Oligogenetic Ecosystem*, de Guyter, Berlin, pp. 29–45.

Cavalier-Smith, T. (1985). Genetic and epigenetic control of the plant cell cycle. In J. A. Bryant and D. Francis (Eds), *The Plant Cell Cycle*, Cambridge University Press, Cambridge, (pp. 179–197).

Commoner, B. (1964). Roles of deoxyribonucleic acid in inheritance. *Nature, Lond.*, **202**, 960–968.

Craigie, R. A. and Cavalier-Smith, T. (1982). Cell volume and the control of the *Chlamydomonas* cell cycle. *J. Cell Sci.*, **54**, 173–191.

Dabrowski, Z. (1968). Quantitative characteristics of some elements of peripheral blood in the crow family. *Acta Biologica Carcoviensia. Series Zoologia*, **11**, 267–283.

Darlington, C. D. (1937). *Recent Advances in Cytology*, 2nd edn, Churchill, London.

Darlington, C. D. (1939). *Evolution of Genetic Systems*, Cambridge University Press, Cambridge.

Dawkins, R. (1982). *The Extended Phenotype: the Gene as the Unit of Selection*, Freeman, Oxford and San Francisco.

Doolittle, W. F. and Sapienza, C. (1980). Selfish genes, the phenotype paradigm and genome evolution. *Nature, Lond.*, **284**, 617–618.

Evans, G. M. (1968). Nuclear changes in flax. *Heredity*, **23**, 25–38.

Fenchel, T. (1974). Intrinsic rate of natural increase: the relationship with body size. *Oecologia*, **14**, 317–376.

Forbes, D. J., Kirschner, M. W. and Newport, J. W. (1983). Spontaneous formation of nuclear-like structures around bacteriophage DNA microinjected into *Xenopus* eggs. *Cell*, **34**, 13–23.

Fredga, K. (1977). Chromosomal changes in vertebrate evolution. *Proc. R. Soc. B*, **199**, 377–397.

Goldschmidt, R. B. (1955). *Theoretical Genetics*, University of California Press, Berkeley and Los Angeles.

Grassé, P. P. (1951). Les Dinoflagellés. In P. P. Grassel (Ed), *Traité de Zoologie*, Masson et Cie, Paris.

Greilhuber, J. and Loidl, J. (1983). On regularities of C-banding patterns, and their possible cause. In P. E. Brandham and M. D. Bennett (Eds), *Kew Chromosome Conference II*, Allen and Unwin, London, p. 344.

Grime, J. P. (1979). *Plant Strategies and Vegetation Processes*, Wiley, Chichester.

Grime, J. P. (1983). Prediction of weed and crop response to climate based upon measurements of nuclear DNA content. In *Aspects of Applied Biology 4. Influence of environmental factors on herbicide performance and crop and weed biology*, pp. 87–98.

Grime, J. P. and Mowforth, M. A. (1982). Variation in genome size—and ecological interpretation. *Nature, Lond.*, **299**, 151–153.

Holm-Hansen, O. (1969). Algae: amounts of DNA and organic carbon in single cells. *Science, N. Y.*, **163**, 87–88.

Horner, H. A. and Macgregor, H. C. (1983). C-value and cell volume: their significance in the evolution and development of amphibians. *J. Cell Sci.*, **63**, 135–146.

Hutchinson, G. E. (1978). *An Introduction to Population Ecology*, Yale University Press, New Haven and London.

Hyman, L. H. (1940). *The Invertebrates: Protozoa through Ctenophora*, McGraw Hill, New York and London.

Ichikawa, S., Sparrow, A. H. Frankton, C., Nauman, A. F., Smith, E. B. and Pond, V. (1971). Chromosome number, volume and nuclear volume relationships in a polyploid series (2×–20×) of the genus *Rumex. Can. J. Genet. Cytol.*, **13**, 842–863.

Ivanov, V. B. (1978). DNA content in the nucleus and rate of development of plants. *Ontogenez*, **9**, 39–53 (translated, Plenum 1978, pp. 38–40).

John, B. and Miklos, G. L. G. (1979). Functional aspects of satellites DNA and heterochromatin. *Int. Rev. Cytol.*, **58**, 1–114.

Kaplan, R. H. and Salthe, S. N. (1979). The allometry of reproduction: an empirical view in salamanders. *Am. Nat.*, **113**, 671–689.

Kenton, A. (1983). Qualitative and quantitative chromosome changes in the evolution of *Gibasis*. In P. E. Brandham and M. D. Bennett (Eds), *Kew Chromosome Conference II*. Allen and Unwin, London, pp. 273–282.

Keyl, H. G. (1965). A demonstrable local and geometric increase in the chromosomal DNA of *Chironomus. Experientia*, **21**, 191–193.

Keyl, H. G. and Pelling, C. (1963). Differentielle DNA-Replication in den Speicheldrusen-Chromosomen von *Chironomus thummi. Chromosoma*, **14**, 347–359.

Keynes, R. D. (1975). In L. Bolis, S. H. P. Madrell and K. Schmidt-Nielsen (Eds), *Comparative Physiology—Functional Aspects of Structural Materials*, North Holland, Amsterdam, pp.00–00.

Lloyd, D., Poole, R. K. and Edwards, S. W. (1982). *The Cell Division Cycle: Temporal Organization and Control of Cellular Growth and Reproduction*, Academic Press, London.

Lima-de-Faria, A. (1954). Chromosome gradient and chromosome field in *Agapanthus*. *Chromosome*, **6**, 330–370.

Lima-de-Faria, A. (1973). Equations defining the position of ribosomal cistrons in the eukaryotic chromosome. *Nature New Biol.*, **241**, 136–139.

Lima-de-Faria, A. (1976). The chromosome field I-V. *Hereditas*, **83**, 1–190; **84**, 19–34.

Lima-de-Faria, A. (1980). Classification of genes, rearrangements and chromosomes according to the chromosome field. *Hereditas*, **93**, 1–46.

Maclean, N. (1973). Suggested mechanism for increase in size of genome. *Nature, New Biol.*, **246**, 205–206.

Macgregor, H. C. (1980). Recent developments in the study of lampbrush chromosomes. *Heredity*, **44**, 3–35.

Macgregor, H. C. (1982). Ways of amplifying ribosomal genes. In E. G. Jordan and C. A. Cullis (Eds), *The Nucleolus*, Cambridge University Press, Cambridge, pp. 129–151.

Malone, T. C. (1980). Algal size. In I. Morris (Ed.), *The Physiological Ecology of Phytoplankton*, Blackwell, Oxford, pp. 433–463.

Marsden, M. P. F. and Laemmli, U. K. (1979). Metaphase chromosome structure: evidence for a radial loop model. *Cell*, **17**, 849–858.

Maynard Smith, J. (1969). Microbial Growth. *Symp. Soc. Gen. Microbiol.*, **19**, 1–13.

McLeish, J. and Sunderland, N. (1961). Measurements of deoxyribonucleic acid (DNA) in higher plants by Feulgen photometry and chemical methods. *Exp. Cell Res.*, **24**, 527–540.

Miklos, G. L. G. (1982). Sequencing and manipulating highly repeated DNA. In G. A. Dover and R. B. Flavell (Eds), *Genome Evolution*, Academic Press, London and New York, pp. 00–00.

Mirsky, A. E. and Ris, H. (1951). The DNA content of animal cells and its evolutionary significance. *J. gen. Physiol.*, **34**, 451–462.

Muntzing, A. (1936). The evolutionary significance of autopolyploidy. *Hereditas*, **21**, 263–378.

Nagl, W. (1978). *Endopolyploidy and Polyteny in Differentiation and Evolution*, North Holland, Amsterdam.

Narayan, R. K. J. (1983). Chromosome changes in evolution of *Lathyrus* species. In P. E. Brandham and M. D. Bennett (Eds), *Kew Chromosome Conference II*, Allen and Unwin, London, pp. 243–250.

Navashin, M. (1931). Chromatin mass and cell volume in related species. *Univ. Calif. Pub. Agr. Sci.*, **6**, 207–230.

Newport, J. and Kirschner, M. (1982). A major developmental transition in early Xenopus embryos: II. Control of the onset of transcription. *Cell*, **30**, 687–696.

Nicholson, G. G. (1983). Studies on the distribution and the relationship between the chromosome races of *Ranunculus ficaria* in S. E. Yorkshire. *Watsonia*, **14**, 321–328.

Oeldorf, E., Nishioka, M. and Bachmann, K. (1978). Nuclear DNA amounts and developmental rate in holarctic anura. *Z. Zool. Syst. Evolut. forsch.*, **16**, 216–224.

Ohno, S. (1972). So much 'junk' DNA in our genome. In H. H. Smith (Ed.), *Evolution of Genetic Systems*, *Brookhaven Symp. Biol.*, **23**, Gordon and Breach, New York, pp. 366–370.

Olmo, E. (1983). Nucleotype and cell size in vertebrates: a review. *Bas. Appl. Histochem.*, **27**, 227–256.

Olmo, E. and Odierna, G. (1982). Relationship between DNA content and cell morphometric parameters in reptiles. *Bas. Appl. Histochem.*, **26**, 27–34.

Orgel, L. E. and Crick, F. H. C. (1980). Selfish DNA: the ultimate parasite. *Nature, Lond.*, **284**, 604–607.

Orgel, L. E., Crick, F. H. C. and Sapienza, C. (1980). Selfish DNA. *Nature, Lond.*, **288**, 645–646.

Peters, R. H. (1983). *The Ecological Implications of Body Size*. Cambridge University Press, Cambridge.

Peterson, S. P. and Berns, M. W. (1978). Chromatin influence on the function and formation of the nuclear envelope shown by laser-induced psoralen photoreaction. *J. Cell Sci.*, **32**, 197–213.

Pianka, E. R. (1970). On r- and K-selection. *Am. Nat.*, **104**, 592–597.

Price, H. J., Bachmann, J., Chambers, K. L. and Riggs, J. (1980). Detection of intraspecific variation in nuclear DNA content in *Microseris douglasii*. *Bot. gaz.*, **141**, 195–198.

Price, H. J., Chambers, K. L. and Bachmann, K. (1981). Geographic and ecological distribution of genomic DNA content variation in *Micoseris douglasii* (Asteraceae). *Bot. gaz.*, **142**, 415–426.

Price, H. J., Sparrow, A. H. and Nauman, A. F. (1973). Correlations between nuclear volume, cell volume and DNA content in meristematic cells of herbaceous angiosperms. *Experientia*, **29**, 1028–1029.

Rees, H. and Jones, R. N. (1972). The origin of the wide species variation in nuclear DNA content. *Int. Rev. Cytol.*, **32**, 53–92.

Rees, H., Shaw, D. D. and Wilkinson, P. (1978). Nuclear DNA variation among Acridid grasshoppers. *Proc. Roy. Soc. and Lond. B.*, **202**, 517–525.

Reynolds, C. S. (1982). Phytoplankton periodicity: its motivation, mechanisms and manipulation. *Rep. Freshwat. Biol. Ass.*, **50**, 60–75.

Reynolds, C. S. (1984). Phytoplankton periodicity: the interactions of form, function and environmental variability. *Freshwat. Biol.*, **14**, 111–142.

Rogers, J. (1983). Retroposons defined. *Nature, Lond.*, **101**, 460.

Sachsenmaier, W. (1981). The mitotic cycle in *Physarum*. In P. C. L. John (Ed.) *The Cell Cycle*, Cambridge University Press, Cambridge, pp. 139–160.

Scheer, U., Kleinschmidt, J. A. and Franke, W. W. (1982). Transcriptional and skeletal elements in nucleoli of amphibian oocytes. In E. G. Jordan and C. A. Cullis (Eds), *The Nucleolus*, Cambridge University Press, Cambridge, pp. 25–42.

Schopf, J. W. and Oehler, D. Z. (1976). How old are the eukaryotes? *Science*, **193**, 47–49.

Seal, A. G. (1983). The distribution and consequences of changes in nuclear DNA amount. In P. E. Brandham and M. D. Bennett (Eds), *Kew Chromosome Conference II*. Allen and Unwin, London, pp. 225–232.

Shuter, B. (1978). Size dependence of phosphorus and nitrogen subsistence quotas in unicellular microorganisms. *Limnol. Oceanogr.*, **23**, 1248–1255.

Shuter, B. J., Thomas, J. E., Taylor, W. D. and Zimmerman, A. M. (1983). Phenotypic correlates of genomic DNA content in unicellular eukaryotes and other cells. *Amer. Nat.*, **122**, 26–44.

Sinnott, E. W. (1960). *Plant Morphogenesis*, McGraw Hill, New York, 550 pp.

Smayda, T. J. (1970). The suspension and sinking of phytoplankton in the sea. *Oceanogr. Mar. Biol. Ann. Rev.*, **8**, 353–414.

Smith, H. M. (1925). Cell size and metabolic activity in amphibia. *Biol. Bull.*, **48**, 347–378.

Sompayrac, L. and Maaløe, O. (1973). Autorepressor model for control of DNA replication. *Nature, New Biol.*, **241**, 133–135.

Southwood, T. R. (1976). Bionomic strategies and population parameters. In R. M.

May (Ed.), *Theoretical Ecology: Principles and Applications*, Blackwell, Oxford, pp. 26–48.

Stahl, A. (1982). The nucleolus and nucleolar chromosomes. In E. G. Jordan and C. A. Cullis (Eds), *The Nucleolus*, Cambridge University Press, Cambridge, pp. 1–24.

Stebbins, G. L. (1938). Cytological characteristics associated with the different growth habits in the dicotyledons. *Amer. J. Bot.*, **25**, 189–198.

Stebbins, G. L. (1966). Chromosomal variation and evolution. *Science, N. Y.*, **152**, 1463–1469.

Stebbins, G. L. (1971). *Chromosomal Evolution in Higher Plants*, Edward Arnold, London.

Stebbins, G. L. (1976). Chromosome, DNA and plant evolution. *Evolutionary Biology*, **9**, 1–34.

Stebbins, G. L. (1979). Fifty years of plant evolution. In O. T. Solbrig, S. Jain, G. B. Johnson and P. H. Raven (Eds), *Topics in Plant Population Biology*, Columbia University Press, New York.

Strasburger, E. (1893). Über die Wirkungssphäre der Kerne und die Zellgrösse. *Histol. Beitr.*, **5**, 97–124.

Szarski, H. (1968). Evolution of cell size in lower vertebrates. In T. Orvig (Ed.), *Current Problems of Lower Vertebrate Phylogeny*, *Nobel Symposium* **4**, Almqvist and Wiksell, Stockholm, pp. 445–453.

Szarski, H. (1970). Changes in the amount of DNA in cell nuclei during vertebrate evolution. *Nature, Lond.*, **226**, 651–652.

Szarski, H. (1976). Cell size and nuclear DNA content in vertebrates. *Int. Rev. Cytol.*, **44**, 93–111.

Szarski, H. (1983). Cell size and the concept of wasteful and frugal evolutionary strategies. *J. theor. Biol.*, **105**, 201–209.

Taylor, W. D. and Shuter, B. J. (1981). Body size, genome size, and intrinsic rate of increase in ciliated protozoa. *Amer. Nat.*, **118**, 160–172.

Teoh, S. B. and Rees, H. (1976). Nuclear DNA amounts in populations of *Picea* and *Pinus* species. *Heredity*, **36**, 123–137.

Thomson, K. S. (1972). An attempt to reconstruct evolutionary changes in the cellular DNA content of lungfish. *J. exp. Zool.*, **180**, 363–372.

Thomson, K. S. and Muraszko, K. (1978). Estimation of cell size and DNA content in fossil fishes and amphibians. *J. exp. Zool.*, **205**, 315–320.

Trombetta, V. V. (1942). The cytonuclear ratio. *Bot. Rev.*, **8**, 317–36.

Van't Hof, J. and Sparrow, A. H. (1963). A relationship between DNA content, nuclear volume, and minimum mitotic cycle time. *Proc. Natl. Acad. Sci. USA*, **49**, 897–902.

Vialli, M. (1957). Volume et contenu en ADN par noyau. *Exp. Cell Res.*, Suppl., **4**, 284–293.

Vivier, E. (1982). Réflexions et suggestions à propos de la systematique des Sporozoaires: création d'une classe des Hematozoa. *Protistologica*, **18**, 449–457.

White, M. J. D. (1973). *Animal Cytology and Evolution*, 3rd edn. Cambridge University Press, Cambridge.

Wilson, E. B. (1925). *The Cell in Development and Heredity*, 3rd edn, Macmillan, New York.

Young, J. Z. (1962). *The Life of Vertebrates*, 2nd edn, Clarendon Press, Oxford.

Zuckerkandl, E. (1976). Gene control in eukaryotes and the C-value paradox: 'excess' DNA as an impediment to transcription of coding sequences. *J. Mol. Evol.*, **9**, 73–104.

The Evolution of Genome Size
Edited by T. Cavalier-Smith
© 1985 John Wiley & Sons Ltd

CHAPTER 5

The Genetic Control of Cell Volume

PAUL NURSE

*ICRF, P.O. Box 123, Lincolns Inn Fields,
London, WC2A 3PX*

INTRODUCTION

The haploid DNA content or DNA C-value of living organisms varies over 1 000 000-fold (see review by Cavalier-Smith, 1978). Even in groups of organisms of similar complexity the range of DNA C-values can be very large; for example, in the vascular plants it is around 600-fold and in the Chordates 500-fold. A possible explanation for this wide range of DNA C-values is that the optimal cell volume for different organisms varies, and that cell volume is related to the DNA content of the cell. This hypothesis makes two assumptions. Firstly, that cell volume is adaptive and thus is subject to selection; this topic was discussed by Cavalier-Smith in Chapter 4. Secondly, that cell volume is genetically controlled by the DNA in the cell (Commoner, 1964). This possibility is considered here.

A variety of experimental approaches has been used to investigate the genetic control of cell volume. One is the study of cells with different DNA ploidies, derived either from organisms which are naturally polyploid or which have been artificially constructed. This approach is useful for establishing the effects of changing gene copy number, and total DNA content. Another approach is to isolate mutants altered in cell volume, which can be used to identify specific genes involved in the determination of cell volume. A third approach is to look for correlations between DNA content and cell volume in a wide variety of organisms. In these experiments cell volume has not always been measured directly. Other measurements such as wet weight, dry mass, protein content and carbon content have been used. It has been assumed that these various measurements are related to cell volume at least in cells that are not extensively vacuolated. However, for the rest of this chapter cell size rather than volume will often be referred to since this vaguer term seems more appropriate to the variety of measurements used.

INFLUENCE OF DNA PLOIDY ON CELL SIZE

The early work on the influence of DNA ploidy was carried out at the beginning of the century and is well summarized in Wilson (1924). One of the first experiments was carried out by Gerassimoff with *Spirogyra bellis*. He compared cell volumes from haploid *Spirogyra* filaments with those from artificially produced diploids and found the latter to be considerably larger. A more extensive range of ploidies was studied in the moss *Amblystegium repens* by the Marchals. They compared the normal haploid gametophyte with diploid and tetraploid gametophytes derived by regeneration of diploid and tetraploid sporophytes, respectively. The cell volumes of a variety of tissues, such as leaves, antheridia, spores and eggs were all found to be directly proportional to the DNA ploidy level. Similar results were found by Gates and Winkler when cells from diploid and tetraploid plants of *Oenothera* and *Solanum* were compared. The earliest experiments carried out on animal cells were by Boveri on the sea urchin *Paracentrotus*. Haploid eggs were produced by artificial parthogenesis, triploid eggs by dispermic fertilization, and tetraploid eggs by shaking a normal diploid egg near the time of first cleavage which resulted in chromosome doubling without division of the nucleus. Cell sizes in the resultant gastrulae were directly related to the cell DNA ploidy level.

These early results have generally been confirmed by later work. For example, Dobzhansky (1929) has shown that the cross-sectional area of wing epidermal cells in *Drosophila* is related to DNA ploidy. In *Aspergillus nidulans* conidial cell volume, and the hyphal cell volume associated with a single nucleus, are directly proportional to ploidy (Pontecorvo, 1953; Clutterbuck, 1969). Similarly, cell volumes and protein contents per cell are increased in diploid strains of the budding yeast *Saccharomyces cerevisiae* and the fission yeast *Schizosaccharomyces pombe*, compared with haploid strains (Adams, 1977; Thuriaux *et al.*, 1978; Nurse and Thuriaux, 1980). Usually cell volumes increase directly in proportion with DNA ploidy as can be seen for a variety of organisms summarized by Ycas *et al.* (1965). However, in some studies, such as those performed with the yeasts mentioned above, increase in cell size is less than increase in DNA ploidy, especially at slower growth rates.

Although ploidy level in the prokaryote *Escherichia coli* cannot be varied in the same way as in eukaryotes, an analogous experiment can be carried out by making use of the variable DNA content/cell at different growth rates (Cooper and Helmstetter, 1968). Because DNA replication occurs in overlapping cycles, faster growing cells have more chromosomal copies.

These faster growing cells are also bigger and it can be calculated that cell mass increases proportionally with the numbers of chromosomal origins (Donachie, 1968). Therefore *E. coli* cells with more chromosomes, equivalent to higher ploidy in eukaryotic cells, are of an increased cell size.

These experiments establish that cell size increases with increase in DNA ploidy level. This can occur in two ways. The first is that there are specific genes which determine cell size, and that their effect is controlled by their dosage in such a way that the two copies present in a diploid result in an approximate doubling in cell size. The second is that cell size is determined by the total DNA content of the cell which is obviously double in a diploid compared with a haploid. These possibilities are considered in the next two sections dealing with the influence of specific genes and of total DNA content on cell size.

INFLUENCE OF SPECIFIC GENES ON CELL SIZE

One problem with investigating the influence of specific genes on cell size is that it is likely to be affected by a variety of factors, such as variations between tissues in multicellular organisms, differences in genetic backgrounds and variations in growth and environmental conditions. This problem can be overcome by using unicellular micro-organisms which can be grown in defined and reproducible culture conditions, and by comparing mutants which are isogenic apart from the gene under study. The two organisms most extensively studied in this way are the yeasts *Schizosaccharomyces pombe* and *Saccharomyces cerevisiae*.

Fission yeast *Schizosaccharomyces pombe*

Mutants have been isolated in this organism which are reduced in cell size but are not influenced in relative growth rate compared with wild type (Nurse, 1975). It was reasoned that small sized mutants would be more useful than large sized ones for investigating the genetic control of cell size because any mutant which was partially defective in the process of cell division could result in large sized cells (for a fuller discussion of this point see Nurse and Thuriaux, 1980 and Nurse, 1981). The first mutant isolated had about half the cell volume, protein, and RNA contents of a wild type cell (Nurse, 1975). It was originally called *cdc* 9–50 but was renamed *wee* 1–50. A further 51 *wee* mutants were isolated by selection of cells which divided at a small size (Thuriaux *et al.*, 1978; Nurse and Thuriaux, 1980), and another 110 *wee* mutants were isolated by Fantes (1981) who made use of the property that *wee* mutants suppress conditional lethal mutations of *cdc* 25 (Fantes, 1979). The mutants define two genes *wee* 1 and *cdc* 2, and genetic analysis suggests that *wee* 1 codes for a negative element or inhibitor concerned with the control of cell size and that *cdc* 2 codes for a positive element or activator (Nurse and Thuriaux, 1980).

Physiological studies indicate that *wee* mutants influence average cell size by altering the size at which a cell undergoes mitosis (Nurse, 1975; Fantes and Nurse, 1978). Growth to a critical cell size is required before a cell can initiate mitosis (Fantes, 1977; Fantes and Nurse, 1977), and this size is altered in the *wee* 1 and *cdc* 2 mutants. Thus *wee* 1 and *cdc* 2 are involved in a control acting in G2 which coordinates mitosis with attainment of a particular cell size. Further studies with the *wee* mutants have revealed an additional second cell size control called start. This functions in G1 and results in commitment to the cell cycle leading to the initiation of DNA replication (Nurse, 1975; Nurse and Thuriaux, 1977; Nurse and Bissett, 1981). This second G1 size control also involves *cdc* 2. Therefore cell size in *S. pombe* is determined, at least in part, by two genes which coordinate progress through the cell cycle with increase in cell size.

Having identified two genes which specifically influence cell volume the next question is whether their gene dosage properties can explain the doubling in cell size in a diploid. This can be answered by comparing a homozygous wild type diploid with a heterozygous diploid (either *wee* 1$^+$/*wee* 1$^-$ or *cdc* 2$^+$/*cdc* 2$^-$), when gene dosage of the relevant gene is reduced from two to one (Nurse and Thuriaux, 1980). Such a reduction does not halve cell size, but only reduces it by 18% for *wee* 1 and increases it by 8% for *cdc* 2. In addition, the cell size of a *wee* 1 mutant haploid is about half a *wee* 1 mutant diploid, and thus a defective *wee* 1 does not disturb the relationship between cell volume and DNA ploidy. Therefore the gene dosage properties of these genes cannot account for the increase in cell size of a diploid over a haploid.

Budding yeast *Saccharomyces cerevisiae*

Analogous mutants to *wee* have been isolated in the budding yeast *S. cerevisiae* where they are called *whi* (Sudbery *et al.*, 1980). *Whi* 1 mutants have about half the cell volume of wild type because they initiate bud emergence and undergo cell division at a reduced cell volume. There is a cell cycle control called start, which governs commitment to the mitotic cycle and determines cell volume at bud emergence (Johnston *et al.*, 1977), and it is likely that *whi* 1 is altered in this control. The gene dosage effects of *whi* 1 are very similar to *wee* 1 in *S. pombe*; reducing *whi* 1 gene dosage from two to one only reduces cell volume by 25%. Also the increase in cell volume of a homozygous *whi* 1 mutant diploid over a haploid *whi* 1 mutant is double. Therefore, although *whi* 1 controls cell volume by co-ordinating commitment to the cell cycle with increase in mass, its properties cannot explain the relationship between cell volume and DNA ploidy.

A second gene *whi* 2 has little effect on cell volume in exponential growth but reduces cell volume to about half wild type in stationary phase. When *whi* 2 mutants are starved, they do not arrest properly in G1 before commitment to the cell cycle, but continue to divide for a while and then arrest at all stages of the cell cycle. The effects of *whi* 1 and 2 in stationary phase cells are additive since the double mutant *whi* 1⁻ *whi* 2⁻ is about a quarter the volume of wild type cells. The gene dosage relations of *whi* 2 have not yet been studied.

Escherichia coli

A *wee* mutant has recently been described in *E. coli* (Martinez-Salas and Vicente, 1980). This is an amber nonsense mutant which is rather different from the yeast *wee* mutants since it is defective in the process of cell elongation. As a consequence cell density increases, cells divide at increasingly smaller sizes and eventually the cells become inviable. Since the mutant is lethal and is defective in growth I think this gene is unlikely to be specifically involved in the control of cell volume.

Multicellular organisms

Multicellular organisms cannot be investigated so easily as the unicellular micro-organisms considered above but there are a variety of studies which suggest that there are specific genes which influence cell volume. Early work by Keeble on *Primula sinensis*, Tischler on *Phragmites communis*, and Tolber on mosses, which is summarized in Sinnott (1960), showed that various closely related races or mutants have cells of different sizes. In the moss study, a factor influencing the differential increase in cell size of the diploid sporophyte over the haploid gametophyte segregated as a single gene. Two studies have shown that the addition of extra specific chromosomes can alter cell size. In *Crepis tectorum* (Schkwarnikow, 1934), addition of an extra 'B' chromosome (*not* a supernumerary B-chromosome) to the normal diploid of 4 pairs of chromosomes increased primary dermatogen cell area by 14%, whilst an extra 'D' chromosome decreased it by 17%. Similarly, the addition of certain extra chromosomes but not others in wheat increased the dry mass of mid-pachytene sporocytes by about 25% (Longwell and Svihla, 1960). Therefore different chromosomes have differing effects on cell size, suggesting that the presence of specific genes present on the chromosomes is influencing cell volume.

A more subtle argument concerning specific genes has been made by Wettstein (1937) working with the moss *Bryum caespiticum*. A haploid with an average cell volume of some 16 900 μm^3 was induced to form a diploid with an average cell volume of 37 800 μm^3. The diploid was cultured for 11 years during which time cell volume gradually reduced to 18 600 μm^3, even though the plant remained diploid. This result can be explained if it is assumed that there are a number of specific genes which determine cell volume, and that it is advantageous for the moss cells to have a particular volume. During the 11 year culture there was gradual selection for the reduced optimal cell volume of the haploid. This interpretation is supported by the fact that a natural diploid has a stable cell volume of 21 900 μm^3, a value similar to the haploid. A similar argument has been used for plants by Trombetta (1942), who has noted that polyploids of ancient origin tend to have smaller cells than those that have arisen more recently, and for fish by Pedersen (1971).

The most complete investigation of specific genes influencing cell volume in animal cells has been carried out in *Drosophila melanogaster* (Dobzhansky, 1929; Brehme, 1941). Large numbers of minute mutants have been isolated which map on all the chromosomes. They are lethal when homozygous, and have a prolonged developmental period when heterozygous. In the latter, the cells which make up the wing epidermis are 10–20% smaller in area. This was established by counting the numbers of surface hairs in a given area of wing since each cell gives rise to a single hair. In addition the cells which make up the eye facets are reduced in size. Thus minute mutants appear to have an effect on cell size although this may be an indirect consequence of the fact that growth rate in the flies is reduced.

Another study of insect wing epidermal cell size has used the parasitic wasp *Habrobracon* (Whiting, 1961). In this wasp, sex is determined by a complex of linked genes located in a sex-determining region. Crossing-over within this region is supressed and so the whole complex is inherited like a single gene. If a particular organism is heterozygous for this region then it is female, but if it is homozygous or hemizygous, then it is male. In nature diploids are females and haploids are males, but it is possible to construct diploids which are homozygous for the region and consequently male. The volumes of the wing epidermal cells (calculated as 3/2 power of cross-sectional area) of the diploid female were only × 1.02 that of the haploid male while those of the diploid male were × 1.91 of the haploid male. It can be concluded that there is a specific gene or set of genes in the sex-determining region which influence cell volume. When the region is heterozygous, cell volume is half that found when the region is homozygous.

Conclusions

These experiments have established that there are genes which specifically influence cell volume. Although the evidence is clearest for unicellular microorganisms there are good indications that such genes also occur in multicellular organisms. The effect of these specific genes is quite limited since none of the mutants identified so far has a cell size more than about 2-fold different from wild type. Gene dosage relations have only been fully investigated using the *wee* and *whi* mutants of the yeasts. In these cases a doubling in gene dosage does not result in a doubling in cell size and thus the action of these genes cannot be responsible for the observed relationship between cell size and DNA ploidy. It is possible that there are other genes responsible for this relationship which have yet to be identified though it should be noted that in the case of *S. pombe* 161 *wee* mutants have been isolated and thus the system may be reaching saturation. Also many other genes will have indirect effects on cell size by altering other factors such as growth rate. The minute mutants of *Drosophila* are probably of this type. Thus there are specific genes which determine cell size but those investigated so far cannot explain the correlation of cell size and DNA ploidy.

INFLUENCE OF TOTAL DNA CONTENT ON CELL VOLUMES

The influence of total DNA content on cell volume can be investigated by making use of organisms which have accessory or B chromosomes. Such chromosomes are not essential and their numbers can vary between strains. Stomatal cell length was found to be increased in strains of rye containing B chromosomes compared to strains lacking them (Muntzing and Akdik, 1948). A fuller study of rye was carried out by John and Jones (1970) who compared the average wet weight of root meristematic cells in a series of strains containing different numbers of B chromosomes. A clear relationship was seen, with cell wet weight increasing in organisms with higher numbers of B chromosomes and higher DNA contents, although the addition of only 1–2 extra chromosomes increasing DNA content by about 25% had little effect. Thus the presence of extra DNA in the cells generally results in an increase in cell volume.

Another approach is to investigate the relationship between cell size and DNA content in a wide range of organisms with different amounts of DNA. Allfrey *et al.* (1955) and Commoner (1964) have pointed out the remarkable correlation between erythrocyte mass and volume and the DNA content of the vertebrates from which they have been derived. For a variety of birds, reptiles, amphibians, and mammals ranging in DNA C-values from 2 to 200 pg/nucleus there is a direct proportional relationship between DNA content and cell size. A similar relationship was found between the fresh weight of root meristemic cells and DNA content for a number of flowering plants whose DNA C-values varied over 10-fold (Martin, 1966). An even larger range of DNA C-values approaching 10^4-fold in 10 different unicellular algae was examined by Holm-Hansen (1969). This author found that the weight of carbon per cell was very closely correlated with the DNA content of the cells. He also noted that although DNA C-values vary by 10^5 in organisms as diverse as bacteria, fungi, protozoans, sponges and vertebrates, the proportion of DNA to cell mass varies by less than 20. These observations indicate that there is a very good relationship between the total DNA content of a cell and its size. (See also Chapter 4 and figure 4.2).

CELL CYCLE CONTROLS AND THE KARYOPLASMIC RATIO

The results of the previous section suggest that cell size is determined by an interaction of the function of specific genes with the total DNA content of the cell. A particular cell size set by the action of specific genes can be altered by changing the DNA content whether by changes in ploidy or by introducing B chromosomes into the cell. Such an interactive system can be best understood in terms of cell cycle controls which coordinate progress through the cell cycle with increase in cell mass. The presence of such controls has already been described in the yeasts which have to attain particular cell sizes to start a cell cycle and to undergo mitosis, and in *E. coli* which initiates DNA replication at a critical cell mass to replicon origin ratio. Similar controls have been reported in a variety of other organisms such as *Physarum polycephalum* (Sudbery and Grant, 1975), *Neurospora crassa* (Alberghina and Sturani, 1981), and mammalian cells (Killander and Zetterberg, 1965; Shields *et al.*, 1978). It is probable that these cell cycle size controls are present in all cells although they may not always be expressed (Nurse and Thuriaux, 1977). The molecular mechanisms of the controls are not understood but at least two of the models proposed involve counting or titrating out particular amounts of a controlling molecule. In the initiator accumulation model a certain amount of an initiator protein is accumulated and then initiates DNA replication (Bleecken, 1971; Sompayrac and Maaløe, 1973). In the nuclear sites model an effector is titrated onto nuclear sites and then initiates mitosis (Sachsenmaier *et al.*, 1972). A likely candidate for measuring an amount of a molecule or for acting as a fixed number of nuclear sites is the DNA itself. For example, origins of replication could act as binding sites for the DNA replication initiator protein (Cavalier-Smith, 1978). In this case the numbers of sites would be measuring the amount of the molecule. Another possibility would be that a number of chromatin sites have to be modified say by phosphorylation (Bradbury *et al.*, 1974) before mitosis can take place. Such molecular mechanisms involve an interaction between specific genes producing controlling molecules, and the total DNA content of the cell which provides a fixed number of sites. Mutations in the specific genes altering the synthesis or action of the controlling molecules will alter cell size at division as will changes in the total DNA content of the cell which will change the numbers of nuclear sites.

These cell cycle controls will set cell size at division and thus determine cell volume in proliferating cells. When proliferation ceases and differentiation takes place, cell size usually does not decrease although there can be massive increases. Examples of the latter are development of egg and xylem cells. Therefore the cell cycle size controls can be considered as establishing the minimum size possible for a cell, but will have only an indirect effect on the maximum possible. In my opinion it is the former which is most likely to be relevant to the DNA C-value paradox since the latter can vary very much indeed, especially during development in multicellular organisms.

Although the above hypothesis can explain how cell size may be determined by an interaction of the function of specific genes with the total DNA content of the cell it offers no explanation as to why these cell cycle controls result in a particular size for a particular cell. The remarkable correlation noted earlier between a cell's DNA content and its size in a wide range of organisms suggests that there are evolutionary constraints relating the nuclear DNA content to cytoplasmic mass. It is possible that a certain cytoplasmic mass may be required to support the demands made by DNA replication and mitosis and the other events which make up the cell cycle. Since the demands of DNA replication and mitosis will be at least partially dictated by the nuclear DNA content, there may be a minimum cytoplasmic mass beneath which it is disadvantageous for a cell to fall for a particular DNA C-value, because of competition of these cell cycle functions with other activities of the cell. The suggestion that a certain cytoplasmic mass is required to 'support' a nucleus with a particular DNA content is of course only a restatement of the karyoplasmic ratio hypothesis. This hypothesis, which was developed by Hertwig and Boveri at the turn of the century, stated that there is a fixed proportion between nuclear volume and protoplasmic volume (Wilson, 1924). This can be restated as a particular minimal cytoplasmic mass (i.e. protoplasmic volume) being required for a nucleus with a particular DNA content (i.e. nuclear volume). Evolutionary constraints such as this (see also Cavalier-Smith, Chapter 4) would produce a selective pressure acting on the specific genes functioning in the cell cycle size control system, resulting in a particular size at division for a cell with a particular DNA content.

RELEVANCE TO THE DNA C-VALUE PARADOX

I have argued that cell cycle controls involving an interaction between the functions of specific genes and the total DNA content of the cell will determine cell size, and that evolutionary constraints will establish the absolute value of cell size that the controls will produce. This can offer at least a partial explanation of the DNA C-value paradox. For example, it has been suggested that a small volume may be of selective advantage to cells which metabolize actively (Szarski, 1974), and certainly the rate of oxygen exchange is more rapid in small erythrocytes than in large ones (Commoner, 1964). The advantage may be related to the increased surface area to volume ratio of small cells. Vertebrates with higher rates of metabolism such as small birds and mammals have smaller and less variable cells, while more sluggish vertebrates such as the Urodela and Dipnoi have a much broader range of cell volumes which can be very large (Szarski, 1974). Therefore, there may be selection in metabolically active organisms for a low DNA C-value which would result in small cell volumes (Cavalier-Smith, 1978). Conversely, in less active organisms there may be a relaxation of selection, leading to drift with respect to DNA content per cell. Without selective pressure acting on the cell cycle control system any organism which drifted to a higher level of DNA content per cell would also have bigger cells. An observation which can be interpreted in this way is the gradual increase in osteocyte cell volumes that has taken place since Devonian times in the lungfish (Thomson, 1972). Cell volume has increased about 30-fold in these organisms which could be the consequence of a relaxation of selection for cell volume, combined with a gradual increase in DNA content. Obviously such 'selective scenarios' are of limited value but at least they can give an indication of how part of the DNA C-value paradox may be resolved.

REFERENCES

Adams, J. (1977). The interrelationship of cell growth and division in haploid and diploid cells of *Saccharomyces cerevisiae*. *Exp. Cell Res.*, **106**, 267–275.

Alberghina, L. and Sturani, E. (1981). Control of growth and the nuclear division cycle in *Neurospora crassa*. *Microbiological Reviews*, **45**, 99–122.

Allfrey, V. G., Mirsky, A. E. and Stern, J. (1955). The chemistry of the cell nucleus. *Advances in Enzymology*, **16**, 411–500.

Bleecken, S. (1971). Replisome controlled initiation of DNA replication. *J. Theor. Biol.*, **32**, 81–92.

Bradbury, E. M., Inglis, R. J., Matthews, H. R. and Langan, T. A. (1974). Molecular basis of control of mitotic cell division in eukaryotes. *Nature, Lond.*, **249**, 553–556.

Brehme, K. S. (1941). Development of the minute phenotype in *Drosophila melanogaster*. A comparative study of the growth of three minute mutants. *J. Exptl. Zool.*, **88**, 135–160.

Cavalier-Smith, T. (1978). Nuclear volume control by nucleoskeletal DNA, selection for cell volume and cell growth rate, and the solution of the DNA C-value paradox. *J. Cell Sci.*, **34**, 247–278.

Clutterbuck, A. J. (1969). Cell volume per nucleus in haploid and diploid strains of *Aspergillus nidulans*. *J. gen. Microbiol.*, **55**, 291–299.

Commoner, B. (1964). Roles of deoxyribonucleic acid in inheritance. *Nature*, **202**, 960–968.

Cooper, S. and Helmstetter, C. (1968). Chromosome replication and the division cycle of *Escherichia coli* B/r. *J. Molec. Biol.*, **31**, 519–540.

Dobzhansky, Th. (1929). The influence of the quantity and quality of chromosomal material on the size of the cells in *Drosophila melanogaster*. *Arch. f. Entw. mech. Org.*, **115**, 363–379.

Donachie, W. (1968). Relationship between cell size and time of initiation of DNA replication. *Nature, Lond.*, **219**, 1077–1079.

Fantes, P. (1977). Control of cell size and cycle time in *Schizosaccharomyces pombe*. *J. Cell. Sci.*, **24**, 51–67.

Fantes, P. (1979). Epistatic gene interactions in the control of division in fission yeast. *Nature, Lond.*, **279**, 428–430.

Fantes, P. A. (1981). Isolation of cell size mutants of a fission yeast by a new selective method: characterization of mutants and implications for division control mechanisms. *J. Bacteriol.*, **146**, 746–754.

Fantes, P. and Nurse, P. (1977). Control of cell size at division in fission yeast by a growth-modulated size control over nuclear division. *Exp. Cell Res.*, **107**, 377–386.

Fantes, P. A. and Nurse, P. (1978). Control of the timing of cell division in fission yeast. *Exp. Cell Res.*, **115**, 317–329.

Holm-Hansen, O. (1969). Algae: amounts of DNA and organic carbon in single cells. *Science*, **163**, 57–88.

John, P. C. L. and Jones, R. N. (1970). Molecular heterogeneity of soluble proteins and histones in relationship to the presence of B-chromosomes in rye. *Exp. Cell Res.*, **63**, 271–276.

Johnston, G. C., Pringle, J. R. and Hartwell, L. H. (1977). Co-ordination of growth with cell division in the yeast *Saccharomyces cerevisiae*. *Exp. Cell Res.*, **20**, 294–312.

Killander, D. and Zetterberg, A. (1965). A quantitative cytochemical investigation of the relationship between cell mass and initiation of DNA synthesis in mouse fibroblasts. *Exp. Cell Res.*, **40**, 12–20.

Longwell, A. C. and Svihla, G. (1960). Specific chromosomal control of the nucleus and of the cytoplasm in wheat. *Exp. Cell Res.*, **20**, 294–312.

Martin, P. G. (1966). Variation in the amounts of nucleic acids in the cells of different species of higher plants. *Exp. Cell Res.*, **44**, 84–94.

Martinez-Salas, E. and Vicente, M. (1980). Amber mutation affecting the length of *Escherichia coli* cells. *J. Bact.*, **144**, 532–541.

Muntzing, A. and Akdik, S. (1948). The effect on cell size of accessory chromosomes in rye. *Hereditas*, **34**, 248–250.

Nurse, P. (1975). Genetic control of cell size at cell division in yeast. *Nature, Lond.*, **256**, 547–551.

Nurse, P. (1981). Genetic analysis of the cell cycle. In W. S. Glover and D. A. Hopwood (Eds), *Genetics as a tool in Microbiology*, Cambridge University Press, Cambridge.

Nurse, P. and Bissett, Y. (1981). Gene required in G_1 for commitment to cell cycle and in G_2 for control of mitosis in fission yeast. *Nature, Lond.*, **292**, 558–560.

Nurse, P. and Thuriaux, P. (1977). Controls over the timing of DNA replication during the cell cycle of fission yeast. *Exp. Cell Res.*, **107**, 365–375.

Nurse, P. and Thuriaux, P. (1980). Regulatory genes controlling mitosis in the fission yeast *Schizosaccharomyces pombe*. *Genetics*, **96**, 627–637.

Pedersen, R. A. (1971). DNA content, ribosomal gene multiplicity, and cell size in fish. *J. Exptl. Zool.*, **177**, 65–78.

Pontecorvo, G. (1953). The genetics of *Aspergillus nidulans*. *Advances in Genetics*, **5**, 141–238.

Sachsenmaier, W., Remy, U. and Plattner-Schobel, R. (1972). Initiation of synchronous mitosis in *Physarum polycephalum*. A model of the control of cell division in eukaryotes. *Exp. Cell Res.*, **73**, 41–48.

Schkwarnikow, P. K. (1934). Uber die Grosse der meristematischen Zellen von trisomen Pflanzen von *Crepis tectorum*. *Planta*, **22**, 375–392.

Shields, R., Brooks, R. F., Riddle, P. N., Capallaro, D. F. and Delia, D. (1978). Cell size, cell cycle and transition probability in mouse fibroblasts. *Cell*, **15**, 469–474.

Sinnott, E. W. (1960). *Plant Morphogenesis*, McGraw-Hill, New York.
Sompayrac, L. and Maaløe, O. (1973). Autorepressor model for control of DNA replication. *Nature, Lond.*, **241**, 133–135.
Sudbery, P. E., Goodey, A. R. and Carter, B. L. A. (1980). Genes which control cell proliferation in the yeast *Saccharomyces cerevisiae*. *Nature, Lond.*, **288**, 401–404.
Sudbery, P. and Grant, W. (1975). The control of mitosis in *Physarum polycephalum*. The effect of lowering the DNA: mass ratio by UV irradiation. *Exp. Cell Res.*, **95**, 405–415.
Szarski, H. (1974). Cell size and nuclear DNA content in vertebrates. *Int. Rev. Cytol.*, **44**, 93–111.
Thomson, K. S. (1972). An attempt to reconstruct evolutionary changes in the cellular DNA content of lungfish. *J. Exptl. Zool.*, **180**, 363–372.
Thuriaux, P., Nurse, P. and Carter, B. (1978). Mutants altered in the control co-ordinating cell division with cell growth in the fission yeast *Schizosaccharomyces pombe*. *Molec. gen. Genet.*, **161**, 215–220.
Trombetta, V. (1942). The cytonuclear ratio. *Botanical Review*, **8**, 317–336.
Wettstein, F. (1937). Experimentelle Untersuchungen zum Artbildungs-problem I. Zell grossenregulation und Fertilwerden einer polyploiden *Bryum* Sippe. *Zeitsch. ind. Abst. Vererb.*, **74**, 34–53.
Whiting, A. R. (1961). Genetics of *Habrobracon*. *Advances in Genetics*, **10**, 295–348.
Wilson, E. (1924). *The Cell in Development and Heredity*, 3rd edn, Macmillan Co., New York.
Yčas, M., Sugita, M. and Bensam, A. (1965). A Model of cell size regulation. *J. Theor. Biol.*, **9**, 444–470.

The Evolution of Genome Size
Edited by T. Cavalier-Smith
© 1985 John Wiley & Sons Ltd

CHAPTER 6

Experimentally Induced Changes in Genome Size

C. A. CULLIS

John Innes Institute, Colney Lane, Norwich, NR4 7UH, U.K.

INTRODUCTION

The interdependence of the genotype and of the environment in which it develops to give the observed phenotype has long been recognized. However, in the absence of selection, the response to environmental factors in one generation is not, as a rule, transmitted to subsequent generations. However, there are exceptions and specific environmental factors have been found to affect subsequent generations (Durrant, 1962; Hill, 1965). In these exceptional cases most of the individuals subjected to a particular set of environmental conditions survive and produce offspring all of which show similar heritable changes. This is in contrast to the effect when the environment exerts a selection pressure. Where selection is operating the response is transmitted to the progeny but in this instance the greater proportion of the original population does not contribute to the subsequent generations. Under these conditions the question arises as to how the genome is altered as a result of selection. Is the change limited to a small specific DNA sequence, for example, the amplification or deletion of the gene for a particular enzyme, or is the specific DNA sequence responsible for the selective advantage just a fraction of a much larger change? A further consideration of this latter possibility is, if the selected sequence is only a small part of the change, how specific are the sets of sequences involved in the responses to different selection pressures? The same considerations arise in the cases of environmentally induced heritable changes in terms of the sets of DNA sequences involved in response to particular environmental factors.

Two experimental systems, namely the selection of methotrexate-resistant cell lines and the environmentally induced heritable changes in flax, will be considered in detail. An environmentally induced change in soya bean cells in culture will also be considered briefly. The first of these systems represents a change which can be selected while the second two are examples of the unselected induction of DNA changes. The findings from the three systems will be compared with findings from other systems and the possible contribution to genome evolution considered.

GENE AMPLIFICATION AND DRUG RESISTANCE

Methotrexate is an analogue of folic acid which binds to the enzyme dihydrofolate reductase (DHFR), preventing the conversion of dihydrofolate to tetrahydrofolate. This latter compound is required for the generation of key precursors to DNA and protein synthesis (Schimke, 1980). Normally, animal cells are sensitive to methotrexate, being killed by low concentrations of the drug. However, cells resistant to high levels of methotrexate can be obtained. There are three ways in which this resistance can occur:

1. by alterations in the transport of the drug so that intracellular levels are minimal (Fisher, 1962);
2. by a mutation of the enzyme DHFR so that it has a lowered affinity for methotrexate (Albrecht *et al.*, 1972);
3. by an increase in the level of the enzyme in the cells (Hakala *et al.*, 1961).

Only the last of these three mechanisms will be considered further here.

Resistant cells containing a higher level of DHFR have been obtained by a stepwise selection procedure (Schimke *et al.*, 1978). This is by initially growing the cells on low concentrations of methotrexate which kills virtually all the cells. Occasionally, some cells survive (approximately 1 in 10^5 in certain systems (Schimke, 1980)) and these are resistant to this level of methotrexate. The resistant cells have a higher level of dihydrofolate reductase than the starting cells. These cells are now grown in a somewhat higher concentration of methotrexate. Again the majority of the cells are killed but the survivors are resistant to the higher concentration of methotrexate. These cells have a higher concentration of the enzyme than those resistant to lower levels of methotrexate. In this fashion cells can be selected which are resistant to much higher concentrations of methotrexate and can have a level of dihydrofolate reductase which can be 400 times higher than in normal methotrexate-sensitive cells.

By the use of a cDNA probe complementary to the dihydrofolate reductase mRNA, it has been shown that the number of copies of the gene for DHFR has been increased in resistant cells, as has the amount of mRNA for DHFR (Schimke *et al.*, 1978). It has also been shown that the level of resistance to methotrexate is dependent on the number of copies of the DHFR gene present.

Two types of methotrexate resistance with increased DHFR can be found

(Dolnick *et al.*, 1979; Kaufman *et al.*, 1979). In one case the resistant cells retain their resistance whether the selective agent (i.e. methotrexate) is present or absent. In the other case the resistance is unstable so that, in the absence of methotrexate, resistance is lost over a number of cell generations.

In the case of the stably resistant cell lines it has been shown that the amplified genes have been integrated into the chromosome(s) and so segregate normally at each division. In a methotrexate-resistant murine lymphoblastoid cell line the amplified DHFR genes have been shown by *in situ* molecular hybridization to be located in the large homogeneously staining region of chromosome number 2. (Dolnick *et al.*, 1979). This segment probably consists of tandem repeats of a basic unit approximately 800 kilobases (kb) long. As the coding regions of the DHFR gene span approximately 40 kb (Kaufman *et al.*, 1978), the increase in DHFR gene number only accounts for about 5% of the total amplified DNA in this case.

The cell lines in which the methotrexate resistance is unstable have the amplified copies of the DHFR gene present as extrachromosomal elements in double-minute chromosomes. Because it is unlikely that double minutes contain centromeric DNA and because double minutes do not associate with the spindle apparatus at mitosis, the double minutes would segregate randomly and unequally. This would generate heterogeneity in the DHFR gene number in daughter cells, and since cells with lower DHFR gene numbers have a more rapid generation time, the reversion in the absence of selection could be accounted for by the faster growth of cells with fewer double minutes. It has also been reported that double minutes tend to aggregate at mitosis with the subsequent formation of micronuclei which can be expelled (Levan and Levan, 1978). This process may account for the single step loss of DHFR genes. It is clear that the amount of DNA in each double minute cannot be accounted for by the amplified DHFR genes which once again only contribute a small fraction of the amplified DNA.

Two other selection systems have been shown to select cell lines with increasing numbers of specific genes, namely, the increase in the numbers of aspartate transcarbamylase genes (Wahl *et al.*, 1979) and the increase in metallothionin-I gene in cadmium-resistant mouse cells (Beach and Palmiter, 1981). In both these cases the piece of DNA amplified was much larger than the gene for the corresponding mRNA. Thus in these selection systems the amplification of a particular gene only accounted for a small fraction of the total amplified DNA. The size of the amplification units associated with duplications of particular selectable loci has consequences for the evolution of genome size. If all the amplification units were similar in size to that of the methotrexate gene (approximately 800 kb) then the selection in a natural environment for amplified genes at a small number of loci would be associated with a significant increase in genome size.

ENVIRONMENTALLY INDUCED HERITABLE CHANGES IN FLAX

Heritable changes can be induced in some flax varieties when they are grown in certain environments (Durrant, 1962, 1971). The initial observation was that stable forms (termed genotrophs) differing in plant weight (large, L and small, S genotrophs) were produced in subsequent generations following the growth of the original (plastic, Pl) variety in particular environments. These two stable genotrophs then behaved as distinct genetic types in most respects.

In addition to the difference in plant weight the two stable genotrophs differed from one another in a number of other characters. These were plant height (L was taller than S), nuclear DNA amount (L had 15% more than S; Evans *et al.*, 1966), ribosomal DNA (rDNA) amount (L had 50% more than S; Cullis, 1975, 1976), isozyme band pattern (Cullis and Kolodynska, 1975) and seed capsule septa hair number (Durrant and Nicholas, 1970).

Nuclear DNA differences

The two stable genotrophs were shown to have different nuclear DNA amounts as determined by Feulgen cytophotometry (Evans *et al.*, 1966). L had the highest value, S the lowest with Pl having an intermediate amount. The nuclear DNA from flax can be separated into three components on neutral CsCl gradients, a light satellite, the main band and a heavy shoulder. There is no detectable variation in the relative amounts of these between genotrophs (Cullis, 1975). In order to characterize the DNA sequences involved in this induced DNA difference, the DNA from a number of genotrophs was analysed by renaturation kinetics. It was shown that the L and S genotrophs differed from one another mainly in the intermediate repetitive sequences (Cullis, 1975). This analysis has since been extended to include a number of other genotrophs, and a detailed comparison was made of the renaturation kinetics of the DNA from the two genotrophs which showed the greatest difference. These two genotrophs have been termed L^H, a large genotroph, and L_6, a small genotroph (Cullis, 1977).

The reassociation curves for the two lines, both for the homologous reaction, and where the DNA of one was driven by that of the other, are shown in Fig. 6.1. It can be seen that when L_6 DNA was driven by either L_6 or L^H DNAs the curves obtained were very similar. However, when L^H DNA was driven by L_6 a very different curve from that for the homologous reaction was obtained. These curves were analysed by a non-linear regression computer program and were fitted to three components. These are given in Table 6.1. It can be seen that L^H and L_6 appear to vary in sequences in both the highly repetitive fractions and in the intermediately repetitive fractions. In order to determine the extent of the difference between L^H and L_6 and the way in

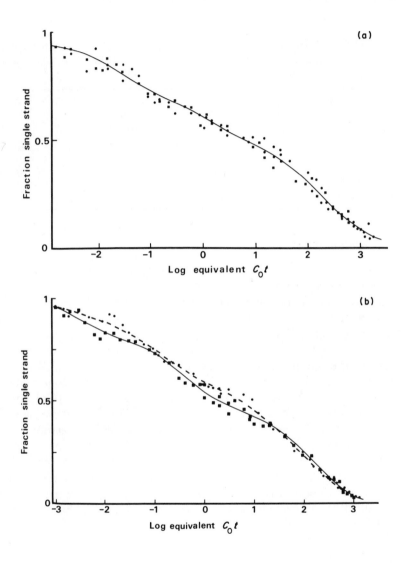

Fig. 6.1. Reassociation kinetics by hydroxyapatite fractionation of L_6 and L^H DNAs sheared to 300 base pairs.

(a) L_6 tracer, L_6 driver ●—●
    L_6 tracer, L^H driver ■—■
(b) L^H tracer, L^H driver ■—■
    L^H tracer, L_6 driver ●—●

which they differed from the original line from which they had both been derived, they were characterized with respect to a number of specific DNA sequences.

Table 6.1. Kinetic components of L_6 and L^H DNAs in homologous and heterologous driver/tracer experiments. The points in Figs. 6.1(a) and (b) were fitted to three components by a non-linear regression computer program.

Driver	Tracer	Fraction %	Rate constant mol l^{-1} s^{-1}	Fraction %	Rate constant mol l^{-1} s^{-1}	Fraction %	Rate constant mol l^{-1} s^{-1}
L^H	L^H	0.24	1900	0.31	8.9	0.42	0.021
L^H	L_6	0.16	1000	0.23	34	0.55	0.056
L_6	L_6	0.16	1400	0.30	30	0.50	0.021
L_6	L^H	0.26	250	0.23	4.2	0.42	0.024

rDNA variation

The number of ribosomal RNA (rRNA) cistrons per genome of a number of flax genotrophs has been determined and shown to vary between genotrophs (Cullis, 1975, 1976). In particular, the number of rRNA genes in L^H was 2570 while that in L_6 was 1050. Thus part of the difference in the highly-repetitive fraction of DNA between L^H and L_6 was contributed by the rRNA genes and this represents about 0.5% of the total DNA. However, there was no significant difference in the number of these genes between L^H and Pl.

The rDNA repeat unit from flax has been cloned and the cloned material used to compare the rDNA sequences in various genotrophs (Goldsbrough and Cullis, 1981). It was shown that the rDNA from all the genotrophs investigated showed a homogeneous set of rRNA genes, arranged in tandem arrays with a repeat length of 8.6 kb. There was neither detectable variation in the size of this repeat nor variation in the position of a number of restriction enzyme sites in the rDNAs from various genotrophs. Thus, whatever the basis of the variation in rDNA amount in the small genotrophs, it was not by the selection of what can presently be recognized as a subfraction of the total rDNA.

5S RNA gene number variation

The 5S RNA genes of flax are arranged as tandem arrays of a 0.35–0.37 kb repeating sequence (Goldsbrough *et al.*, 1981). In contrast to the rDNA the 5S DNA sequences exhibit both length and sequence heterogeneity. In common with the rDNA, however, there is variation in the copy number of the 5S RNA genes in the genotrophs. These vary from 117 000 genes per

2 C nucleus in Pl to 49 600 genes per 2 C nucleus in L^H. There is approximately the same number of genes in L_6 and L^H (Cullis, unpublished). Thus there is no association between the numbers of rRNA and 5S genes.

Cloned nuclear DNA sequences

A number of cloned nuclear DNA sequences have been investigated to find other sequences which vary between genotrophs.

pCL21

This plasmid contains an insert of 4.2 kb, which is a highly-repetitive sequence in flax. It has been used to hybridize to Southern blots of DNA extracted from Pl, L_6 and L^H which was digested with BamHI and equal amounts of each DNA loaded in separate tracks. It can be seen that there is a similar amount of DNA complementary to pCL21 in each of the three genotrophs (Fig. 6.2(a)).

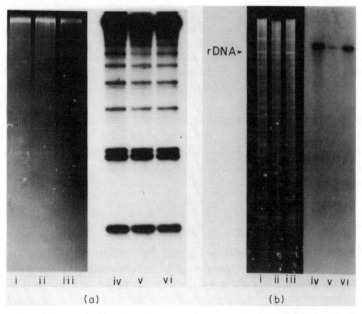

Fig. 6.2.   (a) Equal loadings of Pl (i, iv), L_6 (ii, v) and L^H (iii, vi) DNAs, digested with BamHI, eletrophoresed on a 1.4% agarose gel, transferred to nitrocellulose and hybridized with pCL21. (i, ii, iii) Ethidium bromide stained gel (iv, v, vi) Autoradiograph of filter after hybridization. (b) Equal loadings of Pl (i, iv), L_6 (ii, v) and L^H (iii, vi) DNAs, digested with BamHI, electrophoresed on 1% agarose gel, transferred to nitrocellulose and hybridized to pCG8. (i, ii, iii) Ethidium bromide stained gel (iv, v, vi) Autoradiograph of filter after hybridization.

pCG8

This plasmid contains an insert of 12.9 kb and is an intermediately-repetitive sequence. When this plasmid was radioactively labelled and hybridized to a Southern blot, equivalent to that used for pCL21, it can be seen that there is variability for the number of copies of this sequence between genotrophs (Fig. 6.2(b)). For this sequence there is the highest number of copies in Pl, a lower number in L^H (80% of the number in Pl) and an even lower number in L_6 (30% of the number in Pl).

The four different DNA sequences described above illustrate some of the changes, and the independence of these changes, which can be observed in the genome of the genotrophs. The combinations observed are as follows:

1. The same number of copies in the two induced lines as that in the original line (Pl) from which they were derived (pCL21).
2. Pl and the large genotroph (L^H) having the same number of copies but the small genotroph (L_6) having a reduced number (rDNA).
3. Both the large and small genotrophs having the same number of copies, but this number is less than that of Pl (5S DNA).
4. The original line, the large and small genotrophs all having different numbers of copies of a particular DNA sequence (pCG8).

The plant material (the lines Pl, L_6 and L^H) used as the source of DNA for these experiments has been propagated for a number of generations since they were obtained from the original plastic seed by Durrant. Because of this it is possible that variations have arisen by chance since the separation of the lines and that these are not connected with the original induction phenomenon. However, as is described below, it has been shown that one of these sequences (rDNA) actually varies during the growth in the inducing environment.

THE INDUCTION OF rDNA CHANGES

The rDNA amount, as determined by the hybridization of labelled purified 25S and 18S RNAs to filter bound DNA, has been determined during the growth of the plastic line under two sets of inducing conditions (Cullis and Charlton, 1981). The plants were grown under two nutrient regimes and at various intervals the main stems were harvested. DNA was prepared from the top 1 cm and the remainder of the stem surface sterilized and plated onto solid nutrient medium. After 2–4 weeks the stimulated meristems were harvested and the DNA prepared from the pooled material obtained from regions of the stem. The rDNA amounts in these DNA preparations were determined. It was shown that the rDNA at the base of the stem remained constant in both environments while that at the apex varied (Fig. 6.3).

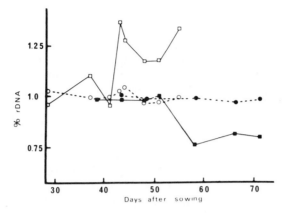

Fig. 6.3. Time course for the induction of rDNA differences in flax. Treatment (1): —□—□— DNA in shoots from top quarter of stem; ○·····○····· DNA from shoots on bottom quarter of the stem. Treatment (2): —■—■— DNA from shoots from top quarter of the stem; ·····●·····●····· DNA from shoots from bottom quarter of the stem (from Cullis, 1982).

Thus the plants growing under inducing conditions were chimeric for rDNA amount, having different amounts at the base and the apex. The rDNA amounts in the progeny from plants grown in one of the environments were determined and found to have similar values for the apical regions of the parent plants after the rDNA had declined. Thus the reduction in the rDNA amount observed during growth under an inducing environment was transmitted to the progeny.

Association of induced changes

The DNA differences, as determined by Feulgen cytophotometry, the plant weight and isozyme differences, induced during the production of the stable genotrophs, were not stable under all subsequent conditions (Joarder *et al.*, 1975) and could vary further. This subsequent variation was not necessarily associated with other phenotypic changes. Lines were produced which had different phenotypes (that is a large genotroph and a small genotroph) and differing rDNA amounts but no detectable nuclear DNA difference. The DNA measurements were made by Feulgen cytophotometry, which is a relatively insensitive technique when compared to the gene counting methods available with cloned probes. Thus the total DNA change which was observed to occur during induction was not necessary for the expression of the phenotypic characters being followed in the genotrophs. The question which arises, as with the amplification of the DHFR genes, is to what extent the change in DNA amount occurred in specific sequences relating to the production of a specific phenotype and how much was fortuitously altered by virtue of the fact that the size of the unit accompanying the selected locus may well be much larger than the locus itself.

The observations of subsequent variation in DNA amount without obvious correlated phenotypic change is consistent with only a small fraction of the altered DNA sequences being directly involved with the determination of the phenotypic characters that have been investigated.

rDNA VARIATION IN SOYA BEAN SUSPENSION CULTURES

Soya bean suspension cultures grown on sucrose as the carbon source have a doubling time of 24 hours (Jackson, 1980). When these cells are supplied with maltose as the carbon source the doubling time rises to 196 hours. Concomitant with this reduction in growth rate is a reduction in rDNA amount of about 30% (Jackson, 1980). In contrast with the flax situation, however, this reduction involved the loss of a specific subset of the ribosomal genes. Reamplification of the rRNA genes did not occur when these cells were returned to sucrose medium. In addition, when these cells were returned to sucrose medium they grew at the same rate as those continuously propagated on sucrose and so containing approximately 30% more rRNA genes. Thus the rate of growth on sucrose is not simply dependent on the number of rRNA genes.

The reduction in rRNA genes did not occur by the selection of cells containing fewer genes but by the reduction of the rRNA genes in most of the cells. This reduction occurred over a small number of cell generations, mostly occurring over two generations and being complete in less than ten generations. The rRNA genes were not the only DNA sequences to vary after the change of growth medium, certain highly repetitive sequences were also lost (Jackson, 1980). Thus, once again, a number of sequences appeared to vary in response to the environmental conditions, and in this case the ones characterized may not have been responsible for the phenotypic effects observed.

POSSIBLE MECHANISMS OF DNA VARIATION

The mechanisms by which DNA variation occurs is not clear. However, a number of possibilities have been discussed for the various systems described here. The initial DHFR gene duplication could occur by a number of mechanisms including unequal crossing-over, or the uptake of a DNA segment from killed cells, which could be either integrated into the chromosome or remain as an extrachromosomal element, or by disproportionate replication (Schimke *et al.*, 1978; Kaufman *et al.*, 1979). Once this initial duplication had occurred further amplification could be produced by more unequal crossing-over, the uptake of additional DNA from lysed cells or the generation of extrachromosomal sequences from rolling circle replication. Various combinations of these would result in different genomic organizations of the amplified genes. The loss of the DHFR genes would occur by a reversal of the amplification processes, namely a loss of the extrachromosomal copies or unequal crossing-over with selection for the cells with reduced gene numbers. A cell population undergoing either the amplification or reduction of genes should be a heterogeneous mixture with different numbers of copies of the gene and this has been observed (Schimke, 1980).

For the metallothionein-I gene, Beach and Palmiter (1981) suggest that random chromosomal breaks occur continuously at a low frequency. Occasionally, a daughter cell will receive a fragment containing the gene and so be at a selective advantage. If this fragment has a suitable origin of replication it could be maintained as a double minute chromosome. Alternatively, it could be integrated into the chromosome.

In flax, it has been suggested that the DNA amplification/diminution occurs via an extrachromosomal mechanism and a model has been proposed (Cullis, 1977). In this case it was suggested that in the appropriate environment, extrachromosomal copies could be generated whose replication was no longer necessarily synchronized with that of the chromosomes. In this way various sequences could be lost or amplified, with both processes possibly occurring in the same cell. However, the data do not exclude unequal crossing-over as the possible mechanism (Cullis, 1982). Since the flax genome is organized in a long period interspersion pattern (Cullis, 1981) and the members of the same families of repeated sequences are also clustered, the opportunity for unequal crossing-over must be considerable.

Irrespective of the details of the mechanisms involved, do these types of gene amplification occur continuously and can they involve any DNA sequence? For both the DHFR gene amplification and the metallothionein-I gene this has been assumed (Schimke, 1980; Beach and Palmiter, 1981). For the flax system the range of sequences involved certainly indicates that most, if not all, the sequences can be involved. Thus if these DNA variations are the product of selection of certain individuals from a variable population produced by whatever mechanism, then does the selective agent in any way enhance the process? That is, is it possible that the environments or selective agents affect the frequency of DNA variation. This does seem to be the case in both flax and soya bean where there was no change in the number of ribosomal genes in certain environments, and large changes in very few cell generations (without massive cell death being observed) in other environments (Cullis and Charlton, 1981; Jackson, 1980). Thus the intriguing possibility remains that the agents of selection may themselves be generating the variation on which the selection then acts.

CONSEQUENCES FOR GENOME EVOLUTION

In all the systems considered here the amount of DNA involved in the amplification or deletion events is very much larger than that required for the response being considered. If this is a general phenomenon in the selection of any trait it has certain consequences for the evolution of genome size.

Consider an organism in a particular environment in which a duplication of a locus is at a selective advantage. If the locus concerned only constitutes a small fraction of the duplicated region, as in the case of the DHFR gene, then the concomitant DNA increase will be relatively large. If the selection is transient then obviously there will be a probability that the amplified segment will be deleted, for example be unequal crossing-over. However, if the sequences have been rearranged before the removal of the selection then the probability of loss may be reduced. Thus the rate of stabilization of DNA increases may be correlated with the rate of DNA rearrangements, so that organisms with large genomes may show higher rates of transposition than those with small genomes. This may also have a bearing on the correlation between genome size and DNA interspersion pattern. The fact that small genomes have a longer period interspersion than large genomes could be due to the lowered rates of transposition and so a lowered probability of fixing any particular amplification. Thus an elucidation of the mechanisms of experimentally induced DNA variation may indicate the process by which genomes expand and contract.

ACKNOWLEDGEMENTS

I wish to thank Professor D. R. Davies, Dr N. Ellis and P. B. Goldsbrough for critically reading the manuscript, and Dr P. Jackson for supplying preprints.

REFERENCES

Albrecht, A., Biedler, J. L. and Hutchinson, D. J. (1972). Two different species of dihydrofolate reductase in mammalian cells differentially resistant to amethopterin and methasquin. *Cancer Research*, **32**, 1539–1546.

Beach, L. R. and Palmiter, R. D. (1981). Amplification of the metallothionein-I gene in cadmium-resistant mouse cells. *Proc. Natl. Acad. Sci. USA*, **78**, 2110–2114.

Cullis, C. A. (1975). Environmentally induced DNA differences in flax. In R. Markham, D. R. Davies, D. A. Hopwood and R. W. Horne (Eds), *Modification of the Information Content of Plant Cells*, North Holland, Amsterdam, pp. 27–36.

Cullis, C. A. (1976). Environmentally induced changes in ribosomal RNA cistron number in flax. *Heredity*, **36**, 73–79.

Cullis, C. A. (1977). Molecular aspects of the environmental induction of heritable changes in flax. *Heredity*, **38**, 124–154.

Cullis, C. A. (1981). DNA sequence organization in the flax genome. *Biochim. Biophys. Acta*, **652**, 1–15.

Cullis C. A. (1982). Quantitative variation of the ribosomal RNA genes. In E. G. Jordan and C. A. Cullis (Eds), *Nucleolus*, Cambridge University Press, Cambridge.

Cullis, C. A. and Charlton, L. (1981). The induction of ribosomal DNA changes in flax. *Plant Sci. Lett.*, **20**, 213–217.

Cullis, C. A. and Kolodynska, K. (1975). Variations in the isozymes of flax (*Linum usitatissimum*) genotrophs. *Biochem. Genet.*, **13**, 687–697.

Dolnick, B. J., Berenson, R. J., Bertino, J. R., Kaufman, R. J., Nunberg, J. H. and Schimke, R. (1979). Correlation of dihydrofolate reductase elevation with gene amplification in a homogeneously staining chromosomal region C5178Y cells. *J. Cell Biol.*, **83**, 394–402.

Durrant, A. (1962). The environmental induction of heritable changes in *Linum*. *Heredity*, **17**, 27–61.

Durrant, A. (1971). Induction and growth of flax genotrophs. *Heredity*, **27**, 277–298.

Durrant, A. and Nicholas, D. B. (1970). An unstable gene in flax. *Heredity*, **25**, 513–527.

Evans, G. M., Durrant, A. and Rees, H. (1966). Associated nuclear changes in the induction of flax genotrophs. *Nature, Lond.*, **212**, 697–699.

Fisher, G. A. (1962). Defective transport of amethopterin (methotrexate) as a mechanism of resistance to the antimetabolite in L5178Y leukemia cells. *Biochemical Pharmacology*, **11**, 1233–1234.

Goldsbrough, P. B. and Cullis, C. A. (1981). Characterization of the genes for ribosomal RNA in flax. *Nucl. Acids. Res.*, **9**, 1301–1309.

Goldsbrough, P. B., Cullis, C. A. and Ellis, T. H. N. (1981). Organization of the 5S RNA genes in flax. *Nuc. Acids. Res.*, **9**, 5895–5904.

Hakala, M. T., Zakrzewski, S. F. and Nichol, C. A. (1961). Relation of folic acid reductase to amethopterin resistance in cultured mammalian cells. *J. Biol. Chem.*, **236**, 952–958.

Hill, J. (1965). Environmental induction of heritable changes in *Nicotiana rustica*. *Nature, Lond.*, **207**, 732–734.

Jackson, P. J. (1980). Characterization of the ribosomal DNA of soybean cells. *Fedn. Proc. Fedn. Am. Socs. exp. Biol.*, **39**, 1878.

Joarder, I. O., Al-Saheal, Y., Begum, J. and Durrant, A. (1975). Environments including changes in amount of DNA in flax. *Heredity*, **34**, 247–253.

Kaufman, R. J., Bertino, J. R. and Schimke, R. T. (1978). Quantitation of dihydrofolate reductase in individual parental and methotrexate-resistant murine cells. *J. Biol. Chem.*, **253**, 5852–5860.

Kaufman, R. J., Brown, P. C. and Schimke, R. T. (1979). Amplified dihydrofolate reductase genes in unstably methotrexate-resistant cells are associated with double minute chromosomes. *Proc. Natl. Acad. Sci. USA*, **76**, 5669–5673.

Levan, A. and Levan, G. (1978). Have double minutes functioning centromeres? *Hereditas*, **88**, 81–92.

Schimke, R. T. (1980). Gene amplification and drug resistance. *Scientific American*, **243** (5), 50–59.

Schimke, R. T., Kaufman, R. J., Alt, F. W. and Kellems, R. F. (1978). Gene amplification and drug resistance in cultured murine cells. *Science*, **202**, 1051–1055.

Wahl, G. A., Padgett, R. A. and Stark, G. R. (1979). Gene amplification causes overproduction of the first three enzymes of UMP synthesis in N-(phosphoneacetyl)-L-aspartate-resistant hamster cells. *J. Biol. Chem.*, **254**, 8679–8689.

CHAPTER 7

DNA Replication and the Evolution of Genome Size

T. CAVALIER-SMITH

SUMMARY

This chapter outlines the basic features of DNA replication in bacteria and in eukaryotes, and discusses their importance for the evolution of genome size. Eukaryote replicons are approximately constant in size despite vast differences in genome size, whereas in bacteria replicon size increases in direct proportion to genome size. This means that the major mechanism for genomic increase in eukaryotes is by increasing the number of replicons, and that the capacity of the eukaryote cell cycle to replicate and segregate tandemly linked replicons is the fundamental reason why they are able to evolve so much larger genomes than prokaryotes.

I argue that in eukaryotes, unlike prokaryotes, larger genomes do not necessarily take any longer to replicate than smaller genomes. Comparison of the replication rates in somatic, meiotic and embryonically cleaving cells shows that the time taken for replication is subject to major physiological regulation and probably depends simply on the number of replisomes per unit amount of DNA. The universal positive correlation between the length of S-phase and genome size is not an inevitable consequence of increased genome size but the indirect consequence of the positive correlation between genome size and cell volume. Since larger cells grow more slowly than smaller ones, for reasons having nothing to do with the duration of DNA replication, it would be wasteful for them to synthesize so many replisomes that they completed replication and division preparations in very much less time than the minimum length of the cell cycle determined by these growth limitations; they can economize on replisomes by allowing S-phase to lengthen in proportion to the overall cell cycle as cell volume increases.

The universal correlation between eukaryote genome size and cell volume could in principle be simply and directly explained by a cell-volume-dependent control over the initiation of DNA replication at the beginning of S-phase of the cell cycle, if such a mechanism depended on the titration of a controlling molecule against the total nuclear DNA content. Present evidence concerning S-phase initiation is insufficient to decide between this explanation for the genome size/cell volume correlation and a more indirect evolutionary explanation along the lines discussed in Chapter 4.

The nature of eukaryote replicon origins, their involvement in the initiation of replication, their possible attachment to the nuclear matrix, and the selective forces preventing or favouring their evolutionary divergence, need to be understood in much more detail because of their great importance in relation to several other aspects of genomic evolution discussed in this chapter: the significance of heterochromatin and its late-replication and its under-replication during polyteny in Diptera; the possible role of intragenomic selection in genomic evolution; the nature of the quantitative constraints on the spacing and properties of replicon origins during evolutionary changes in genome size; and the evolutionary origin of the characteristic eukaryote pattern of many replicons per chromosome.

INTRODUCTION

DNA replication is important for understanding the evolution of genome size for five reasons.

1. The fundamental reason why eukaryotes can evolve genomes that are larger—often by thousands of times—than those of prokaryotes probably lies in their basically different pattern of DNA replication, in which each chromosome consists of many independent units of replication (replicons), and not just one as in bacteria.
2. When different eukaryote species are compared, the time taken for DNA replication (that is the length of the S-period in the cell cycle), and the overall length of the cell cycle, are both found to be strongly positively correlated with genome size; we cannot expect to understand the evolutionary forces influencing the evolution of different genome sizes unless we can determine the causes of these correlations; in particular, does genome size itself directly determine the length of S-phase, and of the whole cell cycle, or is the connection between these variables of a more indirect evolutionary nature?
3. The average size of each replicon is fairly constant in different eukaryote species differing greatly in genome size. This shows that major evolutionary changes in genome size in eukaryotes have predominantly involved changes in the number rather than the length of replicons. We therefore need to understand (a) the mechanisms of change in replicon number, (b) the evolutionary constraints that determine replicon size, (c)

the evolutionary reasons for the great diversity in replicon numbers in different eukaryotes.

4. There is growing evidence that the initiation of DNA replication at the beginning of S-phase depends on the attainment of a critical cell volume in both bacteria and eukaryotes. In bacteria the mechanism of initiation of replication apparently involves the detection of a shift in the ratio between (a) the number of replicon origins and (b) the cell volume, by means of controlling molecules that bind to replicon origins; replication restores this ratio to its original value. In eukaryotes cell volume has long been known to be positively correlated with genome size when different species are compared (see Chapter 4); if the control of DNA replication in eukaryotes also depended on the ratio

$$\frac{\text{total number of replicon origins per cell}}{\text{total cell volume}}$$

then evolutionary changes in replicon numbers might automatically cause corresponding changes in cell volume. Since replicon numbers are closely correlated with genome size, conservation of the replicon number/cell volume ratio throughout eukaryote evolution could in principle explain the correlation between genome size and cell volume (Cavalier-Smith, 1978b). An even more direct explanation of the C-value paradox, and the correlation between genome size and cell volume, would be possible if the volume-dependent control of S-phase initiation depended on the titration of controlling molecules by the total nuclear DNA content. Clearly, it is essential to consider whether or not eukaryote DNA replication is controlled in either of these ways.

5. Natural selection is best thought of as non-random differences in the rate of multiplication of competing 'replicators' (Dawkins, 1978, 1982). Since most 'replicators' consist of DNA, selection at its most fundamental amounts to differences in the multiplication rate of different DNA replicons. Therefore a proper understanding of the nature of intragenomic competition between competing replicons and its possible significance for the evolution of genome size (Chapter 8) depends in part on a good understanding of the mechanisms of control of DNA replication and how they may mutate.

Many of the enzymes and detailed molecular mechanisms that mediate DNA replication are also intimately involved in the mutational processes that increase or decrease genome size, which are discussed in Chapter 12 by Dyson and Sherratt. For this reason, and because they will be unfamiliar to many evolutionary biologists, I shall outline the most fundamental features of DNA replication, before discussing and trying to rationalize the complexities of eukaryote DNA replication in relation to the problem of the evolution of genome size.

THE OVERALL PATTERN OF DNA REPLICATION

In cellular organisms each non-replicating chromosome consists of a single double-stranded DNA molecule to which various proteins are attached. It may be linear, with two free ends, or circular with the two ends covalently joined. The chromosomes of DNA viruses also consist of circular or linear DNA molecules. Study of the simplest virus chromosomes, which can be readily isolated without breakage, and which parasitize the host cell's replication machinery, has made it possible to isolate most components of the replication machinery of the bacterial host cell and to study replication *in vitro* (see Kornberg, 1980, 1982 and Nossal, 1983, for detailed reviews and references); progress along these lines has been slower in eukaryotes but sufficient to suggest that the fundamental processes of replication are common to both superkingdoms.

The chromosomes of bacteria, viruses, mitochondria and chloroplasts are

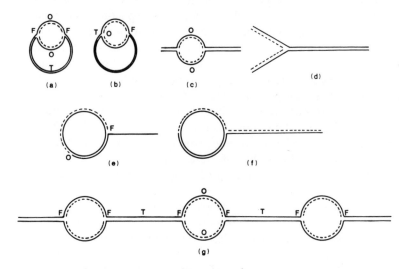

Fig. 7.1.   The major patterns of DNA replication. (a) Symmetrical bidirectional replication of double-stranded circles as in bacteria, chloroplasts and many plasmids and viruses. The two replication forks (F) move in opposite directions away from the replicon origins (O) towards the replicon terminus (T). Parental DNA is shown with a continuous line and newly made DNA with a dashed line. (b) Unidirectional replication of double-stranded circles as in some viruses and some mitochondria. (c), (d) Linear molecules that are single replicons, as in phage T7. (e) Rolling circle replication without replication on the tail generates single-stranded DNA as in certain single-stranded bacteriophages. (f) Rolling circle replication with replication on the tail as in several viruses and amplified nucleolar plasmids. (g) Linear molecules that are polymers of numerous separate replicons, as in eukaryotes; each replicon replicates bidirectionally.

single units of replication—replicons—each characterized by a single starting point for DNA replication known as the replicon origin (Fig. 7.1(a)). Replicon origins have a specific nucleotide sequence and a unique location on the genetic map. DNA replication is initiated only at the origin and proceeds, usually in both directions, in a sequential fashion away from it until the whole chromosome is replicated. A partially replicated chromosome that replicates bidirectionally from a unique origin has two branch points, the replication forks (Fig. 7.1(a)). In a circular chromosome that replicates bidirectionally the replication forks eventually meet each other at a unique chromosomal site—the replication terminus. If the chromosome is linear (e.g. bacteriophage T7) the two replication forks move apart until they reach the chromosome ends which both serve as the replication termini (Fig. 7.1(e)). In a circular chromosome replicating unidirectionally, e.g. phage P_2 (Fig. 7.1(b)), only one branch point is a replication fork, whereas the other marks the location of the adjacent origin and terminus. (In a few viruses replication does not terminate after the synthesis of the whole genome, instead the replication fork continues around the chromosome so as to generate a much longer DNA molecule that is subsequently (or concurrently) cut up into genome-sized pieces by endonucleases: this pattern is known as rolling circle replication (Fig. 7.1(e) and (f)). In a linear chromosome that replicates unidirectionally (e.g. adenovirus) the single branch point is the replication fork which moves from the origin at one end of the chromosome to the terminus at the other (Fig. 7.1(d)). Despite these differences between linear and circular, and uni- and bidirectionally replicating chromosomes, there are only three universal and fundamental processes involved in DNA replication:

1. initiation of replication at the replication origin;
2. movement of the replication fork whilst the replication machinery synthesizes the daughter DNA strands;
3. termination of replication.

In bacteria, which can grow at widely varying rates, the overall rate of DNA replication is controlled by varying the frequency of initiation of replication: the rate of movement of the replication fork remains constant (except at extremely slow growth rates under starvation conditions when paucity of nucleotide precursors slows it down). In wild-type bacteria termination of replication is an essential prerequisite for cell division: the advantage of not dividing until two copies of the chromosome are available for segregation into the two daughter cells are obvious.

Replication of the chromosomes of eukaryote nuclei is more complex because each consists of several or many distinct units of replication, each with a single origin of replication (Fig. 7.1(g)). Each DNA chromosome is a linear DNA molecule and, from the point of view of DNA replication, may be considered as a linear polymer of numerous linear replicons joined end to end by covalent bonds. This change from a single replicon per chromosome to plural replicons per chromosome is one of the most profound in the long history of DNA replication and had, as will be discussed below, profound consequences for the evolution of genome size. It also means that in eukaryotes, unlike bacteria, DNA initiation events must occur at many separate chromosomal loci every cell cycle. Furthermore, a mechanism must exist to ensure that every one of several hundreds or thousands of separate replicons have been replicated before division can commence: a simple signal triggered by the replication of a single replicon terminus would suffice for bacteria but not for eukaryotes. Despite the complexities, each replicon is comparable to the whole bacterial chromosome in that it has a single origin at which initiation occurs and from which two replication forks move away towards their respective termini, which they presumably share with adjacent replicons. The assumption here is that origins and termini are both at definite loci in eukaryotes as in bacteria; to date there is no evidence for this for eukaryote termini; the evidence for origins is discussed later.)

DNA SYNTHESIS AND THE MOVEMENT OF THE REPLICATION FORK

The replication of double-stranded DNA can occur only if the two strands are first pulled apart to form two single-stranded templates. This requires energy; in bacteria this is provided by the hydrolosis of ATP or other nucleoside triphosphates and is catalysed by enzymes known as DNA helicases. DNA helicases bind specifically to single-stranded DNA and are able to use the free energy of hydrolysis of ATP to prise apart the double-stranded DNA, and thus move a replication fork towards the terminus. Some DNA helicases (e.g. *Escherichia coli* DNA helicase I and *E. coli rep* protein), once attached to a region of single-stranded DNA, remain bound to it, and slide unidirectionally along it, whilst they prise apart the double-stranded DNA; this so-called processive movement is probably the basis for the polarized movement of replication forks from origin to terminus. The key event for initiation, discussed in the next section, might therefore be the pulling apart of the parental DNA strands specifically at the replicon origin so as to allow the initial binding of such processive DNA helicases.

The processive movement of a DNA helicase will tend to twist the DNA ahead of the moving fork more tightly, which would throw it into positive supercoils. If the molecule is circular, or, as in linear eukaryote chromosomes, attached to rigid cellular structures, this positive supercoiling would soon become so tight as to resist any further separation of the parental strands by the DNA helicase. Continued fork movement can therefore occur only if covalent bands in one or both of the polynucleotide chains ahead of the fork are enzymatically broken, the chains allowed to unwind relative to each other at the break, and the broken ends resealed by new covalent bonds. Enzymes known as DNA topoisomerases catalyse this breakage and resealing of the phosphodiester backbone of double-stranded DNA to create such a swivel, and have been isolated from prokaryotes and eukaryotes. Topoisomerases are probably essential for the replication of circular and linear double-stranded DNA. They may either break only 1 strand (type I topoisomerase) or both strands (type II topoisomerases).

Neither DNA topoisomerases nor DNA helicases are needed for the first stage of the replication of single-stranded DNA viruses, where the DNA is already single-stranded, and therefore ready to serve directly as a template; no replication fork is needed for the synthesis of a complementary strand to form a double-stranded replicative intermediate, which then replicates like other double-stranded DNA with the aid of a fork-moving DNA helicase and a swivel-making topoisomerase. In both single and double-stranded DNA the actual template is single-stranded and is held in a conformation suitable for the action of the replication machinery by aggregates of protein known as single-stranded DNA binding proteins (SSBs: sometimes also known as helix-destabilizing proteins because when mixed with double-stranded DNA they facilitiate the separation of its two strands).

The replication machinery consists of the enzyme DNA polymerase and a variety of other proteins that are thought to exist in the cell as a complex, the replisome; however, unlike ribosomes and other multienzyme complexes, intact replisomes have not been isolated. In bacteria replisomes are probably attached to the plasma membrane, and in eukaryotes to the proteinaceous nuclear matrix (Smith and Berezney, 1980); the replicating DNA therefore slides like a conveyor belt past the fixed replisome.

Deoxynucleoside triphosphates (dATP, dGTP, dCTP, dTTP) can form base pairs with the single-stranded template and be polymerized by the DNA polymerase to form a complementary polynucleotide chain (Fig. 7.2), with the liberation of one pyrophosphate molecule for the addition of each nucleotide to the growing daughter strand. DNA polymerase, however, unlike RNA polymerases, can add nucleotides covalently only to the end (always the 3'OH end) of an existing polynucleotide or oligonucleotide chain, which is referred to as a primer. Therefore, chain elongation by DNA polymerase cannot begin until suitable primers are formed, initially at the origin of the replicon. This can be done in three ways:

1. The first way is to use pre-existing DNA as a primer. This occurs in rolling circle replication, for example of single-stranded bacteriophages where a DNA topoisomerase makes a nick (i.e. a break in the phosphodiester backbone of one strand only) in the old DNA that ends in a 3'OH group to which DNA polymerases can add further nucleotides. The same

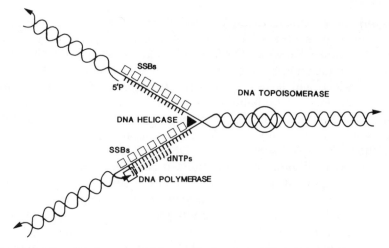

Fig. 7.2.   Simplified structure of a replication fork. The DNA helicase uses
energy derived from ATP hydrolysis to prise apart the parental strands, and
the single-stranded-binding proteins (SSBs) hold them apart to allow base
pairing with the deoxyribonucleoside triphosphates (dNTPs). This tends to
twist up the parental DNA ahead of the fork more tightly, the resulting
strain being relieved by the reversible cutting of the backbone by the DNA
topoisomerase which allows further unwinding to occur. The base-paired
dNTPs are cleaved to liberate pyrophosphate from the nucleotide residues
which are simultaneously covalently attached to the growing 3'OH end of
the lower daughter DNA strand by the DNA polymerase, which is thereby
continuously elongated in the same direction as the moving replication fork.
The upper, lagging, daughter strand elongates discontinuously by the more
complex processes shown in Fig. 7.4. In reality the topoisomerase, helicase
and polymerase may be associated together in a single macromolecular
assembly, the replisome.

principle is found in hairpin priming (Cavalier-Smith, 1974, 1983, 1985a)
where a nuclease (or topoisomerase) creates a 3'OH end, as in parvovi-
ruses (Astell *et al.*, 1983).

2. A second mechanism so far found only in certain single-stranded viruses
    having an origin at one end of the molecule is to attach the first nucleotide
    covalently to a priming protein which remains covalently attached to the
    5' end of the virion DNA.
3. The third and most general and widespread mechanism, found in bacteria,
    mitochondria, eukaryote nuclei and many viruses, is the use of oligon-
    ucleotide primers consisting of a few ribo- or a mixture of ribo- and
    deoxyribonucleotides. In bacteria and most bacteriophages these RNA
    primers are synthesized by a polymerizing enzyme, called DNA primase,
    that is distinct from both DNA and RNA polymerases; but in one group
    of single-stranded viruses (phage M13 and its relatives) RNA polymerase
    itself makes the primer.

The chromosomal sites at which the primers are made is determined not by DNA primase itself but by a prepriming reaction mediated by a protein called *dna B* protein (because it is coded by gene *dna B* that was identified because in mutant form it makes *E. coli* replication heat sensitive). The *dna B* protein, in a definite complex with other proteins called a primosome, can become bound to phage, and probably bacterial, replicon origins and stimulate the synthesis of RNA primers by DNA primase (Fig. 7.3).

In single-stranded viruses at the first stage where only one daughter polynucleotide chain is being synthesized the 3'OH end of this primer can be extended by the DNA polymerase until it meets the 5' tail end. The primer can be excised by a repair exonuclease specific for 5' ends, and the resulting gap (Fig. 7.3) filled by further extension of the 3'OH end by the DNA polymerase. When the gap left by primer excision is completely filled in this way, there still remains a nick (bounded by a 3'OH group on one side and a 5' phosphate group on the other) that can be covalently sealed by the enzyme DNA ligase.

In replication of double-stranded DNA only one of the two daughter strands can be made by such continuous extension of the primers formed at the origin; this is because polymerases can covalently join nucleotides only to the 3'OH end of the primers, and because of the antiparallel arrangement of the two parental strands and of the complementary daughter strands. Therefore the other daughter strand is synthesized discontinuously, new RNA primers being made at intervals as the replication fork advances and

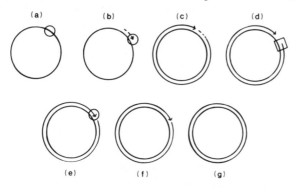

Fig. 7.3. Replication of the complementary strand in a circular single-stranded virus. (a) The origin of replication is recognized either by a primosome (in polyhedral phages) or by RNA polymerase (in filamentous phages). (b) The RNA primer is synthesized either by the RNA polymerase or by the primosome-directed DNA primase. (c) DNA polymerase extends the primer with new DNA until it meets the tail of the primer. (d) An exonuclease removes the primer. (e) A repair DNA polymerase fills in the resulting gap. (f) A DNA ligase covalently seals the remaining nick yielding the double-stranded replicative intermediate (g), the new complementary strand of which serves as a template for the replication of more viral strands identical to the original one.

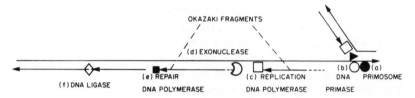

Fig. 7.4. The mechanism of discontinuous DNA replication of the lagging daughter strand. (a) Primosome recognition sequences, occurring at intervals along the DNA, are periodically exposed by the advance of the replication fork by the DNA helicase. (b) The binding of the primosome to its recognition site stimulates the DNA primase to make an RNA (or RNA/DNA) oligonucleotide primer (dashed lines). (c) The replication DNA polymerase covalently attaches deoxynucleotides to the primer's 3'OH end to form a polynucleotide, known as an Okazaki fragment, that grows away from the replication fork. (d) A repair exonuclease removes the RNA primer from the 5' end of the Okazaki fragment. (e) A repair DNA polymerase elongates its 3'OH end until it meets the 5' P tail of the preceding fragment from which the RNA primer has by then been removed. (f) The nick that remains is covalently sealed by the DNA ligase to make a continuous daughter DNA strand.

exposes more single-stranded DNA able to serve as template (Fig. 7.4). These additional primers are elongated by the replisomal DNA polymerase in the direction opposite to the movement of the replication fork, thus forming short polynucleotides called Okazaki fragments. The 5' RNA primers are rapidly excised by a repair exonuclease, the resulting gap is filled by nucleotides that become covalently joined to the 3'OH of the adjacent daughter strand by a repair DNA polymerase and the remaining nick sealed by DNA ligase. The Okazaki fragments, which are about 1000 base pairs in length in bacteria (i.e. about equal to one gene) but only about 200 base pairs in eukaryotes (i.e. about equal to one nucleosome), are thus joined together within a few seconds to yield much longer daughter DNA strands. At each replication fork the strand that grows in this discontinuous fashion is referred to as the lagging strand, in contrast to its continuously synthesized sister—the leading strand. However, it must be remembered that in bidirectional replication one end of each daughter strand is synthesized continuously at one replication fork while the other end is being synthesized discontinuously at the sister replication fork.

INITIATION OF REPLICATION AND ITS CONTROL IN BACTERIA

In single-stranded phages (e.g. M13, ϕ X174) initiation consists simply in the processes discussed above that lead to the synthesis of an RNA primer at the origin. The enzymes responsible are already present in the cell before infection by the phage, and presumably can bind to the phage DNA and initiate replication as soon as it enters the cell. In the double-stranded main chromosome of *E. coli* and its plasmids, however, initiation is more tightly controlled so as to allow one replication, no more and no less, every cell cycle. Initiation in *E. coli* requires a protein coded by gene *dna A* that maps close to the origin. Although the *E. coli* origin, *ori C*, has been cloned into a plasmid and some progress made with an *in vitro* system for its replication, the mechanism of initiation is still not understood. Current ideas of how it may occur are based on physiological experiments and on the study of initiation in plasmids.

Physiological studies show that normal initiation in *E. coli* requires the *de novo* synthesis of at least two proteins (one might be the *dna A* protein) and of polyunsaturated fatty acids that probably form part of the specific membrane site to which the new replisomes become attached. Transcription by RNA polymerase is also necessary even after the last time at which protein synthesis is needed: clearly the transcribed RNA is not needed as a messenger. One possibility is that it serves as an RNA primer as in initiation of phage M13. This is unlikely since it is not clear how the DNA polymerase could extend such a primer, because, unlike in M13 the region ahead of it would be double-stranded and not single-stranded. Single-stranded template DNA adjacent to the 3'OH end of an RNA primer seems essential for the action of DNA polymerases: if this were not so they might 'mistakenly' add DNA to the 3' end of any newly made mRNA, tRNA or rRNA prior to their displacement from the template by reassociation of the parental DNA strands, with lethally disastrous results for the cell.

It therefore seems probable that transcription *per se* at the origin, not the product of transcription, is the essential factor. The sequence of the *E. coli* origin is such that if one DNA strand were transcribed the complementary displaced and non-transcribed DNA strand would be able to fold up in a complex secondary structure with numerous hairpins and several short single-stranded regions. The creation of a specific secondary structure could serve as a signal for initiation proteins, or provide single-stranded regions suitable as substrates for DNA helicases that could cause further unwinding and generate two replication forks. Binding of SSBs to the newly exposed single-stranded DNA would stabilize them and create a substrate for the prepriming reactions that attach the primosome to the origin. It is noteworthy that all bacterial, plasmid, and viral replicon origins so far sequenced have regions capable of forming hairpin loops (this is apparently not the case for eukaryotic origins: Broach *et al.*, 1983). Moreover, there is evidence in several plasmids (see below) for transcriptional activation of this sort as the effective initiator of replication. The *E. coli* origin contains two promoters of transcription that are arranged back-to-back; this arrangement would be ideal for promoting the synthesis of RNA primers for the two leading strands (one for each replication fork), apart from the 'changeover' problem mentioned above.

If replication is initiated by specific transcription at the origin then the problem of the control of replication amounts to the control of this specific transcription. From the beginning, Jacob *et al.* (1964), who first proposed the idea of a replicon, drew a parallel with the operon model for the control of transcription. Because of the behaviour of certain F′ plasmids they proposed a positive control with an initiator protein that could bind to a specific *replicator* sequence at the origin, and so switch on replication. Subsequent studies on the timing of replication in the cell cycle of bacteria growing at different rates, and of the concomitant changes in mean cell mass, allowed Donachie (1968) to refine this model. He pointed out that even under these varying conditions initiation occurred whenever cells reached integral multiples of a basic cell mass. He showed that this would automatically occur if the initiator protein was synthesized as a constant fraction of total cell mass, and if a certain fixed number of initiator molecules had to bind to the replicator to initiate a round of replication. If N_o is the number of replicon origins per cell and N_i is the number of initiators that must bind to each of them to initiate replication, then another round of replication cannot begin until a further $N=N_oN_i$ initiator molecules are synthesized. Sompayrac and Maaløe (1973) proposed an autorepressor model that would enable initiators to be synthesized by the cell as a constant fraction of total cell mass; computer simulation shows that a model like this can fit the experimental data (Margalit *et al.*, 1984).

An alternative model for the coupling of the rate of replication to variable cell growth rates is that of Pritchard *et al.* (1969), who proposed a negative control by a repressor of initiation which bound reversibly to the origin and was diluted out by cell growth until it fell below a critical concentration at which an initiator that was always present could switch on replication. An attractive feature of their repressor model is that it can explain the phenomenon of plasmid incompatibility: related plasmids would have a similar repressor and so inhibit each other's replication but unrelated plasmids would have a different repressor and so be able to coexist in a cell. If a constitutively made positive initiator is also needed the model is consistent with Jacob *et al.*'s early observations on F′ plasmids. An essential feature of the Pritchard model is that repressor synthesis depends on the number of replicon origins in the cell, so that when a replication doubles the number of origins it also increases the concentration of repressor, thus preventing another round of replication until cell volume has doubled; this can occur if (a) the repressor gene is located at the replicon origin and (b) either the gene is transcribed only at the time of its replication and a constant number of transcripts and of protein molecules are made per gene, or the repressor is unstable and its synthesis rate is constant per gene; doubling the number of origins would double the rate of synthesis without affecting the rate of breakdown and therefore increase the concentration of repressor.

A control mechanism remarkably similar to the Pritchard model has been found in certain plasmids. The main difference is that the repressors and initiators are RNA rather than protein molecules. One reason for this may be that control of the number per cell of a particular molecule should be easier and more precise for RNA than for proteins because there is no second translation step to introduce further statistical variation (Cavalier-Smith, 1978a).

Plasmids of *E. coli* are excellent tools for studying control of replication during the cell cycle because, like the main chromosome, they are maintained in constant numbers per cell under standard conditions of growth. Since the genes controlling the rate of their replication are located on the plasmid it is relatively easy to isolate mutants in which the copy number (number of plasmids per cell divided by the number of main chromosomes) is altered. Wild-type plasmids fall into two categories:

1. low copy number plasmids whose numbers are stringently controlled at 1–2 per main chromosome;
2. high copy number plasmids with often a copy number of about 20.

Mutations can be isolated that convert low copy number plasmids to high copy number plasmids or that increase the number of copies of the latter. The simplest cause of such mutants would be a defective repressor. This has been shown to be the case in plasmids R1 and Col El.

In R1 a small RNA molecule coded by a gene *cop A*, located near the plasmid origin, acts as a replication inhibitor (Stougaard *et al.*, 1981). The *cop A* RNA is about 80 nucleotides long and from the DNA sequence coding it appears to have one short and one long hairpin loop and to be non-translatable. It is unstable, with a half-life of a few minutes, and so could act as the unstable repressor of Pritchard's inhibitor-dilution model for replication control. A point mutation in the $RNA_{cop\ A}$ promoter prevents its synthesis and leads to an increase in its copy number. Three mutations that abolish *cop A* RNA's repressor activity map in or close to the single-stranded loop at the end of the large hairpin, suggesting that this region is important for this activity.

Studies on a mutant of plasmid Col El with a copy number of 200–300 instead of the usual 10–15 suggest how this may occur (Muesing, *et al.* 1981). This mutation is a single base-pair GC TA transversion that maps about 400 base pairs from the origin of replication, and alters the nucleotide sequence of two different RNA molecules transcribed from opposite strands of the DNA. One of these RNAs is a precursor of the primer RNA required for initiation of DNA replication *in vitro*: the other is a small non-translated RNA called RNA 1. RNA 1 has 110 nucleotide residues and is probably folded into three hairpins, and the mutation is in the stem of the third hairpin which would considerably reduce its stability. The primer RNA is produced *in vitro* from the primer precursor by partial digestion by RNAse H which is specific for RNA base paired to DNA. Since addition of RNA 1 to the reaction mixture inhibits primer formation, this may be its role *in vivo*. Because it is complementary to the primer it might do this by forming an RNA/RNA complex with it or by binding to the complementary strand at the origin.

IMPORTANCE FOR GENOME SIZE EVOLUTION OF THE DIFFERENT REPLICATION PATTERNS OF BACTERIA AND EUKARYOTES

I have argued (Cavalier-Smith, 1981) that the fundamental reason why eukaryotes can have up to 100 000 times as much DNA per genome as bacteria is that eukaryote chromosomes consist of numerous separate units of replication. As each replicon is far shorter than a prokaryote chromosome it can be replicated in much less time, even though the rate of fork movement is much lower. This slower movement in eukaryotes is usually attributed to the slowing down of parental strand separation by the presence of the nucleosomal histones. Simultaneous replication of numerous replicons allows the whole genome to be replicated in a relatively short period of the cell cycle, the S-phase, irrespective of the total amount of DNA. Genome size can be increased indefinitely simply by increasing the number of replicons which, in principle, need not lead to any increase in the length of S-phase or of the whole cell cycle. Measurements of replicon length in eukaryotes of widely differing genome size (Figs. 7.5 and 7.6) show that the large interspecific variation in eukaryote genome size evolved mainly by altering the number of replicons, changes in replicon length being of only minor importance.

Although simultaneous replication of all replicons is possible in principle, eukaryotic S-phases are in general much longer than would be the case with synchronous replication of all replicons at the observed rates of fork movement. Therefore, different replicons are replicated at different stages in S-phase. A variety of labelling studies have shown that sequences that are replicated at a particular time in one S-phase are replicated at a similar time in subsequent S-phases (Barlow, 1972), so there is a regular program of replication such that at any one time during S-phase only a particular subset of replicons is undergoing replication. The overall time taken for S-phase will depend on the number of sequential subsets of replicons in the genome and the length of the replicons in each.

S-phase is always temporally separate from mitosis and therefore occupies only part of the eukaryote cell cycle. This markedly contrasts with rapidly growing bacteria where chromosome replication occurs continuously throughout the cell cycle and can occur simultaneously with segregation. A consequence of this basic difference is that in normal eukaryote cell cycles successive rounds of DNA replication never overlap: each DNA replication

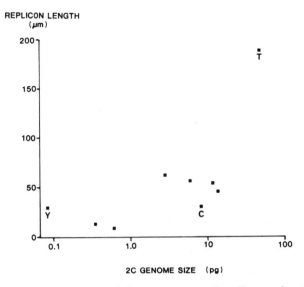

Fig. 7.5. Replicon lengths in relation to genome size. Except for the very large genome of *Triturus* (T) which does have distinctly larger replicons (188 μm) than other eukaryotes (mean 38 μm) there is no clear overall tendency for larger genomes to have larger replicons. Note that yeast (Y) and Chinese hamster (C) which differ about 100-fold in genome size have the same replicon size. Yeast data from Carter (1979), other data from Buongiorno-Nardelli *et al.* (1982).

is terminated and followed by mitosis before the next round is initiated. Cell fusion studies using cells in different phases of the cell cycle show that after the beginning of S-phase reinitiation is blocked until after the next division. Conversely, since inhibition of DNA replication prevents subsequent cell division it follows that cell division can occur only if replication is completed. Some mechanism must exist to ensure that all replicons have finished replication before division is allowed.

S-phase usually occupies only a fraction of interphase, being followed by a period with no DNA replication (G$_2$). The G$_2$ period is a period of preparation for cell division comparable to the D period in the bacterial cell cycle; it is often relatively constant in length in any particular species and is absent only in the polyploid macronuclei of certain large heterokaryote protozoa, whose micronuclei have a normal G$_2$ period. S-phase may begin either immediately after mitosis or more usually after a period of cell growth, the G$_1$ phase.

Since it has been proposed that the mechanism of initiation of S-phase and the time taken for S-phase may be directly connected with genome size both topics will be discussed in detail. A basic question is whether DNA replication in eukaryotes is controlled by means of sequence-specific origins, as in proka

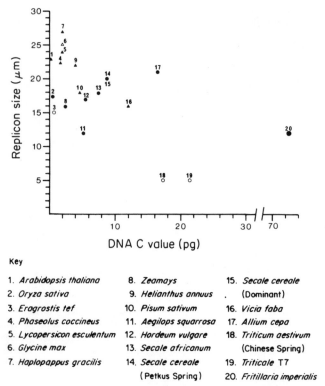

Key

1. *Arabidopsis thaliana*	8. *Zea mays*	15. *Secale cereale*
2. *Oryza sativa*	9. *Helianthus annuus*	. (Dominant)
3. *Eragrostis tef*	10. *Pisum sativum*	16. *Vicia faba*
4. *Phaseolus coccineus*	11. *Aegilops squarrosa*	17. *Allium cepa*
5. *Lycopersicon esculentum*	12. *Hordeum vulgare*	18. *Triticum aestivum*
6. *Glycine max*	13. *Secale africanum*	(Chinese Spring)
7. *Haplopappus gracilis*	14. *Secale cereale*	19. *Triticale* T7
	(Petkus Spring)	20. *Fritillaria imperialis*

Fig. 7.6. Independence of replicon size in root meristem cells and genome size in 19 species of flowering plants differing up to 362-fold in genome size. ▲ Diploid dicotyledons; △ tetraploid dicotyledon; ● diploid monocotyledons; ○ polyploid monocotyledons. From Francis, Kidd and Bennett (1985) by permission of Cambridge University Press.

ryotes, or by a more generalized mechanism that is independent of specific sequences (Harland, 1981). The latter has been suggested because any DNA sequences injected into *Xenopus* eggs can be replicated with the same temporal control as are the *Xenopus* chromosomes during embryonic cleavage.

Since the separate replication units in eukaryotes were initially detected by autoradiography and electron microscopy and were called replicons in the absence of direct evidence that they have unique sequence-specific origins, it is first necessary to outline the evidence for specific origins.

EVIDENCE FOR SPECIFIC REPLICON ORIGINS IN EUKARYOTES

Replicon origins have been shown by electron microscopy of replication bubbles to be located at specific points on cloned ribosomal DNA (rDNA) replicons in the sea urchin *Lytechinus* (Botchan and Dayton, 1982) and in the frog *Xenopus* (Hines and Benbow, 1982). The same is true of the nuclear 2 μm circular plasmid found in many yeast strains (Gunge, 1983). Compelling evidence that this is generally true for eukaryotic DNA comes from recombinant DNA studies of the replication of chimaeric plasmids in yeast.

Plasmids from *E. coli* can be used to transform yeast cells, but cannot replicate autonomously in them. They can be replicated by yeast enzymes only if they become integrated into a yeast chromosome (i.e. become linked to a yeast replicator). However, chimaeric plasmids constructed by inserting fragments of the yeast 2 μm plasmid into an *E. coli* plasmid are able to replicate autonomously (specifically in early S-phase) if, and only if, the fragments contain the 2 μm replication origin ('replicator'). This provides a functional test for yeast replicator sequences. When non-plasmid yeast nuclear DNA is broken into lengths that are short compared with the length of a replicon, and the fragments are inserted into an *E. coli* plasmid, the majority of them do not enable it to replicate autonomously in yeast, and therefore do not contain replicators. But a certain fraction of yeast DNA fragments do facilitate autonomous replication; they are called *ars* (short for autonomously replicating sequences), and presumably contain specific yeast replicators since there are about 400 *ars* and about 400 replicons per yeast genome. Some *ars* sequences exist in multiple copies in the yeast genome, whereas others are unique. Most *ars* sequences show short regions (11 bp) of homology with the 2 μm yeast plasmid replicator and with each other (Broach *et al.*, 1983), but at least one shows no such homology (Gunge, 1983); they differ in transformation efficiency, presumably as a result of evolutionary divergence of their replicator regions.

DNA from a wide variety of other eukaryotes, whether fungi, protozoa, animals or plants, contains *ars* sequences that can be replicated in yeast (Gunge, 1983); some of these are known to be replicon origins, whereas certain other known origins do not function as *ars* in yeast. If the DNA fraction that replicates in the first 1% of S-phase of the highly synchronous cell cycle of the slime mould *Physarum* is used, about 75% of them confer autonomous replication on the yeast plasmid (Waterborg and Shall, 1985). Since these DNA fragments must be highly enriched in replicon origins this, in conjunction with the low incidence of autonomous replication when total DNA is used, clearly shows that specific DNA sequences are important for the initiation of eukaryote DNA replication in normal cell growth–division cycles. These facts, and sequence studies of known eukaryote origins, show not only that there must be considerable conservation in the structure and properties of many eukaryote replicators, but also that marked divergences have also occurred and there cannot be a single category of identical replicators.

TEMPORAL CONTROL OF EUKARYOTE DNA REPLICATION

Several authors have suggested that the temporal control of replication is fundamentally the same in eukaryotes as in prokaryotes (Cavalier-Smith, 1978b, 1980; Cooper, 1979). However, in eukaryotes a fundamental distinction must be made between replicon initiation and the initiation of S-phase. There is no reason to suppose that the sequential initiation of different replicons is mediated by the same mechanism as the initiation of S-phase, and several reasons for thinking that it is not. Understanding both mechanisms is essential for a proper understanding of the evolution of genome size. Let me first consider the initiation of S-phase.

Cell volume and the initiation of S-phase

Two very different situations must be distinguished: (1) typical binary fission cell cycles where DNA replication is preceded and accompanied by cell growth, (2) multiple fission cell cycles, like those during blastula formation in animals and multiple fission in protozoa and algae, where growth does not occur but replication and division repeatedly and rapidly alternate so as to divide one large cell into numerous much smaller ones. Though no growth occurs during multiple fission, it is always preceded by an extensive growth period during which DNA replication does not occur. The basic difference between binary and multiple fission can simply be attributed to a difference between the control of S-phase initiation. In binary fission *each* S-phase is preceded by a period of growth and is triggered by the growth of the cell above a critical volume (Fantes and Nurse, 1981), whereas in multiple fission successive S-phases are not triggered by growth but continue repeatedly until the volume of the resulting daughter cells falls below a critical cell volume, after which no further replication or division occurs until after growth resumes (Craigie and Cavalier-Smith, 1982). A unified model for both binary fission and multiple fission cell cycles (Cavalier-Smith, 1985b) takes into account both the similarities and differences between them.

It has five basic postulates:

1. Chromatin can be switched between two states: the G-state which cannot be replicated and the S-state which can.
2. Replication automatically switches it from the S-state to the G-state so that it cannot be replicated until it is switched back to the S-state.
3. The G to S switch occurs only if (a) mitosis has occurred *and* (b) the DNA/cell volume ratio is below a certain threshold. In binary fission cell cycles with a G_1 this threshold is reached only by growth to a critical volume, but in multiple fission cell cycles and binary fission cycles lacking a G_1 (e.g. the mycetozoan *Physarum*) daughter cells are already above the critical volume.

4. During the growth phase that precedes multiple fission an additional block is imposed on the G to S switch. The stimuli that relieve this block may differ in different types of multiple fission; in animal eggs it is by activation, whether artificial or by natural fertilization, while in *Chlamydomonas* a light- or energy-dependent process appears to be involved (Craigie and Cavalier-Smith, 1982; Cavalier-Smith, 1985b).

5. A round of multiple-fission cell cycles lacking a G_1 will cease as soon as the DNA/cell volume ratio rises above the critical level and a G_1 phase will then become apparent.

The important feature of this model in relation to genome size evolution is the role of the DNA/cell volume ratio and its precise nature. Whether this control actually operates in any particular cell cycle depends on the actual volume of the cell; where genetically programmed developmental factors, as in animal eggs, or environmental factors such as certain growth conditions for yeast (Fantes, 1977) cause cells to grow to such an extent that daughter cells are already above the threshold volume this volume-dependent control is cryptic. It is reasonable to infer that such control is cryptic in cell cycles that lack a G_1 phase, where S-phase automatically follows mitosis. But in most eukaryotes cell cycles exhibit a G_1 during which growth occurs and the volume-dependent mechanism would operate.

There is also much evidence for the view (Van't Hof and Kovacs, 1972; Bryant, 1976; Cavalier-Smith, 1978, 1980, 1985b; Cooper, 1979; Lloyd *et al.*, 1982) that the G_1/S transition is the major control point in the cell cycle and that once cells have completed DNA synthesis they usually proceed automatically to divide after a period of time, the G_2 phase, that is relatively constant for any particular cell type. The cell volume at the time of division will depend on the volume at the G_1/S transition and on the amount of growth during the $S+G_2$ phase; since in binary fission cell cycles the increase during this period will be <2 times, it follows that the million-fold variation in the volume of proliferating eukaryote cells (see Chapter 4) must be chiefly determined by the nature of the DNA/volume control over the G_1/S phase transition. Therefore, evolutionary changes in proliferating cell volume must be brought about primarily by mutations that affect this mechanism.

The fundamental question therefore is whether this control mechanism depends on the total amount of DNA *per se* or on one or more specific genes: in either case replication will double the amount of DNA in question and therefore raise its concentration above the critical threshold and so prevent a further round of replication until the cell volume has doubled. But the implications for genome evolution are radically different for these two variants of the basic model.

If it is the total amount of DNA that is important, and in effect titrated against cell volume, then the total amount of DNA would be a major determinant of cell volume. Cell volume could only increase or decrease in evolution by changing the C-value or by changing the nature or concentration of the titrating molecule(s) that acted as the link between DNA amount and cell volume (models for such mechanisms are discussed in detail in standard works on the cell cycle such as John (1981), Lloyd *et al.* (1982); perhaps the most plausible is the unstable activator model of Wheals and Silverman (1982), where the rate of degradation of the activator is proportional to the number of genome equivalents per cell — this mechanism might be either genic or nucleotypic. Some such mechanism therefore could provide a very simple explanation for the existence of the direct proportionality between C-value and genome size discussed in Chapter 4; such a correlation would inevitably result if evolutionary changes in the cell volume at the G_1/S phase transition were always caused by changing the C-value, whereas changes in the nature or concentration of the titrating molecule did not occur. As Nurse points out (Chapter 5) such a direct nucleotypic control over cell volume cannot be ruled out. But note that the quantitative predictions of different nucleotypic models differ, and that only some would fit the actual genome size data (see Chapter 4).

However, the existence of a universal, unalterable DNA/cell volume ratio at the G_1/S transition, and therefore of such nucleotypic control of cell volume as the sole explanation of the correlation between C-value and cell volume, is called into question by the order of magnitude lower DNA/volume ratio in unicellular eukaryotes compared with multicellular ones as shown in Fig. 4.2. Moreover, in flowering plants vascular cambial meristem cells are an order of magnitude larger than apical meristem cells, even though both cell types have the same diploid DNA content and both are actively proliferating. Though not disproving the nucleotypic hypothesis of cell volume control by C-value, this shows that mutations can alter the DNA critical cell volume ratio at least 10-fold; there is no obvious reason why a hundred-, thousand- or even million-fold variation in this ratio could not occur if there were sufficient selective advantage for it, if so there would be no *mechanistically* necessary reason connected with the volume-dependent control of the G_1/S phase transition why cell volume has to be as closely correlated with C-value as is in fact observed.

A second reason for not previously favouring the nucleotypic hypothesis is that it is most economical to postulate that the volume-dependent control of replication is fundamentally similar in bacteria and eukaryotes (Cavalier-Smith, 1978b). In bacteria, as discussed in an earlier section, this mechanism apparently depends not on molecules that titrate the total cellular DNA content but on those that interact specifically with the replicon origin. I therefore postulated that a similar mechanism operated in eukaryotes (Cavalier-Smith, 1978b, 1980) and that the controlling molecules interacted not just with a single replicon origin (and its daughter copies), as in bacteria, but with the hundreds or thousands of replicon origins in eukaryote euchromatin in such a way that increases in replicon number would cause corresponding increases in cell volume. However, two objections to this theory suggest that it was oversimplified and must be modified. Firstly, it implies that most replicon origins are involved in the control of the G_1/S phase transition, whereas the evidence reviewed above shows that in most cell types only a minority of replicons actually initiate replication at that time; this objection, however, is not compelling since one could argue that all replicons are switched to the S-state at the G_1 transition by replicon-origin-titrating molecules, but that only some of them can be replicated at any one time because, as argued above, the number of replisomes is much smaller than the number of replicon origins. The second objection, pointed out to me by R. A. Craigie, is more compelling. This is that if the G_1/S switch involved the independent titration of a large number of identical replicon origins by identical initiator and/or repressor molecules it is hard to see how they could be synchronously switched to the S-state; it would appear to be inevitable that some by chance would either accumulate initiators or lose repressors long before others and the entry into S-phase of different replicons would follow multi-hit, rather than single-hit kinetics. Though a rigorous study of the kinetics of initiation of the first batch of replicons to be replicated at the G_1/S transition appears not to have been made, the data that are available suggest that it is more synchronous than would be expected if the several hundred replicons were independently controlled.

I therefore now suggest that in eukaryotes, unlike bacteria, S-phase initiation and replicon initiation involve fundamentally distinct mechanisms and that only S-phase initiation involves volume-dependent controls. There are two contrasting types of explanation for such volume-dependent controls, which have fundamentally different implications for the evolution of genome size. First, consider genic control mechanisms in which a single master gene per genome (i.e. one in haploid cells; two in diploid cells) is involved in the volume-dependent control of S-phase initiation: its activation would cause the production of one or more molecules that synchronously convert all the replicons into the S state so that replication can begin. The volume-dependent sensing component of this control mechanism could be fundamentally the same as that in the bacterial cell cycle and conserved during the prokaryote–eukaryote transition. The synchronous switching of the replicons to the S-state might also involve a mechanism found in bacteria, for example DNA methylation, or might involve new mechanisms not found in bacteria such as modifications of histones and/or conformational changes in chromatin. What is important for the present subject is not the exact mechanism but the fact that this type of genic control involves neither the titration of total DNA amounts, nor of large numbers of replicon origins, and therefore there would be no necessary connection between total DNA amounts or replicon numbers and cell volume. In this case the correlation between cell volume and C-values would have nothing to do with the control of the initiation DNA replication but must have an indirect evolutionary explanation as discussed in Chapter 4.

By contrast, with nucleotypic control, involving the titration of a controlling molecule by the total amount of DNA, any mutation changing genome size would proportionally change cell volume. Newport and Kirschner (1982a, b) have recently provided persuasive experimental evidence for such nucleotypic control over the timing of the midblastula transition in *Xenopus* embryos; the midblastula transition is the time of changeover from the rapid cleavage cell cycles without a G_1 or G_2 to move typical cell cycles with both G_1 and G_2 and a longer S, as well as of the switching on of transcription by RNA polymerase II and III. By means of polyspermy and partial egg constriction experiments, Newport and Kirschner (1982a) confirmed earlier evidence that the transition occurs when the nucleus/cytoplasmic volume ratio rises above a critical level. That the important factor was not nuclear volume *per se*, but total DNA amount, was shown by microinjection of foreign DNA of the bacterial plasmid pBR 322 which caused initiation of transcription. Newport and Kirschner (1982b) did not determine whether or not the injected DNA also induced the cell cycle changes observed at the midblastula transition; if future work shows that it does, this would be the first direct demonstration that the DNA amount/cell volume ratio can actually control the initiation of replication. If also the cessation of a round of G_1-less multiple fission is controlled by exactly the same DNA/cell volume titration mechanism as initiates the G_1 to S-phase transition in normal binary fission cell cycles (Cavalier-Smith, 1985), then nucleotypic control over S-phase initiation would be general in eukaryotes, unlike in bacteria (Harland, 1981); but would still not solve the C-value paradox (p. 118).

Genome size and the length of S-phase

It has been shown in flowering plants that the length of S-phase varies systematically with genome size. As Fig. 7.7 indicates, larger genomes tend to have longer S-phases. What are the reasons, both mechanistic and evolutionary, for this? It has frequently been suggested or implied (e.g. Van't Hof and Sparrow, 1963; Van't Hof, 1965; Yeoman, 1981) that the duration of S-phase determines the overall cell cycle time and therefore may be evolutionarily important because of this. However, the balance of evidence argues against this. More extensive evidence for 27 different plant species (Van't Hof, 1974) shows that the regression lines for the relationship between mitotic cycle time and genome size and for the duration of S-phase and genome size are not parallel as first suggested (Van't Hof, 1965); the mitotic cycle lengthens with increasing genome size twice as much as the length of

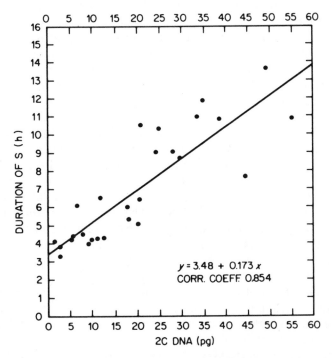

Fig. 7.7. The duration of chromosomal DNA synthesis (S) of root meristem cells of 27 different plant species in relation to their nuclear DNA content. The line is for the regression line $y = 3.48 + 0.173 x$ (correlation coefficient 0.854). Reprinted from Van't Hof (1974) by permission of Plenum Press.

S-phase, showing that other cell cycle phases (mainly G_1) are also longer: therefore the extra cell cycle length is not simply the result of a longer S.

The dispensability of the G_1 period and the exceptional shortness of S-phases in animal embryos show that the length of typical cell cycles with a G_1 is determined primarily by the growth rate of the cell, not by the DNA synthesis rate. The only case where the length of S-phase probably does directly limit the length of the cell cycle is in many cleaving animal embryos where there is no cell growth; in these atypical cell cycles S-phases can be a hundred times shorter than in ordinary somatic cell cycles and take up most, or perhaps sometimes all, of interphase. Since the amount of DNA to be replicated is the same as in typical somatic cell cycles this shows that genome size *per se* does not fundamentally and unavoidably limit the overall rate of replication; the idea that having large genomes directly prevents rapid multiplication just because of the extra DNA to be replicated is therefore untenable. It is more reasonable to suppose that the rate of replication depends simply on the relative amount of the replication machinery compared with the genome size.

Let us consider such a model for the duration of S-phase. The rate of replication will depend simply on the number of replication forks at any instant plus the average rate of travel at each fork. If we suppose that a specific replisome is always associated with each fork then the number of forks that can exist in a nucleus will depend simply on the number of replisomes. One may also suppose that the fork travel rate, which is known to vary several fold, will depend primarily on the concentration of soluble cofactors and/or substrates such as nucleoside triphosphates. The rate of fork movement either remains constant throughout S-phase or increases about 3-fold in early S-phase, perhaps as a result of the accumulation of such soluble factors, and is thereafter constant, in both plants (Van't Hof, 1976), and animals (Kapp and Painter, 1982). The constancy of fork movement rates during most or all of S-phase suggests that precursor concentrations are not rate limiting and that each replisome is functioning at its maximal rate. The linear accumulation of DNA during S-phase implies a fixed number of sites for DNA synthesis (Yeoman and Aitchison, 1976). I suggest that replication proteins are synthesized at the G_1/S transition so as to produce a certain number of replisomes that remains constant throughout S, and that the overall rate of synthesis is controlled simply by the number of replisomes, not the number of potential origins.

The rapid cell cycles lacking G_1 found in many early animal embryos could be sustained if the replication machinery was synthesized not during blastulation itself but in many-fold excess during the growth of the oocyte before fertilization. The increase in the length of S-phase during embryonic development can be accounted for by the reduction in replisome number/ DNA ratio that would result from the repeated DNA synthesis in the absence of (or in the presence of a low rate of) synthesis of replisome proteins. Conversely, the exceptionally low rate of DNA synthesis in premeiotic S-phase (Callan, 1972), could result from a lower number of replisomes being synthesized than in normal cell cycles. It is well known that these major developmental changes in DNA synthesis rates are caused not by changes in the rate of movement of individual replication forks but by changes in the total numbers of forks per nucleus. The reason for such variations in S-phase may simply be a different trade-off between selection for economy and for high speed of replication. Where growth rates are limiting, and there is no point in speeding up replication, economy in replication machinery wins out, making S-phase longer; since growth during premeiotic cell cycles takes far longer than in normal cell cycles greater economy is possible since S-phases can be greatly extended without affecting the overall rate of development. In early embryos, however, where overall growth is absent, replication rates do limit the developmental rate and therefore must be increased by having extra replisomes. It might be objected against this hypothesis that if, as I have suggested, mature oocytes have a massive store of replisomes, why do they not use them earlier to achieve a rapid premeiotic S-phase. However, premeiotic S-phase is completed early in oogenesis before the major period of growth during which one would expect the extra replisomes to be synthesized, and in any case, the activity of those extra replisomes must in some way be inhibited during oocyte growth in order to prevent DNA synthesis leading to polyploidy. Moreover, this objection could not apply to male premeiotic S-phase which is also much longer than in normal mitotic cycles.

It might also be argued that the lengthening of premeiotic S-phase is related not to the lengthening cell cycle but to the concomitant recombination. However, the lengthening is not a sudden event but occurs over several preceding cell cycles which get progressively longer as meiosis approaches (Bennett, 1973). This supports my view that the length of S-phase is simply adapted in the most economic way to the time available for it. As larger cells grow more slowly (for possible reasons see pp. 144–157), they can have longer S-phases. As they also have large genomes, genome size will appear to correlate with the length of S-phase even though it does not determine it.

THE ORGANIZATION, FUNCTION AND EVOLUTIONARY ORIGIN OF EUKARYOTE REPLICONS

The approximate constancy of replicon lengths in normal cell cycles despite large changes in genome size (Fig. 7.5) requires an evolutionary explanation. If replisomes could initiate replication with equal probability at any point on eukaryote DNA then replicon lengths would depend simply on the number of replisomes per nucleus; replisome numbers alone would need to be changed as different genome sizes evolved. However, the evidence discussed earlier in this chapter suggests that replisomes initiate synthesis only at specific replicon origins; therefore the number of specific replicon origins must increase and decrease approximately in step with evolutionary changes in genome size. This means that evolutionary changes in genome size must not be thought of simply as non-specific increases or decreases in any type of DNA sequence. Instead replicon size, and presumably some elements of replicon organization such as origin sequences, must be conserved during such changes: the simplest view of the mechanism of genome size changes consistent with this constraint is that they come about mainly by the duplication or deletion of whole replicons. Our understanding of eukaryote replicon organization is still so primitive that this section must be rather speculative.

However, there are indications that transcriptional activation at replicon origins may be important for replicon initiation both in bacteria and eukaryotes. I suggest that this will turn out to be the case, and that the basic difference between the control of replication in bacteria and eukaryotes lies in the dissociation in eukaryotes of the replicon initiation and S-phase initiation mechanism. However, in bacteria the two mechanisms are obligately coupled by linkage and interactions at the single origin of replication: a volume-dependent control mechanism involving binding of control molecules at the origin initiates a round of replication by leading to the transcriptional activation of the origin at an adjacent (or even overlapping site). When the eukaryote plural replicon mechanism evolved, the sites at which transcriptional activation occurred became highly multiplied, so as to produce numerous replicon origins; these origins would need to be capable of being controlled in *trans* by diffusible controlling molecules, such activation being controlled either by a cell-volume-dependent mechanism, either involving titration of a molecule by a single gene, perhaps directly derived from that of bacteria, or by a new sequence-independent nucleotypic titration mechanism depending on the total DNA content (as discussed on pp. 228–232).

This new hypothesis fits the facts reviewed in this chapter better than the previous suggestion (Cavalier-Smith, 1975, 1981, 1982) that *entire* replicon origins underwent a massive saltatory duplication during the origin or eukaryotes. The numerous sites of transcriptional activation might have originated by the duplication and transposition of the corresponding part of the prokaryote replication origin; but a perhaps simpler idea is that they evolved directly from the promoters of genes coding for mRNA, rRNA and tRNA. Perhaps in the first eukaryote, replication of each gene simply became coupled to transcription of that gene: while the template strand was being transcribed the complementary strand would automatically be exposed to allow the binding of a DNA primosome so as to initiate replication. The volume-dependent and mitosis-dependent G-state to S-state switch and the replication-dependent S to G switch discussed above would ensure that this happened only once per cell cycle.

Though I have suggested that in the original eukaryote there was a 1:1 relationship between gene promoters and replicon origins, this need no longer be universally true in modern eukaryotes. Clearly, as discussed in Chapter 3, species with very large genomes but with relatively few protein-coding or RNA-specifying genes must have many more replicon origins than either type of gene, whereas species like yeast have more genes than replicon origins. Such a dissociation between gene number and replicon number could have arisen secondarily by duplication, deletion and mutational divergence of promoters so as to make some promoters concerned specifically with gene transcription and others specifically with transcriptional activation of DNA replication.

Duplication of promoters is well shown in the spacer region of *Xenopus* rDNA which Moss and Birnsteil (1982) postulated to be a replicon origin since it contains two functional promoters for RNA polymerase I which might be used for the synthesis of an RNA primer; this is the region shown by electron microscopy (Hines and Benbow, 1982) actually to serve as a replicon origin; moreover 140 base pairs downstream is a sequence similar to the origins of replication of polyoma and SV40 viruses.

If replicon origin sequences are conserved they would constitute a class of repetitive DNA sequences that would be interspersed between other sequences at a mean spacing that corresponds with the spacing of replicon origins. I therefore proposed (Cavalier-Smith, 1978b) that a subfraction of interspersed middle repetitive DNA might consist of replicon origins, and that a subclass of the small nuclear RNAs (sn RNAs) are transcripts of replicon origins involved in the control of initiation of DNA replication. The finding that *Alu* sequences, which constitute a large fraction of mammalian middle repetitive DNA, have homologies with the replicon origins of SV40 virus (Jelinek *et al.*, 1980) that they are transcribed by RNA polymerase III which synthesizes uncapped sn RNAs and that some of these snRNAs have homologies with *Alu* sequences might seem to support this hypothesis. However, it is well known that retroviruses uses small RNA molecules as primers for DNA replication, and these facts are equally consistent with the alternative idea that these sequences are integrated proviruses (or former proviruses) that have spread by transposition (Cavalier-Smith, 1978b; Doolittle and Sapienza, 1980); the evidence that at least some *Alu* sequences are transposable elements is steadily increasing (see Chapter 15 by Doolittle; Jelinek and Haynes, 1983; Fox *et al.*, 1983).

Alu sequences are 20-fold more closely spaced than replicon origins in mammalian somatic cells: this could be reconciled with a function as cellular replicon origins only if most are inactive in any one cell cycle, or are mutant and non-functional pseudo-origins. If this were so, the close spacing could not be explained by supposing that most of them are needed only during rapid embryonic replication, since mammals are an exception to the general rule of exceptionally rapid S-phases during embryonic cleavage (Graham, 1973), perhaps because the protection of the cleaving egg from predators by its location in the oviduct removes the selective advantage of rapid development found in eggs that develop externally.

If eukaryote replicon origins are about 250 bp long, like *E. coli ori C*, only about 0.17% of mammalian DNA would consist of replicon origins, given a mean spacing of 150 kb for replicon origins. Although many fungi (Timberlake, 1978) may have as little dispersed repetitive DNA as this, in animals, plants and even protozoa (Allen *et al.*, 1975) this fraction of the genome is commonly 10–100 times as abundant. This quantitative argument supports my earlier suggestion that most middle-repetitive DNA in organisms with very large genomes consists of virus-like transposable genetic elements (Cavalier-Smith, 1978b, pp. 268–273) rather than serves as cellular replicon origins or for transcriptional control during cell differentiation: it also suggests that this is true for species with medium-sized genomes (Doolittle, Chapter 15).

The homologies between parts of the *Alu* sequences and viral origins of replication might simply reflect the fact that duplicative transposons require their own replicon origins; perhaps the pol III transcripts serve as primers for their duplicative transposition just as pol III transcripts serve as primers for retroviruses. Ultimately of course, the most likely source for such 'selfish' sequences would be by mutation from normal cellular components with which they may still share homology. Perhaps RNA polymerase III is also used either to make primers for the DNA replication or for the transcriptional activation of those cellular replicons containing genes that it transcribes (tRNA, 5sRNA and most snRNA).

The idea that in eukaryotes each unit of transcription is also a unit of replication, derived from studies of polytene chromosomes (Pelling, 1966; Rudkin, 1973; Whitehouse, 1978), can be extended by proposing that all three eukaryotic RNA polymerases (pol I, II and III) are each responsible for initiation of their own specific replicons, whether by primer synthesis or by transcriptional activation to expose a template for the attachment of a DNA primase; for any particular replicon the same type of polymerase would be responsible both for ordinary transcription and for initiation of replication. In addition to the evidence already cited for ribosomal genes, transcribed by pol I, that the replicon origin contains promoters for pol I, there are exciting indications of an intimate relationship between transcription and replication for the majority of genes, i.e. protein-coding genes transcribed by polymerase II: Seidman *et al.* (1979) have provided strong evidence (though not conclusive proof), in both SV40 virus and cultured chicken cells, that the leading strand of all replication forks is the coding strand for essentially all the stable nuclear RNA transcripts.

This remarkable finding suggests an obligate connection between the polarity of transcription and replication: *a priori* in the absence of such a connection one would have expected half the genes to be transcribed from leading strands and half from lagging strands. The simplest explanation is that every promoter for mRNA transcription by pol II also serves as a promoter for RNA primer transcription by pol II during DNA replication, and that this primer is used only for leading strand replication; the lagging strand must be primed by a different enzyme. As suggested above transcription of the leading strand template by RNA polymerase II would separate the lagging strand from the nucleosome and expose a primosome entry site like that for the bacterial *dna* B protein (Kornberg, 1980, 1982; Zipursky and Marians, 1981).

Further evidence that replicons are the basic units of interphase chromosomal organization in eukaryotes comes from the growing evidence that they correspond with the structural domains or supercoiled loops of chromatin seen by electron microscopy in both interphase (Cook and Brazell, 1975; Worcel and Benyajati, 1977) and mitosis (Marsden and Laemmli, 1979). Not only is there a close correspondence between the size of the loops and of replicons in a wide variety of eukaryotes (Buongiorno-Nardelli *et al.*, 1982), but replicon origins are specifically attached to the nuclear matrix in interphase in both S and G_2 (Wanka *et al.*, 1982) and replication forks are attached to the matrix during S (Berezney and Coffey, 1975; Pardoll *et al.*, 1980; Goldberg *et al.*, 1983).

THE TEMPORAL SEQUENCE OF REPLICON INITIATION

How is one to explain the reproducible sequence with which different replicons are initiated during S-phase in somatic cells? Barlow (1972) suggested: (1) that replication depended on initiator molecules, (2) that replicons differed in the number of copies of them needed to initiate replication, and (3) that those needing the fewest copies would initiate early in S-phase and those needing more copies would initiate later. However, the mechanistic basis for such a requirement for several initiators per replicon was not made clear, nor is it obvious that it would provide sufficient reproducibility from one cycle to the next since sometimes a replicon requiring several initiators could acquire them by chance at the same time as one requiring only a single initiator.

A simpler explanation for the temporal control is that it depends primarily on the relative strength of the transcriptional activation promoters, together with there being a limiting number per cell of RNA polymerases and primosomes. Promoters with the highest affinity for RNA polymerase would be replicated early, and those with the lowest affinity would be replicated later. The spread of affinities could simply have arisen through random genetic drift rather than by being positively selected in evolution.

The absence of such a temporal sequence in the egg cleavage S-phases of many (but *not* all) animals, such as *Drosophila*, sea urchins and *Xenopus*, would simply result from the vast excess of RNA polymerase and primosomes synthesized during oocyte growth, which would be sufficient to saturate even the weakest promoters and to ensure synchronous replication of all replicons. If cleavage replication used promoters too weak to be used in somatic cells the observed shorter origin to origin spacing would also result. Use of weak promoters for transcriptional activation of replication would also be facilitated by the suppression of transcription that is usual during cleavage and the consequent lack of competition between replication and transcription. In this connection it may be significant that, at least in sea urchins, a different set of histone genes are used during cleavage from those in somatic cells (Maxson *et al.*, 1983), which means that cleavage chromatin may have a structure that differs in significant details for somatic chromatin. Certainly it would be expected that such a separation in gene usage would allow somatic chromatin to become specialized for rapid transcription and cleavage chromatin for rapid replication: it is therefore possible that it is not simply the larger number of polymerases and primosomes but also a modified chromatin structure that leads to the rapid synchronous cleavage replications.

In somatic S-phase the observation that adjacent replicons usually replicate more or less synchronously has led to the proposal of a third level of replication control (Hand, 1980) intermediate between S-phase initiation and replicon initiation, i.e. at the level of clusters of replicons. One possibility for such control is a modification of the supercoiling or folding of a chromosome domain large enough to include several adjacent replicons. However, it is also possible that there is really no intermediate level of control at all: on my promoter-strength hypothesis of temporal control replicon clustering would occur simply if adjacent replicons tended to have promoters of similar strength, which might come about very simply if adjacent replicons have often arisen by tandem duplication. One would simply suppose either that this happened too recently for them to have diverged from each other as much as more distant ones, and/or that closely adjacent tandemly repeated replicon origins are more easily prevented by gene conversion (see Dyson and Sherratt, Chapter 12) from mutual divergence than are distantly linked ones.

A specialized situation is found in the macronuclei of hypotrich heterokaryote protozoa which contain many copies each of from 12–24 000 gene-sized DNA fragments instead of the ordinary chromosomes, which are largely destroyed during the chromatin diminution that occurs during macronuclear formation (see Ammermann, Chapter 14); in contrast to normal eukaryote cell cycles, specific fragments are not replicated in the same order during successive cell cycles. It has been proposed (Cavalier-Smith, 1978b) that this is because the specific sequences necessary for such control—possibly specific replicon origins—are among those lost during chromatin diminution. The mechanism that ensures that each fragment is replicated only once per cell cycle may be a non-sequence-specific one, akin to that postulated for cleaving animal eggs (Harland, 1981). More detailed study of DNA replication in hypotrich macronuclei would be very worth while.

Even heterokaryote protozoa whose macronuclei do not undergo massive chromatin diminution, e.g. *Tetrahymena*, replicate DNA of their micro- and macronuclei at different times during the cell cycle (Mitchison, 1971); this important exception to the rule that nuclei in the same cell replicate in synchrony deserves further study. In the heterokaryote macronucleus there is evidence for a volume-dependent control of the initiation of S-phase (Rasmussen and Berger, 1982, but see also Seyfert and Cleffmann, 1982). The high degree of polyploidy of the macronucleus might be expected to complicate the operation of a gene concentration control mechanism (unless in a diploid or polyploid all the copies of the controlling gene except one are inactivated, for example, by methylation?), but there would be no such problem if the mechanism were nucleotypic, which is possible (p. 178).

LATE-REPLICATION OF CONSTITUTIVE HETEROCHROMATIN

Constitutive heterochromatin and the satellite DNAs often associated with it, are especially prone to undergo large evolutionary changes in amount, and there is long-standing and extensive evidence that it is relatively inert genetically and lacking in protein-coding genes and major effects on the phenotype (for a thorough review see John and Miklos, 1979). How can one account for the distinctive replicative, staining and differential condensation properties of constitutive heterochromatin compared with euchromatin? It is not sufficient merely to label it 'inert', 'junk' or 'selfish' DNA, for these labels can apply equally to euchromatin as is well shown by the fact that B chromosomes may be either heterochromatic or euchromatic. The difference between constitutive heterochromatin and euchromatin are heritable and primarily determined by their sequence differences, though the differences in condensation are probably mediated through the binding of specific proteins.

There are two basic possibilities concerning the origin of constitutive heterochromatin. One is that it is basically homologous in all eukaryotes and arose early in eukaryote evolution (presumably in a protozoan since it is present in both animals and plants) and has been lost (or reduced to an undetectable level) in those animals, plants, fungi and protozoa that lack cytologically detectable constitutive heterochromatic. The second is that it has arisen independently in different groups and the common features such as allocycly, C-banding and late-replication are simple convergences. We shall be unable to choose between these possibilities until the molecular basis of constitutive heterochromatin is known and a more thorough study of its distribution made in protozoa. (By classical criteria most protozoa show no interphase condensed chromatin and therefore no heterochromatin. But in some, notably euglenids, the entire chromosome set is condensed and visible in interphase and might be considered to be heterochromatic: however, since these 'condensed' chromosomes are clearly not genetically inert this would be a misleading use of the term.)

We may also ask how the division between heterochromatin and euchromatin is related to the division of chromosomes into separate replicons. The under-replication of heterochromatin in *Drosophila* polytene chromosomes and its general late-replication strongly suggest that it represents a discrete class of replicons. Because of the late-replication of constitutive heterochromatin it has been suggested (Cavalier-Smith, 1978b) that its replication also is controlled independently of that of bulk euchromatin, even in non-polytene cells.

If, as I have argued, replicons are the basic functional units of eukaryote chromosome organization then each replicon must contain DNA sequences that label it euchromatic or heterochromatic. Differential intragenomic multiplication (by any mechanism including the activities of transposable elements) either of the two categories of replicon or of their 'label' sequences would lead to changes in the ratio of heterochromatin and euchromatin. The erratic distribution of heterochromatin in different species suggests that such an intragenomic flux may be largely uncoupled from organismic selection and therefore from classical ideas of function.

Even if the exact amount of heterochromatin may have no functional significance it does not follow that this is also true of the origin of the basic distinction between euchromatin and constitutive heterochromatin. If constitutive heterochromatinization depends not only on label sequences in the heterochromatic replicons but also on proteins synthesized by other replicons it follows that the origin, and maintenance of the proteins in the face of deleterious mutations, must depend on positive selection for those genes or the organisms bearing them. Since these genes must be transcribed and translated they are probably in euchromatic replicons: so what is their function? An intriguing possibility is that the proteins involved in constitutive and facultative heterochromatinization are basically the same and that the difference between them lies simply in the label sequences. The origin of heterochromatinization proteins and label sequences therefore could be explained in terms of the benefits of switching off one or more protein-coding genes at certain stages of the life cycle.

Replicons with heterochromatin label sequences could then spread by intragenomic duplication. They would inevitably diversify by mutation. Harmful mutant variants would be selected against by organismal stabilizing selection. But mutant replicons whose label sequences mutated to a form allowing permanent, constitutive heterochromatinization that prevented transcription throughout the life cycle, would be selected against much less strongly. Their effects on the phenotype would generally be less than for duplications of transcribed euchromatin, enabling them to multiply (or be deleted) more freely than euchromatic or facultatively heterochromatic sequences.

The preceding hypothesis seems compatible with the distribution of constitutive heterochromatin. It may be argued that the frequently observed influence of large blocks of heterochromatin on chiasma frequencies or distribution may be a simple mechanical side-effect; such effects do not explain why in some animals up to 65% of the DNA is constitutive heterochromatin whereas in other species none is, and there is no reason to think that they are adaptively significant or are evolutionary causes of the origin or maintenance of constitutive heterochromatin.

But why is heterochromatin so often late-replicating? It cannot be simply because heterochromatic replicons are so long that even if replication began at the beginning of S-phase it would not be completed till the end of S-phase, for in cleaving embryos even heterochromatin can be rapidly replicated. The fact that facultative heterochromatin is also late-replicating suggests that late-replication is determined not by the DNA sequence but by secondary modification (e.g. methylation) of the local chromatin structure. The regularity of late-replication suggests that there must be some underlying functional reason for it. Once again, however, this need not primarily concern the heterochromatin which might inevitably be 'parasitizing' a system having more basic significance. For example, it has often been suggested that the final splitting of the centromere might depend on the late replication of certain DNA sequences there. Although this cannot be true of the microtubule-binding sequences themselves that were discussed in Chapter 3 (since these are clearly double before splitting occurs), and although the centromere is by no means always visibly heterochromatic, it might conceivably always contain at least one heterochromatic-type replicon whose replication might necessarily follow that of bulk DNA and precede the onset of mitosis. As suggested in Chapter 3 much heterochromatin might arise by the multiplication of such replicons to the level where they become cytologically detectable.

The mechanism of late-replication of heterochromatin might be connected with its usual absence of transcription. If transcriptional activation of genes transcribed by RNA polymerase II commonly uses the same promoter for replication as for transcription then stabilizing selection will weed out mutations that make it weaker than necessary to allow the minimum frequency of transcription per cell cycle needed for that gene: presumably this would be at least once per cell cycle. But in the case of non-transcribed constitutive heterochromatin such selection could be much weaker. If it originally arose by the tandem duplication of promoter-containing DNA there would be no disadvantage in the *average* strength of promoters dropping below the level where each would be used once per cell cycle, since in a given length of such DNA containing several potential promoters it would not matter if only some of them were used in any one cell cycle. A relative weakness of promoters in constitutive heterochromatin compared with euchromatin might also explain its lack of replication in *Drosophila* polytene chromosomes. Alternatively, or in addition, late-replication might result from the tight folding of the heterochromatin and the relative inaccessibility of the promoter sites, which presupposes a specific mechanism of unfolding in late S-phase.

EVOLUTIONARY CHANGES IN GENOME SIZE IN RELATION TO
REPLICON ORGANIZATION

In most organisms it is not known how overall changes in genome size are
related to changes in the size or number of specific replicons. Such data are
available only for ribosomal RNA genes and polytene chromosomes.

Ribosomal RNA genes

The repeat size of the nucleolar tandemly repeated ribosomal genes varies
only about 5-fold in eukaryotes over the range 9–44 kb (Lewin, 1980). In
view of the evidence from electron microscopy (Botchan and Dayton, 1982;
Hines and Benbow, 1982) and from ribosomal gene amplification
(Macgregor, 1982) that each repeat unit is a separate replicon, it is significant
that the range in repeat size is within that for the average replicon lengths
for the same eukaryotes (Buongiorno-Nardelli *et al.*, 1982), which suggests
that the evolutionary constraints on replicon lengths are similar for all genes.
Part of this variation results from a 2-fold variation in length of the rRNA
transcript, and the rest from a 20-fold variation in the length of the non-
transcribed spacer. The exact length of the non-transcribed spacer is not
critically important since it varies even within a species. However, the overall
repeat length varies only 5-fold in eukaryotes in which genome size varies
about 400-fold, and there is no systematic correlation between rDNA replicon
length and genome size. Thus the variations in length of the non-transcribed
spacer are evolutionary 'noise' unrelated to the C-value paradox. On the
other hand, the 5000-fold variation in the numbers of rRNA replicons per
genome does show a highly significant positive correlation with genome size
(see Fig. 3.1); the reason for this is probably that larger cells require more
ribosomal genes than small ones to maintain a sufficient overall rate of
ribosome synthesis, and because large cells have large genomes (see Chapter
4). Ribosomal gene amplification in animal oocytes (Macgregor, 1982) is
an obvious adaptation to a temporary need for rapid rRNA synthesis in
exceptionally large cells (Gall, 1968): polyploid nurse cells are an alternative
way of achieving this. The fact that large-genomed higher plants have more
copies of rDNA per genome than animals of similar genome size may be
attributed to the absence of gene amplification or nurse cells despite the
need for large reproductive cells.

    The numbers of ribosomal genes are known to vary widely within a species
(see Cullis, Chapter 6), many of which seem to be polymorphic for ribosomal
gene number (Flavell *et al.*, 1983). This suggests that duplication and deletion
of whole rDNA replicons is very easy and frequent. The studies of the
artificial amplification of protein-coding genes reviewed by Cullis (Chapter
6) also suggest that replicon duplication is a very frequent mode of gene

duplication, since the duplicated unit is generally much larger than an individual protein-coding gene but the same order of magnitude as eukaryote replicons.

Polytene chromosomes

The classic studies of the polytene chromosomes of *Chironomus thummi* by Keyl (1965) strongly suggested that doubling of entire replicons caused the 40% increase in the genome size of subspecies *thummi* compared with subspecies *piger*. The extra DNA in *C. thummi thummi* is found in only certain polytene bands, each of which has 2, 4, 8 or 16 times as much DNA as the homologous band in *C. thummi piger*. Such a regular increase could be produced by errors in replication causing the successive doubling of the DNA content of each band, but not by duplicative transposition or unequal crossing-over, the other main possible methods of increase (Cavalier-Smith, 1978b; Dyson and Sherratt, Chapter 12). It is reasonable to treat this as a case of successive whole-replicon doubling, since each band is a distinct unit of replication and bands correspond in length with replicons in non-polytene cells. In *Drosophila* the DNA replications that cause polytenization start simultaneously in each band and continue for varying periods: bands with more DNA take longer to complete their replication. This suggests that each band has only one functional origin of replication.

Hägele (1976) studied the labelling frequency and calculated the time taken for the replication of homologous bands differing in DNA content in *C. thummi thummi* and *piger*, and found that bands take longer to replicate in *thummi*: doubling the DNA content doubles the time taken, but further doublings cause less increase in replication time. If replicon origins, rather than replisome numbers, control the frequency of initiation during polytenization as is implied by its synchrony, then the time taken to replicate an evolutionarily duplicated replicon will depend on whether or not the origin itself is duplicated (or whether both copies of a duplicated origin are subsequently retained). Hägele's results could be explained by supposing that for a single duplication there is no selective advantage in retaining a duplicated origin, whereas for higher levels of duplication selection favours the retention of more than one origin.

REPLICATION, INTRAGENOMIC SELECTION AND GENOME SIZE

The idea of intragenomic selection depends on the assumption that certain DNA sequences can mutate so as to be able systematically to replicate, either by normal replication or by duplicative transposition, at a greater rate than others in the same cell. In the former case this means that they must free themselves from the controls that normally ensure that all replicons replicate once and only once per cell cycle: therefore to understand how such 'selfish' DNA sequences might arise by mutation we need a thorough understanding of the nature of these controls.

Dawkins (1982) referred to such intragenomic competition between independently replicating DNA sequences as 'competition between replicators'. This terminology is potentially confusing since Dawkins' novel use of the word replicator is not the same as that used by students of DNA replication. Jacob *et al.* (1964) coined the term to refer to a specific region at the origin of the bacterial chromosomes to which initiator would be bound and which was essential for the initiation of replication; the rest of the DNA is replicated purely passively by the movement of the replication fork and is *not*, in the original sense, a replicator. Dawkins, however, uses replicator much more vaguely to mean 'anything in the universe, of which copies are made'. Dawkins' term 'active replicator' (one whose nature influences its probability of being copied) is closer in concept to the replicator *sensu* Jacob *et al.* The bulk of the DNA, which is replicated by virtue of being linked to replicator sequences, would fall into Dawkins category of a 'passive replicator': but it is thoroughly confusing to label such DNA as 'selfish DNA' (Orgel and Crick, 1980) since if it has no value to the organism it is merely a 'hitchhiker' on the replication mechanism (Dawkins, 1976), and is not directly being subjected to intragenomic selection. Intragenomic selection could occur only between replicators *sensu* Jacob *et al.*

CONCLUSIONS AND QUESTIONS FOR THE FUTURE

The C-value paradox has often been discussed in relation to transcription and gene expression, but the thesis of this chapter has been that an understanding of eukaryote DNA replication is much more important than is transcription for clarifying the problem.

The fact that eukaryote chromosomes consist of a tandem arrangement of replicons that can be simultaneously replicated allows much larger amounts of DNA to be replicated in a short period than in prokaryotes. I have suggested that this is the basic reason why eukaryotes are able to evolve so much larger genomes than do prokaryotes.

The comparative evidence discussed in this chapter and in Chapter 4 strongly suggests that neither eukaryote genome size itself, nor the length of time taken to replicate eukaryote DNA, can be fundamental determinants of the length of the cell cycle, contrary to what is sometimes supposed. Instead it appears that length of S-phase is secondarily adapted to the length of the cell cycle, and can vary adaptively at different developmental stages in a multicellular organism. It also seems clear that the variation in genome size of different eukaryotes is mainly due to the variation in the number of replicons rather than in their length.

Though the above generalizations provide a useful framework for future research, a more thorough understanding of the problem will depend on detailed research into the structure and organization of individual replicons and how their replication is controlled and modulated in different developmental phases. It will be important to establish the basic similarities and differences between the initiation mechanism of the apparently different categories of eukaryote replicon (i.e. polymerase I, II and III replicons, and constitutive heterochromatic and euchromatic replicons), and to what degree their initiation mechanisms are homologous with replicon initiation or Okazaki fragment initiation of bacteria. A better understanding of the way in which replicons are packed into chromosomes and attached to the nuclear matrix will probably be essential for understanding the evolutionary constraints that keep the mean size of eukaryote replicons fairly similar.

The key question in relation to the C-value paradox is why does the overall number of replicons per genome increase approximately in direct proportion to cell volume. The fundamental issue is whether or not the number of replicons (or genome size itself) causally determines cell volume or whether it is simply correlated with it for secondary evolutionary reasons. A firm decision on this question requires the elucidation of the exact mechanism of the cell-volume-dependent control over the initiation of S-phase.

REFERENCES

Allen, J. R., Roberts, T. M., Loeblich, A. R. III and Klotz, L. C. (1975). Characterisation of the DNA from the dinoflagellate, *Crypthecodinium cohnii* and implications for nuclear organisation. *Cell*, **6**, 161–169.

Astell, C. R., Thomson, M., Chow, M. B. and Ward, D. C. (1983). Structure and replication of minute virus of mice DNA. *Cold Spring Harbor Symp. Quant. Biol.*, **47**, 751–762.

Barlow, P. W. (1972). The ordered replication of chromosomal DNA: a review and a proposal for its control. *Cytobios*, **6**, 55–80.

Bennett, M. D. (1973). The duration of meiosis. In M. Balls and F. S. Billett (Eds.), *The Cell Cycle in Development and Differentiation*. Cambridge University Press, Cambridge, pp. 111–131.

Berezney, R. and Coffey, D. S. (1975). Nuclear protein matrix: association with newly synthesized DNA. *Science*, **189**, 291–292.

Botchan, P. M. and Dayton, A. I. (1982). A specific replication origin in the chromosomal rDNA of *Lytechinus variegatus*. *Nature*, **299**, 453–456.

Broach, J. R., Li, Y.-Y., Feldman, J., Jayaram, M., Abraham, J., Nasymyth, K. A. and Hicks, J. B. (1983). Localization and sequence analysis of yeast origins of DNA replication. *Cold Spring Harbor Symp. Quant. Biol.*, **47**, 1165–1173.

Bryant, J. A. (1976). The cell cycle. In J. A. Bryant (Ed.), *Molecular Aspects of Gene Expression in Plants*, Academic Press, London, pp. 177–216.

Buongiorno-Nardelli, M., Micheli, G., Carrì, M. T. and Marilley, M. (1982). A relationship between replicon size and supercoiled loop domains in the eukaryotic genome. *Nature*, **298**, 100–102.

Callan, H. G. (1972). Replication of DNA in the chromosomes of eukaryotes. *Proc. Roy. Soc. B*, **181**, 19–41.

Carter, B. L. A. (1979). The yeast nucleus. *Advances in Microbiol Physiol.* pp. 243–2.

Cavalier-Smith, T. (1974). Palindromic base sequences and replication of eukaryote chromosome ends. *Nature, Lond.*, **250**, 467–470.

Cavalier-Smith, T. (1975). The origin of nuclei and of eukaryotic cells. *Nature, Lond.*, **256**, 463–468.

Cavalier-Smith, T. (1978a). The evolutionary origin and phylogeny of microtubules, mitotic spindles and eukaryote flagella. *BioSystems*, **10**, 93–114.

Cavalier-Smith, T. (1978b). Nuclear volume control by nucleoskeletal DNA, selection for cell volume and cell growth rate, and the solution of the DNA C-value paradox. *J. Cell Sci.*, **34**, 247–268.

Cavalier-Smith, T. (1980). r- and K-tactics in the evolution of protist developmental systems: cell and genome size, phenotype diversifying selection, and cell cycle patterns. *BioSystems*, **12**, 43–59.

Cavalier-Smith, T. (1981). The origin and early evolution of the eukaryotic cell. In M. J. Carlile, J. F. Collins and B. E. B. Moseley (Eds), *Molecular and Cellular Aspects of Microbial Evolution, Symp. Soc. Gen. Microbiol.* Cambridge University Press, Cambridge, **32**, 33–84.

Cavalier-Smith, T. (1982). The evolution of the nuclear matrix and envelope. In G. G. Maul (Ed.), *The Nuclear Envelope and Matrix*, Alan R. Liss Inc., New York, pp. 307–318.

Cavalier-Smith, T. (1983). Cloning chromosome ends. *Nature, Lond.*, **301**, 112–113.

Cavalier-Smith, T. (1985a). The biochemistry of the replication of chromosome ends. *Biochem. Biophys. Acta. Rev.*, In preparation.

Cavalier-Smith, T. (1985b). A unified model for multiple and binary fission cell cycles and the control of eukaryote DNA replication. In preparation.

Cook, P. R. and Brazell, I. A. (1975). Supercoils in human DNA. *J. Cell Sci.*, **19**, 261–279.

Cooper, S. (1979). A unifying model for the G_1 period in prokaryotes and eukaryotes. *Nature, Lond.*, **280**, 17–19.

Craigie, R. A. and Cavalier-Smith, T. (1982). Cell volume and the control of the *Chlamydomonas* cell cycle. *J. Cell Sci.*, **54**, 173–191.

Dawkins; R. (1976). *The Selfish Gene*, Oxford University Press, London.

Dawkins, R. (1978). Replicator selection and the extended phenotype. *Z. Tierpsychol.*, **47**, 61–76.

Dawkins, R. (1982). *The Extended Phenotype: The Gene as the Unit of Selection*, Freeman, San Francisco.

Donachie, W. D. (1968). Relationship between cell size and the time of initiation of DNA replication. *Nature, Lond.*, **259**, 1077–1079.

Doolittle, W. F. and Sapienza, C. (1980). Selfish genes, the phenotype paradigm and genome evolution. *Nature, Lond.*, **284**, 601–603.

Fantes, P. A. (1977). Control of cell size and cycle time in *Schizosaccharomyces pombe*. *J. Cell Sci.*, **24**, 51–67.

Fantes, P. A. and Nurse, P. (1981). Division timing: controls, models and mechanisms. In P. C. L. John (Ed.), *The Cell Cycle*, Cambridge University Press, Cambridge.

Flavell, R. B., O'Dell, M. and Thompson, W. F. (1983). Cytosine methylation of ribosomal RNA genes and nucleolus organiser activity in wheat. In P. E. Brandham and M. D. Bennett (Eds), *Kew Chromosome Conference II*, Allen and Unwin, London, pp. 11–17.

Fox, G. M., Hess, J. F., Shen, C.-K. J. and Schmid, C. W. (1983). *Alu* family members in the human α-like globin-gene cluster. *Cold Spring Harbor Symp. Quant. Biol.*, **47**, 1131–1139.

Francis, D., Kidd, A. D. and Bennett, M. D. (1985). *DNA replication in relation to DNA C-values*. In J. A. Bryant and D. Francis (Eds.), *The Cell Division Cycle in Plants*. Cambridge University Press, Cambridge, pp. 61–82.

Gall, J. (1968). Differential synthesis of the genes for ribosomal RNA during amphibian oogenesis. *Proc. Natn. Acad. Sci. USA*, **60**, 553–560.

Goldberg, G. I., Collier, I. and Cassel, A. (1983). Specific DNA sequences associated with the nuclear matrix in synchronized mouse 3T3 cells. *Proc. Natl. Acad. Sci. USA*, **80**, 6887–6891.

Graham, C. F. (1973). The cell cycle during mammalian development. In M. Balls and F. S. Billett (Eds), *The Cell Cycle in Development and Differentiation*, Cambridge University Press, Cambridge, pp. 293–310.

Gunge, N. (1983). Yeast DNA plasmids. *Ann. Rev. Microbiol.*, **37**, 253–276.

Hägele, K. (1976). Prolongation of replication time after doublings of the DNA content of polytene chromosome bands of *Chironomus*. *Chromosome*, **55**, 253–258.

Hand, R. (1980). Initation of DNA synthesis in S-phase mammalian cells. In G. L. Whitson (Ed.), *Nuclear-Cytoplasmic Interactions in the Cell Cycle*, pp. 167–179.

Harland, R. H. (1981). Initiation of DNA replication in eukaryotic chromosomes. *Tr. Biochem. Sci.*, **6**, 71–74.

Hines, P. J. and Benbow, R. M. (1982). Initiation of replication at specific origins in DNA molecules microinjected into unfertilized eggs of the frog *Xenopus laevis*. *Cell*, **30**, 459–468.

Jacob, F., Brenner, S. and Cuzin, F. (1964). On the regulation of DNA replication in bacteria. *Cold Spring Harbor Symp. Quant. Biol.*, **28**, 329–348.

Jelinek, W. R., Toomey, T. P., Leinwand, L., Duncan, C. H., Biro, P. A., Choudary, P. V., Weissman, S. M., Rubin, C. M., Houck, C. M., Deininger, P. L. and Schmid, C. W. (1980). Ubiquitous interspersed repeated sequences in mammalian genomes. *Proc. Natl. Acad. Sci. USA*, **77**, 1398–1402.

Jelinek, W. R. and Haynes, S. R. (1983). The mammalian *Alu* family of dispersed repeats. *Cold Spring Harbor Symp. Quant. Biol.*, **47**, 1123–1130.

John, P. C. L. (1981). *The Cell Cycle*, Cambridge University Press, Cambridge.

John, B. and Miklos, G. L. (1979). Functional aspects of satellite DNA and hetero-chromatin. *Int. Rev. Cytol.*, **58**, 1–114.

Kapp, L. N. and Painter, R. B. (1982). DNA replication fork movement rates in mammalian cells. *Int. Rev. Cytol.*, **80**, 1–25.

Keyl, H. G. (1965). Duplikationen von Untereinheiten der chromosomalen DNS während der Evolution von *Chironomus thummi*. *Chromosoma*, **17**, 139–180.

Kornberg, A. (1980). *DNA Replication*, Freeman, San Francisco.

Kornberg, A. (1982). *1982 Supplement to DNA Replication*, Freeman, San Francisco.

Lewin, B. (1980). *Gene Expression*, vol. 2, *Eukaryotic Chromosomes*, 2nd edn, Wiley, New York.

Lloyd, D., Poole, R. K. and Edwards, S. W. (1982). *The Cell Cycle*, Academic Press, London.

Macgregor, H. C. (1982). Ways of amplifying ribosomal genes. In E. G. Jordan and C. A. Cullis (Eds), *The Nucleolus*, Cambridge University Press, Cambridge, pp. 129–151.

Margalit, H., Rosenberger, R. F. and Grover, N. B. (1984). Initiation of DNA replication in bacteria: analysis of an autorepressor model. *J. theor. Biol.*, **111**, 183–199.

Marsden, M. P. F. and Laemmli, U. K. (1979). Metaphase chromosome structure: evidence for a radial loop model. *Cell*, **17**, 849–858.

Maxson, R., Mohun, T. and Kedes, L. (1983). Histone genes. In N. Maclean, S. Gregory and R. A. Flavell (Eds), *Eukaryotic Genes: Their Structure, Activity and Regulation*, Butterworth, London, pp. 277–298.

Mitchison, J. M. (1971). *The Biology of the Cell Cycle*, Cambridge University Press, Cambridge.

Moss, T. and Birnsteil, M. (1982). The structure and function of the ribosomal gene spacer. In E. G. Jordan and C. A. Cullis (Eds), *The Nucleolus*, Cambridge University Press, Cambridge, pp. 73–85.

Muesing, M., Tamm, J., Shepard, H. M. and Polisky, B. (1981). A single base-pair alteration is responsible for the DNA overproduction phenotype of a plasmid copy-number mutant. *Cell*, **24**, 235–242.

Newport, J. and Kirschner, M. (1982a). A major developmental transition in early *Xenopus* embryos. I. Characterization and timing of cellular changes at the midblastula stage. *Cell*, **30**, 675–686.

Newport, J. and Kirschner, M. (1982b). A major developmental transition in early *Xenopus* embryos. II. Control of the onset of transcription. *Cell*, **30**, 687–696.

Nossal, N. G. (1983). Prokaryotic DNA replication systems. *Ann. Rev. Biochem.*, **52**, 581–615.

Orgel, L. E. and Crick, F. H. C. (1980). Selfish DNA: the ultimate parasite. *Nature, Lond.*, **284**, 604–607.

Pardoll, D. M., Vogelstein, B. and Coffey, D. S. (1980). A fixed site of DNA replication in eukaryotic cells. *Cell*, **19**, 527–536.

Pelling, C. (1966). A replicative and synthetic chromosomal unit in the modern concept of the chromomere. *Proc. Roy. Soc. B*, **164**, 279–289.

Pritchard, R. H., Barth, P. T. and Collins, J. (1969). Control of DNA synthesis in bacteria. *Symp. Soc. Gen. Microbiol.*, **19**, 263–297.

Rasmussen, C. D. and Berger, J. D. (1982). Downward regulation of cell size in *Paramecium tetraaurelia*: effects of increased cell size, with or without increased DNA content, on the cell cycle. *J. Cell Sci.*, **57**, 315–329.

Rudkin, G. T. (1973). Cyclic synthesis of DNA in polytene chromosomes of Diptera. In M. Balls and F. S. Billett (Eds), *The Cell Cycle in Development and Differentiation*, Cambridge University Press, Cambridge, pp. 279–292.

Seidman, M. M., Levine, A. J. and Weintraub, H. (1979). The asymmetric segregation of parental nucleosomes during chromosome replication. *Cell*, **18**, 439–449.

Seyfert, H.-M. and Cleffmann, G. (1982). Mean macronuclear DNA contents are variable in the ciliate *Tetrahymena*. *J. Cell Sci.*, **58**, 211–223.

Smith, H. C. and Berezney, R. (1980). DNA polymerase α is tightly bound to the nuclear matrix of actively replicating liver. *Biochem. Biophys. Res. Commun.*, **97**, 1541–1547.

Sompayrac, L. and Maaløe, O. (1973). Autorepressor model for control of DNA replication. *Nature New Biology*, **241**, 133–135.

Stougaard, P., Molin, S. and Nordström, K. (1981). RNAs involved in copy-number control and incompatibility of plasmid R1. *Proc. Natl. Acad. Sci. USA*, **78**, 6008–6012.

Timberlake, W. E. (1978). Low repetitive DNA content in *Aspergillus nidulans*. *Science*, **202**, 973–974.

Van't Hof, J. (1965). Relationships between mitotic cycle duration, S period duration and the average rate of DNA synthesis in the root meristem cells of several plants. *Exp. Cell Res.*, **39**, 48–58.

Van't Hof, J. (1974). The duration of chromosomal DNA synthesis, of the mitotic cycle, and of meiosis of higher plants. In R. C. King (Ed.), *Handbook of Genetics*, vol. 2, *Plants, Plant Viruses and Protists*, Plenum Press, New York, pp. 181–200.

Van't Hof, J. (1976). Replicon size and rate of fork movement in early S of higher plant cells (*Pisum sativum*). *Exp. Cell Res.*, **103**, 393–403.

Van't Hof, J. and Kovacs, C. J. (1972). Mitotic cycle regulation in the meristem of cultured roots: the principal control point hypothesis. *Adv. exp. Med. Biol.*, **18**, 15–32.

Van't Hof, J. and Sparrow, A. K. (1963). A relationship between DNA content, nuclear volume, and minimum mitotic cycle time. *Proc. Natl. Acad. Sci. USA*, **49**, 897–902.

Wanka, F., Pieck, A. C. M., Bekers, A. G. M. and Mullenden, L. H. F. (1982). The attachment of replicating DNA to the nuclear matrix. In G. G. Maul (Ed.), *The Nuclear Envelope and Nuclear Matrix*, Liss, New York, pp. 199–211.

Waterborg, J. H. and Shall, S. (1984). The organization of replicons. In J. A. Bryant and D. Francis (Eds), *The Plant Cell Division Cycle*, Cambridge University Press, Cambridge, pp. 15–35.

Wheals, A. and Silverman, B. (1982). Unstable activator model for size control of the cell cycle. *J. theor. Biol.* **97**, 505–510.

Whitehouse, H. L. K. (1973). *Towards an Understanding of the Mechanism of Heredity*, 3rd edn, Edward Arnold, London.

Worcel, A. and Benyajati, C. (1977). Higher order coiling of DNA in chromatin. *Cell*, **12**, 83–100.

Yeoman, M. M. (1981). The mitotic cycle in higher plants. In P. C. L. John (Ed.), *The Cell Cycle*, pp. 161–184.

Yeoman, M. M. and Aitchison, P. A. (1976). Molecular events of the cell cycle: a preparation for division. In M. M. Yeoman (Ed.), *Cell Division in Higher Plants*, Academic Press, London, pp. 111–132.

Zipursky, S. L. and Marians, K. J. (1981). *Escherichia coli* factor Y sites of plasmid, pBR 322 can function as origins of DNA replication. *Proc. Natl. Acad. Sci. USA*, **78**, 6111–6115.

The Evolution of Genome Size
Edited by T. Cavalier-Smith
© 1985 John Wiley & Sons Ltd

CHAPTER 8

Selfish DNA, Intragenomic Selection and Genome Size

T. Cavalier-Smith

'According to our hypothesis, gemmules multiply by self-division and are transmitted from generation to generation'.
'Though a vast number of active and long-dormant gemmules are diffused and nourished in each living creature, yet there must be some limit to their number.'

(Darwin, 1868, p. 396.)

SUMMARY

The concept of intragenomic selection, or the non-random differential multiplication of different genes or DNA sequences within the genome, goes back to Darwin. Intragenomic selection can take three different forms: differential segregation, differential replication and duplicative transposition. DNA sequences that spread intragenomically in this way may be referred to as genetic symbionts or, potentially more confusingly, 'selfish DNA'. The purpose of this chapter is to argue that the main bulk of the secondary DNA of cells does not consist of genetic symbionts and that the occurrence of intragenomic selection, though important for other evolutionary problems, does not provide an adequate explanation of the C-value paradox.

INTRODUCTION

Since Darwin's 'gemmules' later 'mutated' into the 'pangenes' of de Vries (1889) and then, by deletion of a single triplet, into the 'genes' of Johanssen (1909), one may regard the second quotation from Darwin as the first reference to the problem of the evolutionary determination of genome size. The passage from which it is taken shows that he countenanced the presence in the genome of what are now called pseudogenes (Jeffreys, 1982), i.e. ex-genes that have lost their function and are now useless parasites of the

253

replication machinery. Yet his insistence that 'there must be some limit to their number' shows that he would not have made the mistake of supposing that the mere invention of terms such as 'junk' DNA (Ohno, 1972) or parasitic 'selfish' DNA (Crick, 1979; Orgel and Crick, 1980) solves the problem of genome size. Anyone can see that replication gives particular DNA sequences the potential to spread within the genome ('selfish' DNA) or to be passively transmitted ('junk' DNA). But such elementary *qualitative* considerations concerning DNA sequences do not solve the C-value paradox, which is essentially a quantitative problem: i.e. what determines the 'limit to their number' as Darwin put it.

To deal rigorously with the problem we would have to list all the possible evolutionary forces tending to increase genome size as well as those tending to decrease it, construct a population genetic model of their interactions, make quantitative estimates or preferably actual measurements of their magnitude, substitute these values into the theoretical equations and compare the results with observations of genome size. The aim of this chapter is more modest: to clarify the qualitative aspects of the problem. The appropriate basic model is still the classic mutation-selection equilibrium (Fisher, 1922; Haldane, 1927, 1932) despite the discovery of transposable elements which were not included in the classical theory of population genetics. It will be argued that the widespread occurrence of transposable elements in eukaryotes is not, as Doolittle and Sapienza (1980) suggested, the fundamental cause but only the indirect consequence (Cavalier-Smith, 1978) of their large genome size. But this can be done more clearly after first discussing the concept of intragenomic selection (Cavalier-Smith, 1980).

INTRAGENOMIC SELECTION AND GENETIC SYMBIONTS

'Each living creature must be looked at as a microcosm—a little universe, formed of a host of self-propagating organisms, inconceivably minute and as numerous as the stars in heaven'. (Darwin, 1868, p. 404.) 'Gemmules, derived from the same cells after modification, naturally go on increasing under the same favouring conditions, until at last they become sufficiently numerous to overpower and supplant the old gemmules'. (Darwin, 1868, p. 395.)

'What the world is to organisms in general, each organism is to the molecules of which it is composed. Multitudes of these having diverse tendencies, are competing with one another for opportunity to exist and multiply; and the organism, as a whole, is as much the product of the molecules which are victorious as the Fauna, or Flora of a country is the product of the victorious organic beings in it'. (Huxley, 1873; Critiques and Addresses, p. 209.)

'I am very glad that you have been bold enough to give your idea about Natural Selection amongst the molecules.' (Darwin, 1869; see Darwin, 1887.)

'One possibility is that they [inverted-repeat DNA sequences] have no function for the organism as a whole; they may even be disadvantageous 'selfish genes' maintained by strong positive selection at the level of the individual gene balanced by negative selection at the level of the organism; this would be entirely consistent with their tendency to translocate and cause mutations.' (Cavalier-Smith, 1977.)

In writing the above I had no idea that the idea of 'natural selection amongst the molecules' went back to Darwin and Huxley and I was unaware that I was using Dawkins' (1976) phrase 'selfish genes' for an entirely different concept from his. After later reading 'The Selfish Gene' (Dawkins, 1976) it became clear that he was almost invariably discussing competition between alleles at the same locus, for which the term allelic selection (Cavalier-Smith, 1980) is preferable to the classic term genic selection. The latter can easily be confused with selection arising from competition between different genes at different loci in the genome which is what I had in mind in 1977 and which I (Cavalier-Smith, 1980) named intragenomic selection in order to minimize further confusion of this kind.

Apart from the special case of meiotic drive, allelic selection results from the death and reproduction of individual organisms. Intragenomic selection, by contrast, is the differential multiplication of different DNA sequences within a single organism and is independent of the death and reproduction of the organisms containing them. But note that a sequence that is accidentally duplicated or deleted is not thereby undergoing intragenomic selection, but mutation. Only if the sequence mutates in such a way that it is then systematically multiplied by normal replication or segregation, or by duplicative transposition, at a different rate from other cellular DNA sequences, does it undergo intragenomic selection. When such differential intragenomic multiplication increases the relative number of copies per cell of the new sequence it may aptly be called a 'selfish gene' (Cavalier-Smith, 1977) or 'selfish DNA' (Crick, 1979); it is in this precise sense that Doolittle and Sapienza (1980) and Doolittle (1982) used the phrase 'selfish DNA'. Orgel and Crick (1980) apparently used it in a vaguer and broader sense to include also functionless DNA sequences that have undergone duplication in the past and are now merely passively replicated at the same rate as ordinary DNA. Such DNA is not actively spreading by intragenomic selection, and is no different from the classical idea of purely non-functional inert DNA (Darlington and La Cour, 1941), now commonly called 'junk' DNA (Ohno, 1972). Since the population dynamics of merely inert genes is quite different from 'selfish genes' *sensu* Cavalier-Smith (1977) and 'selfish DNA' *sensu* Doolittle and Sapienza, it is important not to confuse them as did Ohta (1981) when he called purely neutral DNA 'selfish'; Orgel's term 'selfish DNA' was originally stated (Crick, 1979) as 'If some sequences are preferentially replicated in evolution they will spread over the genome in evolution even if they have no particular function, provided they do not harm the "host" organism too much', and this more precise definition was reaffirmed by Orgel et al. (1980).

Such sequences, whether plasmids, viruses, viroids, transposable elements or B chromosomes, may in principle be harmful, neutral or even beneficial to the host cell, so it is better to think of them as symbionts (Cavalier-Smith, 1983), and not necessarily as parasites (Östergren, 1945; Orgel and Crick, 1980); because of their capacity for autonomous replication and non-cellular structure they can be called genetic symbionts (Cavalier-Smith, 1983). Basic features of their population genetics are discussed in Chapter 16.

GENETIC SYMBIONTS AND THE C-VALUE PARADOX

'From the point of view of the selfish genes themselves, there is no paradox. The true "purpose" of DNA is to survive, no more and no less. The simplest way to explain the surplus DNA is to suppose that it is a parasite, or at best a harmless but useless passenger, hitching a ride in the survival machines created by the other DNA'. (Dawkins, 1976, p. 47.)

'The existence of massive amounts of DNA, whose sequence is unconstrained by stabilising selection acting via the phenotype, provides a potent source for the origin of DNA-virus-like self-replicating elements, as well as a superb environment for their increase by direct genic selection (Williams, 1966). Mutations in a section of S-DNA [skeletal DNA] enabling it to replicate at a greater rate than the host chromosome and to insert into other chromosomes will automatically be selected'. (Cavalier-Smith, 1978.)

The quotations reveal two fundamentally contrasting views. Dawkins, followed by Doolittle and Sapienza (1980) and Orgel and Crick (1980), suggested that the existence of genetic symbionts is the main cause of high C-values, whereas I suggested, conversely, that large C-values are a major stimulus to the origin of genetic symbionts, but that the existence of genetic symbionts does not provide a sufficient explanation for the origin of large genomes and the C-value paradox. However, I do not go so far as Campbell (1982) who argues that plasmids, and by implication other non-virulent genetic symbionts, must always be beneficial to the organism. The existence of genetic symbionts must inevitably put some upward evolutionary pressure on genome size since any mutations enabling them to multiply faster than host DNA will automatically increase by intragenomic selection (Cavalier-Smith, 1978). Because host DNA sequences cannot be reduced in concentration below one per genome it follows that, in the absence of any countervailing forces, even a minute intragenomic selective advantage by a genetic symbiont would soon lead to an indefinite and very rapid explosive increase in total nuclear DNA content. Since this does not occur countervailing forces must be very strong. Overall the evidence indicates that genome sizes are not *greatly* increased by the activities of genetic symbionts, though they may be marginally greater than they otherwise would be. The three main modes of intragenomic selection will now be considered in turn.

Differential segregation

Examples of selective elimination and preferential transmission of chromosomes have been reported for a variety of organisms, both in somatic and germ line tissues.

Selective elimination in the germ line (e.g. in *Hordeum* hybrids: Finch and Bennett, 1983) will tend to decrease, rather than increase, genome size. Selective chromosome elimination in somatic cells (e.g. in sciarid and cecidomyid flies: White, 1973) will not in itself alter genome size. If, however, selection against extra inert DNA were much weaker in the germ line than in somatic cells, evolution of an elimination mechanism would make it easier for genetic symbionts to multiply in the somatically eliminated chromosomes and/or for new inert chromosomes supernumerary to the normal haploid set (B chromosomes: Jones and Rees, 1982) to evolve and parasitize the elimination mechanism; this is because elimination from somatic cells would reduce selection against them because of harmful somatic effects. Some B chromosomes, which Östergren (1945) suggested were useless parasitic chromosomes, do undergo preferential elimination in some, but not other, somatic tissues.

Preferential accumulation of chromosomes by differential segregation will affect genome size only if it occurs in the germ line, as in the 'cuckoo' chromosomes of *Aegilops* and *Aegilops*/wheat hybrids (Maan, 1976; Miller, 1983) or many B chromosomes (Jones and Rees, 1982). Jones and Matthews (1983) and Jones (Chapter 13) have persuasive evidence in the case of rye B chromosomes for Östergren's view that B chromosomes are 'selfish' parasites maintained in populations though a balance between preferential intragenomic accumulation and negative selection at the organismic level. They interpret the varying frequency of B chromosomes in terms of a varying strength of organismic selection against them. B chromosomes are known to have phenotypic effects, notably increasing nuclear and cell volume and reducing developmental rates; but only in rare instances (e.g. Teoh and Jones, 1978) is there evidence of a positive organismic selective advantage for organisms possessing Bs compared with those lacking them. While this is probably at least partly because very few quantitative studies of this type have been made, the spread of those B chromosomes possessing an accumulation mechanism, as in rye, can be sufficiently explained by the existence of such a mechanism, even if the Bs have no function or are even mildly harmful to the organism. The experimental evidence suggests that possession of a large number of B chromosomes is always harmful and that organismic selection keeps their numbers low in a population.

The lack of homology of B chromosome with the normal A chromosomes, their dispensibility, the lack of qualitative phenotypic effects comparable to those of trisomic A chromosome, the frequency with which they are entirely heterochromatic, and lack of evidence for transcription, are consistent with the traditional view that they are physiologically and transcriptionally inert. It is reasonable to suggest that their harmful effect is purely from the extra bulk of DNA that they add to the genome. If this is so, the fact that they rarely contribute even as much as 10% extra DNA to the genome means that organismic counterselection against extra DNA of this amount is strong enough to balance the intragenomic accumulation mechanisms. As Jones and Matthews point out, such organismic selection probably eliminates B chromosome mutations that make their accumulation mechanisms too efficient, thus preventing B chromosomes from becoming increasingly 'selfish'.

Though B chromosomes have been reported in hundreds of species, they appear to be absent from the majority, as well as from many or most individuals of the species in which they do occur. Since there is no reason to suppose that the mutational mechanisms that generate B chromosomes are rare, this implies that for most eukaryote species organismic selection against extra DNA is sufficiently strong to prevent even a 5–10% increase in genome size over that now present. Since most of even the highest C-value eukaryotes do not have B chromosomes, and since there is no correlation between genome size and chromosome number (Levin and Funderburg, 1979), it is clear that intragenomic selection of 'selfish' inert chromosomes, though it almost certainly does occur, in no way helps to explain the C-value paradox.

If the harmful effect of B chromosomes is entirely the result of the non-sequence-specific effects of their extra DNA then one could safely conclude from the above argument that other types of non-functional selfish DNA will also be subject to equally strong organismic counter-selection. Though this argument is somewhat weakened by the possibility that the harmful effect of Bs is at least partly due to unidentified sequence-specific or genic effects, the paucity of plasmids in eukaryotes leads to a similar conclusion.

Differential replication

A DNA molecule can be replicated only if it contains one or more replicon origins (see Chapter 7): the unit consisting of a replication origin and the DNA covalently linked to it, which is passively replicated, is referred to as a replicon (Chapter 7). Since eukaryote chromosomes consist of a linear polymer of covalently bonded replicons, the individual replicons can be replicated at systematically different rates over several generations only if they evolve the capacity for duplicative transposition (see following section), or become detached from the rest of the chromosome to form separate plasmids. (See fig. 12.3b p. 361.)

The best known nuclear plasmids in eukaryotes are rDNA plasmids, which occur in the macronuclei of heterokaryote protozoa (*Tetrahymena, Paramecium*), the mycetozoan *Physarum* and temporarily in amphibian oocytes (Macgregor, 1982); though the presence of rDNA plasmids is clearly positively maintained by organismic selection, the 2 μm circular plasmid of yeast has no known functions and may well be a useless genetic symbiont established purely by intragenomic selection. The 2 μm plasmid is present only in a minority of yeast strains (Heitman *et al.*, 1980), and though present in 50–100 copies per nucleus adds only a total of 1–2% of extra DNA to the cell. Though now it is replicated only once per cell cycle, in S-phase, it must once have undergone differential replication and intragenomic increase to reach its present copy number.

The studies on gene amplification in response to drug and other drugs discussed by Cullis (Chapter 6) strongly suggest that mutations to generate plasmids are relatively easy in eukaryotes. Therefore the fact that most eukaryotes appear to lack plasmids altogether cannot be explained in terms of the difficulty of their origin. It is highly probable that they are arising all the time but do not become established or multiply to a detectable level. This inability to become established cannot simply be because most plasmids would lack a centromere and therefore fail to undergo regular segregation. Irregular segregation would tend in itself merely to increase their numbers in some lines and reduce them in others.

The rarity of eukaryote plasmids, and the fact that when apparently useless plasmids are present, as in some yeast strains, they only make up 1–2% of the total DNA strongly implies a universal strong organismic selection against useless genetic symbionts ('selfish DNA'). There is experimental evidence for such negative organismic selection in bacteria (Dale and Smith, 1979; Zünd and Lebek, 1980) for plasmids of 80 kb or more (2% total genome size), and in mammalian cells (Cullis p. 199).

Duplicative transposition and the evolution of moderately repetitive DNA

The main difference between genetic symbionts that spread by duplicative transposition, rather than ordinary replication, is that they are linked to centromeres. Therefore they will undergo regular segregation and be less likely to be lost if they have a very low copy number. But centromere-linkage will not in itself increase their copy number or favour intragenomic spread. Therefore the preceding argument derived from the rarity of plasmids ought also to apply to transposable genetic elements. It is significant that in the best-established candidates for transposable genetic symbionts in eukaryotes copy numbers are kept relatively constant in a particular strain, and that transposition is subject to negative feedback self-control, like replication (Dyson and Sherratt, Chapter 12). This implies that although such genetic symbionts have the evolutionary potential to undergo mutations that would enable them to multiply like wild fire, or like cancer cells, and therefore to greatly inflate genome sizes, this potential must normally be held in check by organismic counterselection arising from the harm to the organism that this would cause. The evidence reviewed by Bouchard (1982) and by Doolittle (Chapter 15) suggests that at least some middle repetitive DNA sequences are genetic symbionts that have arisen by mutation and intragenomic selection. But even if this proves true for most middle repetitive DNA sequences, it would not mean that intragenomic selection of genetic symbionts was the dominant force in the evolution of large genome sizes; intragenomic selection is probably not a continuous force but a very transient phenomenon that rapidly adjusts copy numbers to a new equilibrium level following a mutation in the mechanism that controls copy number.

One must also stress that although repetitive DNA is generally present in greater amounts in larger genomes (Bouchard, 1982) this is also true for non-repetitive DNA. However, the distinction between repetitive and non-repetitive DNA is not absolute and depends on the empirical conditions of the renaturation experiments used to differentiate the various fractions (Bouchard, 1982). Careful studies on plants of differing genome size show that as the stringency of the renaturation conditions is relaxed a greater and greater fraction of DNA shows repetitive properties; at the highest stringencies very little DNA appears to be repetitive, but at the lowest stringencies the vast majority of it does (see also Chapter 15). Only two simple assumptions are needed to explain this:

1. The majority of the DNA of high C-value plants has evolved in the past by repeated duplications; this is expected on *any* hypothesis of genome evolution, whether the extra DNA is 'junk', 'selfish' or 'useful'.
2. Since the repeated sequences evolved they have undergone considerable divergence, the degree of divergence for any subset of repeated sequences depending on how long ago the repeats originated and on the subsequent rate of divergence. This divergence has been much more rapid than for protein-coding sequences.

Though rapid sequence divergence does not necessarily imply an absence of genic functions, and can in principle occur through very strong selective forces, in this case there is no actual evidence for any genic function for the vast majority of the extra DNA in high C-value species (Chapter 3), and the classical genetic load arguments make it exceedingly improbable that strong natural selection could be causing such divergence: such selection would have to be far stronger than selection for genes that code for proteins and also affect many more sequences. Therefore it can be concluded from this great sequence divergence that most of the sequences in non-coding DNA, including a high proportion of moderately repetitive DNA (Bouchard, 1982; Singer, 1982), are under little or no selective constraint with respect to *sequence*, and are free to diverge by mutation and random drift.

Since known genetic symbionts such as viruses, plasmids and transposons have highly conserved genic sequences, and little or no non-genic DNA, it is improbable that more than a small fraction of the repeated DNA of high C-value species can consist of active genetic symbionts. It appears to be truly non-genic, consisting of neither organismic genes nor genetic symbiont genes. It must mainly be either true 'junk', or have a positive *quantitative* function that does not depend on particular sequences as I argued in Chapter 4.

Moderately repetitive DNA, far from being a uniform category of DNA sequences, has been proven to vary greatly in its frequency and properties in different organisms (Bouchard, 1982). I previously stressed (Cavalier-Smith, 1978) that these data do not support the idea that its primary function is in gene regulation. On the gene-regulation hypothesis one would expect its abundance to be related to the numbers of protein-coding genes and to the developmental complexity of the organism. But this is simply not the case. Unicellular organisms such as protozoa can have even higher levels of middle repetitive DNA than mammals (Allen *et al.*, 1975); conversely, in some complex organisms such as chickens half the genome consists of long stretches (< 17 kb) of protein-coding DNA devoid of repeated DNA. The amounts of moderately repetitive DNA show a much better relationship with genome size than they do with organismic complexity. Not only do organisms with larger genomes usually have more moderately repeated DNA but there is a marked tendency for those with the largest genomes (e.g. some plants) to have relatively larger and those with the smallest genes (notably fungi) to have relatively much smaller amounts.

This is readily understood if, as previously proposed (Cavalier-Smith, 1978), the large amounts of non-conserved moderately repetitive DNA in high C-value species like salamanders and grasses are simply the inevitable left-overs of duplications and transpositions that occurred during past increases in genome size. By contrast with such rapidly diverging sequences, there are also evolutionary conserved moderately repetitive DNA sequences (Bouchard, 1982); one possible function for them is to act as replicon origins (Cavalier-Smith, 1978; Jelinek *et al.*, 1980; Georgiev *et al.*, 1981; see discussion in Chapter 7), but they must also include true genetic symbionts of no value to the host, which, as previously emphasized (Cavalier-Smith, 1978) can be expected to find a more favourable habitat in nuclei containing large amounts of non-coding DNA. Some non-conserved repetitive sequences may have originated from genetic symbionts by mutations abolishing their capacity for independent multiplication. But a non-functional sequence that originated thus from a genetic symbiont is no more entitled to be called a genetic symbiont or 'selfish DNA' than is a functionless pseudogene that originated from a cellular protein-coding genome to be regarded as a functional cellular gene. The former is no more *maintained* by intragenomic, than is the latter by organismic, selection. It is thus essential to distinguish the mechanism of *origin* of such sequences from that of their *maintenance*. Neither is maintained by sequence-dependent intragenomic or organismic selection. Both are maintained by the same force: either actively, by sequence-independent selection for the overall mass of DNA (Chapter 4), in which case, though inert, they are *not* junk; or passively by 'hitching' (i.e. simply by virtue of their linkage to replicon origins and centromeres) in which case they are *both* junk DNA.

The great diversity in the types of middle repetitive DNA and their interspersion pattern in different organisms (Bouchard, 1982) is most simply explained as the incidental consequence of differing propensities for the generation and spread of such sequences, probably by a variety of different passive and active mechanisms, coupled with different tolerances to such spread (Bouchard, 1982); organismic selection for genome size (Cavalier-Smith, 1978) is almost certainly a major, but probably not the only, factor that influences both the propensity for, and tolerance to, such spread.

CONCLUSION

Chapter 3 concluded that the greater proneness of large genomes to lethal and mutagenic damage by mutagens made it unreasonable to argue that the extra DNA of high C-value species was purely junk DNA, and that it must be being actively maintained by some positive evolutionary force. The evidence and arguments discussed in this chapter strongly suggest that intragenomic selection, though it almost certainly plays an important role in the evolution of B chromosomes, plasmids and transposable elements, is not strong enough to provide such a positive force. Merely to demonstrate a basic flaw in the preceding argument would be insufficient to turn the intragenomic selection of genetic symbionts into a satisfactory solution to the C-value paradox. Also, it would be necessary to explain quantitatively why intragenomic selection should result in genomes that are directly proportional to cell volume in size.

NOTE: 'SELFISH DNAS WITH SELF-RESTRAINT'

Since the completion of this chapter Doolittle *et al.* (1984) have explicitly recognized that genetic symbionts are usually subject to strong intracellular population controls. However, some of the anthropomorphic language they use to describe this phenomenon is potentially confusing. 'Selfish DNA with self-restraint' is almost a contradiction in terms and obscures the fact (which *they* are perfectly aware of) that the restraint is exercised not by intragenomic selection but by classical cellular or organismic selection. To call such selection group selection, as they do, merely because it is at a higher level than intragenomic selection, is confusing because 'group selection' usually refers to interdeme selection. Though interdeme selection is, as they say, 'comparatively weak', organismic selection is very strong and fully capable of selecting mutations in genetic symbionts that favour the cell and therefore are *not* selfish *sensu* Orgel and Crick (1980). Since cellular or organismic selection causes the restraint it is misleading as to the mechanism to say 'the better strategy for the transposon in the long run is to exercise self-restraint'; evolutionarily speaking it is not *self*-restraint—an accurate statement is that 'their tendency to increase will be held in check by selection acting . . . via the fitness of the organism carrying them' (Cavalier-Smith, 1978, p. 269).

It is not because interdeme selection is at a higher level than organismic selection that it is so much weaker, but because demes (unlike organisms) are seldom genetically discrete units with high rates of death and reproduction. Intragenomic selection, despite being at a lower level than organismic selection, is in general a less important and weaker evolutionary factor than organismic selection because unrestricted intragenomic multiplication of a mutant symbiont would normally automatically lead to its own extinction, which is not the case with a conventional organism; only when the genetic symbiont evolves the capacity to survive outside cells and to infect new ones (i.e. becomes a virus) can intragenomic selection become of overriding evolutionary importance (Cavalier-Smith, 1978; 1983).

REFERENCES

Allen, J. R., Roberts, T. M., Loeblich, A. R. III and Klotz, L. C. (1975). Characterisation of the DNA from the dinoflagellate, *Crypthecodinium cohnii* and implications for nuclear organisation. *Cell*, **6**, 161–169.

Bouchard, R. A. (1982). Moderately repetitive DNA in evolution. *Int. Rev. Cytol.*, **76**, 113–193.

Campbell, A. (1982). Evolutionary significance of accessory DNA elements in bacteria. *Ann. Rev. Microbiol.*, **35**, 55–83.

Cavalier-Smith, T. (1977). Visualising jumping genes. *Nature, Lond.*, **270**, 10–12.

Cavalier-Smith, T. (1978). Nuclear volume control by nucleoskeletal DNA, selection for cell volume and cell growth rate, and the solution of the C-value paradox. *J. Cell Sci.*, **34**, 247–278.

Cavalier-Smith, T. (1980). How selfish is DNA? *Nature, Lond.*, **285**, 617–618.

Cavalier-Smith, T. (1983). Genetic symbionts and the origin of split genes and linear chromosomes. In H. E. A. Schenk and W. Schwemmler (Eds), *Endocytobiology II: Intracellular Space as Oligogenetic Ecosystem*, Walter de Gruyter, Berlin and New York, pp. 29–45.

Crick, F. H. C. (1979). How to live with a golden helix. *Miami Winter Symposium*, vol. 16, 1–13.

Dale, J. W. and Smith, J. T. (1979). The effect of a plasmid on growth and survival of *E. coli. Antonie van Leewenhoek*, **45**, 103–111.

Darlington, C. D. and La Cour, L. (1941). The detection of inert genes. *J. Hered.*, **32**, 114–121.

Darwin, C. (1868). *The Variation of Animals and Plants under Domestication*, vol. II, Murray, London.

Darwin, F. (ed) (1887). *The Life and Letters of Charles Darwin*, vol. 3, p. 119, Murray, London.

Dawkins, R. (1976). *The Selfish Gene*, Oxford University Press, Oxford.

de Vries, H. (1889). *Intracelluläre Pangenesis*, Fischer, Jena.

Doolittle, W. F. (1982). Selfish DNA after 18 months. In G. A. Dover and R. B. Flavell (eds), *Genome Evolution*, Academic Press, London, pp. 3–28.

Doolittle, W. F. and Sapienza, C. (1980). Selfish genes, the phenotype paradigm and genome evolution. *Nature, Lond.*, **284**, 617–618.

Doolittle, W. F., Kirkwood, T. B. L. and Dempster, M. A. H. (1984). Selfish DNAs with self-restraint. *Nature*, **307**, 501–502.

Finch, R. A. and Bennett, M. D. (1983). The mechanism of somatic chromosome elimination in *Hordeum*. In P. E. Brandham and M. D. Bennett (Eds), *Kew Chromosome Conference II*, Allen and Unwin, London, pp. 147–154.

Fisher, R. A. (1922). On the dominance ratio. *Proc. Roy. Soc. Ed.*, **42**, 321–341.

Georgiev, G. P., Ilyin, Y. V., Chmelianskaite, V. G., Ryskov, A. P., Kramerov, D. A., Skryabin, K. G., Krayev, A. S., Lukanidin, E. M. and Grigoryan, M. S. (1981). Mobile dispersed genetic elements and other middle repetitive DNA sequences in the genomes of *Drosophila* and mouse: transcription and biological significance. *Cold Spring Harbor Symp. Quant. Biol.*, **45**, 641–654.

Haldane, J. B. S. (1927). A mathematical theory of natural and artificial selection. Part V. *Proc. Camb. Phil. Soc.*, **23**, 838–844.

Haldane, J. B. S. (1932). *The Causes of Evolution*, Longmans Green, London.

Heitmann, C., Mannhaupt, G., Michaelis, G., Pratje, E., Adolf, M. (1980). The distribution of the 2 μm DNA plasmid in the genus *Saccharomyces*. In H. E. A. Schenk and W. Schwemmler (Eds), *Endocytobiology: Endosymbiosis and Cell Biology*, de Gruyter, Berlin and New York, pp. 791–796.

Jeffreys, A. J. (1982). Evolution of globin genes. In G. A. Dover and R. B. Flavell (Eds), *Genome Evolution*, Academic Press, London, pp. 157–176.

Jelinek, W. R., Toomey, T. P., Leinwand, L., Duncan, C. H., Biro, P. A., Coudray, P. V., Weisman, S. M., Rubin, C. M., Houck, C. M., Deininger, P. L. and Schmid, C. W. (1980). Ubiquitous, interspersed repeated sequences in mammalian genomes. *Proc. Natl. Acad. Sci. USA*, **77**, 1398–1402.

Johanssen, W. (1909). *Elemente der exakten Erblichkeitslehre*, Gustav Fischer, Jena.

Jones, R. N. and Matthews, R. B. (1983). Selfish B chromosomes in rye. In P. E. Brandham and M. D. Bennett (Eds), *Kew Chromosome Conference II*, Allen and Unwin, London, pp. 183–190.

Jones, R. N. and Rees, H. (1982). *B chromosomes*, Academic Press, London.

Levin, D. A. and Funderburg, S. W. (1979). Genome size in angiosperms: temperate versus tropical species. *Amer. Nat.*, **114**, 784–795.

Maan, S. S. (1976). Alien chromosome controlling sporophytic sterility in common wheat. *Crop Sci.*, **16**, 580–583.

Macgregor, H. C. (1982). Ways of amplifying ribosomal genes. In E. G. Jordan and C. A. Cullis (Eds), *The Nucleolus*, Cambridge University Press, Cambridge, pp. 129–151.

Miller, T. E. (1983). Preferential transmission of alien chromosomes in wheat. In P.

E. Brandham and M. D. Bennett (Eds), *Kew Chromosome Conference II*, Allen and Unwin, London, pp. 173–182.

Ohno, S. (1972). So much 'junk' DNA in our genome. *Brookhaven Symp. Biol.*, **23**, 366–370.

Ohta (1981) Population genetics of selfish DNA. *Nature, Lond.*, **292**, 648–649.

Orgel, L. E. and Crick, F. H. C. (1980). Selfish DNA: the ultimate parasite. *Nature, Lond.*, **284**, 604–607.

Orgel, L. E., Crick, F. H. C. and Sapienza, C. (1980) Selfish DNA. *Nature, Lond.*, **288**, 645–646.

Östergren, G. (1945). Parasitic nature of extra fragment chromosomes. *Botan. Not.*, **2**, 157–163.

Singer, M. F. (1982). Highly repeated sequences in mammalian genomes. *Int. Rev. Cytol.*, **76**, 67–112.

Teoh, S. B. and Jones, R. N. (1978). B chromosome selection and fitness in rye. *Heredity*, **41**, 35–48.

White, M. J. D. (1973). *Animal Cytology and Evolution*, 3rd edn, Cambridge University Press, Cambridge.

Williams, G. C. (1966). *Adaptation and Natural Selection*. Princeton University Press, Princeton.

Zünd, P. and Lebek, G. (1980). Generation time-prolonging R plasmids: correlation between increases in the generation time of *Escherichia coli* caused by R plasmids and their molecular size. *Plasmid*, **3**, 65–69.

The Evolution of Genome Size
Edited by T. Cavalier-Smith
© 1985 John Wiley & Sons Ltd

CHAPTER 9

Genome Size and Natural Selection: Observations and Experiments in Plants

KONRAD BACHMANN

Hugo de Vries Laboratorium, University of Amsterdam, Kruislaan 318, 1098 SM Amsterdam, The Netherlands

KENTON L. CHAMBERS

Herbarium, Department of Botany, Oregon State University, Corvallis, Oregon, 97331, USA

H. JAMES PRICE

Genetics, Department of Plant Sciences, Texas A & M University College Station, Texas, 77843, USA

SUMMARY

Certain basic morphological and physiological parameters (cell size, mitotic cycle time) seem to be directly dependent on the total amount of DNA per haploid chromosome set. These primary effects find their expression in the phenotype of the organism, in plants more directly than in animals. In plants, the basic mutational event causing interspecific DNA amount differences seems to be 'saltational', i.e. relatively large parts of the genome are gained or lost in one step. The concomitant phenotypical changes should expose genome size to natural selection. Experimental studies of crosses of plant species with high and low DNA contents have failed to show a correlation between the F2 segregation of genome sizes and that of phenotypic characters. Reasons for this failure are discussed and it is suggested that a selective effect on genome size might best be looked for after strong selection on the phenotype under stressful environmental conditions.

INTRODUCTION

This chapter will discuss some recent observations in plants bearing on the selective role of the nuclear DNA amount, including those DNA sequences that serve no sequence-dependent function. If such sequences are purely 'selfish DNA' (Doolittle and Sapienza, 1980; Orgel and Crick, 1980) they could be removed from the genome with no detectable effect on the organism, and therefore with no selective consequences. In contrast to this view, we propose that the brute bulk amount of all DNA in the genome exerts an influence on the phenotype so that selection under certain circumstances permits or even favours the accumulation of nuclear DNA more or less irrespective of its nucleotide sequence, while under different circumstances the loss of sequences not necessary to survival may bring a considerable adaptive and reproductive advantage.

During the last several years we have begun to look at this problem in the plant genus *Microseris*. Our results agree well with results from other laboratories on other plant taxa. They agree in principle with comparable results obtained in animal groups, but there are some differences due to the drastically different developmental patterns of plants and animals.

VARIATION OF GENOME SIZE IN *MICROSERIS*

Microseris is a genus of about 20 species of dandelion-like composites (Chambers, 1955; Bachmann *et al.*, 1979), all but two of which occur in Western North America. Most are diploid with $n=9$ chromosomes. There are two major distinct groups of species. One consists of relatively large, perennial species that are self-sterile and depend on cross-fertilization. These species, for which *M. laciniata* is an example, grow in coastal northwestern California and in western Oregon and Washington as well as some higher elevations outside that area (Chambers, 1957; Mauthe *et al.*, 1982). The other group consists of smaller annual species that are self-fertile and seem to reproduce mainly by selfing; *M. douglasii* is a typical example. Their distribution centres roughly on the San Francisco Bay area. They occur in much drier locations than the perennials, down to Baja California (Chambers, 1955).

The two groups have strikingly different genome sizes. The haploid genomes of the perennials contain about 3.4 pg of DNA, those of the annuals about 1.4 pg (Price and Bachmann, 1975). Since the annuals are derived and specialized compared to the perennials, their evolution probably involved the loss of about 60% of the genomic DNA. The alternative possibility that the perennials have gained additional DNA after the annuals had evolved from them is unlikely for several reasons. Those features of the perennials that make them unspecialized morphologically, and ecologically adapted to the mesic conditions from which the dry-adapted annuals have originated, are the features that are correlated with (and, we argue here, determined by) the high genome size. This has been shown directly for the duration of the mitotic cycle (Price and Bachmann, 1976) and for cell size (pollen grains: Feuer and Tomb, 1977). Thus, the large genome size is an integral part of the ancestral condition of the perennials. This is supported by comparative biochemical fractionation of *Microseris* genomes (Bachmann and Price, 1977; Hemleben *et al.*, 1978). While the annuals have closely similar total DNA amounts, these equal-sized genomes contain very different proportions of highly repetitive, middle repetitive and ribosomal DNA fractions. There is no common small genome. Rather, the different small genomes can easily be interpreted as various remnants of the reduction of a large genome such as that of *M. laciniata*.

Several aspects of this evolutionary pattern in *Microseris* appear to be typical for other plant groups. Massive decreases in the nuclear DNA amount have been postulated, for instance for *Lathyrus* (Rees and Hazarika, 1969; Rees and Narayan, 1981), for *Crepis* (Jones and Brown, 1976), for *Scilla* (Greilhuber, 1979), and for *Atriplex* (Belford and Thompson, 1981). In these cases, as in *Microseris*, the nuclear DNA amounts of the existing species are not randomly distributed between the highest and the lowest. In these genera, and in *Anemone* (Rothfels *et al.*, 1966), *Vicia* (Martin and Shanks, 1966) and *Clarkia* (Rees and Narayan, 1981) genome sizes are distributed in discrete groups. This is somewhat different from the typical pattern in animals, where the nuclear DNA amounts in groups of related species often show a continuous log-normal distribution (Bachmann *et al.*, 1972; Hinegardner, 1976).

A SALTATORY MODE OF INCREASE?

We presume that the discrete DNA amount distributions found in many plant groups represent the basic mutational pattern rather than the result of specific selection. The remarkable differences in genome size could arise by a 'saltatory' process possibly in a single individual or certainly within a lineage of very few generations. A molecular mechanism that cuts out homologous repetitive sequences from all chromosomes simultaneously, together with any other sequences embedded in them, is easier to imagine today than a few years back, and it would explain some of the patterns found in species comparisons (Bachmann and Price, 1977; Rees and Narayan, 1977). The event could be rare, or rarely leave a viable genome. The population resulting from such a change, if it survives at all, may constitute a new species with new adaptations subject to stabilizing selection at or near the new level.

Generally, all specimens of a species have very closely identical genome sizes. At a much smaller level of variation, each individual probably has its own genome size. Apparently, the specific genome size is controlled closely by stabilizing (normalizing) selection, and no appreciable pool of different genome sizes is maintained in populations which is available for directional selection. Directional selection may be a typical feature after speciation by polyploidy. Many polyploid species have nuclear DNA contents that differ from the sum of the component genome sizes (Bachmann and Bogart, 1975; Grant, 1969). Saltatory changes in genome size without polyploidization may be followed by a period of genome size adjustment through directional selection. Since we cannot estimate original DNA values in cases other than those involving polyploidization, a saltational event will have to be observed (or produced) before this point can be settled. In any case, the usual constancy of specific genome sizes compared to the large differences among species argues against slow continuous changes in genome sizes through directional selection. Moreover, intergenomic recombination, usually the major factor in the generation of genetic diversity, hardly ever takes place among organisms with greatly differing genome sizes. Such recombination has been achieved experimentally, for instance in *Lolium* (Hutchinson *et al.*, 1979). Since the genome size differences usually involve all chromosomes, a wide continuous distribution between the parental.extremes should be seen in the resulting population. No such case is known from nature.

These considerations lead to the conclusion that in plants the 'mutational effects' play a considerable role in the establishment of genome sizes. If this is the case, it will greatly influence the mode of selection. While a steady selection pressure for a certain genome size is necessary for the accumulation of small mutational changes, the selective factors coming into play after a large saltatory change may be completely new. The product of saltatory evolution may not be recognized by the agents of natural selection as a deviation from its parental species norm. Thus it may not be in competition with its parental population but with some other species. This depends on the kind and magnitude of the phenotypic changes correlated with a genome size change.

PHENOTYPIC CORRELATES WITH GENOME SIZE

Comparisons of phenotypic differences of closely related species with differing genome sizes have been made in several groups of plants and animals (Bachmann and Rheinsmith, 1973; Bachmann *et al.*, 1972, 1979; Conner *et al.*, 1972; Hinegardner, 1976; Price *et al.*, 1974; Rees, 1972). These have shown that a few basic properties of nuclei and cells seem to be affected directly by the DNA content of the genome, cell size and cell cycle time being the most obvious (Bachmann *et al.*, 1972; Bennett, 1972; Cavalier-Smith, 1978 and this volume (Chapter 4); Evans *et al.*, 1972; Martin, 1966; Olmo and Morescalchi, 1975; Price *et al.*, 1973; Price and Bachmann, 1976; Szarski, 1976; Van't Hof, 1965). The relationship between genome size and the minimum duration of the mitotic cycle is very close, at least in plants. The observation that different major plant groups have different such close correlations clearly shows, however, that even this relationship is not entirely determined by the brute amount of DNA (Rees *et al.*, 1982).

Based on these correlations, derived characteristics of species should show similar correlations: sizes of structures made up of cells, growth rates, generation times. Such correlations have been found. It must be noted, however, that with increasingly complex interactions between the primary cellular feature and the final phenotype the correlations can become increasingly imprecise. Body size in animals, for instance, has virtually no relationship with cell size and neither has the size difference between trees and related herbaceous plants (El Lakany and Dugle, 1972; Lack *et al.*, 1978). The relationship between the rate of embryonic development and genome size in frogs which was considered to be direct and linear (Bachmann, 1972) turned out to be more complex, when additional data were obtained (Oeldorf *et al.*, 1978): obviously development can be slowed down independently of genome size but genome size sets a limit to the genetical acceleration of development. This is comparable to the relationship between genome size and the minimum generation time in higher herbaceous plants, where only species with less than 30 pg of DNA per nucleus can be annuals or ephemerals while species with more nuclear DNA than that appear to be limited to a perennial style (Bennett, 1972). In spite of the indirect relationship, these features ultimately seem to depend on the brute amount of DNA present in the genome. Bennett (1972) has termed these quantitative effects of the DNA amount 'nucleotypic effects' as compared to the genotypic effects coded for by sequence information. We have previously characterized these nucleotypic effects as setting the quantitative scale, both in time and space, for the qualitative DNA effects (Bachmann *et al.*, 1979).

As convincing as these correlations may be, an appraisal of the effectiveness of selection will finally depend on experimental evidence: how closely is a selection for genome size followed by changes in phenotypic characteristics, or how closely is selection in supposedly nucleotypically determined characters followed by genome size changes.

GENETICS OF GENOME SIZE DIFFERENCES

An experimental genetic approach to selection for genome size has been hampered by the nucleotypic effects on size and generation time: selection experiments depend on large numbers of individuals with short generation times. All experimentally suitable organisms have such small genomes that genome size changes can not be routinely determined by the usual spectrophotometric methods. Any pair of related species with an appreciable difference in genome size diverges so much in life cycle parameters as to make comparisons difficult. These problems are more easily circumvented in plants, and it is in plants that recently the first relevant experiments have been performed. These experiments have not shown a single reliable case of a correlated change in nuclear DNA amount and any phenotypic parameter. At first glance this result was surprising, but on looking again it would have been surprising if they had. A short description of the experiments will show why.

In the grass genus, *Lolium*, inbreeding species have about 40% more nuclear DNA than the outbreeders (Rees and Jones, 1967). Despite the large DNA difference, the inbreeding and outbreeding species of *Lolium* can be hybridized. Most of the F1 hybrids, moreover, are fertile, producing F2s by selfing or intercrossing, and backcross progenies with either 'high' or 'low-DNA' parents. The DNA amount difference between the inbreeding and outbreeding *Lolium* species is not due to a difference in chromosome number. All species are diploids with $2n=14$, and the DNA amount difference involves all chromosomes. In the hybrids, intermediate DNA amounts are found, and there is segregation of DNA amounts in their offspring. For instance, in the cross of *L. perenne* (4.16 pg DNA per diploid nucleus) with *L. temulentum* (6.23 pg), an F1 with 5.33 pg DNA was obtained (expected 5.20). The backcross of this to the low DNA parent resulted in offspring with DNA amounts varying between that of the hybrid and that of the low DNA parent. In the parent and backcross families 19 morphological and other phenotypic characters were investigated. For most of these characters there are substantial and highly significant differences between the parents. The scores for these phenotypic values also segregated in the backcross offspring. However, there was no correlation between the DNA amount and the score for any of these characters that would indicate a direct relationship (Hutchinson *et al.*, 1979). We have recently performed intraspecific crosses between strains of *Microseris douglasii* with different DNA amounts, and have found no correlation between DNA amount segregation and that of any other character (unpublished data).

In both the *Lolium* and *Microseris* experiments at least some of the phenotypic characters examined were known to be under polygenic control. Certainly in both cases the hybrid individuals were heterozygous for many structural genes. Thus, the phenotypic segregation involves a part that is due to the segregation of loci with a normal genetic influence via their transcriptional and translational products in addition to any segregation of quantitative nucleotypic effects. Since the polygenic segregation need not parallel but may even counteract any nucleotypic effects, any detection of a correlation between DNA amount and phenotypic characters would virtually have demanded a very strange explanation: it would have meant that the supposed 'polygenes' are not genes at all, but (transcribed or non-transcribed) DNA sequences acting quantitatively via their amount. Such relationships between some 'polygenes' and the nuclear DNA content have been suggested, but can certainly never account for all polygenic variation. In any case, the analysis of the data obtained by Hutchinson *et al.* (1979) in terms of quantitative genetics is illuminating. The data given in Fig. 3 of that paper together with estimates on the statistical variation involved in the measurements suggest that a correlation could hardly have been detected even if it has existed. This is even more the case for our own data where the differences in DNA amount between the parental strains are much smaller than for *Lolium*. The total range of DNA values involved in the *Lolium* segregation is about 1 pg of DNA, that in *Microseris* 0.1 pg. Such small genome size differences are hardly accompanied by significant phenotypic differences in interspecies correlations.

Segregation analysis in a cross between species differing by very large amounts of DNA may not be possible, though it is worth a few tries. Surprisingly, a considerable difference in DNA content does not seem to be an obstacle to proper meiotic pairing between chromosomes (Togby, 1943; Rees *et al.*, 1982). Usually, though, a large change in nuclear DNA content is accompanied by sufficient restructuring of the chromosomes to interfere with meiosis.

ARTIFICIAL SELECTION AND GENOME SIZE

As a different, easier experimental approach would be welcome, we suggest that a very strong selection on the phenotype should be applied for several generations, and the product of such a selection should be checked for any DNA amount differences. This eliminates the need for a large number of very precise DNA amount determinations which soon become a limiting factor. Also, much stronger selection can be exerted on the phenotype than on the very small DNA amount differences. Lastly, an effect on genome size via phenotypic selection would correspond to the natural selection process that we postulate.

Recent analysis of intraspecific DNA amount variation in *Microseris douglasii* (Price *et al.*, 1981) gives some useful hints for such experiments, especially concerning relevant selective factors. We have looked at DNA amounts in plants of 24 geographically, ecologically and morphologically diverse populations of this very variable species in California. The plants with the largest genomes had 20% more DNA per genome (of about 1.5 pg haploid) than those with the smallest genomes. This variation was not correlated with morphological traits. Mean DNA contents of populations varied about 14%. In most populations genome size was constant even when there were many morphologically diverse biotypes. Populations with lower genome sizes apparently can grow throughout the range of the species, while those with higher DNA content are restricted to sites with well-developed soil and a fair amount of moisture. At a single site near Jolon, California, the mean genome size was 9% higher in plants grown from seed collected in 1973 compared with plants grown from the 1962 and 1977 harvests. The low-DNA years each followed major California droughts. All this points to environmental stress as a major selective factor: when the environment selects for those specimens that can survive in dry poor soil by quickly forming diminutive plants with hardly any structures besides reduced but normal flowering heads, there may be an enormously strong selection for any genetic factor that helps fast growth and small stature. Such situations are not rare in nature. Any selection experiment in normal or luxurious greenhouse circumstances may never come near the selective differential of these natural circumstances. Selection under nutrient and moisture stress may be the ideal way to set up the next experiments. We have recently buried in natural *Microseris* habitats in California mixed lots of 'high'- and 'low-DNA' seed of *M. douglasii*, recognizable by morphologic markers. We shall follow their fate over the coming seasons.

Experimental proof of a change of genome size brought about by selection on the phenotype still is lacking and there is nothing known about the dynamics of the supposed correlations between phenotypic characters and the DNA amount during experimental selection. On the other hand, the much discussed genome size–phenotype correlations are as valid as ever. It is presumed that an experimental demonstration of a genome size decrease will be feasible under strong selection pressure for rapid reproduction under stressful environmental conditions. This may take a few years. Any change in genome size in response to selection would necessarily also constitute 'intragenomic selection' among the sequences for those that can be lost in order to achieve the quantitative response without qualitative side effects. It would be interesting to check if mobile transposable elements (the ideal 'selfish DNA') are preferentially preserved or if they are especially vulnerable when there is a premium on small genome size.

NOTE ADDED IN PROOF:

Work performed since the completion of this chapter has shown striking differences between *Microseris* and the reported results on *Lolium* in crosses between plants differing in genome size. Such crosses in *Microseris* lead to all sorts of unpredictable genomic readjustments during the development of the F1 plant, with or without effects on chromosome number and meiosis. Different genome sizes are found in offspring families derived from different heads of the same plant by selfing (Price *et al.*, 1983, *Amer. J. Bot.* **70**, 1133). Phenotypic effects are genetically stable or unstable and may involve profound alterations of developmental regulatory processes and "macromutations".

REFERENCES

Bachmann, K. (1972). Nuclear DNA and developmental rate in frogs. *Quart. J. Florida Acad. Sci.*, **35**, 225–231.

Bachmann, K. and Bogart, J. P. (1975). Comparative cytochemical measurements in the diploid-tetraploid species pair of hylid frogs, *Hyla chrysoscelis* and *versicolor*. *Cytogenet. Cell Genet.*, **15**, 186–194.

Bachmann, K., Chambers, K. L. and Price, H. L. (1979). Genome size and phenotypic evolution in *Microseris* (Asteraceae, Cichorieae), *Plant Syst. Evol. Suppl.*, **2**, 41–66.

Bachmann, K., Goin, O. B. and Goin, C. J. (1972). Nuclear DNA amounts in vertebrates. *Brookhaven Symp. Biol.*, **23**, 419–450.

Bachmann, K. and Price, H. J. (1977). Repetitive DNA in Cichorieae (Compositae). *Chromosoma*, **61**, 267–275.

Bachmann, K. and Rheinsmith, E. L. (1973). Nuclear DNA amounts in Pacific Crustacea. *Chromosoma*, **43**, 225–236.

Belford, H. S. and Thompson, W. F. (1981). Single copy DNA homologies in *Atriplex* I. Cross reactivity estimates and the role of deletions in genome evolution. *Heredity*, **46**, 91–108.

Bennett, M. D. (1972). Nuclear DNA content and minimum generation time in herbaceous plants. *Prco. R. Soc. B*, **181**, 109–135.

Chambers, K. L. (1955). A biosystematic study of the annual species of *Microseris*. *Contrib. Dudley Herb*, **4**, 207–312.

Chambers, K. L. (1957). Taxonomic notes on some compositae of the western United States. *Contrib. Dudley Herb.*, **5** 57–67.

Cavalier-Smith, T. (1978). Nuclear volume control by nucleoskeletal DNA, selection for cell volume and cell growth rate, and the solution of the DNA C-value paradox. *J. Cell Sci.*, **34**, 247–278.

Conner, W., Hinegardner, R. and Bachmann, K. (1972). Nuclear DNA amounts in polychaete annelids. *Experientia*, **28**, 1502–1504.

Doolittle, W. F. and Sapienza, C. (1980). Selfish genes, the phenotype paradigm and genome evolution. *Nature, Lond.*, **284**, 601–603.

El-Lakany, M. H. and Dugle, J. R. (1972). DNA content in relation to phylogeny of selected boreal forest plants. *Evolution*, **26**, 427–434.

Evans, G. M., Rees, H., Snell, C. L. and Sun, S. (1972). The relationship between nuclear DNA amount and the duration of the mitotic cycle. *Chromosomes Today*, **3**, 24–31.

Feuer, S. and Tomb, A. S. (1977). Pollen morphology and detailed structure of family Compositae, tribe Cichorieae. II. Subtribe Microseridinae. *Amer. J. Bot.*, **64**, 230–245.

Grant, W. F. (1969). Decreased DNA content of birch (*Betula*) chromosomes at high ploidy as determined by cytophotometry. *Chromosoma*, **26**, 326–336.

Greilhuber, J. (1979). Evolutionary changes of DNA and heterochromatin amounts in the *Scilla bifolia* group (Liliaceae). *Plant Syst. Evol.*, Suppl. **2**, 263–280.

Hemleben, V., Bachmann, K. and Price, H. J. (1978). Ribosomal gene numbers in *Microseridinae* (*Compositae*: Cichorieae). *Experientia*, **34**, 1452–1453.

Hinegardner, R. (1976). Evolution of genome size. In F. J. Ayala (Ed.), *Molecular Evolution*, Sinauer, Sunderland.

Hutchinson, J., Rees, H. and Seal, A. G. (1979). An assay of the activity of supplementary DNA in *Lolium*. *Heredity*, **43**, 411–421.

Jones, R. N. and Brown, L. M. (1976). Chromosome evolution and DNA variation in *Crepis*. *Heredity*, **36**, 91–104.

Lack, H. W., Sack, G. and Bachmann, K. (1978). The genome of *Dendroseris litoralis*, an arboreal insular endemic of the tribe Lactuceae (Asteraceae). *Beitr. Biol. Pflanzen*, **54**, 425–441.

Martin, P. G. (1966). Variation in the amounts of nucleic acids in the cells of different species. *Exp. Cell Res.*, **44**, 84–90.

Martin, P. G. and Shanks, R. (1966). Does *Vicia faba* have multistranded chromosomes? *Nature, Lond.*, **211**, 650–651.

Mauthe, S., Bachmann, K., Chambers, K. L. and Price, H. J. (1982). Variability of the inflorescence among populations of *Microseris laciniata* (Asteraceae, Lactuceae). *Beitr. Biol. Pflanzen*, **56**, 25–52.

Oeldorf, E., Nishioka, M. and Bachmann, K. (1978). Nuclear DNA amounts and developmental rate in holarctic anura. *J. Zool. Syst. Evolutionsfschg.*, **16**, 216–224.

Olmo, E. and Morescalchi, A. (1975). Evolution of the genome and cell sizes in salamanders. *Experientia*, **31**, 804–806.

Orgel, L. E. and Crick, F. H. C. (1980). Selfish DNA: the ultimate parasite. *Nature, Lond.*, **284**, 604–607.

Price, H. J. and Bachmann, K. (1975). DNA content and evolution in the Microseridinae. *Amer. J. Bot.*, **62**, 262–267.

Price, H. J. and Bachmann, K. (1976). Mitotic cycle time and DNA content in annual and perennial Microseridinae (Compositae Cichoriaceae). *Plant Syst. Evol.*, **126**, 323–330.

Price, H. J., Chambers, K. L. and Bachmann, K. (1981). Geographic and ecological distribution of genomic DNA content variation in *Microseris douglasii* (Asteraceae). *Bot. Gaz.*, **142**, 415–426.

Price, H. J., Sparrow, A. H. and Nauman, A. F. (1973). Correlations between nuclear volume, cell volume and DNA content in meristematic cells of herbaceous angiosperms. *Experientia*, **29**, 1028–1029.

Price, H. J., Sparrow, A. H. and Nauman, A. F. (1974). Evolutionary and developmental considerations of the variability of nuclear parameters in higher plants I. Genome volume, interphase chromosome volume and estimated DNA content of 236 gymnosperms. *Brookhaven Symp. Biol.*, **25**, 390–421.

Rees, H. (1972). DNA in higher plants. *Brookhaven Symp. Biol.*, **23**, 394–418.

Rees, H. and Hazarika, M. H. (1969). Chromosome evolution in *Lathyrus*. *Chromosomes Today*, **2**, 158–165.

Rees, H., Jenkins, G., Seal, A. G. and Hutchinson, J. (1982). Assays of the phenotypic effects of changes in DNA amounts. In G. A. Dover and R. B. Flavell (Eds), *Genome Evolution*, Academic Press, London.

Rees, H. and Jones, G. H. (1967). Chromosome evolution in *Lolium*. *Heredity*, **22**, 1–18.

Rees, H. and Narayan, R. K. J. (1977). Nuclear DNA variation in *Lathyrus*. *Chromosomes Today*, **6**, 131–139.

Rees, H. and Narayan, R. K. J. (1981). Chromosomal DNA in higher plants. *Phil. Trans. R. Soc. Lond. B*, **292**, 569–578.

Rothfels, K., Sexsmith, E., Heimburger, M. and Krause, M. O. (1966). Chromosome size and DNA content of species of *Anemone* L. and related genera (Ranunculaceae). *Chromosoma*, **20**, 54–74.

Szarski, H. (1976). Cell size and nuclear DNA content in vertebrates. *Int. Rev. Cytol.*, **44**, 93–109.

Togby, H. A. (1943). A cytological study of *Crepis fuliginosa*, *C. neglecta*, and their F1 hybrid, and its bearing on the mechanism of phylogenetic reduction in chromosome number. *J. Genetics*, **45**, 67–111.

Van't Hof, J. (1965). Relationship between mitotic cycle duration, S period duration and the average rate of DNA synthesis in the root meristem cells of several plants. *Exp. Cell Res.*, **39**, 48–58.

The Evolution of Genome Size
Edited by T. Cavalier-Smith
© 1985 John Wiley & Sons Ltd

CHAPTER 10

Basis of Diversity in Mitochondrial DNAs

G. D. CLARK-WALKER

Department of Genetics, Research School of Biological Sciences, Australian National University, Canberra, Australia

SUMMARY

As well as differing many-fold in size, mitochondrial genomes differ in the amount of non-coding DNA, in the number and arrangement of protein-coding genes, in the number of DNA molecules per genome, in whether their genes are split or not by non-coding sequences, and even in the genetic code. This chapter discusses the cause of these evolutionary differences.

INTRODUCTION: THE DIVERSITY OF MITOCHONDRIAL DNA STRUCTURE

Before the discovery that eukaryotic genes are both bounded and interrupted by non-coding sequences, it seemed likely that each nucleotide in a genome would have a specific purpose. Surprisingly, mitochondrial DNA (mtDNA) in multicellular animals may be the only example of frugal sequence architecture in eukaryotes. By contrast, mtDNAs from various fungi can have genes interrupt with non-coding sequence and bounded by larger spacer regions (Borst and Grivell, 1981; Mahler, 1981; Dujon, 1981; Wallace, 1982; Grivell, 1983) which is reminiscent of nuclear DNA from vertebrates and higher plants.

Diversity of mtDNA structure is not confined to the characteristics just mentioned as the genome can be either circular, linear (Goddard and Cummings, 1977; Goldbach *et al.*, 1977; Wesolowski and Fukuhara, 1981), perhaps even dispersed between several molecules, e.g. plant mtDNAs (Leaver *et al.*, 1982; Dale, 1982; Wallace, 1982), or incorporated into a network like kinetoplast DNA (Borst and Hoeijmakers, 1979). Sizes of mtDNAs can vary between closely related organisms (Clark-Walker *et al.*,

1981a; Ward *et al.*, 1981; McArthur and Clark-Walker, 1983) and the size of the coding portion of individual genes is also changeable (e.g. ribosomal RNA genes, Grant and Lambowitz, 1981; Kuntzel and Kochel, 1981). Within mtDNAs there are rearrangements of genes (Macino *et al.*, 1980; Clark-Walker and Sriprakash, 1981), presence or absence of genes (Tzagoloff *et al.*, 1979; Borst and Grivell, 1981) and fragments of genes (Clark-Walker *et al.*, 1981b; Brown *et al.*, 1983a). Furthermore, some molecules contain tandem duplications (Manella *et al.*, 1979), inverted duplications (Goldbach *et al.*, 1977; Clark-Walker *et al.*, 1981b; Hudspeth *et al.*, 1983), multiple copies of short sequences (e.g. PstI palindromic sites in *Neurospora crassa*: Yin *et al.*, 1981; G + C rich regions in *Saccharomyces cerevisiae*: Prunell and Bernardi, 1977; Cosson and Tzagoloff, 1979; Sor and Fukuhara, 1982a) or even a chloroplast DNA sequence (Stern and Lonsdale, 1982). In addition, differences have been found in the genetic code (Barrell *et al.*, 1979; Hensgens *et al.*, 1979; Bonitz *et al.*, 1980a; Heckman *et al.*, 1980; Anderson *et al.*, 1981), and the mtDNA of mammals has a high base substitution rate characterized by a predominance of transitions over transversions (Brown *et al.*, 1982). Finally, it must be mentioned that plasmids have been found in mitochondria of both fungi (Stohl *et al.*, 1982; Nargang *et al.*, 1983) and higher plants (Pring *et al.*, 1977; Kemble and Bedbrook, 1980; Palmer *et al.*, 1983) but their relationship to the mitochondrial genome remains unclear.

From the above list it is apparent that the mitochondrion has proved to be an excellent cocoon for the metamorphosis of DNA. In this article my intention is to try to deduce how some of the diversity in mtDNA structure has arisen and to highlight gaps in our knowledge that more data may help to bridge.

ORIGIN OF MITOCHONDRIA

In trying to account for some aspects of mtDNA diversity it is pertinent to ask whether any differences can be traced to possible multiple origins for mitochondria. Proposals for the acquisition of mitochondria and mtDNA centre on either the endosymbiont hypothesis or the membrane sequestration of episomal DNA (Margulis, 1970; Raff and Mahler, 1972; Gray and Doolittle, 1982; Wallace, 1982 and articles in *Ann. N.Y. Acad. Sci.* vol 361, 1981. Evolution of organelle genomes and protein synthesizing systems). Amongst adherents of the endosymbiont hypothesis, the author included, there is debate concerning the possibility that mitochondria may have arisen on more than one occasion. This debate stems from two observations. In the first instance, it has been proposed that mitochondria of *Euglena* and *Crithidia*, *Tetrahymena* and collectively plants, animals and fungi were established by three separate symbioses (Dayhoff and Schwartz, 1981). This proposal is based on cytochrome c sequence comparisons and it is assumed for every organism that this nuclear encoded gene has been derived from the endosymbiont. However, it is not essential to the endosymbiont hypothesis that all genes concerned with mitochondrial function should be derived from the organelle as it has been suggested that the host at the time of colonization could have had a respiratory system (Hall, 1973) instead of being a fermentative organism as originally envisaged (Margulis, 1970). Hence, in the absence of knowledge as to whether the cytochrome c gene is derived from the host or the endosymbiont in each instance, it is impossible to make a decision based on the sequence of a nuclear gene as to the mono- or polyphyletic origin of mitochondria.

The second set of observations provoking discussion on possible multiple origins of mitochondria comes from sequence comparisons of small ribosomal RNA (S rRNA) genes. It has been found that bacterial and fungal mtDNA sequences of S rRNA show greater similarity than the latter does with S rRNA of mammalian mtDNA (Kuntzel and Kochel, 1981). These observations have led the authors to propose separate origins for fungal and mammalian mtDNAs. Their interpretation is weakened by data showing that mammalian mtDNA undergoes a higher base substitution rate than nuclear sequences (Brown *et al.*, 1982). This latter result raises the possibility that different organisms' mtDNA may have different rates of base substitution and that the rates may have changed with time. Hence it will be important to determine whether high base substitution rate is a universal attribute of the mitochondrial genome if we are to attempt to use sequence comparisons for constructing phylogenetic trees linking distantly related organisms.

High base substitition rate in mtDNA may also help to explain differences in the genetic code, not only from the universal code but also between mitochondrial genomes. For example, both fungi and mammals use the stop codon UGA for tryptophan whereas in plants this amino acid is coded by CGG (normally arginine) as well as the usual UGG (Fox and Leaver, 1981). At first glance this information could be taken to support separate origins for the plant mitochondrial genome from those of animals and fungi. However, it has also been found that some ascomycetes differ in their genetic code, for instance in the mtDNA of *Saccharomyces cerevisiae* the CUN family codes for threonine in place of leucine and AUA is used for methionine, while in the mitochondrial genomes of *Aspergillus nidulans* and *Neurospora crassa* CUN is retained for leucine and AUA is used for isoleucine. This knowledge together with the possibility that mtDNA in general may have a high base substitution rate makes it likely that the codon changes listed above and additional variations (Wallace, 1982) have occurred since the mitochondrial genome originated.

In conclusion there is at present no compelling evidence to support a polyphyletic origin for the mitochondrial genome. In the absence of such evidence the simplest postulate is that all aspects of mtDNA diversity must have occurred subsequent to the single event leading to the establishment of this organelle. Some steps in diversification can be deduced in outline from the available information on mtDNA structure and central to this analysis is mtDNA size.

DIVERSITY IN mtDNA SIZE

The most notable aspect of mtDNA size is that molecules from animals, excluding protists, have sizes of 16–19 kbp, whereas wide variations occur in other groups even amongst related forms (Gillham and Boynton, 1981; Wallace, 1982). Striking examples of size variations amongst related forms are the 7-fold range of melon mtDNAs from 330–2400 kbp (Ward *et al.*, 1981) and the 4-fold range in the mtDNA of the *Dekkera/Brettanomyces* yeasts from 28–100 kbp (McArthur and Clark-Walker, 1983). Sizes of mtDNAs are puzzling on two counts because we have to explain not only how variation occurs within a set of mtDNAs but, conversely, the absence of extensive size variation in the animal mtDNAs. Moreover, not only do plant mitochondrial genomes vary in size between species (as determined by renaturation analysis), but the genome within individuals appears to consist of many separate molecules as observed by both electron microscopy and gel electrophoresis (Wallace, 1982). Whether this heterogeneity reflects a truly polyoecious* genome or recombination intermediates (Sederoff *et al.*, 1981) remains to be demonstrated. However, recombination products may exist as cross hybridization has been demonstrated between some of the different size classes of DNA (Dale, 1981).

In attempting to account for size diversity between mtDNAs it needs to be emphasized that closely related organisms such as members of the melon family or yeast genus mentioned above show some of the greatest variation. This suggests that most of the intrageneric variability is unlikely to be due to the differential presence of genic sequences. Nevertheless, before examining possible causes for intrageneric variability it is now clear that separate groups do differ in the genes encoded by their mtDNAs.

DIFFERENTIAL PRESENCE OF GENES

A teleological explanation has been advanced to account for the retention of the mitochondrial genome as a separate entity. This argument is based on the observation that most of the encoded proteins are highly hydrophobic. According to this proposal, *in situ* production of hydrophobic proteins has been selected for in the face of a counter pressure to transfer genes to the nucleus. Implicit in this argument is the need for constancy of the mtDNA

---

* A neologism derived from *poly* (many) *oecious* (house).

genetic composition; if exceptions have to be made the proposal becomes less convincing. Many such exceptions have been found. In the first instance, the hydrophobic subunit 9 protein of the ATPase complex has been found to be encoded by the mitochondrial genome of yeast but in mammals this gene is in the nucleus (Tzagoloff *et al.*, 1979; Anderson *et al.*, 1981). On the other hand, filamentous fungi, specifically *Neurospora crassa*, contain ATPase subunit 9 sequences in both nuclear and mtDNA although it is the nuclear sequence that is transcribed and translated (Van den Boogaart *et al.*, 1982). The conventional explanation advanced to account for the dual location of this sequence is that the mitochondrial gene has been translocated to the nucleus but it remains a possibility that the nuclear sequence is a host gene as discussed earlier. However, the former explanation has been strengthened by the observation that mtDNA sequences are known to have been translocated to the nucleus in yeast (Farrelly and Butow, 1983), sea urchin (Jacobs *et al.*, 1983) and filamentous fungi (Wright and Cummings, 1983).

Other examples of genic differences between mitochondrial genomes are the presence of seven unidentified reading frames (URFs) in mammalian mtDNAs (Anderson *et al.*, 1981; Bibb *et al.*, 1981) and the absence of similar sequences from the mtDNA of the yeast *Torulopsis glabrata*. The evidence that the mammalian mtDNA URF sequences are active genes rests on the identification of transcripts for these regions (Battey and Clayton, 1978 and 1980; Montoya *et al.*, 1981) and the conservation of sequences between these regions in human, bovine and mouse mitochondrial genomes (*op. cit.*).

Absence from the 18.9 kbp *T. glabrata* mtDNA of at least the larger representatives of the mammalian URF sequences can be deduced from the transcript map of this yeast's mitochondrial genome (Clark-Walker and Sriprakash, 1983a) (Fig. 10.1). Transcripts have been identified for cytochrome b; cytochrome oxidase subunits 1, 2 and 3; ATPase subunits 6 and 9 and the large and small rRNAs. Although a region of the mtDNA has been shown to hybridize to the variant 1 protein gene of *S. cerevisiae* mtDNA no transcript has been found for this region. Other gaps in the map are occupied by tRNA genes. Additionally, the ATPase associated protein (aapl) gene (Novitski *et al.*, 1983), which is thought to be equivalent to the mammalian URFA6L sequence, has been found to be co-transcribed with the ATPase subunit 6 sequence. Hence it appears that there is insufficient unaccounted sequence in this genome to specify the larger URF sequences found in mammalian mtDNAs.

By contrast the 32 kbp mitochondrial genome of *Aspergillus nidulans* has been found to contain eight unidentified reading frames, two of which show homology to the mammalian mtDNA sequences URF1 and 4 (Netzker *et al.*, 1982; Brown *et al.*, 1983b). A further difference between the *A. nidulans* and *T. glabrata* mtDNAs may lie in the absence of the variant 1 protein

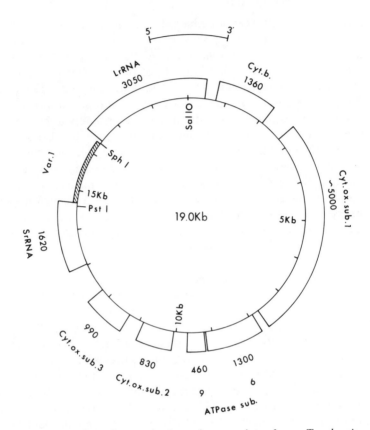

Fig. 10.1. Location and size of transcripts from *Torulopsis glabrata* mtDNA. The hatched region labelled var.1 hybridizes to the variant 1 sequence of *Saccharomyces cerevisiae* mtDNA but a transcript of this region in *T. glabrata* mitochondria has not been found. The divisions on the circle represent 1000 bp and the length of each transcript is given in nucleotides. All transcripts are produced from the same strand according to the orientation at the top of the figure. Reproduced with permission from Clark-Walker and Sriprakash (1983a).

sequence from the genome of the fungus as this segment has not yet been reported despite sequence data having been obtained for 27 kbp of the molecule (Brown *et al.*, 1983b). As noted above this sequence may not be an active gene in *T. glabrata* mtDNA whereas in mitochondria of *S. cerevisiae* this sequence is transcribed and translated (Hudspeth *et al.*, 1982).

Characterization of plant mtDNAs in terms of genic content is less advanced than for fungi or animals. In general plant mitochondrial genomes, ranging from around 100 kbp to over 2000 kbp (Wallace, 1982), might be thought to contain many more genes than the mtDNAs of fungi or animals.

This may not necessarily be true as only 18–20 polypeptides have been found to be made in isolated mitochondria from maize (Leaver *et al.*, 1982). From amongst the sequences that have been identified in plant mtDNAs, which include some tRNA genes (Wallace, 1982), the cytochrome oxidase subunit 2 gene (Fox and Leaver, 1981) and large, small and 5s-rRNAs (Bonen and Gray, 1980), the 5s-rRNA sequence is unique to plants. Other differences can be anticipated because in plants there may be interaction between the chloroplast and mitochondrial genomes which could require specific regulatory genes.

From these results it is clear that mtDNAs from separate groups of organisms can differ in the genes they possess. Missing mtDNA genes have either been transferred to the nucleus (before codon changes occurred) or lost from the organism. In yeasts, candidates for the latter possibility are the equivalents of the URF genes of mammalian mtDNA as it remains a possibility that these sequences are absent from yeast nuclear DNA. However, as already alluded to, differential gene retention is not the explanation for mtDNA size diversity amongst related organisms. At present, the best illustration that factors other than genic sequences contribute to changes in mtDNA size, comes from analysis of mitochondrial genomes from yeasts.

SIZE VARIATION IN mtDNAs OF YEASTS

The size of mtDNA in yeasts ranges from 18.9 kbp in *T. glabrata* (O'Connor *et al.*, 1976) and *Schizosaccharomyces pombe* (O'Connor *et al.*, 1975) to around 100 kbp in *B. custersii* (McArthur and Clark-Walker, 1983) with the most frequent size occurring between 25 and 40 kbp (Clark-Walker *et al.*, 1981a; McArthur and Clark-Walker, 1983 and Clark-Walker, unpublished observations). Even within genera there is considerable size diversity as noted above for the *Dekkera/Brettanomyces* yeasts. In addition, sizes in the *Saccharomyces* genus range from 23.7 kbp for *S. exiguus* to 81 kbp for some strains of *S. cerevisiae* (Clark-Walker *et al.*, 1981a and unpublished observations). This size diversity raises a question as to the direction of change since yeasts diverged. For example, are the smaller genomes derived from a larger form or vice-versa? In my view neither of these explanations is correct. It is proposed that both larger and smaller yeast mtDNAs are derived from one of intermediate size. This view is based on information obtained from gene mapping and sequencing studies on the 68–81 kbp mtDNA of *Saccharomyces cerevisiae* and on other mitochondrial genomes from both yeasts and the filamentous ascomycete *Aspergillus nidulans*. It is therefore my intention in the following section to use data from these studies to try to deduce the structure of the ancestral yeast mtDNA.

STRUCTURE OF mtDNAs

The first evidence that the mtDNA from *S. cerevisiae* has some unusual structures came from studies on base composition heteregeneity where it was shown that over 50% of the 68–81 kbp molecule is composed of A + T rich sequences with less than 5% G + C (Prunell and Bernardi, 1974). *S. cerevisiae* mtDNA also contains 70–100 regions of 30–50 bp that are rich in G + C (Prunell and Bernardi, 1977). Some of these G + C clusters contain HpaII and HaeIII sites (CCGG and GGCC) and, as discussed below, some of them occur in the genome more than once. The presence of these unusual sequence elements has been confirmed by sequence analysis whereby coding regions are found to be bounded by stretches of A + T rich non-coding spacers and G + C clusters have been found both within genes and between them (Cosson and Tzagoloff, 1979; Tzagoloff *et al.*, 1980; Borst and Grivell, 1981; Bernardi, 1982; Sor and Fukuhara, 1982a and 1982b; Farrelly *et al.*, 1982).

Sequence analysis has also confirmed earlier genetic evidence that some genes contain intervening sequences. For example, it has been found that the cytochrome b gene can exist in two forms due to the presence of five introns in the long version (Lazowska *et al.*, 1980) and the absence of the first three introns in the short form (Nobrega and Tzagoloff, 1980). Introns that cause such polymorphism have been termed optional (Borst and Grivell, 1981) and other examples are found in the large rRNA gene (Bos *et al.*, 1978; Dujon, 1980) and the cytochrome oxidase subunit 1 gene (Sanders *et al.*, 1977; Bonitz *et al.*, 1980b; Hensgens *et al.*, 1983).

In addition to being 'optional' there are two other notable attributes of intervening sequences in *S. cerevisiae* mtDNA. Firstly, most of the longer sequences contain open reading frames (ORFs) and, secondly, these ORFs show homology. In the first instance there is now substantial evidence that the ORFs located in the second (bI2) and fourth (bI4) introns of the long cytochrome b gene code for polypeptides that are involved in processing the primary transcript of this gene (De La Salle *et al.*, 1982; Mahler *et al.*, 1982; Weiss-Brummer *et al.*, 1982). Of more interest to the present discussion is sequence homology between introns which indicates that some of the ORFs are likely to have arisen by transposition/duplication. Thus it has been shown that the ORF in the fourth intron of the cytochrome b gene has sequence homology with the ORF in the fourth intron of the cytochrome oxidase subunit 1 gene (Bonitz *et al.*, 1980b). Likewise, the first and second introns of the cytochrome oxidase subunit 1 gene contain ORFs that are partially homologous (*op. cit.*). Indeed several other ORFs, and even some introns that lack open reading frames, have been found by computer search to show either sequence or structural homology (Michel *et al.*, 1982; Hensgens *et al.*, 1983). Such studies have enabled the first authors to construct dendrograms which show that introns can be placed in one of two broad groups. This implies that introns within a group share an ancestral sequence and at different times there have been duplications followed by sequence divergence. Whether these transposition/duplication events took place before or after yeasts originated is unknown. However, evidence for an ancient origin of one intron in yeast mtDNA comes from the observation that the intervening sequences in the L rRNA genes of *S. cerevisiae* mtDNA and *Physarum polycephalum* nuclear DNA have surrounding sequences that are strikingly similar (Nomiyama *et al.*, 1981).

Other events which have enlarged the *S. cerevisiae* mtDNA concern the short G + C rich clusters mentioned above and the 300 bp *ori/rep* sequence that confers replicative advantage on segments of mtDNA found in petite mutants (Blanc and Dujon, 1980, 1982; de Zamaroczy *et al.*, 1981). There are at least 70 G + C rich clusters of between 30 and 50 bp distributed around the molecule (Prunell and Bernardi, 1977). Many regions containing these clusters have been sequenced and 17 of them have been listed by Sor and Fukuhara (1982b). The points to note from this analysis are that each of the G + C rich regions is flanked by A + T rich sequence and bounded on the 5' end by TAGT and at the 3' end by AAGGAG. In four separate cases, near perfect repeats of a G + C element have been described (Cosson and Tzagoloff, 1979; Sor and Fukuhara, 1982b; Farrelly *et al.*, 1982; Bernardi, 1982; Butow *et al.*, 1983) and the S rRNA gene has been found to contain an optional G + C element (Sor and Fukuhara, 1982b). These results have led Sor and Fukuhara to suggest that the G + C element is propagated in the mtDNA and that the constant terminal nucleotides may play a part in the insertion of this element into A + T rich regions of the genome. It is important to add that these G + C elements appear to be confined to the mtDNA of *S. cerevisiae* and its close relatives as they have not been found in the mitochondrial genomes of other yeasts (Sor and Fukuhara, 1982a; McArthur and Clark-Walker, 1983). In this context it is interesting to note that the 18 nucleotide palindromic sequence found in multiple copies in *N. crassa* mtDNA (Yin *et al.*, 1981) has not been observed in the mitochondrial genome of *A. nidulans*. Again the suggestion has been made that this sequence element is propagated (*op. cit.*) and this proposal is supported by the observation that it occurs on a mitochondrial plasmid in one strain of *N. crassa* (Nargang *et al.*, 1983). The origins of the 18 nucleotide palindromic sequence and the G + C rich element in *S. cerevisiae* mtDNA are unknown but the possibility cannot be excluded that they have arisen from outside the mitochondrion like the chloroplast DNA sequence found in plant mtDNA (Stern and Lonsdale, 1982).

Propagation of the 300 bp *ori/rep* sequence in *S. cerevisiae* mtDNA also appears to have occurred. Some strains of this yeast have three *ori/rep* sequences in mtDNA (Blanc and Dujon, 1980) while others apparently have as many as seven (de Zamaroczy *et al.*, 1981). These sequences share extensive but not complete homology and again the suggestion has been made that they have arisen by duplication (Dujon, 1981; Bernardi, 1982).

In addition to optional introns there is now evidence for an optional sequence in a spacer region as it has been found that a polymorphism of 2000 nucleotides exists near the 3' end of the ATPase subunit 6 sequence (Cobon *et al.*, 1982). As the location of the 2000 nucleotide sequence is in the vicinity of *ori*7 it is possible that this polymorphism is associated with the gain or loss of an *ori/rep* element.

From the foregoing data on G + C and *ori/rep* structures it can be seen that the mitochondrial genome of *S. cerevisiae* has undergone a series of transposition/duplications that have led to expansion of the molecule. The question now becomes, have the *S. cerevisiae* mtDNA molecules that lack optional sequences arisen by loss of such structures or have they never gained them? For some optional introns a strong case can be made for their loss. Consider, for example, the short and long forms of the cytochrome b gene. The short form lacks the first three introns (bI1, bI2 and bI3) yet in the mtDNA of *Aspergillus nidulans* the cytochrome b gene contains a single intron that has homology to bI3 in the long form of the yeast gene (Waring *et al.*, 1982). Furthermore, the intron is located in the same place in the cytochrome b gene in each instance, thereby strengthening the idea that this sequence was present in the mtDNA of the shared ancestor before yeasts and filamentous fungi diverged. Therefore, it is reasonable to conclude that the absence of bI3 from the short form of the gene in *S. cerevisiae* mtDNA has resulted from loss.

Likewise the absence of the intervening sequence in the L rRNA gene of some *S. cerevisiae* mtDNAs can be explained as a loss. This is deduced from observations that similar flanking sequences and location have been found for the intervening sequence in the L rRNA gene of mtDNA from *N. crassa* (Yin *et al.*, 1981), *A. nidulans* (Kochel and Kuntzel, 1982) and *S. cerevisiae* (Dujon, 1980; Bos *et al.*, 1980). Hence it appears that an ancestral yeast mitochondrial genome had an intervening sequence in both the cytochrome b and L rRNA genes and that these introns, where absent in present forms of *S. cerevisiae* mtDNA, have been lost.

Similar reasoning can be applied to explain the uninterrupted structure of the cytochrome b and L rRNA genes in other yeast mtDNAs. For example, it has been shown from an analysis of transcripts of *T. glabrata* mtDNA that the cytochrome b and L rRNA genes lack intervening sequences (Clark-Walker and Sriprakash, 1983a; Fig. 1). As it is known that *S. cerevisiae* and *T. glabrata* have 80–90% sequence homology in genic regions of mtDNA (Clark-Walker, unpublished), and are therefore less divergent than either is to *A. nidulans*, it can be concluded that the absence of introns from the two genes of *T. glabrata* mtDNA is due to loss. Absence of the L rRNA intron from *T. glabrata* mtDNA as well as from some *S. cerevisiae* mtDNAs must also mean that elimination of this sequence has occurred on more than one occasion.

From the foregoing discussion the generalization is that yeast mtDNAs have undergone multiple independent events, both expansionary and contractionary, since yeasts diverged from a common progenitor. Another type of event that has complicated the task of establishing the form of the ancestral yeast mtDNA is sequence rearrangement.

SEQUENCE ARRANGEMENT IN YEAST mtDNAs

From studies on gene mapping of mtDNAs from the yeasts *T. glabrata* and *Kloeckera africana* (27.1 kbp) it is apparent that there is only one instance of two juxtaposed genes sharing the same order and orientation (Clark-Walker and Sriprakash, 1981 and 1982). Likewise, gene topography in *S. cerevisiae* mtDNA bears little resemblance to either of the genomes mentioned above. Nevertheless we have been fortunate to discover a common linkage group of five genes in mtDNAs of *T. glabrata* and *Saccharomyces exiguus* which may represent the order of these sequences in an earlier yeast mtDNA. A starting point for this investigation was the attractive observation that the ATPase subunit 6 and 9 genes in *T. glabrata* mtDNA have the same order and orientation as analogous genes in the *Escherichia coli atp* or *unc* operon (Downie *et al.*, 1981; Gay and Walker, 1981). There-fore, bearing in mind the possible endosymbiotic origin of mitochondria, the question arose as to whether the juxtaposition of the ATPase genes in *T. glabrata* mtDNA is present in other yeast mtDNAs.

To answer this question, a number of yeast mtDNAs were surveyed for juxtaposition of the ATPase subunit 6 and 9 genes by determining whether the two probes would hybridize to a common DNA fragment released by restriction endonuclease digestion. From amongst eight previously untested yeast mtDNAs, with sizes ranging from 23.7 kbp in *Saccharomyces exiguus* to around 100 kbp in *Brettanomyces custersii*, common fragment hybridization was found in only the smallest mitochondrial genome from *S. exiguus*, thereby suggesting that in the other mtDNAs the two ATPase genes are unlikely to be closely positioned (Clark-Walker and Sriprakash, unpublished observations). We therefore analyzed mtDNA from *S. exiguus* in more detail to see if further similarities exist between this molecule and that from *T. glabrata*.

The results of this study are summarized in Fig. 10.2 and details have been published elsewhere (Clark-Walker and Sriprakash, 1983b). Starting with the cytochrome b region, there is a cluster of five genes that is common to both *S. exiguus* and *T. glabrata* mtDNAs. The cytochrome oxidase subunit 1 gene in each case is mosaic and contains a sequence hybridizing to the fourth intron of the cognate gene in *S. cerevisiae* mtDNA and there is tentative

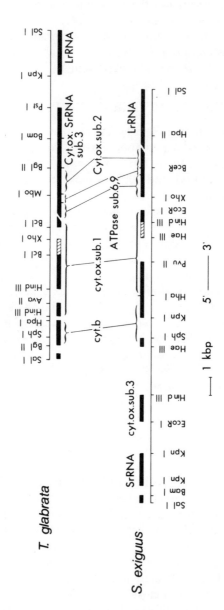

Fig. 10.2. Location of regions in mtDNA of *T. glabrata* (18.9 kbp) and *S. exiguus* (23.7 kbp) hybridizing to specific probes from *S. cerevisiae* mtDNA. The molecules have been opened at the unique SalI site in the L rRNA gene and aligned from the SphI site in the cytochrome b gene. Abbreviated symbols stand for L and S rRNA (large and small ribosomal RNAs); cyt.ox.sub1-3 (cytochrome oxidase subunits 1–3); cyt.b (cytochrome b); ATPase sub.6 and 9 (adenosine triphosphatase subunits 6 and 9). Additionally, sequences hybridizing to the variant 1 gene of *S. cerevisiae* mtDNA have been found in both mtDNAs. For *T. glabrata*, variant 1 is located in the segment between the S rRNA and L rRNA genes (see Fig. 10.1) whereas in *S. exiguus*, variant 1 occurs between S rRNA and cyt.ox. sub.3.

evidence that the cytochrome b gene of *S. exiguus* mtDNA contains an intervening sequence.

A point that serves to emphasize the remarkable nature of the five gene cluster in the mtDNAs of *S. exiguus* and *T. glabrata* is that these yeasts do not appear to be closely related by other criteria. The buoyant densities of their nuclear DNAs are 1.692 and 1.700 g/cm³ for *S. exiguus* and *T. glabrata*, respectively (Yarrow and Nakase, 1975; O'Connor *et al.*, 1976), and there are few physiological similarities due to the former yeast being a mesophilic soil saprophyte (Van der Walt, 1970) while the latter is a thermophilic inhabitant of mammalian intestinal tracts (Travassos and Cury, 1971). Therefore the truth seems to be that the common five gene cluster, with the two ATPase genes in the same order and orientation as analogous genes in the *E. coli unc* operon, reflects gene order in an earlier yeast mtDNA.

Further comparative studies are necessary to try to extend such observations to cover the remainder of the mitochondrial genome but one question arises from these results that needs examination. Both *T. glabrata* and *S. exiguus* mtDNAs are small and at least in the former there is little space between the genes. On the other hand, in the large mtDNA from *S. cerevisiae*, genes are bounded by A + T rich non-coding sequences which can extend for several hundred nucleotides. The question then arises as to which type of mtDNA more closely resembles the ancestral molecule?

Again the truth may lie between these two representatives. The compromise view is that the ancestral molecule would have contained separate linked clusters bounded by spacer sequences but there would have been relatively little spacer sequence between the genes of each cluster. Candidates for the linked clusters are genes of like function such as the ATPase 6 and 9 sequences mentioned above, the S and L rRNA genes, tRNAs and the cytochrome oxidase subunit genes. Based on this tentative model, the mtDNA of *T. glabrata* would have undergone some sequence rearrangement and elimination of spacer sequences while the *S. cerevisiae* mtDNA would have undergone much more rearrangement and acquired extra spacer sequence so that few or none of the ancestral linkage groups are now extant. Indeed it can be imagined that expansion of spacer regions by transposition/ duplication of short sequences would predispose the molecule to rearrangement by homologous site recombination.

In general it can be seen that since yeasts diverged from a common progenitor their mtDNAs have been in a state of flux so that each contemporary molecule has been established by separate processes leading to rearrangement, insertion of sequences (in *S. cerevisiae* mtDNA, G + C rich sequences, *ori/rep* sequences and possibly some introns and spacers) and loss of sequences (introns, spacers and genes transferred to the nucleus or lost from the organism). These processes are seen as independent of each other so that a molecule could be gaining some of the sequence elements mentioned above and losing others. The mechanisms of these changes are largely unknown as are the pressures leading to their establishment.

STRUCTURE OF ANIMAL mtDNAs

Perhaps the most striking contrast in structure between mtDNAs of mammals and other organisms is that the former molecules have genes that not only lack introns but are juxtaposed so closely that in some cases the stop codon is formed by addition of poly A to the RNA transcript (Anderson *et al.*, 1981). Indeed such a tight arrangement of genes means that every nucleotide has a purpose; in this case it is a coding function.

One consequence of such a frugal architecture is that bounding promoter and terminator sequences do not exist. Apparently each DNA strand is transcribed into a polycistronic RNA from one or only a few promoters followed by processing of the RNA generally at tRNA sequences interspersed throughout the genome (Battey and Clayton, 1980; Montoya *et al.*, 1981).

There is, moreover, a region in animal mtDNAs that is an exception to the frugal structure of the remainder. Size changes of a few hundred base pairs have been found between human, cow and mouse mtDNAs near the origin of heavy strand replication (Anderson *et al.*, 1981; Bibb *et al.*, 1981). However, larger changes in this region have been found in mtDNAs of *Drosophila* species as an A + T rich sequence can vary from 0.9 to 5 kbp (Fauron and Wolstenholme, 1976).

In a previous section, mention was made that animal (multicellular invertebrate and vertebrate) mtDNAs do not show large size variation. Furthermore, sequencing studies have shown matching topography between the mammalian mtDNAs of human, cow and mouse (Anderson *et al.*, 1981; Bibb *et al.*, 1981). This concordance does not extend to the mtDNA of *Drosophila yakuba* where a segment of the genome containing the S and L rRNA genes, an interspersed valine tRNA and URF1 is inverted relative to this region in mammalian mtDNA (Clary *et al.*, 1982). Tight genic structure may also be correlated with lack of sequence rearrangements in mammalian mtDNA and it is perhaps significant that one end of the inverted segment mentioned above is located in the A + T rich variable region.

Viewed in the context of knowledge about the ancient ancestry of some intervening sequences in fungal mtDNAs, it seems justified to suppose that the absence of introns from the cytochrome b, L rRNA and possibly the cytochrome oxidase subunit 1 genes of mammalian mtDNAs is due to loss. Likewise, spacer sequences may have been eliminated but disruption of supposed linkage groups may have preceded this change. One linkage group, namely the S and L rRNA region, appears to have been preserved much more than in other organism's mtDNAs as this structure has the same order and orientation as in bacteria, although the rRNAs are considerably smaller.

SIGNIFICANCE OF mtDNA DIVERSITY

There is a hierarchy of changes that can happen to DNA, and mtDNA is no exception. Most prevalent are silent base substitutions (Brown *et al.*, 1982) but in addition mtDNAs of yeasts can have, in roughly decreasing frequency of occurrence, optional introns, sequence rearrangements, expansion or contraction in size of spacers and, in *S. cerevisiae*, optional genic sequences (e.g. G + C sequence elements) while in *K. africana* there is an inverted duplication. Additionally, it must be emphasized that within two yeast genera namely *Saccharomyces* and *Dekkera/Brettanomyces* there has been independent dispersal in size of the mitochondrial genome. This diversity presents two problems. Firstly, we have to resolve the interrelationships of the molecules together with their mechanisms of change and, secondly, we need to ascertain the significance of the diversity in neo-Darwinian terms.

Some aspects of the first problem have been dealt with above, but in considering the second issue it would be useful to know whether a change in DNA structure (even a single base substitution in a non-coding sequence) can alter the phenotype? A further question bearing on this issue is whether a change in phenotype confers selective advantage or·disadvantage in every instance? While recognizing that these questions cannot be answered in the absence of experimental tests, nevertheless a reasonable assumption would be that not all changes to DNA structure alter the phenotype. This position then enables us to view changes in DNA from a different perspective.

The alterations just described in mtDNAs can be seen as manifestations of *processes* that are sometimes of selective advantage. Thus in DNA molecules in general that are continually undergoing insertions, deletions, duplications and rearrangements there is the possibility of new functions emerging from a background of events that do not change the phenotype.

ACKNOWLEDGEMENTS

I thank W. M. Brown, R. J. Evans, P. Hoeben, G. L. G. Miklos and K. S. Sriprakash, for critical comments and A. J. Bendich, R. A. Butow, G. B. Cox and A. C. Wilson for helpful discussions.

NOTE ADDED IN PROOF:

Transcripts of the variant 1 protein gene of *T.glabrata* mt DNA have been detected (R. A. Butow personal communication) thereby raising the possibility that this is an active gene.

REFERENCES

Anderson, S., Bankier, A. T., Barrell, B. G., de Bruijn, M. H. L., Coulson, A. R., Drouin, J., Eperon, I. C., Nierlich, D. P., Roe, B. A., Sanger, F., Schreier, P. H., Smith, J. H., Staden, R. and Young, I. G. (1981). Sequence and organization of the human mitochondrial genome. *Nature, Lond.*, **290**, 457–464.

Barrell, B. G., Bankier, A. T. and Drouin, J. (1979). A different genetic code in human mitochondria. *Nature, Lond.*, **282**, 189–194.

Battey, J. and Clayton, D. A. (1978). The transcription map of mouse mitochondrial DNA. *Cell*, **14**, 143–156.

Battey, J. and Clayton, D. A. (1980). The transcription map of human mitochondrial DNA implicates transfer RNA excision as a major processing event. *Proc. Natl. Acad. Sci. USA*, **255**, 11599–11606.

Bernardi, G. (1982). Evolutionary origin and the biological function of noncoding sequences in the mitochondrial genome of yeast. In P. P. Slonimski, P. Borst and G. Attardi (Eds), *Mitochondrial Genes*, Cold Spring Harbor Laboratory Press, Cold Spring Harbor, pp. 269–278.

Bibb, M. J., Van Etten, R. A., Wright, C. T., Walberg, M. W. and Clayton, D. A. (1981). Sequence and gene organization of mouse mitochondrial DNA. *Cell*, **26**, 167–180.

Blanc, H. and Dujon, B. (1980). Replicator regions of the yeast mitochondrial DNA responsible for suppressiveness. *Proc. Natl. Acad. Sci. USA*, **77**, 3942–3946.

Blanc, H. and Dujon, B. (1982). Replicator regions of the yeast mitochondrial DNA active *in vivo* and in yeast transformants. In P. P. Slonimiski, P. Borst and G. Attardi (Eds), *Mitochondrial Genes*, Cold Spring Harbor Laboratory Press, Cold Spring Harbor, pp. 279–294.

Bonen, L. and Gray, M. W. (1980). Organization and expression of the mitochondrial genome of plants. I. The genes for wheat mitochondrial ribosomal and transfer RNA: evidence for an unusual arrangement. *Nucleic Acids Res.*, **8**, 319–335.

Bonitz, S. G., Berlani, R. Coruzzi, G., Li, M., Macino, G., Nobrega, M. P., Thalenfeld, B. E. and Tzagoloff, A. (1980a). Codon recognition rules in yeast mitochondria. *Proc. Natl. Acad. Sci. USA*, **77**, 3167–3170.

Bonitz, S. G., Coruzzi, G., Thalenfeld, B. E., Tzagoloff, A. and Macino, G. (1980b). Assembly of the mitochondrial membrane system. Structure and nucleotide sequence of the gene coding for subunit 1 of yeast cytochrome oxidase. *J. Biol. Chem.*, **255**, 11927–11941.

Borst, P. and Grivell, L. A. (1981). Small is beautiful—portrait of a mitochondrial genome. *Nature*, **290**, 443–444.

Borst, P. and Hoeijmakers, J. H. J. (1979). Kinetoplast DNA. *Plasmid*, **2**, 20–40.

Bos, J. L., Heyting, C., Borst, P., Arnberg, A. C. and Van Bruggen, E. F. J. (1978). An insert in the single gene for the large ribosomal RNA in yeast mitochondrial DNA. *Nature*, **275**, 336–338.

Bos, J. L., Osinga, K. A., Van der Horst, G., Hecht, N. B., Tabak, H. F., Van Ommen, G.-J. B. and Borst, P. (1980). Splice point sequence and transcripts of the intervening sequence in the mitochondrial 21 S ribosomal RNA gene of yeast. *Cell*, **20**, 207–214.

Brown, T. A., Davies, R. W., Ray, J. A., Waring, R. B. and Scazzocchio, C. (1983b). The mitochondrial genome of *Aspergillus nidulans* contains reading frames homologous to the human URFs 1 and 4. *EMBO J.*, **2**, 427–435.

Brown, T. A., Davies, R. W., Waring, R. B. and Ray, J. A. (1983a). DNA duplication has resulted in transfer of an amino-terminal peptide between two mitochondrial proteins. *Nature, Lond.*, **302**, 721–723.

Brown, W. M., Prager, E. M., Wang, A. and Wilson, A. C. (1982). Mitochondrial DNA sequences of primates: tempo and mode of evolution. *J. Mol. Evol.*, **18**, 225–239.

Butow, R. A., Ainley, W. M., Zassenhaus, H. P., Hudspeth, M. E. and Grossman, L. I. (1983). The unusual organization of the yeast mitochondrial Var1 gene. In

P. Nagley, A. W. Linnane, W. J. Peacock and J. A. Pateman (Eds), *Manipulation and Expression of Genes in Eukaryotes*, Academic Press, Sydney, pp. 269–277.

Clark-Walker, G. D. and Sriprakash, K. S. (1981). Sequence rearrangements between mitochondrial DNAs of *Torulopsis glabrata* and *Kloeckera africana* identified by hybridization with six polypeptide encoding regions from *Saccharomyces cerevisiae* mitochondrial DNA. *J. Mol. Biol.*, **151**, 367–387.

Clark-Walker, G. D. and Sriprakash, K. S. (1982). Size diversity and sequence rearrangements in mitochondrial DNAs from yeasts. In P. P. Slonimski, P. Borst and G. Attardi (Eds), *Mitochondrial Genes*, Cold Spring Harbor Laboratory Press, Cold Spring Harbor, pp. 349–354.

Clark-Walker, G. D. and Sriprakash, K. S. (1983a). Map location of transcripts from *Torulopsis glabrata* mitochondrial DNA. *EMBO J.* **2**, 1465–1472.

Clark-Walker, G. D. and Sriprakash, K. S. (1983b). Order and orientation of genic sequences in circular mitochondrial DNA from *Saccharomyces exiguus*: implications for evolution of yeast mtDNAs. *J. Mol. Evol.* **19**, 333–341.

Clark-Walker, G. D., McArthur, C. R. and Daley, D. J. (1981a). Does mitochondrial DNA length influence the frequency of spontaneous petite mutants in yeasts? *Current Genetics*, **4**, 7–12.

Clark-Walker, G. D., McArthur, C. R. and Sriprakash, K. S. (1981b). Partial duplication of the large ribosomal RNA sequence in an inverted repeat in circular mitochondrial DNA from *Kloeckera africana*. Implications for mechanisms of the petite mutation. *J. Mol. Biol.*, **147**, 399–415.

Clary, D. O., Goddard, J. M., Martin, S. C., Fauron, C. M.-R. and Wolstenholme, D. R. (1982). *Drosophila* mitochondrial DNA: a novel gene order. *Nucleic Acids Res.*, **10**, 6619–6637.

Cobon, G. S., Beilharz, M. W., Linnane, A. W. and Nagley, P. (1982). Biogenesis of mitochondria: mapping of transcripts from the *oli2* region of mitochondrial DNA in two grande strains of *Saccharomyces cerevisiae*. *Current Genetics*, **5**, 97–107.

Cosson, J. and Tzagoloff, A. (1979). Sequence homologies of (guanosine + cytosine)-rich regions of mitochondrial DNA of *Saccharomyces cerevisiae*. *J. Biol. Chem.*, **254**, 42–43.

Dale, R. M. K. (1981). Sequence homology among different size classes of plant mitochondrial DNAs. *Proc. Natl. Acad. Sci. USA*, **78**, 4453–4457.

Dale, R. M. K. (1982). Structure of plant mitochondrial DNAs. In P. P. Slonimski, P. Borst and G. Attardi (Eds), *Mitochondrial Genes*, Cold Spring Harbor Laboratory Press, Cold Spring Harbor, pp. 471–476.

Dayhoff, M. O. and Schwartz, R. M. (1981). Evidence on the origin of eukaryotic mitochondria from protein and nucleic acid sequences. *Ann. New York Acad. Sci.*, **361**, 92–104.

De La Salle, H., Jacq. C. and Slonimski, P. P. (1982). Critical sequences within mitochondrial introns: pleiotropic mRNA maturase and cis-dominant signals of the *box* intron controlling reductase and oxidase. *Cell*, **28**, 721–732.

deZamaroczy, M., Marotta, R., Faugeron-Fonty, G., Goursot, R., Mangin, M., Baldacci, G. and Bernardi, G. (1981). The origins of replication of the mitochondrial genome of yeast and the phenomenon of suppressivity. *Nature, Lond.*, **292**, 75–78.

Downie, J. A., Cox, G. B., Langman, L., Ash, G., Becker, M. and Gibson, F. (1981). Three genes coding for subunits of the membrane sector (Fo) of the *Escherichia coli* adenosine triphosphatase complex. *J. Bacteriol.*, **145**, 200–210.

Dujon, B. (1980). Sequence of the intron and flanking exons of the mitochondrial

21 S rRNA gene of yeast strains having different alleles at the ω and *rib-1* loci. *Cell*, **20**, 185–197.

Dujon, B. (1981). Mitochondrial genetics and functions. In J. N. Strathern, E. W. Jones and J. R. Broach (eds), *The Molecular Biology of the Yeast Saccharomyces cerevisiae*, Cold Spring Harbor Laboratory Press, Cold Spring Harbor, pp. 505–635.

Farrelly, F., Zassenhaus, H. P. and Butow, R. A. (1982). Characterization of transcripts from the *var1* region on mitochondrial DNA of *Saccharomyces cerevisiae*. *J. Biol. Chem.*, **257**, 6581–6587.

Farrelly, F. and Butow, R. A. (1983). Rearranged mitochondrial genes in the yeast nuclear genome. *Nature*, **301**, 296–301.

Fauron, C. M. R. and Wolstenholme, D. R. (1976). Structural heterogeneity of mitochondrial DNA molecules within the genus *Drosophila*. *Proc. Natl. Acad. Sci. USA*, **73**, 3623–3627.

Fox, T. D. and Leaver, C. J. (1981). The *Zea mays* mitochondrial gene coding cytochrome oxidase subunit II has an intervening sequence and does not contain TGA codons. *Cell*, **26**, 315–323.

Gay, N. J. and Walker, J. E. (1981). The *atp* operon: nucleotide sequence of the promoter and the genes for the membrane proteins, and the δ subunit of *Escherichia coli* ATP-synthetase. *Nucleic. Acids Res.*, **9**, 3919–3926.

Gillham, N. W. and Boynton, J. E. (1981). Evolution of organelle genomes and protein-synthesizing systems. *Ann. New York Acad. Sci.*, **361**, 20–43.

Goddard, J. M. and Cummings, D. J. (1977). Mitochondrial DNA replication in *Paramecium aurelia*. Cross-linking of the initiation end. *J. Mol. Biol.*, **109**, 327–344.

Goldbach, R. W., Arnberg, A. C., Van Bruggen, E. F. J., Defize, J. and Borst, P. (1977). The structure of *Tetrahymena pyriformis* DNA. I. Strain differences and occurence of inverted repetitions. *Biochem. Biophys. Acta*, **447**, 37–50.

Grant, D. M. and Lambowitz, A. M. (1981). Mitochondrial ribosomal RNA genes, in H. Busch and L. Rothblum (Eds), *The Cell Nucleus*, vol. X, Academic Press, New York, pp. 387–408.

Gray, M. W. and Doolittle, W. F. (1982). Has the endosymbiont hypothesis been proven? *Microbiol. Rev.*, **46**, 1–42.

Grivell, L. A. (1983). Mitochondrial DNA. *Scientific American*, **248**, (3), 60–73.

Hall, J. B. (1973). The nature of the host in the origin of the eukaryote cell. *J. Theor. Biol.*, **38**, 413–418.

Heckman, J. E., Sarnoff, J., Alzner-de Weered, B., Yin, S. and RajBhandary, U. L. (1980). Novel features in the genetic code and codon reading patterns in *Neurospora crassa* mitochondria based on sequences of six mitochondrial tRNAs. *Proc. Natl. Acad. Sci. USA*, **77**, 3159–3163.

Hensgens, L. A. M., Grivell, L. A., Borst, P. and Bos, J. L. (1979). Nucleotide sequence of the mitochondrial structural gene for subunit 9 of yeast ATPase complex. *Proc. Natl. Acad. Sci. USA*, **76**, 1663–1667.

Hensgens, L. A. M., Bonen, L., de Haan, M., van der Horst, G. and Grivell, L. A. (1983). Two intron sequences in yeast mitochondrial COX1 gene: homology among URF-containing introns and strain-dependent variation in flanking exons. *Cell*, **32**, 379–389.

Hudspeth, M. E. S., Ainley, W. M., Shumard, D. S., Butow, R. A. and Grossman, L. I. (1982). Location and structure of the *var1* gene on yeast mitochondrial DNA: nucleotide sequence of the 40.0 allele. *Cell*, **30**, 617–626.

Hudspeth, M. E. S., Shumard, D. S., Bradford, C. J. R. and Grossman, L. I. (1983). Organization of *Achlya* mtDNA: a population with two orientations and a large inverted repeat containing the rRNA genes. *Proc. Natl. Acad. Sci. USA*, **80**, 142–146.

Jacobs, H. T., Posakony, J. W., Grula, J. W., Roberts, J. W., Xin, J.-H., Britten, R. J. and Davidson, E. H. (1983). Mitochondrial DNA sequences in the nuclear genome of *Strongylocentrotus purpuratus*. *J. Mol. Biol.*, **165**, 609–632.

Kemble, R. J. and Bedbrook, J. R. (1980). Low molecular weight circular and linear DNA in mitochondria from normal and male sterile *Zea mays* cytoplasm. *Nature, Lond.*, **284**, 565–566.

Kochel, H. G. and Kuntzel, H. (1982). Mitochondrial L-rRNA from *Aspergillus nidulans*: potential secondary structure and evolution. *Nucleic Acids Res.*, **10**, 4795–4801.

Kuntzel, H. and Kochel, H. G. (1981). Evolution of rRNA and origin of mitochondria. *Nature*, **293**, 751–755.

Lazowska, J., Jacq, C. and Slonomski, P. P. (1980). Sequence of introns and flanking exons in wild-type and box 3 mutants of cytochrome b reveals an interlaced splicing protein coded by an intron. *Cell*, **22**, 333–348.

Leaver, C. J., Forde, B. G., Dixon, L. K. and Fox, T. D. (1982). Mitochondrial genes and cytoplasmically inherited variation in higher plants. In P. P. Slonimski, P. Borst and G. Attardi (Eds), *Mitochondrial Genes*, Cold Spring Harbor University Press, Cold Spring Harbor, pp. 457–470.

McArthur, C. R. and Clark-Walker, G. D. (1983). Mitochondrial DNA size diversity in the *Dekkera/Brettanomyces* yeasts. *Current Genetics*, **7**, 29–35.

Macino, G., Scazzocchio, C., Waring, R. B., Berks, M., McP. and Davies, R. W. (1980). Conservation and rearrangement of mitochondrial structural gene sequences. *Nature, Lond.*, **288**, 404–406.

Mahler, H. R. (1981). Mitochondrial evolution: organization and regulation of mitochondrial genes. *Ann. New York Acad. Sci.*, **361**, 53–75.

Mahler, H. R., Hanson, D. K., Lamb, M. R., Perlman, P. S., Anziano, P. Q., Glaus, K. R. and Haldi, M. L. (1982). Regulatory interactions between mitochondrial genes: expressed introns—their function and regulation. In P. P. Slonimski, P. Borst and G. Attardi (Eds), *Mitochondrial Genes*, Cold Spring Harbor Laboratory Press, Cold Spring Harbor, pp. 185–199.

Mannella, C. A., Goewert, R. R. and Lambowitz, A. M. (1979). Characterization of variant *Neurospora crassa* mitochondrial DNAs which contain tandem reiterations. *Cell*, **18**, 1197–1207.

Margulis, L. (1970). *Origin of Eukaryotic Cells*, Yale University Press, New Haven.

Michel, F., Jacquier, A. and Dujon, B. (1982). Comparison of fungal mitochondrial introns reveals extensive homologies in RNA secondary structure. *Biochimie*, **64**, 867–881.

Montoya, J., Ojala, D. and Attardi, G. (1981). Distinctive features of the 5′-terminal sequences of the human mitochondrial mtDNAs. *Nature, Lond.*, **290**, 465–474.

Nargang, F. E., Bell, J. B., Stohl, L. L. and Lambowitz, A. M. (1983). A family of repetitive palindromic sequences found in *Neurospora* mitochondrial DNA is also found in a mitochondrial plasmid DNA. *J. Biol. Chem.*, **258**, 4257–4260.

Netzker, R., Kochel, H. G., Basak, N. and Kuntzel, H. (1982). Nucleotide sequence of *Aspergillus nidulans* mitochondrial genes coding for ATPase subunit 6, cytochrome oxidase subunit 3, seven unidentified proteins, four tRNAs and L-rRNA. *Nucleic Acids Res.*, **10**, 4783–4794.

Nobrega, F. G. and Tzagoloff, A. (1980). Assembly of the mitochondrial membrane system. DNA sequence and organization of the cytochrome b gene in *Saccharomyces cerevisiae* D273-10B. *J. Biol. Chem.*, **255**, 9828–9837.

Nomiyama, H., Sakaki, Y. and Takagi, Y. (1981). Nucleotide sequence of a ribosomal RNA gene intron from slime mould *Physarum polycephalum*. *Proc. Natl. Acad. Sci., USA*, **78**, 1376–1380.

Novitski, C. E., Macreadie, I. G., Maxwell, R. J., Lukins, H. B., Linnane, A. W. and Nagley, P. (1983). Features of nucleotide sequences in the region of the *oli2* and *aapl* genes in the yeast mitochondrial genome. In P. Nagley, A. W. Linnane, W. J. Peacock and J. A. Pateman (Eds), *Manipulation and Expression of Genes in Eukaryotes*, Academic Press, Sydney, pp. 257–268.

O'Connor, R. M., McArthur, C. R. and Clark-Walker, G. D. (1975). Closed circular DNA from mitochondrial-enriched fractions of four *petite*-negative yeasts. *Eur. J. Biochem.*, **53**, 137–144.

O'Connor, R. M., McArthur, C. R. and Clark-Walker, G. D. (1976). Respiratory deficient mutants of *Torulopsis glabrata*, a yeast with circular mitochondrial deoxyribonucleic acid of 6 μm. *J. Bacteriol.*, **126**, 959–968.

Palmer, J. D., Shields, C. R., Cohen, D. B. and Orton, T. J. (1983). An unusual mitochondrial DNA plasmid in the genus *Brassica*. *Nature, Lond.*, **301**, 725–728.

Pring, D. R., Levings, C. S., III, Hu, W. W. L. and Timothy, D. H. (1977). Unique DNA associated with mitochondria in the 'S'-type cytoplasm of male-sterile maize. *Proc. Natl. Acad. Sci. USA*, **74**, 1904–1909.

Prunell, A. and Bernardi, G. (1974). The mitochondrial genome of wild-type yeast cells. IV. Genes and spacers. *J. Mol. Biol.*, **86**, 825–841.

Prunell, A. and Bernardi, G. (1977). The mitochondrial genome of wild-type yeast cells. VI. Genome organization. *J. Mol. Biol.*, **110**, 53–74.

Raff, R. A. and Mahler, H. R. (1972). The non-symbiotic origin of mitochondria. *Science*, **169**, 641–646.

Sanders, J. P. M., Heyting, C., Verbeet, M. P., Meijlink, F. C. P. W. and Borst, P. (1977). The organization of genes in yeast mitochondrial DNA III. Comparison of the physical maps of mitochondrial DNAs from three wild-type *Saccharomyces* strains. *Molec. gen. Genet.*, **157**, 239–261.

Sederoff, R. R., Levings, C. S., III, Timothy, D. H. and Hu, W. W. L. (1981). Evolution of DNA sequence organization in mitochondrial genomes of *Zea*. *Proc. Natl. Acad. Sci. USA*, **78**, 5953–5957.

Sor, F. and Fukuhara, H. (1982a). Nature of an inserted sequence in the mitochondrial gene coding for the 15S ribosomal RNA of yeast. *Nucleic Acids Res.*, **10**, 1625–1633.

Sor, F. and Fukuhara, H. (1982b). Nucleotide sequence of the small ribosomal RNA gene from the mitochondria of *Saccharomyces cerevisiae*. In P. P. Slonimski, P. Borst and G. Attardi (Eds), *Mitochondrial Genes*, Cold Spring Harbor Laboratory Press, Cold Spring Harbor, pp. 255–262.

Stern, D. B. and Lonsdale, D. M. (1982). Mitochondrial and chloroplast genomes of maize have a 12-kilobase DNA sequence in common. *Nature, Lond.*, **299**, 698–702.

Stohl, L. L., Collins, R. A., Cole, M. D. and Lambowitz, A. M. (1982). Characterization of two new plasmid DNAs found in mitochondria of wild-type *Neurospora intermedia* strains. *Nucleic Acids Res.*, **10**, 1439–1458.

Travassos, L. R. R. G. and Cury, A. (1971). Thermophilic enteric yeasts. *Ann. Rev. Microbiol.*, **25**, 49–74.

Tzagoloff, A., Macino, G. and Sebald, W. (1979). Mitochondrial genes and translation products. *Ann. Rev. Biochem.*, **48**, 419–441.

Tzagoloff, A., Nobrega, M., Akai, A. and Macino, G. (1980). Assembly of the mitochondrial membrane system. Organization of yeast mitochondrial DNA in the *oli*1 region. *Current Genetics*, **2**, 149–157.

Van den Boogaart, P., Samallo, J. and Agsteribbe, E. (1982). Similar genes for a mitochondrial ATPase subunit in the nuclear and mitochondrial genomes of *Neurospora crassa*. *Nature, Lond.*, **298**, 187–189.

Van der Walt, J. P. (1970). Genus 16 *Saccharomyces*. In J. Lodder (Ed), *The Yeasts*, North Holland, Amsterdam, pp. 555–718.

Wallace, D. C. (1982). Structure and evolution of organelle genomes. *Microbiol. Rev.*, **46**, 208–240.

Ward, B. L., Anderson, R. S. and Bendich, A. J. (1981). The mitochondrial genome is large and variable in a family of plants (*Cucurbitaceae*). *Cell*, **35**, 793–803.

Waring, R. B., Davies, R. W., Scazzocchio, C. and Brown, T. A. (1982). Internal structure of a mitochondrial intron of *Aspergillus nidulans*. *Proc. Natl. Acad. Sci. USA*, **79**, 6332–6336.

Weiss-Brummer, B., Rodel, G., Schweyen, R. J. and Kaudewitz, F. (1982). Expression of the split gene *cob* in yeast: evidence for a precursor of a maturase protein translated from intron 4 and preceding exons. *Cell*, **29**, 527–536.

Wesolowski, M. and Fukuhara, H. (1981). Linear mitochondrial deoxyribonucleic acid from the yeast *Hansenula mrakii*. *Mol. Cell Biol.*, **1**, 387–393.

Wright, R. M. and Cummings, D. J. (1983). Integration of mitochondrial gene sequences within the nuclear genome during senescence in a fungus. *Nature, Lond.*, **302**, 86–88.

Yarrow, D. and Nakase, T. (1975). DNA base composition of species of the genus *Saccharomyces*. *Antonie van Leeuwenhoek*, **41**, 81–88.

Yin, S., Heckman, J. and RajBhandary, U. L. (1981). Highly conserved GC-rich palindromic DNA sequences flank tRNA genes in *Neurospora crassa* mitochondria. *Cell*, **26**, 325–332.

The Evolution of Genome Size
Edited by T. Cavalier-Smith
© 1985 John Wiley & Sons Ltd

CHAPTER 11

Evolution of Plastid DNA[1]

NICHOLAS W. GILLHAM, JOHN E. BOYNTON and ELIZABETH H. HARRIS
Departments of Zoology and Botany, Duke University, Durham, NC 27706

INTRODUCTION

Chloroplasts, like mitochondria, contain DNA but are not genetically autonomous organelles (cf. Gillham, 1978). While chloroplast genomes encode
certain specific chloroplast components, most other chloroplast components
are determined by the nuclear genome (Bartlett *et al.*, 1981; Gillham, 1978).
Although both non-symbiotic and symbiotic origins have been proposed for
chloroplasts (cf. Bogorad, 1975; Margulis, 1970, 1975; Uzzell and Spolsky,
1974, 1981), most investigators now favour the hypothesis of symbiotic origin,
possibly involving several independent endosymbiotic events (cf. Cavalier-
Smith, 1982; Doolittle and Bonen, 1981; Gibbs, 1981; Gray and Doolittle,
1982; Weeden, 1981; Whatley, 1981). Very little information is available on
chloroplast genomes among primitive eukaryotes (cf. Gillham and Boynton,
1981). For example, the chloroplast genomes of only one chrysophyte, *Olisthodiscus* (Aldrich and Cattolico, 1981; Ersland *et al.*, 1981), and one xantho-
phyte, *Vaucheria* (Herrmann and Possingham, 1980), have been examined,
and chloroplast genomes of red and brown algae have yet to be studied at
all. In fact virtually all chloroplast genomes characterized to date[1] are from
green algae or from higher plants, which supposedly derive from green
algae. In the case of vascular plants, the emphasis has been on angiosperms.
Chloroplast genomes from only three archegoniates have been described
(Herrmann *et al.*, 1980b). These points should be kept in mind in considering
the seeming degree of conservation one sees in conformation, size and structure of chloroplast genomes.

The purpose of this chapter is to review briefly the structure of chloroplast
genomes and the chloroplast components for which they code, and then to
turn to recent work on the organization and structure of specific chloroplast
genes and the evolution of plastid genomes in nature and in the laboratory.

[1] 1982. Please see important note on page 334.

SIZE, NUMBER AND ORGANIZATION OF CHLOROPLAST GENOMES

The properties of chloroplast DNA (cpDNA) and the known functions of this molecule have been reviewed recently (Bartlett *et al.*, 1981; Bedbrook and Kolodner, 1979; Bogorad *et al.*, 1978, 1979, 1980; Coen *et al.*, 1978; Edelman, 1981; Gillham, 1978; Gillham and Boynton, 1981; Gray and Doolittle, 1982; Herrmann and Possingham, 1980; Kirk and Tilney-Bassett, 1978; Rochaix, 1981; von Wettstein, 1981). We will therefore restrict ourselves to a series of general statements about cpDNA and the proteins coded by this molecule before turning to those aspects of cpDNA which at the moment seem to be of the greatest evolutionary interest.

Chloroplast DNA molecules from higher plants and most algae examined to date are circular, have a mean G + C content of 36–38 mole % and range in size from 56×10^6 daltons in the siphonaceous green alga *Codium* to 126×10^6 d in the green alga *Chlamydomonas reinhardtii*. Most higher plants have chloroplast genomes in the range of $90–100 \times 10^6$ d, corresponding to about 135 to 150 kbp (Table 11.1). Little evidence of intermolecular heterogeneity has been found among cpDNA molecules. *Acetabularia* is the only organism thus far described which may be an exception to this general pattern. Reports (Green *et al.*, 1977; Padmanabhan and Green, 1978) show that this siphonaceous alga has a cpDNA molecule that renatures at the same rate as *E. coli* DNA. If so, the cpDNA of *Acetabularia* could have a molecular weight of 1.5×10^9 d and a length of approximately 750 μm. So far the largest cpDNA fragments obtained from chloroplast lysates of this alga are 200 μm linear molecules, about 4.14×10^8 d, or approximately four times the size of other cpDNA genomes (Green, 1976). Because Hedberg *et al.* (1981) have reported that another siphonaceous alga, *Codium*, has the *smallest* chloroplast genome known based on electron microscopy and restriction enzyme digestion, the *Acetabularia* results should probably be rechecked by summation of fragment molecular weights following digestion of cpDNA with restriction enzymes.

All higher plants clearly contain multiple copies of the chloroplast genome per plastid and usually have many plastids per cell as well. The number of copies of cpDNA per plastid can vary from very few to 200, with the normal range being around 20 to 80 molecules per plastid (cf. Herrmann and Possingham, 1980). This number can be influenced by physiological conditions. For example, during differentiation of proplastids into plastids in beet, there is a 25-fold increase in plastid volume that is accompanied by an increase from 10 to over 100 cpDNA molecules per plastid (Herrmann *et al.*, 1974). In pea, Lamppa and Bendich (1979a) estimate an average of 244 molecules per chloroplast in young leaves, but 174 in fully green leaves. Despite this decrease in cpDNA molecules per per plastid, there is an

Table 11.1. Restriction maps of chloroplast genomes.

Organism	Estimated genome size (kb)	Ava I	Bal I	Bam I	Bgl I	Bgl II	Bst EII	Eco RI	Hpa I	Kpn I	Pst I	Pvu I	Pvu II	Sac I	Sal I	Sma I	Xba I	Xho I	Xma I	Reference
Atriplex triangularis	152								*					*						Palmer, 1982
Atropa belladonna	160			*								*								Fluhr and Edelman, 1981b
Chlamydomonas reinhardtii	190		*	*				*												Rochaix, 1978, 1981
Cucumis sativa (cucumber)	155									*		*		*						Palmer, 1982
Euglena gracilis	140	*	*	*			*	*		*	*	*	*	*	*			*		Hallick et al., 1978; El-Gewely et al., 1981
Euoenothera Plastome IV	153				*					*	*									Gordon et al., 1981
Lycopersicon esculentum (garden tomato)	158			*					*	*	*	*				*		*		Palmer and Zamir, 1982
Nicotiana tabacum (tobacco)	160				*					*		*		*	*			*		Fluhr and Edelman, 1981a
Petunia hybrida	160			*				*	*			*		*	*	*	*	*		Seyer et al., 1981; Jurgenson and Bourque, 1980
Petunia parodii	160			*			*	*	*			*		*				*		Bovenberg et al., 1981
Pisum sativum (pea)	120							*		*	*	*	*	*	*			*		Fluhr and Edelman, 1981b; Palmer and Thompson, 1981b
Sinapis alba (mustard)	158							*		*	*	*		*				*		Link et al., 1981
Solanum lycopersicoides	158			*			*		*		*	*		*						Palmer and Zamir, 1982
Spinacia oleracea (spinach)	140			*						*		*	*	*	*			*	*	Herrmann and Possingham, 1980
Spirodela oligorrhiza	180									*		*		*	*			*		Driesel et al., 1980; van Ee et al., 1980a, b
Triticum aestivum (wheat)	135									*	*				*					Bowman et al., 1981
Vicia faba (broad bean)	121									*				*	*					Koller and Delius, 1980
Vigna radiata (mung bean)	150						*			*	*	*	*	*	*	*		*		Palmer and Thompson, 1981b
Zea mays (maize)	132		*					*				*		*						Bedbrook and Bogorad, 1976; Palmer and Thompson, 1982

The Evolution of Genome Size

increase in the number of cpDNA molecules per cell since the average number of plastids per cell increases from 24 to 64 during greening (Lamppa *et al.*, 1980). Lamppa and Bendich (1979a) also found that the relative amount of cpDNA varies from 12% of the total DNA in cells of fully green leaves to only 0.4% in root cells.

Chloroplast DNA molecules are not distributed randomly throughout the plastid, but are organized into structures called nucleoids (Bisalputra and Bisalputra, 1969; Chiang *et al.*, 1981; Coleman, 1978, 1979; Coleman and Heywood, 1981; Coleman *et al.*, 1981; Gibbs *et al.*, 1974; Herrmann and Possingham, 1980; James and Jope, 1978; Kuroiwa and Suzuki, 1980, 1981; Kuroiwa *et al.*, 1981; Lüttke, 1981; Ris, 1961). Studies of nucleoid structure by electron microscopy, by fluorescent light microscopy following staining of cpDNA with the fluorescent dye 4',6-diamidino-2-phenylindole (DAPI) and by autoradiography following labelling of cpDNA with radioactive precursors have revealed three general patterns of chloroplast nucleoid organization:

1. cpDNA appears to be concentrated in a single structure located more or less centrally in proplastids, chloroplasts of certain small algae and chromoplasts.
2. cpDNA is present as a continuous ring which encircles the rim of the chloroplast in five closely related classes of algae (Raphidophyceae, Chrysophyceae, Bacillariophyceae, Xanthophyceae, Phaeophyceae).
3. cpDNA is present in variable numbers of discrete nucleoids in mature plastids of higher plants and certain algae, e.g. *Chlamydomonas*.

Chloroplast genomes of most higher plants consist of one large and one small region of unique sequence DNA separated by two large inverted repeats. Each inverted repeat contains one set of rRNA genes which accounts for about 20% of the sequence. The inverted repeat regions are relatively G + C rich whereas the unique sequences are A + T rich (Herrmann and Possingham, 1980). In the chloroplast genome of *Chlamydomonas reinhardtii*, two unique sequences of more nearly equal length separate the two large inverted repeats and the molecule is correspondingly larger than that in higher plants (Rochaix, 1978). In two closely related genera of legumes, pea (*Pisum*) and broad bean (*Vicia*), there is but one set of rRNA cistrons per cpDNA molecule and no inverted repeat region (Chu *et al.*, 1981; Palmer and Thompson, 1981a, b; Koller and Delius, 1980). Chloroplast genomes of *Euglena gracilis* strain Z and variety *bacillaris* consist of a single large region of unique sequence DNA and a triplet tandem repeat of 5.6 kbp, with each repeat containing one set of rRNA genes (Gray and Hallick, 1978, 1979; Rawson *et al.*, 1978). An additional 16S rRNA gene is found just proximal to the tandem repeat region in these strains as well (Jenni and Stutz, 1979). Wurtz and Buetow (1981) have reported that one laboratory stock of *E. gracilis* strain Z has but a single set of rRNA genes.

Evidence indicating that plastid genomes exhibit prominent intramolecular heterogeneities in their base composition is presented by Crouse *et al.* (1978), Herrmann and Possingham (1980), Schmitt *et al.* (1981) and Tabidze and Beridze (1979). A class of short repetitive sequences has been detected in cpDNA from *Chlamydomonas reinhardtii* (Gelvin and Howell, 1979; Rochaix, 1972, 1978). These sequences are 0.1–0.3 kbp in size, constitute between 4 and 7% of the cpDNA and are interspersed throughout the chloroplast genome. In *Euglena gracilis* (strain Z) Jenni *et al.* (1981) report that the multiple copies of the chloroplast genome are not uniform in size and contain a variable number of short repeats clustered in a specific region.

CHLOROPLAST GENOME FUNCTION

Chloroplast genome function has been reviewed in detail recently (Bartlett *et al.*, 1981; Bedbrook and Kolodner, 1979; Edelman, 1981; Gillham, 1978; Herrmann and Possingham, 1980; Rochaix, 1981; Steinback, 1981; von Wettstein, 1981, 1982). The average cpDNA molecule (approximately 10^8 d) could theoretically code for as many as 300 polypeptides of molecular weight 20 000 d (Gillham, 1978). Chloroplast gene products identified to date include chloroplast rRNAs, probably a complete set of tRNAs, and a number of proteins, most of which are associated either with the photosynthetic lamellae or with the chloroplast protein synthesizing system.

Two or three chloroplast envelope polypeptides appear to be synthesized inside isolated chloroplasts of spinach and pea, but, as discussed by Douce and Joyard (1979), the possibility cannot be overlooked that the envelopes used in these studies were contaminated with stroma or thylakoid components. Among the soluble proteins in the chloroplast stroma, only the CO_2-fixing enzyme ribulose bisphosphate carboxylase (RuBPCase) is known to require chloroplast protein synthesis (cf. Bartlett *et al.*, 1981). The RuBPCase holoenzyme in higher plants and green algae has a molecular weight of about 550 000 d and consists of eight identical large subunits (55 000 d) and eight identical small subunits (14 000 d). The gene for the large subunit of this enzyme is localized in cpDNA (cf. Coen *et al.*, 1977) and its message is translated in the chloroplast. In contrast, the small subunit gene is located in the nucleus (cf. Cashmore, 1979) and its message is decoded in the cytoplasm.

The gene coding for the large subunit has been mapped in the chloroplast genomes of several plants including maize (Link and Bogorad, 1980), mustard (Link, 1981), tobacco (Seyer *et al.*, 1981), spinach (Herrmann *et al.*, 1980a), wheat (Gatenby *et al.*, 1981) and the green alga *Chlamydomonas* (Malnoë *et al.*, 1979; Rochaix and Malnoë, 1978b). The DNA sequences of the maize and spinach large subunit genes have been determined (McIntosh *et al.*, 1980; Zurawski *et al.*, 1981) while the amino acid sequence of barley large subunit has been ascertained directly (Poulsen, 1979; Poulsen *et al.*, 1979). Langridge (1981) has reported that the large subunit in spinach appears to be made as a precursor 1000–2000 d larger than mature large subunit and that this precursor is processed to mature large subunit by a soluble chloroplast factor. Gatenby *et al.* (1981) have observed expression of cloned large subunit genes from maize and wheat in *E. coli*. The nuclear gene for the small subunit from pea has been cloned (Bedbrook *et al.*, 1980; Broglie *et al.*, 1981) and sequenced (Bedbrook *et al.*, 1980). The small subunit polypeptide is made as a precursor with an amino-terminal extension which is then processed (reviewed by Ellis *et al.*, 1980; Schmidt *et al.*, 1980). An interesting problem in regulation of the synthesis of the two subunits of RuBPCase is created by the report of Cashmore (1979) that only one or very few copies of the small subunit gene are present per cell, whereas there are multiple copies of the large subunit gene per plastid and usually multiple plastids per cell as well.

The question remains open whether other stroma proteins are also chloroplast gene products. Ellis and colleagues (1977, 1978, 1980) have reported detection of 80 radioactive spots on two-dimensional gels of stroma proteins following labelling of isolated pea chloroplasts. However, as Bartlett *et al.* (1981) point out, these results must be interpreted with caution since only the large subunit of RuBPCase clearly comigrated with an authentic stroma polypeptide.

The thylakoid membranes of the chloroplast contain many distinct polypeptides making up the light-harvesting pigment-protein complexes, electron transport chain and the ATP synthesizing system. In *Chlamydomonas* Chua and Bennoun (1975) were able to resolve 33 polypeptides from purified thylakoid membranes on SDS gels. Recent reports (von Wettstein, 1981; von Wettstein *et al.*, 1982) show that a minimum of 43 polypeptides can be resolved from purified thylakoid membranes of barley. Of these at least 30 can be related to specific photosynthetic functions. Three functional complexes can be recognized in the thylakoid membranes: photosystem I (PSI), photosystem II (PSII) and the chloroplast coupling factor (CF_1-CF_0 complex) (Bartlett *et al.*, 1981; von Wettstein, 1981; von Wettstein *et al.*, 1982). Two proteins in the PSI complex, the chlorophyll a binding protein and cytochrome f, have been identified as products of chloroplast protein synthesis. Three other proteins of this complex, ferredoxin, ferredoxin NADP reductase and plastocyanin, are synthesized in the cytoplasm and transported into the chloroplast. Three chlorophyll protein complexes associated with PSII (CPII, III and IV) have been characterized. The chlorophyll a/b binding apoprotein of CPII, the principal complex, is synthesized in the cytoplasm while the apoproteins of CPIII and IV are made in the chloroplast. Rochaix (1981) has identified the map positions for the latter apoproteins on the chloroplast genome of *Chlamydomonas*. Recently, Broglie *et al.* (1981)

have cloned the nuclear gene coding for the chlorophyll a/b polypeptide from pea.

Other chloroplast gene products associated with the PSII complex are cytochrome b_{559} and the 32 kd 'photoprotein'. Synthesis of the latter protein, which turns over rapidly, is induced by light. The 32 kd protein (peak D) and the large subunit of RuBPCase were very early identified as major products of protein synthesis in isolated chloroplasts of pea (Blair and Ellis, 1973; Eaglesham and Ellis, 1974; Ellis *et al.*, 1980). Like the large subunit of RuBPCase, the 32 kd protein is made as a precursor 1000–2500 d larger than the mature protein (Edelman and Reisfeld, 1980; Grebanier *et al.*, 1978). Two proteins that turn over rapidly, D-1 and D-2, are made on chloroplast ribosomes in *Chlamydomonas* (Chua and Gillham, 1977). Hoffman-Falk *et al.* (1980) have reported that peak D from pea, peak D-1 from *Chlamydomonas* and the 32 kd proteins from maize and *Spirodela* all appear to be identical on the basis of partial proteolytic digestion patterns. The gene coding for the 32 kd protein has been mapped on the chloroplast genomes of maize (Bedbrook *et al.*, 1978), mustard (Link, 1981), spinach (Driesel *et al.*, 1980) and *Chlamydomonas* (Rochaix, 1981). In *Spirodela*, the 32 kd protein appears to be the site of action of the herbicide diuron (Mattoo *et al.*, 1981).

The chloroplast ATPase consists of peripheral (CF_1) and integral membrane (CF_0) complexes. The α, β and ϵ subunits of CF_1 are products of chloroplast protein synthesis whereas γ and δ are most likely made in the cytoplasm although there is some conflict on this point (see Bartlett *et al.*, 1981; Nechushtai *et al.*, 1981). CF_0 consists of three or four subunits, of which at least two are made in the chloroplast and one in the cytoplasm. The genes coding for the α, β and ϵ subunits of CF_1 have been mapped on the chloroplast genome of spinach (Westhoff *et al.*, 1981). The genes coding for the α and β subunits have been localized in *Chlamydomonas* (Rochaix, 1981) and in *Spirodela* (de Heij and Groot, 1981).

Although chloroplast rRNAs and tRNAs are coded by the chloroplast genome in every species examined to date, many of the proteins of the chloroplast translation system are nuclear gene products. There is no evidence in any plant to suggest that the tRNA synthetases are coded by the chloroplast genome. In *Euglena* there is positive evidence that some of these synthetases are coded by nuclear genes (cf. Gillham, 1978). Certain chloroplast elongation factors appear to be coded by the chloroplast genome in some species and by the nuclear genome in others. The chloroplast elongation factor G ($EF-G_{chl}$) is a product of chloroplast protein synthesis in both *Chlorella* and spinach (Ciferri and Tiboni, 1976; Ciferri *et al.*, 1979), whereas in *Euglena* it is a nuclear gene product (Breitenberger *et al.*, 1979). A nuclear gene conferring fusidic acid resistance on this factor has been identified in *Chlamydomonas* (Carbonera *et al.*, 1981). In spinach (Ciferri *et al.*, 1979) and *Chlamydomonas* (Watson and Surzycki, 1981) elongation factor Tu (EF-Tu$_{chl}$) appears to be a chloroplast gene product, while in *Euglena* elongation factor Ts (EF-Ts$_{chl}$) is a nuclear gene product (Fox *et al.*, 1980). In *E. coli* EF-Ts is normally complexed with EF-Tu (cf. Weissbach, 1980).

In contrast to mitochondria of yeast and *Neurospora* where all but one ribosomal protein is made in the cytoplasm, chloroplasts appear to synthesize a considerable number of their ribosomal proteins. In *Euglena* nine of 57 chloroplast ribosomal proteins were found by Freyssinet (1978) to be synthesized in the chloroplast and 12 in the cytoplasm. The site of synthesis of the remaining proteins was uncertain. Isolated chloroplasts of pea were found to synthesize at least six of the 24 small subunit proteins and at least five of the 32 large subunit proteins (Eneas-Filho *et al.*, 1981). Richardson *et al.* (1982) have determined that in *Chlamydomonas* at least five of the 33 large and 14 of the 31 small subunit proteins are made in the chloroplast, while the remaining proteins of each subunit are products of cytoplasmic protein synthesis. Thus the ribosomal proteins appear to outnumber all other known chloroplast gene products except tRNAs. From the evolutionary viewpoint a comparison of the structure and organization of chloroplast and nuclear genes coding for chloroplast ribosomal proteins should be of great interest. The regulation of these two sets of genes which are present in vastly different copy numbers in the cell poses an even more complicated problem than that discussed previously for the small and large subunit genes of RuBPCase.

ORGANIZATION AND STRUCTURE OF THE rRNA GENES

The organization of the chloroplast rRNA genes has been studied in *Euglena*, *Chlamydomonas* and several higher plants (cf. Dyer and Bedbrook, 1980; Dyer and Leaver, 1981). In a few cases chloroplast rRNA genes have also been sequenced. The striking observation emerging from this work is that, despite some novel embellishments, a close similarity exists between the organization and structure of these chloroplast genes and the genes of the *rrn* operons coding for rRNA in *E. coli* (Fig. 11.1). In fact, sufficient homology exists between bacterial, mitochondrial and chloroplast rRNAs that any of these can be used as probes to identify rRNA genes in other chloroplast genomes (Bohnert *et al.*, 1980; Delius and Koller, 1980). Although eight *rrn* operons have been reported in *E. coli*, Ellwood and Nomura (1982) have found no evidence for the existence of the *rrnF* operon and believe there are only seven. The *rrnA* and *rrnD* operons contain genes coding for isoleucine tRNA (tRNA$_1^{Ile}$) and alanine tRNA (tRNA$_{1B}^{Ala}$) in the spacer between the 16S and 23S rRNA genes (Nomura and Post, 1980). The spacers in the *rrnB*, *rrnC*, *rrnE* and *rrnG* operons all contain a gene coding for glutamine tRNA (tRNA^{Glu}). The *rrnC* operon has distal genes coding for tRNAs for aspartate (tRNA^{Asp}) and tryptophan (tRNA^{Trp}). The *rrnH* operon has a distal tRNA^{Asp} gene and the *rrnD* operon has a distal gene coding for a threonine tRNA (tRNA^{Thr}) (Ellwood and Nomura, 1982).

In most *Euglena gracilis* strains the chloroplast rRNA genes are organized in three tandem repeats of 5600 bp each (Gray and Hallick, 1978, 1979; Graf *et al.*, 1980; Helling *et al.*, 1979; Jenni and Stutz, 1978; Orozco *et al.*, 1980a, b; Rawson *et al.*, 1978). An additional 16S rRNA gene has been identified proximal to the first repeat in strain Z by Jenni and Stutz (1979). The arrangement of the rRNA genes within each repeat is reminiscent of the *E. coli rrnD* operon (Fig. 11.1). Each repeat unit is transcribed into a precursor RNA molecule with a molecular weight of about 1.8×10^6 d (Wollgiehn and Parthier, 1979). Orozco *et al.* (1980b) also report that the 700 base pair spacer between the repeat units contains a pseudo tRNA^{Ile} gene and 54 base pairs of the 3' end of a tRNA^{Trp} gene. Keller *et al.* (1980) have confirmed the assignments of the tRNA^{Ile} and tRNA^{Ala} genes between the 16S and 23S rRNA genes by hybridization of the cognate tRNAs to appropriate restriction fragments of chloroplast DNA. However, they do not report hybridization of tRNA^{Ile} within the 700 base pair spacer despite the presence of the pseudo-gene there.

The organization of the chloroplast rRNA genes in *Chlamydomonas* is very different from *Euglena* and departs further from the bacterial model (Fig. 11.1; Rochaix, 1978, 1981; Rochaix and Malnoë, 1978a). There is one set of rRNA genes in each inverted repeat. The spacer between the 16S and 23S genes contains at least one and possibly more 4S (tRNA) genes (Malnoë and Rochaix, 1978). The 23S rRNA gene appears to be a mosaic of four coding sequences, two short spacers and a large intervening sequence. The 7S, 3S and 23S rRNAs are associated with the large subunit of the chloroplast ribosome and extensive sequence homology exists between the 3S and 7S genes and the 5' end of the 23S rRNA genes of *E. coli*, *Euglena* and maize (Rochaix and Darlix, 1982). The 3S RNA gene is flanked by two A-T rich hairpin structures which are missing from the corresponding region of the *E.*

E. coli rrnD operon
5' — 16S — [68 bp] — tRNA^{Ile} — [42 bp] — tRNA^{Ala} — [174 bp] — 23S — [92 bp] — 5S — 3'

Euglena gracilis
5' — 16S — [87 bp] — tRNA^{Ile} — [9 bp] — tRNA^{Ala} — [16 bp] — 23S — [? bp] — 5S — 3'

Chlamydomonas reinhardtii
5' — 16S — [1700 bp — tRNAs?] — 7S — [23 bp] — 3S — [~80 bp] — 23S — [? bp] — 5S — 3'
　　　　　　　　　　　　　　　　　　　　　　　　　　(includes 870 bp intron)

Zea mays
5' — 16S — [2100 bp, including tRNA^{Ile}, tRNA^{Ala}] — 23S — [78 bp] — 4.5S — [? bp] — 5S — 3'
　　　　　　　　　　　　　　　　　　　　　　　　　　(256 bp in tobacco)

Fig. 11.1. Linear organization of ribosomal RNA genes in *E. coli* and chloroplasts of three organisms. Spacer regions are designated by brackets.

coli 23s rRNA gene. In maize one of these hairpin structures has been maintained while in *Euglena* the other, which is A-T rich as in *C. reinhardtii*, has been preserved (Rochaix and Darlix, 1982). Within the 23S rRNA gene of *Chlamydomonas* there is a large intron (870 bp; Rochaix and Malnoë, 1978a; Rochaix, 1981). A cloned intron probe from the 23S intron does not hybridize elsewhere in the chloroplast genome although the probe shows faint hybridization with several restriction fragments of the nuclear genome (Rochaix, 1981). Partial nucleotide sequence complementarity observed between the 5' ends of the 7S and 3S RNAs and the 23S rRNA sequences which flank the ribosomal intron suggest that these small rRNAs might play a role in the processing of the 23S rRNA precursor.

The organization of the chloroplast rRNA genes has been studied in a number of higher plants with the typical pattern of two inverted repeats, including maize (Fig. 11.1; Bedbrook and Bogorad, 1976; Bedbrook *et al.*, 1977), mung bean (Palmer and Thompson, 1981b), mustard (Link *et al.*, 1981), spinach (Whitfeld *et al.*, 1978), *Oenothera* (Gordon *et al.*, 1981), tobacco (Fluhr and Edelman, 1981a; Kusuda *et al.*, 1980; Jurgenson and Bourque, 1980; Seyer *et al.*, 1981; Sugiura and Kusuda, 1979; Takaiwa and Sugiura, 1980b; Tohdoh *et al.*, 1981) and wheat (Bowman *et al.*, 1981). In addition, the structures of the single sets of rRNA genes in pea (Chu *et al.*, 1981; Palmer and Thompson, 1981b) and broad bean (Delius and Koller, 1980; Koller and Delius, 1980) have been examined. The spacer between the 16S and 23S rRNA genes ranges in size from 1800 base pairs in spinach to 2100 base pairs in maize and tobacco to 2400 base pairs in *Oenothera* and broad bean. Delius and Koller (1980) using heteroduplexing have found homology between the spacer in broad bean and the *rrnD* operon of *E. coli*, which contains the tRNA^{Ile} and tRNA^{Ala} genes, but not between the broad bean spacer and the spacer of the *rrnB* operon, which contains tRNA^{Glu}. Within the spacer a gene coding for tRNA^{Ile} has been identified in spinach (Driesel *et al.*, 1979; Bohnert *et al.*, 1979) while genes coding for both tRNA^{Ile} and tRNA^{Ala} have been found in maize (Koch *et al.*, 1981). While the same two tRNAs are found in *Euglena* and in the *E. coli rrnA* and *rrnD* operons, the maize genes coding for tRNA^{Ile} and tRNA^{Ala} are split by intervening sequences of 949 and 806 base pairs, respectively (Koch *et al.*, 1981). These introns occupy over 80% of the spacer between the 16S and 23S rRNA genes and sequence homology suggests they have a common origin. Obviously, it would be desirable to know whether the same organiz-ation applies within this spacer region in other higher plants and *Chlamydo-monas*. In both maize (Schwarz *et al.*, 1981a) and tobacco (Tohdoh *et al.*, 1981) a tRNA^{Val} gene is located about 300 base pairs proximal to the 5' end of the 16S rRNA gene on the same strand and is probably part of the primary transcript of the rRNA genes (see below). The presence of proximal tRNA genes has not been reported for the *rrn* operons of *E. coli*.

In *E. coli* the 5S rRNA gene is part of the primary transcript of each *rrn* operon, but this probably is not true in the higher plant chloroplast. Experiments with isolated spinach chloroplasts show that a 2.7×10^6 d rRNA is accumulated with sequences homologous to 16S and 23S rRNA (Bohnert *et al.*, 1976; Hartley and Head, 1979). While 4.5S rRNA is probably a part of this precursor, 5S rRNA is not (Hartley, 1979). In tobacco, Takaiwa and Sugiura (1980a) have found sequences in the region between the 4.5S and 5S genes that resemble *E. coli* promoters, suggesting the chloroplast 5S rRNA gene is transcribed separately from the rest of the operon. Whereas the 5s rRNA gene of prokaryotes is transcribed as part of the rRNA operon, separate transcription is the norm for eukaryotes, even where the 5S rRNA genes are closely linked to the genes coding for the high molecular weight rRNAs such as in *Dictyostelium* and *Saccharomyces* (cf. Planta and Meyerink, 1980).

In the *rrn* operons of *E. coli* two tandem promoters initiate transcription approximately 300 (P_1) and 200 (P_2) bases upstream from the 16S rRNA gene (de Boer *et al.*, 1979; Nomura and Post, 1980). The P_1 promoters have a common 15 base sequence including the Pribnow box (Pribnow, 1975) but there is considerable sequence diversity on either side. The P_2 promoters for three of the operons are identical but the fourth is different. Promoter-like sequences have been searched for upstream of the 5' end of the chloroplast 16S rRNA gene in *Euglena* (Orozco *et al.*, 1980b), tobacco (Tohdoh *et al.*, 1981) and maize (Schwarz *et al.*, 1981a). In *Euglena* no promoter-like regions were observed in a sequence extending 400 base pairs upstream from the 5' end of the 16S rRNA gene. However, this leader sequence exhibits a high (70%) homology with the spacer sequence between the 16S and 23S rRNA genes, including a region homologous to the 3' end of a 16S rRNA gene, the adjacent spacer and a 'pseudo' tRNA^{Ile} gene. In addition, the leader region contains two sequences which seem to conform to parts of a tRNA^{Trp} gene. The first is just upstream of the sequence corresponding to the 3' end of a 16S rRNA gene and includes the 5' end of the partial tRNA gene plus half of the D loop. The second sequence, which follows the pseudo tRNA^{Ile} gene, starts with a short precise overlap of the first part of the partial gene and includes the anticodon stem and loop.

Tohdoh *et al.* (1981) have sequenced 627 base pairs upstream of the 5' end of the 16S rRNA gene in tobacco in search of promoters. There is a tRNA^{Val} gene between 229 and 300 base pairs upstream of the 16S rRNA gene. When purified *E. coli* RNA polymerase was used to transcribe this fragment, two major RNA products of 460 and 240 base pairs were obtained, the former of which seems to be an authentic transcript. Transcription begins at what appears to be a promoter region between 320 and 326 base pairs upstream from the 5' end of the 16S gene. A TAGGATT sequence similar to a Pribnow box (Pribnow, 1975) is found 331 to 337 base pairs upstream and a TTG sequence is found about 20 base pairs further upstream. These promoter sequences are similar to promoters of *E. coli* operons except that the TTG sequence is characteristically 35 base pairs upstream of the promoter rather than 20 base pairs upstream (cf. Rosenberg and Court, 1979). Since the tobacco promoter region is upstream of the tRNA^{Val} gene, this gene must be part of the transcript of the chloroplast rRNA operon. Schwarz *et al.* (1981a) have sequenced a 635 base pair region upstream of the 5' end of the maize chloroplast 16S rRNA gene. A tRNA^{Val} gene preceded by TTG and TAGGATT sequences is found in the same relative position as in tobacco. *E. coli* RNA polymerase binds to this region as well as to two other regions proximal to the 16S rRNA gene in maize (Schwarz *et al.*, 1981a). By analogy with tobacco, the most likely promoter is the one proximal to the tRNA^{Val} gene.

In *E. coli* the 16S rRNA gene is flanked by sequences which base pair to form a stem structure containing the RNAse III cleavage site (Young and Steitz, 1978). This enzyme clips the rRNA precursor at this point to generate the immediate precursor of 16S rRNA. So far, similar sequences have not been found flanking chloroplast 16S rRNA genes.

The complete sequence of the chloroplast 16S rRNA gene has been determined in maize by Schwarz and Kössel (1980) who found that 74% of the base pairs were identical to those found in the *E. coli* gene. The longest stretch of identical nucleotides observed was 53 bp and the 16S rRNA gene from maize was found to be 50 nucleotides shorter than the *E. coli* 16S rRNA gene. This reduction in size was due primarily to three deletions of groups of nucleotides, two of which form short hairpin structures in *E. coli*.

In prokaryotes the 3' end of the 16S rRNA contains the sequence which pairs with the complementary sequence ACCUCCUUA$_{OH}$ discovered prior to the initiation codon in mRNA by Shine and Dalgarno (1974). Taniguchi and Weissmann (1978) demonstrated directly that this sequence is required for formation of initiation complexes by blocking it with the complementary oligonucleotide. Steitz (1980) has found extensive homology in the fifty nucleotides of the 3' termini of *E. coli* 16S rRNA and eukaryotic 18S rRNAs. However, the CCUCC part of the Shine-Dalgarno sequence is deleted in rat 18S rRNA. Steitz points out that the CUCC portion of this sequence in the *E. coli* 16S rRNA is the region most often utilized for mRNA binding during initiation. In blue-green algae, the sequence of 16S rRNA terminates in CCUCCUUU$_{OH}$ in one strain of *Synechococcus* (Borbely and Simoncsits, 1981) while in another strain of *Synechococcus* and in several strains of *Aphanocapsa* the sequence CCUCCU$_{OH}$ is found (Bonen *et al.*, 1979). Thus blue-green algae contain the CCUCC sequence as in *E. coli*, but the number and species of nucleotides following this sequence differ. In the chloroplast 16S rRNA gene of maize, the 3' terminal sequence is CCUCCUUU$_{OH}$ as in one of the *Synechococcus* strains (Borbely and Simoncsits, 1981; Schwarz and Kössel, 1979, 1980). In *Euglena gracilis* strain Z, the 3′ terminal sequence of the 16S rRNA gene is ACUCC$_{OH}$ (Orozco *et al.*, 1980b), while in the *bacillaris* strain of this alga there is an additional C at the 3′ end (Steege *et al.*, 1982). In summary, the two 16S rRNA 3′ terminal sequences from chloroplasts are prokaryotic in character and predict the presence of Shine-Dalgarno sequences in chloroplast mRNAs similar to those already identified in the message coding for the large subunit of RuBPCase (see below).

Young *et al.* (1979) have sequenced the spacer between the 16S and 23S rRNA genes in the *rrnD* and *rrnX* operons in *E. coli*. The *rrnX* operon is a hybrid containing the 5′ end of the *rrnH* operon and the 3′ end of the *rrnC* operon (Nomura, 1982). The comparable sequence has been determined for the *Euglena* (Graf *et al.*, 1980; Orozco *et al.*, 1980b) and maize (Koch *et al.*, 1981) chloroplast spacers. The *E. coli* spacer can be divided into five segments (Young *et al.*, 1979). The first segment extends 68 base pairs from the 3′ end of the 16S rRNA gene to the 5′ end of the tRNA$_1^{Ile}$ gene. In *Euglena* the comparable region differs by a single deletion of eight base pairs from a segment of the leader of the 16S rRNA gene (Orozco *et al.*, 1980b). Segment I from *Euglena* also shares limited homology with segment I from *E. coli* (Graf *et al.*, 1980). The repetition of much of the 16S rRNA gene leader sequence in segment I of the rRNA gene of *Euglena* suggests that these sequences will not be able to base pair to form a stem-loop structure in the primary transcript which includes 16S rRNA. However, a stem-loop structure can be formed within segment I itself immediately prior to the 5′ end of the tRNA^{Ile} gene (Graf *et al.*, 1980). Interestingly, a similar sequence can be found just prior to the pseudo tRNA^{Ile} gene in the leader (Orozco *et al.*, 1980b). The similarity of these structures supports the suggestion of Graf *et al.* that they may contain recognition sequences for a processing enzyme. However, neither of these stem-loop structures contains an obvious recognition site for *E. coli* RNAse III. Although segment I is shorter in maize than in *Euglena*, a stem-loop structure is again found just proximal to the 5′ end of the tRNA^{Ile} gene (Koch *et al.*, 1981).

Segments II and IV of the *E. coli* spacer in the *rrnD* operon contain the genes for tRNA$_1^{Ile}$ and tRNA$_{1B}^{Ala}$, respectively. The same two genes are present in the same relative positions in *Euglena* (Graf *et al.*, 1980; Orozco *et al.*, 1980b) and maize (Koch *et al.*, 1981) and share extensive sequence homology with their counterparts in *E. coli*. However, the tRNA^{Ile} and tRNA^{Ala} genes in maize contain intervening sequences of 949 and 806 base pairs, respectively. Koch *et al.* have identified open reading frames in the introns of both genes, each beginning with an AUG codon and preceded by a Shine-Dalgarno-like sequence. The open reading frame in the tRNA^{Ile} gene could code for a protein of 123 amino acids and terminates with a UAA codon, while the open reading frame in the tRNA^{Ala} gene could specify a protein of 45 amino acids and terminates with the codon UAG. Koch *et al.*

have examined sequences common to the 5' and 3' ends of the two introns and propose that they may have a common origin based on their considerable homology.

In *E. coli* segment III, a 42 base pair sequence separates the two tRNA genes (Young *et al.*, 1979), whereas in *Euglena* the intergenic spacer is only six base pairs (Graf *et al.*, 1980). These six base pairs of the *Euglena* spacer are homologous to six contiguous base pairs in the interior of the *E. coli* spacer. In contrast to *Euglena*, the spacer between tRNA^{Ile} and tRNA^{Ala} in maize is 58 base pairs long (Koch *et al.*, 1981). The two *Euglena* tRNAs can form a dimeric tRNA precursor structure (Graf *et al.*, 1980), whereas in maize a comparable structure may not be stable because of the length of segment III.

Segment V in *E. coli* extends from the 3' end of the tRNA$^{Ala}_{1B}$ gene to the 5' end of the 23S rRNA gene and constitutes 40% of the spacer DNA (Young *et al.*, 1979). There is extensive sequence complementarity between segment V and sequences flanking the 3' end of the 23S rRNA gene. Bram *et al.* (1980) showed that these sequences in *E. coli* can pair to form a giant stem-loop structure where the loop is 23S rRNA and the stem contains an RNAse III processing site. In *Euglena* the sequence equivalent to segment V is reduced to a few base pairs (Graf *et al.*, 1980). As in segment III, the retained base pairs in *Euglena* line up precisely with homologous base pairs in the interior of segment V of *E. coli*. These results suggest that the processing enzyme in *Euglena* which makes the immediate precursors of 16S and 23S rRNA recognizes different features of secondary structure than RNAse III from *E. coli*. In maize the equivalent of segment V is 142 base pairs long (Koch *et al.*, 1981). This sequence has not been compared to segment V of *E. coli* nor is it known whether there are complementary sequences at the 3' end of the chloroplast 23S rRNA gene.

The complete sequences of *E. coli* 23S rDNA (Brosius *et al.*, 1980) and rRNA (Branlant *et al.*, 1981; Glotz *et al.*, 1981) have been compared to the 23S rDNA gene in maize chloroplast DNA (Edwards and Kössel, 1981) as well as directly to maize chloroplast 23S rRNA and the rRNAs of the large subunit of the mitochondrial ribosome in mouse and humans (Branlant *et al.*, 1981). Branlant *et al.* (1981) divide *E. coli* 23S rRNA into seven domains. The 5' terminal domain (domain I) encompasses nucleotides 1–530 and remains associated with protein L24 upon digestion of the L24-23S rRNA complex. The 5' end of domain I appears to base pair with a complementary sequence of several bases at the 3' end of the molecule (domain VII). On the basis of sequence homology (Nazar, 1980; Jacq, 1981) and secondary structural considerations (Branlant *et al.*, 1981; Clark and Gerbi, 1982), 5.8S rRNA in eukaryotic ribosomes has been proposed to play the same role as the 5' end of 23S rRNA in prokaryotes. As mentioned earlier, Rochaix and Darlix (1982) reported extensive sequence homology between the 3S and 7S chloroplast rRNAs found associated with the large subunit of the chloroplast ribosome in *Chlamydomonas* and the 5' end of *E. coli* 23S rRNA. Their analysis also revealed a high degree of sequence homology between the 3S and 7S genes and 5.8S RNA. Thus, both 5.8S and 7S rRNA appear to have been split off from the 5' end of ancestral 23S rRNA genes and may share similar functions. In maize the interaction predicted between the 5' and 3' ends of *E. coli* 23S rRNA occurs between the 5' end of the molecule and 4.5S rRNA (Glotz *et al.*, 1981). The latter molecule is derived from the 3' end of 23S rRNA in chloroplasts of higher plants (see below). In *Euglena* no small rRNAs derived from the chloroplast large ribosomal subunit have been reported. In summary, the chloroplast 23S rRNA gene can either be intact, as in *Euglena*, or a small rRNA species may be split off at the 5' end as in *Chlamydomonas* or at the 3' end as in higher plants.

Between the 5' terminal region of *E. coli* 23S rRNA and the so-called central region (domain IV, positions 1196–1650) Branlant *et al.* (1981) define domains II and III. Both of these regions are strongly conserved in maize 23S rRNA. Domain IV together with the 5' terminus (domain I) and the 3' terminus (domain VII) constitute a compact area in the large subunit of the *E. coli* ribosome. In maize 23S rRNA domain IV includes a large insert of 65 bases relative to *E. coli* (Edwards and Kössel, 1981; Glotz *et al.*, 1981). Domains V and VI (bp 1651–2629) are highly conserved in maize 23S rRNA except for a region of 60 base pairs which forms a loop in *E. coli* and is absent in maize. In *E. coli* domain VI includes the binding sites for ribosomal proteins L1, L5, L18 and L25, as well as 5S rRNA (Branlant *et al.*, 1981). In the yeast mitochondrial genome, mutations to chloramphenicol resistance have been found to map in mitochondrial 21S rRNA at positions corresponding to 2447 and 2503 of *E. coli* 23S rRNA (Dujon, 1980). The nucleotides at both these positions are the same in maize 23S rRNA as in *E. coli* and in wild type yeast. The 3' region in *E. coli* 23S rRNA is domain VII (bp 2630–2904). Near the 3' end of domain VII maize 23S rRNA contains a 78 base pair insert compared to *E. coli*. In total there are 450 compensating base pair changes between *E. coli* and maize chloroplast 23S rRNA species (Glotz *et al.*, 1981) out of the total parallel sequence which occupies 2904 base pairs in *E. coli*.

The gene coding for 4.5S rRNA in maize is distal to the gene coding for 23S rRNA. This rRNA, which is more than 60% homologous in primary sequence with the 3' end of *E. coli* 23S rRNA (Edwards *et al.*, 1981; Edwards and Kössel, 1981; Machatt *et al.*, 1981; MacKay, 1981), and shows extensive conservation of secondary structure (Clark and Gerbi, 1982), is a characteristic component of the large subunit of chloroplast ribosomes of flowering plants and occurs in size classes of 77 and 100 nucleotides (Bowman and Dyer, 1979). The nucleotide sequence of 4.5S rRNA has been determined in maize (Edwards *et al.*, 1981) and tobacco (Takaiwa and Sugiura, 1980a). No rRNAs equivalent to 4.5S rRNA have yet been reported from algae.

As mentioned earlier, Rochaix and Malnoë (1978a) found an intervening sequence in the gene coding for 23S rRNA in *Chlamydomonas*. The ends of this intervening sequence are each flanked by an identical three base pair repeat (Allet and Rochaix, 1979). In addition, a five base pair repeat is found adjacent to the distal end of the intervening sequence and again 16 base pairs from the other end in the coding part of the gene. Finally, an inverted repeat reminiscent of those found at the ends of IS elements can be drawn near the ends of the intervening sequence. Rochaix (1981) has shown that the intron can be placed formally between positions 2593 and 2594 in *E. coli* 23S rRNA which is close to the 3' end of domain VI (Branlant *et al.*, 1981). Similarly, Dujon (1980) has shown that an intron present in the 21S rRNA gene of mitochondrial DNA of certain strains of yeast (*Saccharomyces cerevisiae*) can be placed between positions 2447 and 2449 in the *E. coli* 23S rRNA gene.

The 23S-5S intergenic regions of the *rrnD* and *rrnX* operons of *E. coli* are 92 base pairs long and are identical except for two nucleotide substitutions (Bram *et al.*, 1980). The flanking regions on either side of the 23S rRNA gene can base pair to form a processing site for RNAse III. So far, the only equivalent chloroplast intergenic spacer that has been sequenced is the 256 base pair segment between the 4.5S and 5S rRNA genes in tobacco (Takaiwa and Sugiura, 1980a). A promoter-like sequence (TATGCCTT) is proximal to the 5S gene by 32 base pairs and preceded by 25 base pairs by the TTG sequence characteristic of the -35 region close to *E. coli* promoters (Rosenberg and Court, 1979). These results are consistent with the observations of Hartley (1979) and suggest that the 5S rRNA gene is transcribed independently of the rest of the rRNA genes.

Chloroplast 5S rRNA sequences have been reported for a number of higher plants (Delihas *et al.*, 1981; Dyer and Bowman, 1979; Takaiwa and Sugiura, 1980a, 1981). Osawa and Hori (1980) divide 5S rRNAs into four types. The first, to which all eukaryotic cytoplasmic 5S rRNAs belong, is 116–118 nucleotides long in plants and 120 nucleotides in other eukaryotes. This type differs from all bacterial 5S rRNAs by having a well-conserved loop between positions 83 to 94. It also lacks the hairpin structure that exists in the 5S rRNAs of eubacteria (Fox and Woese, 1975) and specific sequences present in prokaryotic 5S rRNA which interact with 23S rRNA.

The other three species of 5S rRNA are of prokaryotic origin. The two eubacterial types are 116 and 120 nucleotides long, respectively. The latter type is found in the cyanobacterium *Anacystis nidulans* whose 5S rRNA sequence has been compared to chloroplast 5S rRNA (Delihas *et al.*, 1981; Dyer and Bowman, 1979). The fourth 5S rRNA species, found in the Archaebacterial genus *Halobacterium*, contains a terminal helix that is less complete than the other three types. The first to third nucleotides from the 5' end do not base pair with the 3' end, and a small loop is seen between positions 87 to 94 reminiscent of the loop found in eukaryotes. Delihas *et al.* (1981) have compared the sequence of chloroplast 5S rRNA from spinach to the sequence of 5S rRNA in *A. nidulans*, *Lemna* (Dyer and Bowman, 1979) and tobacco (Takaiwa and Sugiura, 1980a, 1981). The nucleotide sequences of the spinach and tobacco 5S rRNAs are identical and differ by only two nucleotides from that of *Lemna*. These sequences show extensive homology with *Anacystis* 5S rRNA, and the secondary structure models obtained for all these 5S rRNAs are very similar. Thus chloroplast 5S rRNAs appear very closely related to the eubacterial 120 nucleotide type (Osawa and Hori, 1980).

ORGANIZATION AND STRUCTURE OF THE tRNA GENES

Chloroplast tRNAs have been identified for most of the amino acids (cf. Dyer and Leaver, 1981; Mubumbila *et al.*, 1980; Weil *et al.*, 1980). Although a few tRNAs remain to be identified in each species examined, if the data are taken in aggregate only tRNA^{Cys} has not yet been found in any plant. Multiple isoaccepting tRNAs are found for different amino acids having more than one pair of codons. The genetic code in chloroplasts appears to be read according to the rules of the Wobble hypothesis (Crick, 1966) which demands a minimum of 32 tRNAs, and not according to the modified two letter code proposed by Lagerkvist (1978) which requires only 24 tRNAs and is employed in mitochondria (Heckman *et al.*, 1980; Barrell *et al.*, 1979, 1980; Bonitz *et al.*, 1980).

The organization of individual tRNA genes has been studied most completely in spinach (Burkard *et al.*, 1980; Driesel *et al.*, 1979; Steinmetz *et al.*, 1978, 1980), where a minimum of 21 tRNA genes corresponding to 14 amino acids have been mapped on the chloroplast genome. Of these, 15 genes, corresponding to tRNAs for 12 amino acids, are located in the large single copy region. Each copy of the inverted repeat contains a gene for tRNA^{Ile} in the spacer between the 16S and 23S rRNA genes as would be expected from the sequencing data for maize and *Euglena*, but a tRNA^{Ala} gene has not yet been found in the spinach spacer. The genes for tRNA$^{Leu}_1$, tRNA$^{Leu}_2$ and tRNA$^{Leu}_3$ also map in the inverted repeat, but outside the rRNA genes. Although no tRNA^{Val} gene has been mapped proximal to the 5' end of the 16S rRNA in the spinach inverted repeat, this gene might be expected to be present by analogy with maize and tobacco. Chloroplast 4S RNA genes have also been mapped in *Chlamydomonas* (Malnoë and Rochaix, 1978) and *Euglena* (El-Gewely *et al.*, 1981; Hallick *et al.*, 1979).

Numerous chloroplast tRNAs have now been sequenced directly or their sequences have been deduced from DNA sequencing experiments (Table

Table 11.2. Sequence homology of chloroplast, prokaryotic and eukaryotic tRNAs.

Chloroplast tRNA	Source	% homology with cognate tRNA from				Reference
		Prokaryote	Source	Eukaryote	Source	
tRNA$_f^{Met}$	*Scenedesmus*	81	*E. coli*	64	*Scenedesmus* cytoplasm	McCoy and Jones, 1980
	spinach	81–84	various	64–69	various	Calagan et al., 1980
tRNA$_1^{Ile}$	maize	81	*E. coli*	58	*Torulopsis*	Koch et al., 1981
	Euglena	86	*E. coli*	–	–	Orozco et al., 1980b
tRNA$_{1B}^{Ala}$	maize	77	*E. coli*	63	yeast	Koch et al., 1981
	Euglena	79	*E. coli*	–	–	Orozco et al., 1980b
tRNA$_m^{Met}$	spinach	67	*E. coli*	50–55	various	Pirtle et al., 1981
tRNA^{Val}	spinach	58–65	various	47–53	various	Sprouse et al., 1981
	maize, tobacco	67	*E. coli*	49	yeast	Schwarz et al., 1981a
tRNA^{Phe}	*Euglena*	75	*E. coli*	64	*Euglena* cytoplasm	Chang et al., 1976, 1981
tRNA^{Thr}	spinach	63	*E. coli*	65	yeast	Kashdan et al., 1980
tRNA$_1^{Leu}$	*Phaseolus*	74	*E. coli*	64	*Schizosaccharomyces*	LaRue et al., 1981
tRNA$_2^{Leu}$	*Phaseolus*	68	*A. nidulans*	55	yeast	LaRue et al., 1981
tRNA$_3^{Leu}$	*Phaseolus*	66	*E. coli*	50	yeast	LaRue et al., 1981
tRNA^{Trp}	spinach	68	*E. coli* su+ UGA	–	–	Canaday et al., 1981
tRNA^{His}	maize	60	*E. coli*	–	–	Schwarz et al., 1981b

11.2). These tRNAs invariably exhibit more sequence homology with their cognate prokaryotic tRNAs than they do with the same eukaryotic tRNAs. Chloroplast tRNAs have the 3' CCA$_{OH}$ terminal characteristic of all tRNAs, but this sequence does not appear to be present at the 3' end of the gene (Schwarz *et al.*, 1981a). Thus the CCA$_{OH}$ must be added post-transcriptionally as it is eukaryotes and certain phage T4 tRNAs (Altman, 1978) rather than part of the primary transcript as it is in *E. coli*. Despite the fact that this sequence is encoded by prokaryotic DNA, organisms such as *E. coli* also have an enzyme which can add this sequence post-transcriptionally. This enzyme probably repairs mature tRNAs since there is constant turnover of this end of the molecule (Altman, 1978).

Chloroplast initiator tRNAs have been sequenced directly in *Scenedesmus* (McCoy and Jones, 1980), spinach (Calagan *et al.*, 1980) and the bean *Phaseolus* (Canaday *et al.*, 1980a), and are structurally more similar to prokaryotic than eukaryotic initiator tRNAs (Canaday *et al.*, 1980a). Methionine elongator tRNA (tRNA$_M^{et}$) from spinach chloroplasts has only 49% homology with the initiator species (Pirtle *et al.*, 1981).

The leucine isoaccepting tRNAs of *Phaseolus* have been studied in detail (Osorio-Almeida *et al.*, 1980) and compared to homologous tRNA species from spinach (Canaday *et al.*, 1980b; Weil *et al.*, 1980). Three isoaccepting species were identified in *Phaseolus*: tRNA$_1^{Leu}$ 5'(U*AA)3', tRNA$_2^{Leu}$ 5'(C^{m}AA)3' and tRNA$_3^{Leu}$ 5'(UA^{m7}G)3'. The tRNA$_3^{Leu}$ anticodon is unusual since it is the first tRNA in which a ^{m7}G has been found in the third position in the anticodon. There are six leucine codons. According to the wobble hypothesis two of the six leucine codons, 5'UUA3' and 5'UUG3', should be translatable by one tRNA with the anticodon 5'UAA3'. If this is correct tRNA$_1^{Leu}$ should translate both codons while tRNA$_2^{Leu}$ should translate only 5'UUG3'. There should be at least two more isoaccepting leucyl tRNAs: A tRNA$_3^{Leu}$ has been found which can translate 5'CUA3' and 5'CUG3', but the other isoaccepting tRNA which can read 5'CUU3' and 5'CUC3' has yet to be identified in chloroplasts, although it has been found in *E. coli* tRNA3Leu (Gauss and Sprinzl, 1981). Both the 5'CUU3' and the 5'CUC3' codons are present in the DNA sequence of the large subunit of the enzyme RuBPCase from maize (McIntosh *et al.*, 1980), and the 5'CUC3' codon is found in the large subunit gene from spinach (Zurawski *et al.*, 1981).

The *Phaseolus* tRNA$_1^{Leu}$ reads the 5'UU$_G^A$3' leucine codon, as does the tRNA$_1^{Leu}$ from *Neurospora* mitochondria which has a modified U in the 5' position in the anticodon (Heckman *et al.*, 1980). The other mitochondrial leucyl tRNA (tRNA$_2^{Leu}$) has the anticodon 5'UAG3' with the U in the 5' position being unmodified. Heckman *et al.* suggest that tRNA$_2^{Leu}$ may read the complete 5'CUN3' family of codons because the U in the 5' position is unmodified. The corresponding *Phaseolus* tRNA (tRNA$_3^{Leu}$) also has an unmodified U in this position, but it has the unique methylated G in the 3' position. Whether this tRNA can read the complete family of leucine codons or whether the aforementioned anticodon modification or other structural modifications cause this tRNA to read according to the wobble rules remains to be established. Spinach tRNA$_3^{Leu}$ is practically identical in sequence to the cognate tRNA from *Phaseolus* including the anticodon with the modified G at the 3' position (Canaday *et al.*, 1980b; Weil *et al.*, 1980). Based on sequence comparisons, LaRue *et al.* (1981) have proposed a phylogeny for leucyl tRNAs in prokaryotes and chloroplasts.

Chloroplast phenylalanine tRNAs have been sequenced in *Euglena* (Chang *et al.*, 1976), *Phaseolus* (Guillemaut and Keith, 1977) and spinach (Canaday *et al.*, 1980b; Weil *et al.*, 1980). The codons for phenylalanine are 5'UUU3' and 5'UUC3' and the anticodon for the single acceptor tRNA is 5'GAA3' in most prokaryotes. Chang *et al.*, (1976) compared prokaryotic and eukaryotic phenylalanyl tRNAs with the *Euglena* chloroplast tRNA. At five of seven positions which distinguish tRNA^{Phe} between prokaryotes and eukaryotes, *Euglena* tRNA^{Phe} has the nucleotides found in prokaryotes. Six of these nucleotides are of the prokaryotic type in spinach and bean. The sequence homology between the chloroplast and cytoplasmic tRNA^{Phe} from *Euglena* is only 64%, but the homology between the *Euglena* cytoplasmic tRNA and mammalian cytoplasmic tRNA^{Phe} is very high (95%) (Chang *et al.*, 1981). Since the cytoplasmic tRNA^{Phe} from *Euglena* shows relatively low homology with cytoplasmic phenylalanyl tRNAs from other plants, Chang *et al.* suggest that *Euglena* might better be classified as an animal than a plant!

Chloroplast tRNA^{Val} sequences have been determined in spinach (Sprouse *et al.*, 1981), maize (Schwarz *et al.*, 1981a) and tobacco (Tohdoh *et al.*, 1981). Identical tRNA^{Val} genes occur in the leader region prior to the 16S rRNA gene in maize and tobacco (Tohdoh *et al.*, 1981). The spinach tRNA^{Val} sequenced by Sprouse *et al.* appears to be distinct from the other two chloroplast tRNAs. This tRNA has the anticodon 5'U*AC3' which should recognize the codons 5'GU$_G^A$3' while the tRNA^{Val} genes from maize and tobacco have the anticodon 5'GAC3' which should read the other two valine codons 5'GU$_C^U$3'. Driesel *et al.* (1979) have identified two isoaccepting chloroplast tRNA^{Val} genes in spinach and mapped one of these in the large unique sequence region of the chloroplast genome. If this is the same tRNA^{Val} gene sequenced by Sprouse *et al.* it might be predicted that the other isoaccepting spinach tRNA^{Val} gene would correspond to the valyl tRNAs found in maize and tobacco in the leader region prior to the 16S rRNA gene. Since the cognate pair of tRNAs is found in *E. coli* (Gauss and Sprinzl, 1981), the results suggest that the valine family of codons is probably read according to the rules proposed by the wobble hypothesis. This conjecture is strengthened further by the presence of a modified U in the anticodon of the spinach valyl tRNA sequenced by Sprouse *et al.* In contrast, Heckman *et al.* (1980) found that *Neurospora* mitochondrial tRNA^{Val} has an unmodified U in the anticodon (5'UAC3'), which is consistent with the hypothesis that an unmodified U in the wobble position permits reading of the entire codon family.

The tRNA^{Ile} and tRNA^{Ala} genes residing in the spacer between the 16S and 23S chloroplast rRNA genes have been sequenced in maize (Koch *et al.*, 1981) and *Euglena* (Graf *et al.*, 1980; Orozco *et al.*, 1980b). There are three codons for isoleucine (5'AUU3', 5'AUC3', 5'AUA3'), the fourth codon in this group being the initiator codon 5'AUG3'. In *E. coli*, tRNA$_1^{Ile}$, which maps in the spacer between the 16S and 23S rRNA genes in the *rrnD* and *rrnX* operons (Young *et al.*, 1979), has the anticodon 5'GAU3' which will read the codons 5'AU$_C^U$3'. This is the same anticodon found in both chloroplast tRNAs. The alanine family includes the four codons (5'GCN3'). These are probably read by a single tRNA with the anticodon 5'UCC3' in *Neurospora* mitochondria (Heckman *et al.*, 1980). This same anticodon is found in the tRNA$_{1B}^{Ala}$ gene located in the spacer between the 16S and 23S rRNA genes in *E. coli* (Young *et al.*, 1979), *Euglena* (Graf *et al.*, 1980; Orozco *et al.*, 1980b) and maize (Koch *et al.*, 1981). Although one assumes the U is modified, this does not appear to be established as yet (Gauss and Sprinzl, 1981). If the alanine codons in chloroplast tRNA are read according to the rules of wobble, we would expect to find a second isoaccepting species of alanine tRNA. Two isoacceptors have been identified among spinach chloroplast tRNAs, one of which maps in the large unique sequence part of the chloroplast genome (Driesel *et al.*, 1979). By analogy with maize and *Euglena* the other isoacceptor would be expected to map in the spacer between the 16S and 23S rRNA genes. Koch *et al.* (1981) find a very high degree of sequence homology between the maize tRNA^{Ile} and tRNA^{Ala} genes and the comparable genes in the *Euglena* and *E. coli* spacers.

Three other chloroplast tRNAs have been sequenced. In maize a tRNA^{His} gene overlaps a gene coding for a 1.6 kb RNA by a few bases, but the two genes are transcribed from complementary strands (Schwarz *et al.*, 1981b). The sequence of tRNA^{Thr} from spinach chloroplasts shows essentially equal homology with prokaryotic and eukaryotic tRNA^{Thr} species (Kashdan *et al.*, 1980). A chloroplast tRNA^{Trp} gene from spinach has the anticodon sequence 5'CCA3' which should recognize the tryptophan codon 5'UGG3' and ought not to decode the terminator codon 5'UGA3' (Canaday *et al.*, 1981). This tRNA has a U11:A24 base pair which is distinct from the U11:G24 base pair found in wild type tRNA^{Trp} from *E. coli*. However, an *E. coli* suppressor tRNA^{Trp} strain which reads 5'UGA3' as sense has the base pair U11:A24 (cf. Canaday *et al.*, 1981). Since a 5'UGA3' stop codon terminates the gene coding for the large subunit of RuBPCase in spinach (Zurawski *et al.*, 1981), the spinach tRNA^{Trp} probably does not function as a suppressor.

STRUCTURE OF THE GENE FOR THE LARGE SUBUNIT OF
RuBPCase

Among chloroplast genes specifying polypeptides, only the genes coding for the large subunit of RuBPCase from maize (McIntosh *et al.*, 1980) and spinach (Zurawski *et al.*, 1981) have so far been sequenced. The nucleotide sequences obtained have been compared to the amino acid sequence reported for barley (Poulsen, 1979; Poulsen *et al.*, 1979). The transcript of the large subunit gene in both spinach and maize is preceded by a 5′ untranslated region and followed by a 3′ untranslated region. In spinach transcription of the large subunit mRNA begins 178–179 nucleotides upstream of the 5′AUG3′ codon at which translation is presumed to start, whereas in maize the transcript begins only 59–63 nucleotides upstream of this codon. Like the 16S and 5S rRNA genes in maize, the spinach large subunit gene has a region containing a prokaryotic promoter sequence (TATAAT) and a TTG sequence about 20 base pairs upstream from the promoter sequence (Rosenberg and Court, 1979). In contrast, obvious promoter sequences are not evident immediately upstream of the large subunit gene in maize, although Gatenby *et al.* (1981) report that such sequences can be found considerably further upstream. Since the leader region is 117 nucleotides shorter in maize than spinach, these sequences of the spinach leader probably have little effect on translation of the mRNA for this polypeptide. There is 95% homology between spinach and maize in the 20 nucleotides preceding the AUG initiator codon, whereas there is only 65% homology upstream of this region. A 5′GGAGG3′ Shine-Dalgarno sequence occurs at positions -6 to -10 in both maize and spinach which is complementary to the 5′CCUCC3′ sequence found in maize 16S rRNA.

The AUG codon presumed to be the initiator codon in both the maize and spinach sequences is found 14 codons prior to the alanine residue which occurs at the N terminus of the large subunit polypeptide in barley and wheat (Zurawski *et al.*, 1981). These findings, together with those of Langridge (1981) discussed earlier, suggest that the 14 codons specify a leader sequence which is processed during large subunit maturation. The spinach large subunit mRNA has 82–85 untranslated nucleotides at the 3′ end, and the region immediately preceding the 3′ end of the mRNA is capable of forming a stem-loop structure analogous to those found at prokaryotic transcription termination sites. Within the large subunit gene itself, strong conservation of amino acid sequence is evident. Of the 475 amino acids found in the spinach large subunit gene, 10% are changed in maize. Amino acid substitutions account for 45 of these changes with the remaining four being deletions or additions. These changes are not randomly distributed throughout the molecule. Two of the three tryptic peptides located close to the catalytic site show 97% conservation in both spinach and maize, while there is 34% divergence in the third tryptic peptide.

Analysis of codon usage in the large subunit gene reveals that all but three of the 61 codons specifying amino acids are found in the large subunit gene of maize, whereas all but six are used in the spinach large subunit gene. The codons 5'ACG3' (threonine), 5'CGG3' (arginine) and 5'AGC3' (serine) are not used in either gene, and there is a marked bias towards codons having an A or T in the third position in both genes. No internal termination codons are found in either gene and the stop codon at the 3' end of the spinach gene is UAG while in maize it is UAA. All evidence we have to date indicates that the genetic code of the chloroplast is read according to the rules of the wobble hypothesis in contrast to the modified two letter code used in the mitochondrion.

CHANGES IN CHLOROPLAST GENOME ORGANIZATION DURING EVOLUTION

From the foregoing sections, one can conclude that chloroplast genomes are conserved in terms of the components they code for, e.g. the same set of thylakoid membrane polypeptides, the large subunit of RuBPCase, ribosomal and transfer RNAs. Solution hybridization experiments also indicate that extensive base sequence homologies exist among chloroplast genomes of land plants separated by several hundred million years of evolution (Bisaro and Siegel, 1980; Lamppa and Bendich, 1979b; Palmer and Thompson, 1982). Although the estimates of homology differ somewhat depending on the methods of assay and size of probes used, one sees values of 70 to 100% between genera within dicotyledonous families such as the Leguminoseae or the Solanaceae and about 50% between plants in different dicotyledon families. Slightly less homology was found between chloroplast genomes of monocotyledons and dicotyledons. Two lines of evidence suggest that the conserved sequences are short and are interspersed with divergent sequences. First, hydroxylapatite chromatography gave higher estimates of homology for a given species pair than did S-1 analysis when the same size cpDNA was used as a probe. Second, use of a long cpDNA probe (1860 bp) in the hybridization experiments instead of a short probe (735 bp) gave higher estimates of homology for the same species pair. Both results suggest that heterologous regions might be interspersed with conserved regions. The hybrid duplexes examined had a lower melting temperature than the homo-duplexes, again suggesting that the precision of pairing within homologous regions was less than complete.

Reciprocal Southern hybridizations between cpDNA of maize and pea revealed that the conserved sequences were distributed over 40 to 50% of the chloroplast genome (Lamppa and Bendich, 1981). Heteroduplex analysis by these same investigators showed that the conserved and divergent sequences of pea and maize cpDNA were interspersed, with 50% of the conserved sequences being less than 550 bases long and having a mean length of 250 bases. Sequences this short are unlikely to contain genes other than those coding for tRNAs. The heteroduplexing studies of both Lamppa and Bendich (1981) and Thomas and Tewari (1974) show that the 16S and 23S rRNA genes as well as the spacer between them are conserved in higher plants. Since much of the spacer region in maize is occupied by the introns that split tRNA[Ile] and tRNA[Ala] (Koch *et al.*, 1981; see p. 311), the results predict that the spacer in pea may have the same structure as that in maize. Interestingly, the regions flanking the rRNA gene set in pea are heterologous compared to those in maize.

As detailed information accumulates on restriction maps of chloroplast genomes in different species (Table 11.1), on the structure of specific chloroplast genes and on the arrangement of these genes within the genome, additional observations of evolutionary significance are beginning to emerge. Although the data are still sparse in a phylogenetic sense (Gillham and Boynton, 1981), recent evidence suggests that only a limited set of rearrangements have occurred in the chloroplast genomes of land plants during their evolution (Palmer and Thompson, 1982). The most prominent structural feature of chloroplast genomes in higher plants is the large inverted repeat sequence of 20 to 25 kb, part of which codes for ribosomal RNA (see p. 302). This inverted repeat has been found in all vascular plants examined with the exceptions of pea (Chu *et al.*, 1981; Palmer and Thompson, 1981b) and broad bean (Koller and Delius, 1980), where one of the repeats has been deleted. The linear order of common sequence elements, defined by restriction enzymes making a moderate number of cleavages, is remarkably conserved among angiosperm chloroplast genomes that contain the large inverted repeat (Fluhr and Edelman, 1981b; Palmer, 1982; Palmer and Thompson, 1982; Palmer and Zamir, 1982). Only two rearrangements were found among chloroplast genomes from five species examined by Palmer and Thompson (1982). The chloroplast genomes of spinach, petunia and cucumber appear essentially identical in gross sequence arrangement and differ from that of mung bean by a single inversion of approximately 50 kb. A second smaller inversion differentiates the chloroplast genome of maize from those of spinach, petunia and cucumber. In contrast there is an extremely scrambled arrangement of shared chloroplast gene sequences between the two species of the tribe Viciae of the legume family (pea and broad bean) that have lost the inverted repeat, compared to the tribe Phaseoleae in which the repeated structure is retained (Palmer and Thompson, 1981b, 1982). Furthermore, there appears to be a sizeable rearrangement of shared sequences between pea and broad bean, the two species that lack inverted repeats. As a consequence, one sees more major rearrangements in the chloroplast genomes of one tribe of the legume family than among the remaining monocotyledon and dicotyledon families. Since the chloroplast genome of *Chlamydomonas reinhardtii* possesses the inverted repeat structure found in most angiosperms, but has unique sequence regions of more equal size, comparisons of shared sequence organization between chloroplast genomes of this alga and that of higher plants should be most interesting.

Although colinearity appears to exist between shared sequences of chloroplast genomes of a variety of species of monocotyledons and dicotyledons, one can demonstrate that evolution of the genomes has taken place in the form of deletion or addition of specific regions. For example, Fluhr and Edelman (1981b) find that the chloroplast genome of *Petunia parodii* differs from that of *Nicotiana tabacum* by two separate 100 bp deletions, two separate 50 bp deletions and a 100 bp insertion, while the overall genomes are colinear. Likewise the chloroplast genome of *Atropa belladonna* differs from that of *N. tabacum* by two 100 bp deletions and two 50 bp deletions. One of the 100 bp deletions and both 50 bp deletions appear to correspond to the segments deleted in *Petunia*.

Of particular interest are comparisons of chloroplast genome organization of species within the genera *Nicotiana*, *Oenothera* and *Lycopersicon* where much is known about the biosystematic relationships of the members. Comparative studies of chloroplast DNA restriction patterns have been made between tobacco (*Nicotiana tabacum*) and related species in the genus *Nicotiana* by several groups of investigators (Atchison *et al.*, 1976; Kung *et al.*, 1982; Rhodes *et al.*, 1981; Vedel *et al.*, 1976). Based on extensive taxonomic and cytogenetic analysis Goodspeed (1954) recognized 60 species in this genus which he divided into three subgenera: Rustica, Tabacum and Petunioides. The first two subgenera are exclusively South American in distribution and contain species with either 12 or 24 chromosome pairs. Species of the subgenus Petunioides occur in North and South America, Australia and the islands of the South Pacific. The species of this subgenus have 9 to 24 chromosome pairs. A major evolutionary trend in the genus is the formation of amphiploids as the result of interspecific hybridization. Goodspeed postulates that the ancestral *Nicotiana* species had six chromosome pairs and gave rise by successive hybridization to species with 12 and then 24 chromosome pairs. Since chloroplast DNA is maternally inherited in *Nicotiana* (Frankel *et al.*, 1979), it might be predicted that chloroplast DNAs from *Nicotiana* species derived by amphiploidy should have the same restriction pattern as the primeval maternal parent.

Chloroplast DNA restriction patterns in *Nicotiana* have been examined in terms of the evolutionary patterns perceived by Goodspeed (Atchison *et al.*, 1976; Kung *et al.*, 1982; Rhodes *et al.*, 1981). Rhodes *et al.* classified the species of all three subgenera into four groups based on Eco RI restriction pattern differences in cpDNA, and further divided group IV into three subgroups. In general, the results obtained complement those reported by Goodspeed. For example, Goodspeed postulated that the group I species *N. excelsior, N. gossei* and *N. megalosiphon* were all derived from a cross between *N. suaveolans* and *N. fragrans*. The fact that *N. suaveolans* belongs to group I is consistent with this model. The finding that *N. tabacum* belongs to group III together with the American species of the section Alatae of the Petunioides subgenus is also predictable. A member of this group, *N. sylvestris*, was postulated to be the female parent of *N. tabacum* (Gray *et al.*, 1974). The reason why the restriction patterns of group II and III plants are so similar is not clear. Based on taxonomic criteria, *N. glauca*, the sole species with a group II restriction pattern, is not closely related to the group III plants. In addition, while groups IVa and IVb contain species supposedly of common origin, these two chloroplast DNAs show a high degree of divergence. Kung *et al.* (1982) constructed a phylogenetic tree of chloroplast DNA evolution in the genus *Nicotiana* on the basis of Sal I and Sma I cpDNA restriction patterns which shows both similarities and differences to Goodspeed's phylogeny. In addition, Kung *et al.* have found that both sequential gains and losses of specific Sma I sites have occurred in different *Nicotiana* species. Intraspecific chloroplast DNA polymorphism has been observed in *Nicotiana debneyi* by Scowcroft (1979). This polymorphism involves the presence or absence of a single Eco RI site with each population of plants exhibiting one pattern or the other. Chloroplast DNA alterations have also been observed in male sterile lines (Frankel *et al.*, 1979; Kung *et al.*, 1981).

Oenothera is also a particularly appropriate genus for such studies since it has been extremely well characterized both taxonomically and genetically by Cleland (1972), Renner (1937), Stubbe (1966) and Kutzelnigg and Stubbe (1974). Independent assortment and crossing-over of the seven pairs of chromosomes are effectively suppressed during meiosis by a series of reciprocal translocations which characterize the respective species. Nuclear versus plastome (plastid) compatibility can be studied in interspecific hybrids where certain combinations have normal plastids, others have plastids with defective structure and some combinations are actually lethal. Within the subgenus *Euoenothera*, which consists of 10 North American and 14 European species, there are three basic haploid nuclear genomes (A, B, C) and five different plastome types (I–V) (cf. Kutzelnigg and Stubbe, 1974). Plastome IV was considered to be the most primitive since it was fully compatible with all but one of the six different diploid nuclear genome combinations. The five *Euoenothera* plastome types have distinctive cpDNA restriction patterns which differ from one another by a series of deletions or insertions (Gordon *et al.*, 1981; Herrmann *et al.*, 1980a). No rearrangements were observed and the deletions/insertions appear to be localized either in the regions of the inverted repeats not containing the rRNA cistrons or in the adjacent large single copy DNA. The changes in size of fragments within the inverted repeat region are symmetrical. This has led Herrmann and his colleagues to postulate that a rectification mechanism must exist to keep the sequences of the two repeats identical. Direct evidence of this hypothesis will be discussed shortly for mutations of *Chlamydomonas* having physical alterations in their chloroplast genomes.

Clegg *et al.* (1982) have found that the chloroplast genomes of 12 strains of pearl millet (*Pennisetum*) from different areas of the world and one species of the related genus *Cenchrus* showed no differences in restriction pattern for the inverted repeat region. The only changes found were accounted for by single base pair substitutions in the unique sequence regions examined. No evidence was found for additions, deletions or rearrangements. The rate of intergeneric changes for chloroplast genomes of these species appeared to be considerably lower than that found for intraspecific changes in mitochondrial DNA from pocket gophers or old field mice by Avise *et al.* (1979a, b).

Biosystematic relationships of the garden tomato *Lycopersicon esculentum* and its relatives in the genera *Lycopersicon* and *Solanum* are also exceedingly well known (Rick, 1979). Palmer and Zamir (1982) have demonstrated that the chloroplast genomes of 12 species of *Lycopersicon* and three of *Solanum* differ by only a total of 39 restriction site changes among 484 restriction sites (2800 bp) surveyed using 25 different endonucleases. Of the 19 restriction site changes whose direction can be determined, nine are site gains and ten are site losses. Only one of these changes appears to occur within a fragment included in the inverted repeat region of *L. esculentum*. Furthermore, the low rate of base sequence changes is paralleled by an extremely low rate of convergent change in restriction sites, as only one of the 39 mutations appears to have occurred independently in two different lineages. A significant variation in fragment size was observed in only a single restriction fragment for *Solanum juglandifolium* and *S. pennellii*.

A maternal phylogeny for the chloroplast genomes constructed for the 15 *Lycopersicon* and *Solanum* species examined is generally consistent with relationships based on morphology and hybridization experiments. Two of the *Solanum* species are farthest from the *Lycopersicon* species both in terms of number of cpDNA changes and in their biosystematic relationships. The species *Solanum pennellii*, which forms fertile hybrids with *Lycopersicon esculentum* and other species in this genus, also has cpDNA restriction patterns with few differences from members of the genus *Lycopersicon*. However, *S. pennellii* appears to share a deletion in a specific restriction fragment with *S. juglandifolium* and to differ from all other members of the genus *Lycopersicon* in this respect. A limited amount of intraspecific polymorphism for cpDNA was found among six collections of *L. peruvianum* and this encompasses all the variation found for *L. chilense* and *L. chemielewskii*. Using the chloroplast genome as a sole criterion, Palmer and Zamir

(1982) suggest that the latter species might well be relegated to the *L. peruvianum* complex. Only in the case of *L. esculentum* and *S. lycopersicoides* were the chloroplast genomes mapped and the map locations of the single Bgl I and Pvu II site mutations localized.

Finally, restriction patterns of organelle DNAs have been used as tools to study phylogeny in maize, wheat, cabbage and their relatives. Timothy *et al.* (1979) compared chloroplast and mitochondrial DNA restriction patterns from six annual races of teosinte, perennial teosinte and maize. Three groups of chloroplast DNAs were detected while four groups of mitochondrial DNAs were found. Separation of the teosinte and maize organelle DNAs into five groups approximated the systematic relationships of the taxa. Vedel *et al.* (1978, 1981) have used Eco RI digests of organelle DNA to study wheat phylogeny. One of the basic problems is which pairs of the diploid wheat species combined to produce alloploid wheats. For example, the hexaploid wheat species *Triticum aestivum* (AABBDD) is supposed to have arisen by hybridization of *T. turgidum* (AABB) and *Aegilops squarrosa* (DD) on the basis of meiotic pairing figures. Since the organelle DNAs of *T. aestivum* are identical in restriction pattern to those of *T. turgidum*, but different from *Ae. squarrosa*, *T. turgidum* must have been the female parent in the cross. *Ae. ventricosa*, a tetraploid used to transfer resistance to eyespot disease into *T. aestivum*, contains organelle DNAs different in restriction patterns from *T. aestivum* and other wheats. Therefore, the organelle genomes of *Ae. ventricosa* cannot be responsible for eyespot resistance and the gene(s) for this trait must be nuclear.

Lebacq and Vedel (1981) compared Sal I digests of chloroplast and mito-chondrial DNA from six species of *Brassica* (mustard, cabbage, turnip and relations). Four species were identical in patterns while two differed from each other and from the other four species. Mitochondrial DNAs showed more differences than cpDNAs. So far, these studies have not permitted determination of the maternal parent in three interspecific crosses leading to production of three allotetraploid species (rape, leaf mustard and Abyssinian mustard) within this genus.

The major conclusion to be drawn from the evolutionary studies discussed so far is that the inverted repeat regions of the chloroplast genome are highly conserved as are the rRNA genes included within them, whereas the unique sequence regions appear somewhat more divergent by whatever method is used to measure divergence. Bedbrook *et al.* (1977) have suggested that a possible function of inverted repeats may be to ensure homogeneity of the nucleotide sequences in these regions. Sequence homogeneity could be main-tained by intramolecular recombination or by heteroduplex repair events (Bedbrook and Kolodner, 1979). This general notion has found wide accept-ance (cf. Herrmann *et al.*, 1980a; Rochaix, 1978; Clegg *et al.*, 1982; Palmer and Thompson, 1981b).

Myers *et al.* (1982) have isolated and characterized a series of non-photosynthetic mutants in *Chlamydomonas reinhardtii* having physical alterations in the chloroplast genome. One group of mutants, obtained with the thymidine analogue 5-fluorodeoxyuridine (FdUrd), had simple deletions, whereas the second group, obtained after treatment of cells grown in FdUrd with x-rays, contained deletions plus inversions and/or duplications. All of the physical alterations map within or close to the inverted repeat regions. Only in the case of four deletion mutants mapping in the *ac-u-c* locus has the non-photosynthetic phenotype of the mutants been shown to result from the physical alterations. All four of these mutants have deletions in unique sequence cpDNA adjacent to one of the inverted repeats, and in one of the mutants, *ac-u-c-2-43*, the deletion extends through one of the two inverted repeat regions. This mutant, like pea and broad bean, has only one set of rRNA genes, yet makes normal amounts of chloroplast ribosomes. Most of the other FdUrd-induced mutations involve alterations in both the inverted repeat regions, and these alterations are often symmetrical. For example, mutant *ac-u-ε* has symmetrical deletions plus inversions in the two inverted repeats (Myers *et al.*, 1982), whereas mutant *ac-u-g-2-3* has symmetrical small deletions (Grant *et al.*, 1980).

Several bleached mutants of *Euglena* thought for many years to lack chloroplast DNA have been shown by Heizmann *et al.* (1981) to contain defective chloroplast genomes present in very low copy numbers. The rRNA genes are preferentially retained in these mutants.

CONCLUDING REMARKS ON CHLOROPLAST GENOME ORGANIZATION

Although chloroplast genomes, with the possible exception of *Acetabularia*, appear to be reasonably homogenous in size and probably function, recent physical studies of chloroplast genome structure are beginning to reveal a variety of novel features. Genes coding for the tRNAs associated with the rRNA genes have large intervening sequences in higher plants and the 23S rRNA gene itself may contain intervening sequences. While base pair substitutions and rearrangements can occur within the unique sequence segments of cpDNA, the structure of the inverted repeats is highly conserved, probably because of intramolecular pairing and recombination or repair. Interestingly, many of the differences in cpDNA organization found in different species of plants in nature have been duplicated in the laboratory with *Chlamydomonas*. Codon usage in chloroplast genes appears to be conventional, based on studies of tRNAs and sequencing of the gene coding for the large subunit of RuBPCase. As more chloroplast genes coding for specific proteins are studied, we will begin to understand the evolution of chloroplast genomes better. One wonders, for example, whether the RuBPCase gene is typical in lacking intervening sequences or whether intervening sequences will prove to be a common feature of chloroplast genomes as they are in the case of the tRNA genes in the spacer between the 16S and 23S rRNA genes in maize cpDNA or in the 23S rRNA gene of *Chlamydomonas*.

CONSTRAINTS ON THE EVOLUTION OF THE CHLOROPLAST GENOME

Mutation is the ultimate source of genetic variability in all living organisms. New combinations of nuclear gene mutations are achieved by recombination and independent assortment in meiosis. In contrast, chloroplast and mito-chondrial genomes are inherited maternally in most organisms and therefore the opportunity for recombination and assortment of new mutations may never arise (cf. Gillham, 1978; Kirk and Tilney-Bassett, 1978; Sears, 1980b). Even in those plants in which biparental inheritance of plastids occurs, plastid fusion may be rare, preventing recombination of mutations which have occurred in the genomes of different plastids. Therefore evolution of chloro-plast and mitochondrial genomes must take place largely by the occurrence and fixation of sequential mutations. Recombination of chloroplast genomes has so far been observed only in *Chlamydomonas* where organelle fusion occurs. Recombination and segregation of chloroplast genes in biparental zygotes of *C. reinhardtii* have been used to construct genetic maps of the chloroplast genome (Boynton *et al.*, 1976; Harris *et al.*, 1977; Sager, 1972, 1977).

Fixation of a new chloroplast mutation represents a problem in population genetics at the intraorganelle and intracellular levels (cf. Birky, 1978; Birky *et al.*, 1981). The process by which a single mutant allele becomes homoplasmic against a background of many wild type alleles cannot be studied directly in most plants that contain many plastids per cell and many cells per tissue. Only in suitable models such as *Chlamydomonas*, which contains a single plastid per cell, can mutation, segregation and recombination of chloroplast genomes be measured directly. On the other hand, the process of somatic segregation which establishes cell lines containing only mutant plastids has been a subject for study almost since the time that non-Mendelian inheritance of plastome mutations was first reported.

In *Chlamydomonas reinhardtii* chloroplast mutations resistant to antibiotics are recovered spontaneously at frequencies of 10^{-6} to 10^{-7} under conditions where they can be selected positively (cf. Gillham and Levine, 1962; Lee and Haughn, 1980; Wurtz *et al.*, 1979). This is not greatly different from the frequency of nuclear mutations of similar phenotype. Since the nucleus of *C. reinhardtii* contains one copy of each gene while the chloroplast contains 50 to 100 copies, this observation is paradoxical and cannot be accounted for by random segregation of chloroplast genomes. However, by placing five constraints on the population of genomes in the single chloroplast of *Chlamydomonas* and, by analogy, in one chloroplast of a higher plant cell, it is possible to account for the relatively rapid appearance of new chloroplast mutations.

First, the grouping of chloroplast DNA molecules into nucleoids (see p. 300) effectively reduces the population size at the time of chloroplast division. Second, variability in nucleoid number is known to occur in both *Chlamydomonas* (Coleman, 1978; Matagne and Hermesse, 1980, 1981) and higher plants (Herrmann, 1970; Herrmann and Kowallik, 1970), and mutations occurring in plastids with the smallest number of nucleoids would be fixed most readily. A decrease in the average number of DNA molecules per nucleoid could also increase recovery of chloroplast mutations. Growth of *C. reinhardtii* in the thymidine analogue 5-fluorodeoxyuridine, which selectively reduces the amount of chloroplast DNA (Wurtz *et al.*, 1977) and the number of nucleoids per plastid (Matagne and Hermesse, 1981), greatly enhances the recovery of chloroplast mutants with a variety of phenotypes (Bennoun *et al.*, 1978; Harris *et al.*, 1982; Shepherd *et al.*, 1979; Spreitzer and Mets, 1980, 1981; Wurtz *et al.*, 1979).

Third, segregation of individual DNA molecules or nucleoids themselves in a reductional rather than an equational fashion during chloroplast division could explain both the high frequency with which chloroplast mutations are detected in *C. reinhardtii* and the fact that these mutations are not distributed clonally in fluctuation tests as are nuclear mutations (Birky, 1978; Lee and Haughn, 1980). Reductional segregation of the nucleoids could also explain the rapid segregation of chloroplast genes in crosses of *C. reinhardtii*, if certain assumptions are made regarding the extent of mixing and the nature of recombinational events taking place between nucleoids (VanWinkle-Swift, 1980). Furthermore, the actual patterns of segregation and recombination observed for chloroplast genes in pedigrees of biparental zygotes (Forster *et al.*, 1980) are consistent with the reductional segregation model of Van Winkle-Swift (1980).

Fourth, subjecting the population of chloroplast genomes to the forces of random drift during the period when their numbers are reduced can further enhance the rate of fixation of new chloroplast mutations (cf. Birky, 1978). The combination of random drift with a directional force that favours fixation of alleles from the maternal parent could account for the predominantly maternal pattern of inheritance in *C. reinhardtii* (Birky *et al.*, 1981).

Fifth, recombination of chloroplast genomes within the plastid itself may play a role in the fixation of new mutations. Analysis of progeny from crosses of *C. reinhardtii* strains carrying different chloroplast markers suggests that recombination of chloroplast genes is usually a non-reciprocal process (Forster *et al.*, 1980; Gillham, 1965; VanWinkle-Swift and Birky, 1978). Whether recombination occurs within higher plant plastids has not been ascertained because of the lack of appropriate markers and the infrequent occurrence of plastid fusion (see below).

Although fixation of a mutation within the chloroplast of *Chlamydomonas* is sufficient to generate a new cell line carrying that mutation, higher plant cells usually contain many plastids. These plastids must then segregate and assort in such a way that a cell line is generated in which all of the plastids are mutant. Somatic cell fusion studies with higher plants have been particularly useful in modelling this part of the process.

Analysis of plants regenerated from somatic fusions between cells of higher plant species having distinctive chloroplast DNA restriction patterns or other chloroplast markers reveals that rapid, apparently random sorting out usually occurs to yield approximately equal numbers of regenerated plants of the two parental chloroplast genotypes (Aviv *et al.*, 1980; Aviv and Galun, 1980; Belliard *et al.*, 1978; Chen *et al.*, 1977; Douglas *et al.*, 1981; Medgyesy *et al.*, 1980; Melchers *et al.*, 1978; Melchers, 1980; Scowcroft and Larkin, 1981; Sidorov *et al.*, 1981). In a few cases only plants of one parental chloroplast genotype were recovered (Evans *et al.*, 1980; Maliga *et al.*, 1980). Certain exceptions to this rapid sorting out are found where regenerated plants still contain both plastid types (Chen *et al.*, 1977; Sidorov *et al.*, 1981). Iwai *et al.* (1981) report that while the chloroplast RuBPCase large subunit (LS) marker of only one parent was expressed in leaves of the regenerated hybrid, plants regenerated from individual pollen cells of this hybrid showed a 7:2 ratio for the LS phenotypes of the two parents, with the majority type that of the somatic hybrid. At least in terms of this chloroplast marker, it must be concluded that cells can remain heteroplasmic to a certain degree without expressing one of the two parental alleles. A similar situation has been found in the single chloroplast of *C. reinhardtii* with respect to chloroplast mutations for streptomycin dependence (Bolen *et al.*, 1980).

Most of the foregoing reports involved fusions between protoplasts of different species, between varieties with well known nucleocytoplasmic interactions or between cultivars with mutant and normal plastids. Scowcroft and Larkin (1981) have observed sorting out of plastids in plants regenerated between somatic fusions of two genetically intercompatible strains of *Nicotiana debneyi* having normal chloroplasts with distinctive DNA restriction patterns. Furthermore, their somatic hybrids were identified without selection from among a population of plants regenerated from individual drop cultures. Since the same rapid and random sorting out was seen for the chloroplast types in the six hybrids characterized, these authors argue that neither the interspecific nature of the somatic hybrids studied by previous workers nor the selection schemes imposed to identify them can be responsible for the rapid sorting out of chloroplast genomes observed.

Scowcroft and Larkin (1981) discuss the paradox that we would not expect to obtain only homoplasmic somatic hybrids in the 40–50 cell generations that occur from fusion to regenerated plant when taking into consideration that a total of approximately 400 chloroplasts (200 of each parental type) are present in the initial fused cell. The authors observe that a marked reduction in the number of chloroplasts, to about 20 per cell, occurs by three days after protoplast fusion, comparable to the 10–20 proplastids per cell observed in meristematic cells of plant shoots (Possingham, 1980). Hence random drift during chloroplast segregation at the time of cytokinesis of the fused protoplast and/or its progeny cells having reduced numbers of chloroplasts per cell could well explain the rapid segregation observed both at the callus and regenerated plant level.

No clear evidence is available from the foregoing studies with somatic hybrids for plastid fusion or for the generation of progeny with recombinant chloroplast genotypes. Obviously fusion of genetically dissimilar plastids is a prerequisite for recombination of their genomes. The extent to which plastid fusion occurs in higher plant species is poorly understood, and profiles of plastids in the process of fusing are easily confused with those in the process of division. Chloroplast fusion has been reported in only a few algae, possibly in several mosses, and in one species of higher plant (cf. Sears, 1980b). Vaughn (1981) presents electron microscopic evidence for fusion of wild type chloroplasts with mutant plastids in mixed cells of variegated strain of *Hosta sieboldii* and suggests that such fusions will arrest the rate of sorting out of the two plastid types in variegated tissues of this cultivar. In situations where plastid fusion rarely occurs and where plastid segregation is rapid, rare recombinational events will only be detected through the use of selective markers in the chloroplast genomes of the two parental strains. To date such studies have not been done with either somatic hybrids or sexual hybrids of higher plants.

In the case of vegetative growth of certain plastome (chloroplast) mutations of higher plants that result in variegated pigmentation, mixed cells can be observed to persist for many generations (Keresztes, 1971; Ueda and Wada, 1959; Vaughn and Wilson, 1980; Yasui, 1929). While many variegated cultivars resulting from plastome mutations will give rise to pure white or pure green shoots and leaves by somatic segregation (cf. Kirk and Tilney-Bassett, 1978), others will largely or exclusively continue to produce variegated tissues. Obviously in higher plants we have to deal with the distribution of cells with exclusively mutant or normal chloroplasts as well as mixed cells in different embryonic layers of the meristem in accounting for the different tissue patterns observed. Segregation at the cellular level must be superimposed on the segregation of plastids when mixed cells divide and on the segregation of chloroplast genomes when heteroplasmic plastids divide.

The reason why plastids are so often maternally inherited in plants remains to be elucidated. Sears (1980b) has recently reviewed both the occurrence and mechanisms postulated to be responsible for this phenomenon throughout the plant kingdom. Elimination of plastids from the paternal parent may occur in a number of ways: exclusion from the male gamete during spermatogenesis, loss from the motile sperm, exclusion during fertilization or degradation of the paternal plastid DNA itself. Sears states that the many ways of achieving maternal transmission of plastids among diverse plants probably indicate that parallel or convergent evolution has occurred in those taxa adjusted to different selective pressures associated with sexual reproduction. In addition to the exclusion mechanisms discussed by Sears, Vaughn *et al.* (1980) have presented electron microscopic evidence that paternal chloroplasts physically degenerate during microsporogenesis in certain species of higher plants.

The case of *Chlamydomonas* deserves special comment since this genus is supposedly primitive in the phylogeny of green plants. In the isogamous species *C. reinhardtii* fusion of the two parental chloroplasts has been seen in electron micrographs of sexual zygotes fixed a few hours after gamete fusion (Blank *et al.*, 1978; Cavalier-Smith, 1970, 1976; Gillham, 1978). However, it is not known whether such plastid fusions are isolated events that produce the approximately 5% frequency of spontaneous biparental zygotes or represent normal occurrences in all zygotes in the population. If it is assumed that chloroplast fusion occurs in all zygotes, bringing together approximately 100 copies of the chloroplast genome of each of the two parents, then a strong directional selection against the paternal chloroplast genomes must occur to explain the fact that the majority (<95%) of the zygotes of this alga transmit only the chloroplast genomes of the maternal parent to the meiotic progeny (Gillham, 1978; Sager, 1972, 1977). There is considerable controversy regarding the fate of the two parental chloroplast DNAs during maturation of the zygote and the mechanism(s) responsible for maternal inheritance (Adams *et al.*, 1976; Chiang, 1976; Gillham, 1978; Sager, 1972, 1977; Sears, 1980a). Grant *et al.* (1980) have shown that maternal transmission of chloroplast genetic markers is accompanied by maternal transmission of chloroplast DNA. Similar demonstrations of cotransmission of chloroplast genetic markers and DNA have been made in the interspecific cross of *C. eugametos* and *C. moewusii* (Lemieux *et al.*, 1980; Mets, 1980). Certain F_1 and backcross progeny from this cross have been shown to be recombinant for their cpDNA restriction patterns (Lemieux *et al.*, 1980). Modification of maternal chloroplast DNA by methylation and restriction of unmethylated paternal chloroplast DNA has been invoked as the mechanism responsible for maternal transmission of chloroplast DNA in *C. reinhardtii* (cf. Sager *et al.*, 1981). However, extensive methylation of paternal chloroplast DNA by a nuclear gene mutation is not sufficient to ensure its transmission in crosses (Bolen *et al.*, 1982).

In contrast to the meiotic zygotes of *C. reinhardtii*, spontaneous vegetative diploids from the same cross transmit chloroplast markers from both parents to a high percentage of their progeny (Gillham, 1978; VanWinkle-Swift, 1978). These rare (<5%) spontaneous diploids can be selected on minimal medium by employing parental strains with closely linked auxotrophic markers that complement *in vivo*. Diploids produced by asexual fusion of the same auxotrophic strains show a similar high frequency of biparental transmission of chloroplast markers regardless of the mating types involved (Adams, 1981; Matagne, 1981; Matagne and Hermesse, 1980, 1981). Clearly, the mechanism responsible for the preferential loss of chloroplast markers from the paternal (*mt⁻*) strain is only operational in the meiotic zygotes of a cross. Zygotes that divide mitotically or asexual diploid progeny produced by cell fusion transmit chloroplast markers from both parents despite the presence of both *mt⁺* and *mt⁻* alleles in these cells.

Although biparental transmission of plastids occurs frequently in certain genera of higher plants such as *Oenothera* and *Pelargonium* (cf. Sears, 1980b), there is no evidence at present that the maternal and paternal plastids fuse in the zygote or in mixed cells derived therefrom. Likewise, no bona fide recombinant genotypes have been reported among the progeny of such crosses in spite of the availability of parental strains of *Pelargonium* with physically distinct chloroplast genomes (Metzlaff *et al.*, 1981). The rates and pattern of segregation observed for morphologically distinct markers in *Pelargonium* crosses have been studied extensively (cf. Abdel-Wahab and Tilney-Bassett, 1981; Tilney-Bassett and Birky, 1981).

The nuclear genome may exert a substantial effect on the plastid composition of plants such as *Oenothera* which exhibit biparental inheritance of plastid types (see p. 324 above). Incompatibility between the nuclear and plastid genomes is also seen in the interspecific cross of *Chlamydomonas eugametos* × *C. moewusii*. Lemieux (1981) has shown that the viability of the F₁ progeny from this interspecific cross is very low, but in subsequent backcrosses viability increases markedly. Examination of the restriction patterns of chloroplast DNA from the surviving F₁ progeny show that specific chloroplast DNA segments from each parent are retained. In short, for survival of the hybrid, specific chloroplast DNA fragments from each parent must be present. Obviously, in the absence of recombination of chloroplast genes compatibility of the chloroplast and nuclear genomes can never be achieved in interspecific hybrids. This could be an important factor in determining interspecific compatibility barriers in flowering plants. From the point of view of the plant breeder, the matter of chloroplast fusion and chloroplast gene recombination is central to the development of compatible hybrids between species and selection of chloroplast genomic traits of agricultural importance.

ACKNOWLEDGEMENTS

This work was supported by NIH grant GM-19427. We wish to thank our colleagues for allowing us to cite certain unpublished results.

NOTE ADDED IN PROOF:

The writing of this review was completed in early March 1982 based on our critical review of the literature published through December 1981. In a few cases, papers available to us in preprint form were included and these were subsequently published in 1982 or later. During the long interval between our submission of this manuscript in the spring of 1982 and its publication in 1985, much new information has accrued on the evolution of plastid DNA. Also, in a few instances, findings that appeared to be correct when our review was written has subsequently proved to be in error. For a very recent and critical consideration of the evolution of plastid DNA, the reader is referred to a review by Jeffrey D. Palmer entitled "Evolution of Chloroplast and Mitochondrial DNA in Plants and Algae" that will appear in *Monographs in Evolutionary Biology: Molecular Evolutionary Genetics*, edited by R. J. MacIntyre and published by Plenum. In this review, the reader will find a number of other excellent recent reviews cited.

REFERENCES

Abdel-Wahab, O. A. L. and Tilney-Bassett, R. A. E. (1981). The role of plastid competition in the control of plastid inheritance in the zonal *Pelargonium*. *Plasmid*, **6**, 7–16.

Adams, G. M. W., VanWinkle-Swift, K. P., Gillham, N. W. and Boynton, J. E. (1976). Plastid inheritance in *Chlamydomonas reinhardtii*. In R. A. Lewin (Ed.), *The Genetics of Algae*, University of California Press, Berkeley, pp. 69–118.

Aldrich, J. and Cattolico, R. A. (1981). Isolation and characterization of chloroplast DNA from the marine chromophyte, *Olisthodiscus luteus*: Electron microscopic visualization of isomeric molecular forms. *Plant Physiol.*, **68**, 641–647.

Allet, B. and Rochaix, J.-D. (1979). Structure analysis at the ends of the intervening DNA sequences in the chloroplast 23S ribosomal genes of *C. reinhardtii*. *Cell*, **18**, 55–60.

Altman, S. (1978). Biosynthesis of tRNA. In S. Altman (Ed.), *Transfer RNA*, MIT Press, Cambridge, Massachusetts, pp. 48–77.

Atchison, B. A., Whitfeld, P. R. and Bottomley, W. (1976). Comparison of chloroplast DNAs by specific fragmentation with EcoRI endonuclease. *Molec. Gen. Genet.*, **148**, 263–269.

Avise, J. C., Lansman, R. A. and Shade, R. O. (1979a). The use of restriction endonucleases to measure mitochondrial DNA sequence relatedness in natural populations. I. Population structure and evolution in the genus *Peromyscus*. *Genetics*, **92**, 279–295.

Avise, J. C., Giblin-Davidson, C., Laerm, J., Patton, J. C. and Lansman, R. A. (1979b). Mitochondrial DNA clones and matriarchal phylogeny within and among geographic populations of the pocket gopher, *Geomys pinetis*. *Proc. Natl. Acad. Sci. USA*, **76**, 6694–6698.

Aviv, D., Fluhr, R., Edelman, M. and Galun, E. (1980). Progeny analysis of the interspecific somatic hybrids: *Nicotiana tabacum* (CMS) + *Nicotiana sylvestris* with respect to nuclear and chloroplast markers. *Theor. Appl. Genet.*, **56**, 145–150.

Aviv, D. and Galun, E. (1980). Biochemical and genetic analysis of plants derived from the fusion of X-irradiated male sterile *Nicotiana tabacum* protoplasts and *N.*

sylvestris protoplasts. In L. Ferenczy and G. L. Farkas (Eds), *Advances in Protoplast Research*, Pergamon Press, Oxford and Akademiai Kiado, Budapest, pp. 357–362.

Barrell, B. G., Anderson, S., Bankier, A. T., de Bruijn, M. H. L., Chen, E., Coulson, A. R., Drouin, J., Eperon, I. C., Nierlich, D. P., Roe, B. A., Sanger, F., Schreier, P. H., Smith, A. J. H., Staden, R. and Young, I. G. (1980). Different pattern of codon recognition by mammalian mitochondrial tRNAs. *Proc. Natl. Acad. Sci. USA*, **77**, 3164–3166.

Barrell, B. G., Bankier, A. T. and Drouin, J. (1979). A different genetic code in human mitochondria. *Nature, Lond.*, **282**, 189–194.

Bartlett, S. G., Boynton, J. E. and Gillham, N. W. (1981). Genetics of photosynthesis and the chloroplast. *Symp. Soc. Gen. Microbiol.*, **31**, 379–412.

Bedbrook, J. R. and Bogorad, L. (1976). Endonuclease recognition sites mapped on *Zea mays* chloroplast DNA. *Proc. Natl. Acad. Sci. USA*, **73**, 4309–4313.

Bedbrook, J. R. and Kolodner, R. (1979). The structure of chloroplast DNA. *Ann. Rev. Plant Physiol.*, **30**, 593–620.

Bedbrook, J. R., Kolodner, R. and Bogorad, L. (1977). *Zea mays* chloroplast ribosomal RNA genes are part of a 22,000 base pair inverted repeat. *Cell*, **11**, 739–749.

Bedbrook, J. R., Link, G., Coen, D. M., Bogorad, L. and Rich, A. (1978). Maize plastid gene expressed during photoregulated development. *Proc. Natl. Acad. Sci. USA*, **75**, 3060–3064.

Bedbrook, J. R., Smith, S. M. and Ellis, R. J. (1980). Molecular cloning and sequencing of cDNA encoding the precursor to the small subunit of chloroplast ribulose-1,5-bisphosphate carboxylase. *Nature, Lond.*, **287**, 692–697.

Belliard, G., Pelletier, G., Vedel, F. and Quetier, F. (1978). Morphological characteristics and chloroplast DNA distribution in different cytoplasmic parasexual hybrids of *Nicotiana tabacum*. *Molec. Gen. Genet.*, **165**, 231–237.

Bennoun, P., Masson, A., Piccioni, R. and Chua, N.-H. (1978). Uniparental mutants of *Chlamydomonas reinhardi* defective in photosynthesis. In G. Akoyunoglou and J. H. Argyroudi-Akoyunoglou (Eds), *Chloroplast Development*, Elsevier/North Holland Biomedical Press, Amsterdam, pp. 721–726.

Birky, C. W., Jr. (1978). Transmission genetics of mitochondria and chloroplasts. *Ann. Rev. Genet.*, **12**, 471–512.

Birky, C. W., Jr., VanWinkle-Swift, K. P., Sears, B. B., Boynton, J. E. and Gillham, N. W. (1981). Frequency distributions for chloroplast genes in *Chlamydomonas* zygote clones: Evidence for random drift. *Plasmid*, **6**, 173–192.

Bisalputra, T. and Bisalputra, A. A. (1969). The ultrastructure of chloroplast of a brown alga *Sphacelaria* sp. I. Plastid DNA configuration—the chloroplast genophore. *J. Ultrastruct. Res.*, **29**, 151–170.

Bisaro, D. and Siegel, A. (1980). Sequence homology between chloroplast DNAs from several higher plants. *Plant Physiol.*, **65**, 234–237.

Blair, E. G. and Ellis, R. J. (1973). Protein synthesis in chloroplasts. I. Light-driven synthesis of the large subunit of fraction I protein by isolated pea chloroplasts. *Biochim. Biophys. Acta*, **319**, 223–234.

Blank, R., Grobe, B. and Arnold, C.-G. (1978). Time sequence of nuclear and chloroplast fusions in the zygote of *Chlamydomonas reinhardii*. *Planta*, **138**, 63–64.

Bogorad, L. (1975). Evolution of organelles and eukaryotic genomes. *Science*, **188**, 891–898.

Bogorad, L., Bedbrook, J. R., Coen, D. M., Kolodner, R. and Link, G. (1978). Genes for chloroplast proteins and RNAs. In G. Akoyunoglou and J. H. Argyroudi-Akoyunoglou (Eds), *Chloroplast Development*, Elsevier/North Holland Biomedical Press, Amsterdam, pp. 541–551.

Bogorad, L., Jolly, S. O., Kidd, G., Link, G. and McIntosh, L. (1980). Organization and transcription of maize chloroplast genes. In C. J. Leaver (Ed), *Genome Organization and Expression in Plants*, Plenum Press, New York, pp. 291–303.

Bogorad, L., Link, G., McIntosh, L. and Jolly, S. O. (1979b). Genes on the maize chloroplast chromosome. In D. J. Cummings, P. Borst, I. B. Dawid, S. M. Weissman and C. F. Fox (Eds), *Extrachromosomal DNA*, Academic Press, New York, pp. 113–126.

Bohnert, H. J., Driesel, A. J., Crouse, E. J., Gordon, K., Herrmann, R. G., Steinmetz, A., Mubumbila, M., Keller, M., Burkard, G. and Weil, J. H. (1979). Presence of a transfer RNA gene in the spacer sequence between the 16S and 23S rRNA genes of spinach chloroplast DNA. *FEBS Lett.*, **103**, 52–56.

Bohnert, H. J., Driesel, A. J. and Herrmann, R. G. (1976). Characterization of the RNA compounds synthesized by isolated chloroplasts. In Th. Bücher, W. Neupert, W. Sebald and S. Werner (Eds), *Genetics and Biogenesis of Chloroplasts and Mitochondria*, Elsevier/North-Holland Biomedical Press, Amsterdam, pp. 629–636.

Bohnert, H. J., Gordon, K. H. J. and Crouse, E. J. (1980). Homologies among ribosomal RNA and messenger RNA genes in chloroplasts, mitochondria, and *E. coli*. *Molec. Gen. Genet.*, **179**, 539–545.

Bolen, P. L., Gillham, N. W. and Boynton, J. E. (1980). Evidence for persistence of chloroplast markers in the heteroplasmic state in *Chlamydomonas reinhardtii*. *Current Genetics*, **2**, 159–167.

Bolen, P. L., Grant, D. M., Swinton, D., Boynton, J. E. and Gillham, N. W. (1982). Extensive methylation of chloroplast DNA by a nuclear gene mutation does not affect chloroplast gene transmission in *Chlamydomonas*. *Cell*, **28**, 335–343.

Bonen, L., Doolittle, W. F. and Fox, G. E. (1979). Cyanobacterial evolution: results of 16S ribosomal ribonucleic acid sequence analyses. *Can. J. Biochem.*, **57**, 879–888.

Bonitz, S. G., Berlani, R., Coruzzi, G., Li, M., Macino, G., Nobrega, F. G., Nobrega, M. P., Thalenfeld, B. E. and Tzagoloff, A. (1980). Codon recognition rules in yeast mitochondria. *Proc. Natl. Acad. Sci. USA*, **77**, 3167–3170.

Borbely, G. and Simoncsits, A. (1981). 3'-terminal conserved loops of 16S rRNAs from the Cyanobacterium *Synechococcus* AN PCC 6301 and maize chloroplast differ only in two bases. *Biochem. Biophys. Res. Commun.*, **101**, 846–852.

Bovenberg, W. A., Kool, A. J. and Nijkamp, H. J. J. (1981). Isolation, characterization and restriction endonuclease mapping of the *Petunia hybrida* chloroplast DNA. *Nucl. Acids Res.*, **9**, 503–517.

Bowman, C. M. and Dyer, T. A. (1979). 4.5S ribonucleic acid, a novel ribosome component in the chloroplasts of flowering plants. *Biochem. J.*, **183**, 605–613.

Bowman, C. M., Koller, B., Delius, H. and Dyer, T. A. (1981). A physical map of wheat chloroplast DNA showing the location of the structural genes for the ribosomal RNAs and the large subunit of ribulose 1,5-bisphosphate carboxylase. *Molec. Gen. Genet.*, **183**, 93–101.

Boynton, J. E., Gillham, N. W., Harris, E. H., Tingle, C. L., VanWinkle-Swift, K. and Adams, G. M. W. (1976). Transmission, segregation and recombination of chloroplast genes in *Chlamydomonas*. In Th. Bücher, W. Neupert, W. Sebald and S. Werner (Eds), *Genetics and Biogenesis of Chloroplasts and Mitochondria*, Elsevier/North-Holland Biomedical Press, Amsterdam, pp. 313–322.

Bram, R. J., Young, R. A. and Steitz, J. A. (1980). The ribonuclease III site flanking 23S sequences in the 30S ribosomal precursor RNA of *E. coli*. *Cell*, **19**, 393–401.

Branlant, C., Krol, A., Machatt, M. A., Pouyet, J., Ebel, J.-P., Edwards, K. and Kössel, H. (1981). Primary and secondary structures of *Escherichia coli* MRE 600 23S ribosomal RNA. Comparison with models of secondary structure for maize chloroplast 23S rRNA and for large portions of mouse and human 16S mitochondrial rRNAs. *Nucl. Acids Res.*, **9**, 4303–4324.

Breitenberger, C. A., Graves, M. C. and Spremulli, L. L. (1979). Evidence for the nuclear location of the gene for chloroplast elongation factor G. *Arch. Biochem. Biophys.*, **194**, 265–270.

Broglie, R., Bellemare, G., Bartlett, S. G., Chua, N.-H. and Cashmore, A. R. (1981). Cloned DNA sequences complementary to mRNAs encoding precursors to the small subunit of ribulose-1,5-bisphosphate carboxylase and a chlorophyll a/b binding polypeptide. *Proc. Natl. Acad. Sci. USA*, **78**, 7304–7308.

Brosius, J., Dull, T. J. and Noller, H. F. (1980). Complete nucleotide sequence of a 23S ribosomal RNA gene from *Escherichia coli*. *Proc. Natl. Acad. Sci. USA*, **77**, 201–204.

Burkard, G., Canaday, J., Crouse, E., Guillemaut, P., Imbault, P., Keith, G., Keller, M., Mubumbila, M., Osorio, L., Sarantoglou, V., Steinmetz, A. and Weil, J. H. (1980). Transfer RNAs and aminoacyl-tRNA synthetases in plant organelles. In C. J. Leaver (ed), *Genome Organization and Expression in Plants*, Plenum Press, New York, pp. 313–320.

Calagan, J. L., Pirtle, R. M., Pirtle, I. L., Kashdan, M. A., Vreman, H. J. and Dudock, B. S. (1980). Homology between chloroplast and prokaryotic initiator transfer RNA: nucleotide sequence of spinach chloroplast methionine initiator transfer RNA. *J. Biol. Chem.*, **255**, 9981–9984.

Canaday, J., Guillemaut, P., Gloeckler, R. and Weil, J.-H. (1980b). Comparison of the nucleotide sequence of chloroplast tRNAs^{Phe} and tRNAs$_3^{Leu}$ from spinach and bean. *Plant Sci. Lett.*, **20**, 57–62.

Canaday, J., Guillemaut, P., Gloeckler, R. and Weil, J.-H. (1981). The nucleotide sequence of spinach chloroplast tryptophan transfer RNA. *Nucl. Acids Res.*, **9**, 47–53.

Canaday, J., Guillemaut, P. and Weil, J.-H. (1980a). The nucleotide sequences of the initiator transfer RNAs from bean cytoplasm and chloroplasts. *Nucl. Acids Res.*, **8**, 999–1008.

Carbonera, D., Sora, S., Riccardi, G., Camerino, G. and Ciferri, O. (1981). Characterization of a mutant of *Chlamydomonas reinhardtii* resistant to fusidic acid. *FEBS Lett.*, **132**, 227–230.

Cashmore, A. R. (1979). Reiteration frequency of the gene coding for the small subunit of ribulose-1,5-bisphosphase carboxylase. *Cell*, **17**, 383–388.

Cavalier-Smith, T. (1970). Electron microscopic evidence for chloroplast fusion in zygotes of *Chlamydomonas reinhardii*. *Nature, Lond.*, **228**, 333–335.

Cavalier-Smith, T. (1976). Electron microscopy of zygospore formation in *Chlamydomonas reinhardii*. *Protoplasma*, **87**, 297–315.

Cavalier-Smith, T. (1982). The origins of plastids. *Biol. J. Linnean Soc.*, **17**, 289–306.

Chang, S. H., Brum, C. K., Silberklang, M., RajBhandary, U. L., Hecker, L. I. and Barnett, W. E. (1976). The first nucleotide sequence of an organelle transfer RNA: chloroplastic tRNA^{Phe}. *Cell*, **9**, 717–723.

Chang, S. H., Hecker, L. I., Brum, C. K., Schnabel, J. J., Heckman, J. E., Silberklang, M., RajBhandary, U. L. and Barnett, W. E. (1981). The nucleotide sequence of *Euglena* cytoplasmic phenylalanine transfer RNA. Evidence for possible classification of *Euglena* among the animal rather than the plant kingdom. *Nucl. Acids Res.*, **9**, 3199–3204.

Chen, K., Wildman, S. G. and Smith, H. H. (1977). Chloroplast DNA distribution in parasexual hybrids as shown by polypeptide composition of fraction I protein. *Proc. Natl. Acad. Sci. USA*, **74**, 5109–5112.

Chiang, K. S. (1976). On the search for a molecular mechanism of cytoplasmic inheritance: Past controversy, present progress and future outlook. In Th. Bücher, W. Neupert, W. Sebald and S. Werner (Eds), *Genetics and Biogenesis of Chloroplasts and Mitochondria*, Elsevier/North-Holland Biomedical Press, Amsterdam, pp. 305–312.

Chiang, K.-S., Friedman, E., Malavasic, M. J. Jr., Lin Feng, M.-H., Eves, E. M., Feng, T.-Y. and Swinton, D. C. (1981). On the folding and organization of chloroplast DNA in *Chlamydomonas reinhardtii*. *Ann. New York Acad. Sci.*, **361**, 219–247.

Chu, N. M., Oishi, K. K. and Tewari, K. K. (1981). Physical mapping of the pea chloroplast DNA and localization of the ribosomal RNA genes. *Plasmid*, **6**, 279–292.

Chua, N.-H. and Bennoun, P. (1975). Thylakoid membrane polypeptides of *Chlamydomonas reinhardtii*: Wild-type and mutant strains deficient in photosystem II reaction center. *Proc. Natl. Acad. Sci. USA*, **72**, 2175–2179.

Chua, N.-H. and Gillham, N. W. (1977). The sites of synthesis of the principal thylakoid membrane polypeptides in *Chlamydomonas reinhardtii*. *J. Cell Biol.*, **74**, 441–452.

Ciferri, O., Di Pasquale, G. and Tiboni, O. (1979). Chloroplast elongation factors are synthesized in the chloroplast. *Eur. J. Biochem.*, **102**, 331–335.

Ciferri, O. and Tiboni, O. (1976). Evidence for the synthesis in the chloroplast of elongation factor G. *Plant Sci. Lett.*, **7**, 455–466.

Clark, C. G. and Gerbi, S. A. (1982). Ribosomal RNA evolution by fragmentation of the 23S progenitor: maturation pathway parallels evolutionary emergence. *J. Mol. Evol.*, **18**, 329–336.

Clegg, M. T., Rawson, J. R. Y. and Thomas, K. (1982). DNA sequences variation in plants. I. Chloroplast DNA evolution in pearl millet and related species. In preparation. [Subsequently published as, "Chloroplast DNA Evolution in Pearl Millet and Related Species" *Genetics* **106**, 449–461 (1984)].

Cleland, R. E. (1972). *Oenothera: Cytogenetics and Evolution*, Academic Press, New York.

Coen, D. M., Bedbrook, J. R., Bogorad, L. and Rich, A. (1977). Maize chloroplast DNA fragment encoding the large subunit of ribulosebisphosphate carboxylase. *Proc. Natl. Acad. Sci. USA*, **74**, 5487–5491.

Coen, D. M., Bedbrook, J. R., Link, G. Grebanier, A., Steinback, K., Beaton, A., Rich, A. and Bogorad, L. (1978). Genes and mRNAs for maize chloroplast proteins: Changes during light-induced chloroplast development. In G. Akoyunoglou and J. H. Argyroudi-Akoyunoglou (Eds), *Chloroplast Development*, Elsevier/North-Holland Biomedical Press, Amsterdam, pp. 553–558.

Coleman, A. W. (1978). Visualization of chloroplast DNA with two fluorochromes. *Exp. Cell Res.*, **114**, 95–100.

Coleman, A. W. (1979). Use of the fluorochrome 4′6-diamidino-2-phenylindole in genetic and developmental studies of chloroplast DNA. *J. Cell Biol.*, **82**, 299–305.

Coleman, A. W. and Heywood, P. (1981). Structure of the chloroplast and its DNA in chloromonadophycean algae. *J. Cell Sci.*, **49**, 401–409.

Coleman, A. W., Maguire, M. J. and Coleman, J. R. (1981). Mithramycin- and 4′-6-diamidino-2-phenylindole (DAPI)-DNA staining for fluorescence microspectrophotometric measurement of DNA in nuclei, plastids, and virus particles. *J. Histo-chem. Cytochem.*, **29**, 959–968.

Crick, F. H. C. (1966). Codon-anticodon pairing: The wobble hypothesis. *J. Mol. Biol.*, **19**, 548–555.

Crouse, E. J., Schmitt, J. M., Bohnert, H.-J., Gordon, K., Driesel, A. J. and Herrmann, R. G. (1978). Intramolecular compositional heterogeneity of *Spinacia* and *Euglena* chloroplast DNAs. In G. Akoyunoglou and J. H. Argyroudi-Akoyu-noglou (Eds), *Chloroplast Development*, Elsevier/North-Holland Biomedical Press, Amsterdam, pp. 565–572.

de Boer, H. A., Gilbert, S. F. and Nomura, M. (1979). DNA sequences of promoter regions for rRNA operons *rrnE* and *rrnA* in *E. coli*. *Cell*, **17**, 201–209.

de Heij, J. T. and Groot, G. S. P. (1981). *Spirodela oligorhiza* chloroplast DNA codes for ATPase subunits α and β. Immunological evidence from a coupled transcription-translation system. *FEBS Lett.*, **134**, 6–10.

Delihas, N., Andersen, J., Sprouse, H. M. and Dudock, B. (1981). The nucleotide sequence of the chloroplast 5S ribosomal RNA from spinach. *Nucl. Acids Res.*, **9**, 2801–2805.

Delius, H. and Koller, B. (1980). Sequence homologies between *Escherichia coli* and chloroplast ribosomal DNA as seen by heteroduplex analysis. *J. Mol. Biol.*, **142**, 247–261.

Doolittle, W. F. and Bonen, L. (1981). Molecular sequence data indicating an endo-symbiotic origin for plastids. *Ann. New York Acad. Sci.*, **361**, 248–256.

Douce, R. and Joyard, J. (1979). Structure and function of the plastid envelope. *Adv. in Botanical Res.*, **7**, 1–116.

Douglas, G. C., Wetter, L. R., Keller, W. A. and Setterfield, G. (1981). Somatic hybridization between *Nicotiana rustica* and *Nicotiana tabacum*. IV. Analysis of nuclear and chloroplast genome expression in somatic hybrids. *Can. J. Bot.*, **59**, 1509–1513.

Driesel, A. J., Crouse, E. J., Gordon, K., Bohnert, H. J., Herrmann, R. G., Steinmetz, A., Mubumbila, M., Keller, M., Burkard, G. and Weil, J. H. (1979). Fractionation and identification of spinach chloroplast transfer RNAs and mapping of their genes on the restriction map of chloroplast DNA. *Gene*, **6**, 285–306.

Driesel, A. J., Speirs, J. and Bohnert, H. J. (1980). Spinach chloroplast mRNA for a 32,000 dalton polypeptide. Size and localization on the physical map of the chloroplast DNA. *Biochim. Biophys. Acta*, **610**, 297–310.

Dujon, B. (1980). Sequence of the intron and flanking exons of the mitochondrial 21S rRNA gene of yeast strains having different alleles at the ω and *rib-1* loci. *Cell*, **20**, 185–197.

Dyer, T. A. and Bedbrook, J. R. (1980). The organization in higher plants of the genes coding for chloroplast ribosomal RNA, in C. J. Leaver (Ed), *Genome Organization and Expression in Plants*, Plenum Press, New York, pp. 305–311.

Dyer, T. A. and Bowman, C. M. (1979). Nucleotide sequences of chloroplast 5S ribosomal ribonucleic acid in flowering plants. *Biochem. J.*, **183**, 595–604.

Dyer, T. A. and Leaver, C. J. (1981). RNA: structure and metabolism. In A. Marcus (Ed), *The Biochemistry of Plants*, vol. 6, *Proteins and Nucleic Acids*, Academic Press, New York, pp. 111–168.

Eaglesham, A. R. J. and Ellis, R. J. (1974). Protein synthesis in chloroplasts. II. Light-driven synthesis of membrane proteins by isolated pea chloroplasts. *Biochim. Biophys. Acta*, **335**, 396–407.

Edelman, M. (1981). Nucleic acids of chloroplasts and mitochondria. In A. Marcus

(Ed), *The Biochemistry of Plants*, vol. 6, *Proteins and Nucleic Acids*, Academic Press, New York, pp. 249–301.

Edelman, M. and Reisfeld, A. (1980). Synthesis, processing and functional probing of P-32000, the major membrane protein translated within the chloroplast. In C. J. Leaver (Ed), *Genome Organization and Expression in Plants*, Plenum Press, New York, pp. 353–362.

Edwards, K., Bedbrook, J., Dyer, T. and Kössel, H. (1981). 4.5S rRNA from *Zea mays* chloroplasts shows structural homology with the 3′ end of prokaryotic 23S rRNA. *Biochemistry International*, **2**, 533–538.

Edwards, K. and Kössel, H. (1981). The rRNA operon from *Zea mays* chloroplasts: nucleotide sequence of 23S rDNA and its homology with *E. coli* 23S rDNA. *Nucl. Acids Res.*, **9**, 2853–2869.

El-Gewely, M. R., Lomax, M. I., Lau, E. T., Helling, R. B., Farmerie, W. and Barnett, W. E. (1981). A map of specific cleavage sites and transfer RNA genes in the chloroplast genome of *Euglena gracilis bacillaris*. *Molec. Gen. Genet.*, **181**, 296–305.

Ellis, R. J. and Barraclough, R. (1978). Synthesis and transport of chloroplast proteins inside and outside the cell. In G. Akoyunoglou and J. H. Argyroudi-Akoyunoglou (Eds), *Chloroplast Development*, Elsevier/North-Holland Biomedical Press, Amsterdam, pp. 185–194.

Ellis, R. J., Highfield, P. E. and Silverthorne, J. (1977). The synthesis of chloroplast proteins by subcellular systems. In D. O. Hall, J. Coombs and T. W. Goodwin (Eds), *Proceedings of the Fourth International Congress on Photosynthesis*, The Biochemical Society, London, pp. 497–506.

Ellis, R. J., Smith, S. M. and Barraclough, R. (1980). Synthesis, transport and assembly of chloroplast proteins. In C. J. Leaver (Ed), *Genome Organization and Expression in Plants*, Plenum Press, New York, pp. 321–335.

Ellwood, M. and Nomura, M. (1982). Chromosomal locations of the genes for rRNA in *Escherichia coli* K-12. *J. Bacteriol.*, **149**, 458–468.

Eneas-Filho, J., Hartley, M. R. and Mache R. (1981). Pea chloroplast ribosomal proteins: Characterization and site of synthesis. *Molec. Gen. Genet.*, **184**, 484–488.

Ersland, D. R., Aldrich, J. and Cattolico, R. A. (1981). Kinetic complexity, homogeneity and copy number of chloroplast DNA from the marine alga *Olisthodiscus luteus*. *Plant Physiol.*, **68**, 1468–1473.

Evans, D. A., Wetter, L. R. and Gamborg, O. L. (1980). Somatic hybrid plants of *Nicotiana glauca* and *Nicotiana tabacum* obtained by protoplast fusion. *Physiol. Plant.*, **48**, 225–230.

Fluhr, R. and Edelman, M. (1981a). Physical mapping of *Nicotiana tabacum* chloroplast DNA. *Molec. Gen. Genet.*, **181**, 484–490.

Fluhr, R. and Edelman, M. (1981b). Conservation of sequence arrangement among higher plant chloroplast DNAs: Molecular cross hybridization among the Solanaceae and between *Nicotiana* and *Spinacia*. *Nucl. Acids Res.*, **9**, 6841–6853.

Forster, J. L., Grabowy, C. T., Harris, E. H., Boynton, J. E. and Gillham, N. W. (1980). Behavior of chloroplast genes during the early zygotic divisions of *Chlamydomonas reinhardtii*. *Current Genetics*, **1**, 137–153.

Fox, G. E. and Woese, C. R. (1975). 5S RNA secondary structure. *Nature*, **256**, 505–507.

Fox, L., Erion, J., Tarnowski, J., Spremulli, L., Brot, N. and Weissbach, H. (1980).*Euglena gracilis* chloroplast EF-Ts. Evidence that it is a nuclear-coded gene product. *J. Biol. Chem.*, **255**, 6018–6019.

Frankel, R., Scowcroft, W. R. and Whitfeld, P. R. (1979). Chloroplast DNA variation in isonuclear male-sterile lines of *Nicotiana*. *Molec. Gen. Genet.*, **169**, 129–135.

Freyssinet, G. (1978). Determination of the site of synthesis of some *Euglena* cytoplasmic and chloroplast ribosomal proteins. *Exp. Cell Res.*, **115**, 207–219.

Gatenby, A. A., Castleton, J. A. and Saul, M. W. (1981). Expression in *E. coli* of the maize and wheat chloroplast genes for the large subunit of ribulose bisphosphate carboxylase. *Nature, Lond.*, **291**, 117–121.

Gauss, D. H. and Sprinzl, M. (1981). Compilation of tRNA sequences. *Nucl. Acids Res.*, **9**, r1–r23.

Gelvin, S. B. and Howell, S. H. (1979). Small repeated sequences in the chloroplast genome of *Chlamydomonas reinhardi*. *Molec. Gen. Genet.*, **173**, 315–322.

Gibbs, S. P. (1981). The chloroplasts of some algal groups may have evolved from endosymbiotic eukaryotic algae. *Ann. New York Acad. Sci.*, **361**, 193–208.

Gibbs, S. P., Cheng, D. and Slankis, T. (1974). The chloroplast nucleoid in *Ochromonas danica*. I. Three-dimensional morphology in light- and dark-grown cells. *J. Cell Sci.*, **16**, 557–577.

Gillham, N. W. (1965). Induction of chromosomal and non-chromosomal mutations in *Chlamydomonas reinhardi* with N-methyl-N'-nitro-N-nitrosoguanidine. *Genetics*, **52**, 529–537.

Gillham, N. W. (1978). *Organelle Heredity*, Raven Press, New York.

Gillham, N. W. and Boynton, J. E. (1981). Evolution of organelle genomes and protein-synthesizing systems. *Ann. New York Acad. Sci.*, **361**, 20–43.

Gillham. N. W. and Levine, R. P. (1962). Pure mutant clones induced by ultra-violet light in the green alga, *Chlamydomonas reinhardi*. *Nature, Lond.*, **194**, 1165–1166.

Glotz, C., Zweib, C., Brimacombe, R., Edwards, K. and Kössel, H. (1981). Secondary structure of the large subunit ribosomal RNA from *Escherichia coli*, *Zea mays* chloroplast, and human and mouse mitochondrial ribosomes. *Nucl. Acids Res.*, **9**, 3287–3306.

Goodspeed, T. H. (1954). The genus *Nicotiana*. *Chronica Botanica*, **16**, 1–536.

Gordon, K. H. J., Crouse, E. J., Bohnert, H.-J. and Herrmann, R. G. (1981). Restriction endonuclease cleavage site map of chloroplast DNA from *Oenothera parviflora* (*Euoenothera* Plastome IV). *Theor. Appl. Genet.*, **59**, 281–296.

Graf, L., Kössel, H. and Stutz, E. (1980). Sequencing of 16S-23S spacer in a ribosomal RNA operon of *Euglena gracilis* chloroplast DNA reveals two tRNA genes. *Nature, Lond.*, **286**, 908–910.

Grant, D. M., Gillham, N. W. and Boynton, J. E. (1980). Inheritance of chloroplast DNA in *Chlamydomonas reinhardtii*. *Proc. Natl. Acad. Sci. USA*, **77**, 6067–6071.

Gray, J. C., Kung, S. D., Wildman, S. G. and Sheen, S. J. (1974). Origin of *Nicotiana tabacum* L. detected by polypeptide composition of Fraction I protein. *Nature, Lond.*, **252**, 226–227.

Gray, M. W. and Doolittle, W. F. (1982). Has the endosymbiont hypothesis been proven? *Microbiol. Revs.*, **46**, 1–42.

Gray, P. W. and Hallick, R. B. (1978). Physical mapping of the *Euglena gracilis* chloroplast DNA and ribosomal RNA gene region. *Biochemistry*, **17**, 284–289.

Gray, P. W. and Hallick, R. B. (1979). Isolation of *Euglena gracilis* chloroplast 5S ribosomal RNA and mapping the 5S rRNA gene on chloroplast DNA. *Biochemistry*, **18**, 1820–1825.

Grebanier, A. E., Coen, D. M., Rich, A. and Bogorad, L. (1978). Membrane proteins synthesized but not processed by isolated maize chloroplasts. *J. Cell Biol.*, **78**, 734–746.

Green, B. R. (1976). Covalently closed minicircular DNA associated with *Acetabularia* chloroplasts. *Biochim. Biophys. Acta*, **447**, 156–166.

Green, B. R., Muir, B. L. and Padmanabhan, U. (1977). The *Acetabularia* chloroplast genome: small circles and large kinetic complexity. In C. F. L. Woodcock (Ed), *Progress in Acetabularia Research*, Academic Press, New York, pp. 107–122.

Guillemaut, P. and Keith, G. (1977). Primary structure of bean chloroplastic tRNA^Phe. Comparison with euglena chloroplastic tRNA^Phe. *FEBS Lett.*, **84**, 351–356.

Hallick, R. B., Gray, P. W., Chelm, B. K., Rushlow, K. E. and Orozco, E. M. Jr. (1978). *Euglena gracilis* chloroplast DNA structure, gene mapping, and RNA transcription. In G. Akoyunoglou and J. H. Argyroudi-Akoyunoglou (Eds), *Chloroplast Development*, Elsevier/North-Holland Biomedical Press, Amsterdam, pp. 619–622.

Hallick, R. B., Rushlow, K. E., Orozco, E. M. Jr., Steigler, G. L. and Gray, P. W. (1979). Chloroplast DNA of *Euglena gracilis*: Gene mapping and selective *in vitro* transcription of the ribosomal RNA region. In D. J. Cummings, P. Borst, I. B. Dawid, S. M. Weissman and C. F. Fox (Eds), *Extrachromosomal DNA*, Academic Press, New York, pp. 127–141.

Harris, E. H., Boynton, J. E. and Gillham, N. W. (1982). Induction of nuclear and chloroplast mutations which affect the chloroplast in *Chlamydomonas reinhardtii*, in *Methods in Chloroplast Molecular Biology* (Eds. M. Edelman, R. Hallick and N.-H. Chua), pp. 3–23, Elsevier/North-Holland Biomedical Press, Amsterdam.

Harris, E. H., Boynton, J. E., Gillham, N. W., Tingle, C. L. and Fox, S. B. (1977). Mapping of chloroplast genes involved in chloroplast ribosome biogenesis in *Chlamydomonas reinhardtii*. *Molec. Gen. Genet.*, **155**, 249–265.

Hartley, M. R. (1979). The synthesis and origin of chloroplast low-molecular-weight ribosomal ribonucleic acid in spinach. *Eur. J. Biochem.*, **96**, 311–320.

Hartley, M. R. and Head, C. (1979). The synthesis of chloroplast high-molecular-weight ribosomal ribonucleic acid in spinach. *Eur. J. Biochem.*, **96**, 301–309.

Heckman, J. E., Sarnoff, J., Alzner-DeWeerd, B., Yin, S. and RajBhandary, U. L. (1980). Novel features in the genetic code and codon reading patterns in *Neurospora crassa* mitochondria based on sequences of six mitochondrial tRNAs. *Proc. Natl. Acad. Sci. USA*, **77**, 3159–3163.

Hedberg, M. F., Huang, Y.-S. and Hommersand, M. H. (1981). Size of the chloroplast genome in *Codium fragile*. *Science*, **213**, 445–447.

Heizmann, P., Doly, J., Hussein, Y., Nicholas, P., Nigon, V. and Bernardi, G. (1981). The chloroplast genome of bleached mutants of *Euglena gracilis*. *Biochim. Biophys. Acta*, **653**, 412–415.

Helling, R. B., El-Gewely, M. R., Lomax, M. I., Baumgartner, J. E., Schwartzbach, S. D. and Barnett, W. E. (1979). Organization of the chloroplast ribosomal RNA genes of *Euglena gracilis bacillaris*. *Molec. Gen. Genet.*, **174**, 1–10.

Herrmann, R. G. (1970). Multiple amounts of DNA related to the size of chloroplasts. I. An autoradiographic study. *Planta*, **90**, 80–96.

Herrmann, R. G. and Kowallik, K. V. (1970). Multiple amounts of DNA related to the size of chloroplasts. II. Comparison of electron-microscopic and autoradiographic data. *Protoplasma*, **69**, 365–372.

Herrmann, R. G., Kowallik, K. V. and Bohnert, H. J. (1974). Structural and functional aspects of the plastome. I. The organization of the plastome. *Port. Acta Biol. Ser. A*, **14**, 91–110.

Herrmann, R. G., Palta, H. K. and Kowallik, K. V. (1980b). Chloroplast DNA from three archegoniates. *Planta*, **148**, 319–327.

Herrmann, R. G. and Possingham, J. V. (1980). Plastid DNA—The plastome. In J. Reinert (Ed), *Results and Problems in Cell Differentiation,* vol. 10, *Chloroplasts,* Springer-Verlag, Berlin, pp. 45–96.

Herrmann, R. G., Seyer, P., Schedel, R., Gordon, K., Bisanz, C., Winter, P., Hildebrandt, J. W., Wlaschek, M., Alt, J., Driesel, A. J. and Sears, B. B. (1980a). The plastid chromosomes of several dicotyledons. In T. Bücher, W. Sebald and H. Weiss (Eds), *Biological Chemistry of Organelle Formation,* Springer-Verlag, Berlin, pp. 97–112.

Hoffman-Falk, H., Mattoo, A. K., Marder, J., Fluhr, R. and Edelman, M. (1980). *Proc. 13th FEBS Meeting,* p. 22.

Iwai, S., Nakata, K., Nagao, T., Kawashima, N. and Matsuyama, S. (1981). Detection of the *Nicotiana rustica* chloroplast genome coding for the large subunit of Fraction I protein in a somatic hybrid in which only the *N. tabacum* chloroplast genome appeared to have been expressed. *Planta,* **152,** 478–480.

Jacq, B. (1981). Sequence homologies between eukaryotic 5.8S rRNA and the 5′ end of prokaryotic 23S rRNA: evidences for a common evolutionary origin. *Nucl. Acids Res.,* **9,** 2913–2932.

James, T. W. and Jope, C. (1978). Visualization by fluorescence of chloroplast DNA in higher plants by means of the DNA-specific probe 4′6-diamidino-2-phenylindole. *J. Cell Biol.* **79,** 623–630.

Jenni, B., Fasnacht, M. and Stutz, E. (1981). The multiple copies of the *Euglena gracilis* chloroplast genome are not uniform in size. *FEBS Lett.,* **125,** 175–179.

Jenni, B. and Stutz, E. (1978). Physical mapping of the ribosomal DNA region of *Euglena gracilis* chloroplast DNA. *Eur. J. Biochem.,* **88,** 127–134.

Jenni, B. and Stutz, E. (1979). Analysis of *Euglena gracilis* chloroplast DNA. Mapping of a DNA sequence complementary to 16S rRNA outside of the three rRNA gene sets. *FEBS Lett.,* **102,** 95–99.

Jurgenson, J. E. and Bourque, D. P. (1980). Mapping of rRNA genes in an inverted repeat in *Nicotiana tabacum* chloroplast DNA. *Nucl. Acids Res.,* **8,** 3505–3516.

Kashdan, M. A., Pirtle, R. M., Pirtle, I. L., Calagan, J. L., Vreman, H. J. and Dudock, B. S. (1980). Nucleotide sequence of a spinach chloroplast threonine tRNA. *J. Biol. Chem.,* **255,** 8831–8835.

Keller, M., Burkard, G., Bohnert, H. J., Mubumbila, M., Gordon, K., Steinmetz, A., Heiser, D., Crouse, E. J. and Weil, J. H. (1980). Transfer RNA genes associated with the 16S and 23S rRNA genes of *Euglena* chloroplast DNA. *Biochem. Biophys. Res. Commun.,* **95,** 47–54.

Keresztes, A. (1971). Light microscopic examination of chloroplast mutation in *Tradescantia* leaves. *Acta Botanica Scientiarum Hungaricae,* **17,** 379–389.

Kirk, J. T. O. and Tilney-Bassett, R. A. E. (1978). *The Plastids,* Elsevier/North-Holland Biomedical Press, Amsterdam.

Koch, W., Edwards, K. and Kössel, H. (1981). Sequencing of the 16S-23S spacer in a ribosomal RNA operon of *Zea mays* chloroplast DNA reveals two split tRNA genes. *Cell,* **25,** 203–213.

Koller, B. and Delius, H. (1980). *Vicia faba* chloroplast DNA has only one set of ribosomal RNA genes as shown by partial denaturation mapping and R-loop analysis. *Molec. Gen. Genet.,* **178,** 261–269.

Kung, S. D., Zhu, Y. S., Chen, K., Shen, G. F. and Sisson, V. A. (1981). *Nicotiana* chloroplast genome. II. Chloroplast DNA alteration. *Molec. Gen. Genet.,* **183,** 20–24.

Kung, S. D., Zhu, Y. S. and Shen, G. F. (1982). *Nicotiana* chloroplast genome. III. Chloroplast DNA evolution. *Theor. Appl. Genet.,* **61,** 73–79.

Kuroiwa, T. and Suzuki, T. (1980). An improved method for the demonstration of the *in situ* chloroplast nuclei in higher plants. *Cell Structure and Function*, **5**, 195–197.

Kuroiwa, T. and Suzuki, T. (1981). Circular nucleoids isolated from chloroplasts in a brown alga *Ectocarpus indicus*. *Exp. Cell Res.*, **134**, 457–461.

Kuroiwa, T., Suzuki, T., Ogawa, K. and Kawano, S. (1981). The chloroplast nucleus: Distribution, number, size, and shape, and a model for the multiplication of the chloroplast genome during chloroplast development. *Plant Cell Physiol.*, **22**, 381–396.

Kusuda, J., Shinozaki, K., Takaiwa, F. and Sugiura, M. (1980). Characterization of the cloned ribosomal DNA of tobacco chloroplasts. *Molec. Gen. Genet.*, **178**, 1–7.

Kutzelnigg, H. and Stubbe, W. (1974). Investigations on plastome mutants in *Oenothera*. 1. General considerations. *Sub-Cell Biochem.*, **3**, 73–89.

Lagerkvist, U. (1978). "Two out of three": An alternative method for codon reading. *Proc. Natl. Acad. Sci. USA*, **75**, 1759–1762.

Lamppa, G. K. and Bendich, A. J. (1979a). Changes in chloroplast DNA levels during development of pea (*Pisum sativum*). *Plant Physiol.*, **64**, 126–130.

Lamppa, G. K. and Bendich, A. J. (1979b). Chloroplast DNA sequence homologies among vascular plants. *Plant Physiol.*, **63**, 660–668.

Lamppa, G. K. and Bendich, A. J. (1981). Fine scale interspersion of conserved sequences in the pea and corn chloroplast genomes. *Molec. Gen. Genet.*, **182**, 310–320.

Lamppa, G. K., Elliot, L. V. and Bendich, A. J. (1980). Changes in chloroplast number during pea leaf development. An analysis of a protoplast population. *Planta*, **148**, 437–443.

Langridge, P. (1981). Synthesis of the large subunit of spinach ribulose bisphosphate carboxylase may involve a precursor polypeptide. *FEBS Lett.*, **123**, 85–89.

LaRue, B., Newhouse, N., Nicoghosian, K., and Cedergren, R. J. (1981). The evolution of multi-isoacceptor tRNA families: Sequence of tRNA$^{\text{Leu}}_{\text{CAA}}$ and tRNA$^{\text{Leu}}_{\text{CAG}}$ from *Anacystis nidulans*. *J. Biol. Chem.*, **256**, 1539–1543.

Lebacq, P. and Vedel, F. (1981). Sal I restriction enzyme analysis of chloroplast and mitochondrial DNAs in the genus *Brassica*. *Plant Sci. Lett.*, **23**, 1–9.

Lee, R. W. and Haughn, G. W. (1980). Induction and segregation of chloroplast mutations in vegetative cell cultures of *Chlamydomonas reinhardtii*. *Genetics*, **96**, 79–94.

Lemieux, C. (1981). Transmission, segregation and recombination of non-Mendelian genetic markers and chloroplast DNAs in interspecific crosses between *Chlamydomonas eugametos* and *C. moewusii*. Ph.D thesis, Dalhousie University, Halifax, Nova Scotia.

Lemieux, C., Turmel, M. and Lee, R. W. (1980). Characterization of chloroplast DNA in *Chlamydomonas eugametos* and *C. moewusii* and its inheritance in hybrid progeny. *Current Genetics*, **2**, 139–147.

Link, G. (1981). Cloning and mapping of the chloroplast DNA sequences for two messenger RNAs from mustard (*Sinapis alba* L.). *Nucl. Acids Res.*, **9**, 3681–3694.

Link, G. and Bogorad, L. (1980). Sizes, locations, and directions of transcription of two genes on a cloned maize chloroplast DNA sequence. *Proc. Natl. Acad. Sci. USA*, **77**, 1832–1836.

Link, G., Chambers, S. E., Thompson, J. A. and Falk, H. (1981). Size and physical organization of chloroplast DNA from mustard (*Sinapis alba* L.). *Molec. Gen. Genet.*, **181**, 454–457.

Lüttke, A. (1981). Heterogeneity of chloroplasts in *Acetabularia mediterranea*.

Heterogeneous distribution and morphology of chloroplast DNA. *Exp. Cell Res.*, **131**, 483–488.

Machatt, Mohamed Ali, Ebel, J.-P. and Branlant, C. (1981). The 3'-terminal region of bacterial 23S ribosomal RNA: structure and homology with the 3'-terminal region of eukaryotic 28S rrRNA and with chloroplast 4.5S rRNA. *Nucl. Acids Res.*, **9**, 1533–1549.

MacKay, R. M. (1981). The origin of plant chloroplast 4.5S ribosomal RNA. *FEBS Lett.*, **123**, 17–18.

Maliga, P., Nagy, F., Xuan, Le Thi, Kiss, Zs. R., Menczel, L. and Lázár, G. (1980). Protoplast fusion to study cytoplasmic traits in *Nicotiana*. In L. Ferenczy and G. L. Farkas (Eds), *Advances in Protoplast Research*, Pergamon Press, Oxford and Akademiai Kiado, Budapest, pp. 341–348.

Malnoë, P. and Rochaix, J.-D. (1978). Localization of 4S RNA genes on the chloroplast genome of *Chlamydomonas reinhardii*. *Molec. Gen. Genet.*, **166**, 269–275.

Malnoë, P., Rochaix, J.-D., Chua, N.-H. and Spahr, P.-F. (1979). Characterization of the gene and messenger RNA of the large subunit of ribulose 1,5-diphosphate carboxylase in *Chlamydomonas reinhardii*. *J. Mol. Biol.*, **133**, 417–434.

Margulis, L. (1975). Symbiotic theory of the origin of eukaryotic organelles: criteria for proof. *Symp. Soc. Exp. Biol.*, **29**, 21–38.

Margulis, L. (1970). *Origin of Eukaryotic Cells*, Yale University Press, New Haven.

Matagne, R. F. (1981). Transmission of chloroplast alleles in somatic fusion products obtained from vegetative cells and/or 'gametes' of *Chlamydomonas reinhardi*. *Current Genetics*, **3**, 31–36.

Matagne, R. F. and Hermesse, M.-P. (1980). Chloroplast gene inheritance studied by somatic fusion in *Chlamydomonas reinhardtii*. *Current Genetics*, **1**, 127–131.

Matagne, R. F. and Hermesse, M.-P. (1981). Modification of chloroplast gene transmission in somatic fusion products and vegetative zygotes of *Chlamydomonas reinhardi* by 5-fluorodeoxyuridine. *Genetics*, **99**, 371–381.

Mattoo, A. K., Pick, U., Hoffman-Falk, H. and Edelman, M. (1981). The rapidly metabolized 32,000 dalton polypeptide of the chloroplast is the 'proteinaceous shield' regulating photosystem II electron transport and mediating diuron herbicide sensitivity. *Proc. Natl. Acad. Sci. USA*, **78**, 1572–1576.

McCoy, J. M. and Jones, D. S. (1980). The nucleotide sequence of *Scenedesmus obliquus* chloroplast tRNA$_f^{Met}$. *Nucl. Acids Res.*, **8**, 5089–5093.

McIntosh, L., Poulsen, C. and Bogorad, L. (1980). Chloroplast gene sequence for the large subunit of ribulose bisphosphate carboxylase of maize. *Nature, Lond.*, **288**, 556–560.

Medgyesy, P. Menczel, L. and Maliga, P. (1980). The use of cytoplasmic streptomycin resistance: Chloroplast transfer from *Nicotiana tabacum* into *Nicotiana sylvestris*, and isolation of their somatic hybrids. *Molec. Gen. Genet.*, **179**, 693–698.

Melchers, G. (1980). Protoplast fusion, mechanism and consequences for potato breeding and production of potatoes and tomatoes. In L. Ferenczy and G. L. Farkas (Eds), *Advances in Protoplast Research*, Pergamon Press, Oxford and Akademiai Kiado, Budapest, pp. 283–286.

Melchers, G., Sacristan, M. D. and Holder, A. A. (1978). Somatic hybrid plants of potato and tomato regenerated from fused protoplasts. *Carlsberg Res. Commun.*, **43**, 203–218.

Mets, L. (1980). Uniparental inheritance of chloroplast DNA sequences in interspecific hybrids of *Chlamydomonas*. *Current Genetics*, **2**, 131–138.

Metzlaff, M., Börner, T. and Hagemann, R. (1981). Variations of chloroplast DNAs in the genus *Pelargonium* and their biparental inheritance. *Theor. Appl. Genet.*, **60**, 37–41.

Milner, J. J. and Heishberger, C. L. (1979) *Eugleua gracilis* chloroplast DNA codes for polyadenylated RNA *Plant Physiol* **64**, 818–821.

Mubumbila, M., Burkard, G., Keller, M., Steinmetz, A.,Crouse, E. and Weil, J.-H. (1980). Hybridization of bean, spinach, maize, and *Euglena* chloroplast tRNAs with homologous and heterologous chloroplast DNAs. An approach to the study of homology between chloroplast tRNAs from various species. *Biochim. Biophys. Acta*, **609**, 31–39.

Myers, A. M., Grant, D. M., Rabert, D. K., Harris, E. H., Boynton, J. E. and Gillham, N. W. (1982). Mutants of *Chlamydomonas reinhardtii* with physical alterations in their chloroplast DNA. *Plasmid*, **7**, 133–151.

Nazar, R. N. (1980). A 5.8S rRNA-like sequence in prokaryotic 23S rRNA. *FEBS Lett.*, **119**, 212–214.

Nechushtai, R. Nelson, N., Mattoo, A. K. and Edelman, M. (1981). Site of synthesis of subunits to photosystem I reaction center and the proton-ATPase in *Spirodela*. *FEBS Lett.*, **125**, 115–119.

Nomura, M. (1982). Personal communication.

Nomura, M. and Post, L. E. (1980). Organization of ribosomal genes and regulation of their expression in *Escherichia coli*. In G. Chambliss, G. R. Craven, J. Davies, K. Davis, L. Kahan and M. Nomura (Eds), *Ribosomes: Structure, Function and Genetics*, University Park Press, Baltimore, pp. 671–691.

Orozco, E. M. Jr., Gray, P. W. and Hallick, R. B. (1980a). *Euglena gracilis* chloroplast ribosomal RNA transcription units. I. The location of transfer RNA, 5S, 16S and 23S ribosomal RNA genes. *J. Biol. Chem.*, **255**, 10991–10996.

Orozco, E. M. Jr., Rushlow, K. E., Dodd, J. R. and Hallick, R. B. (1980b). *Euglena gracilis* chloroplast ribosomal RNA transcription units. II. Nucleotide sequence homology between the 16S-23S ribosomal RNA spacer and the 16S ribosomal RNA leader regions. *J. Biol. Chem.*, **255**, 10997–11003.

Osawa, S. and Hori, H. (1980). Molecular evolution of ribosomal components. In G. Chambliss, G. R. Craven, J. Davies, K. Davis, L. Kahan and M. Nomura (Eds), *Ribosomes: Structure, Function and Genetics*, University Park Press, Baltimore, pp. 333–335.

Osorio-Almeida, M. L., Guillemaut, P., Keith, G., Canaday, J. and Weil, J. H. (1980). Primary structure of three leucine transfer RNAs from bean chloroplast. *Biochem. Biophys. Res. Commun.*, **92**, 102–108.

Padmanabhan, U. and Green, B. R. (1978). The kinetic complexity of *Acetabularia* chloroplast DNA. *Biochim. Biophys. Acta*, **521**, 67–73.

Palmer, J. D. (1982). Physical and gene mapping of chloroplast DNA from *Atriplex triangularis* and *Cucumis sativa*. *Nucl. Acids Res.*, **10**, 1593–1605.

Palmer, J. D. and Thompson, W. F. (1981a). Clone banks of the mung bean, pea and spinach chloroplast genomes. *Gene*, **15**, 21–26.

Palmer, J. D. and Thompson, W. F. (1981b). Rearrangements in the chloroplast genomes of mung bean and pea. *Proc. Natl. Acad. Sci. USA*, **78**, 5533–5537.

Palmer, J. D. and Thompson, W. F. (1982). Chloroplast DNA rearrangements are more frequent when a large inverted repeat sequence is lost. *Cell*, **29**, 537–550.

Palmer, J. D. and Zamir, D. (1982). Chloroplast DNA evolution and phylogenetic relationships in *Lycopersicon*. *Proc. Natl. Acad. Sci. USA*, **79**, 5006–5010.

Pirtle, R., Calagan, J., Pirtle, I., Kashdan, M., Vreman, H. and Dudock, B. (1981). The nucleotide sequence of spinach chloroplast methionine elongator tRNA. *Nucl. Acids Res.*, **9**, 183–188.

Planta, R. J. and Meyerink, J. H. (1980). Organization of the ribosomal RNA genes in eukaryotes. In G. Chambliss, G. R. Craven, J. Davies, K. Davis, L. Kahan and M. Nomura (Eds), *Ribosomes: Structure, Function and Genetics*, University Park Press, Baltimore, pp. 871–887.

Possingham, J. V. (1980). Plastid replication and development in the life cycle of higher plants. *Ann. Rev. Plant Physiol.*, **31**, 113–129.

Poulsen, C. (1979). The cyanogen bromide fragments of the large subunit of ribulose-bisphosphate carboxylase from barley. *Carlsberg Res. Commun.*, **44**, 163–189.

Poulsen, C. Martin, B. and Svendsen, I. (1979). Partial amino acid sequence of the large subunit of ribulosebisphosphate carboxylase from barley. *Carlsberg Res. Commun.*, **44**, 191–199.

Pribnow, D. (1975). Nucleotide sequence of an RNA polymerase binding site at an early T7 promoter. *Proc. Natl. Acad. Sci. USA*, **72**, 784–788.

Rawson, J. R. Y., Kushner, S. R., Vapnek, D., Alton, N. K. and Boerma, C. L. (1978). Chloroplast ribosomal RNA genes in *Euglena gracilis* exist as three clustered tandem repeats. *Gene*, **3**, 191–209.

Renner, O. (1937). Zur Kenntnis der Plastiden- und Plasmavererbung. *Cytologia (Tokyo)*, **2**, 644–653.

Rhodes, P. R., Zhu, Y. S. and Kung, S. D. (1981). *Nicotiana* chloroplast genome. I. Chloroplast DNA diversity. *Molec. Gen. Genet.*, **182**, 106–111.

Richardson, C. B., Schmidt, R. J., Boynton, J. E. and Gillham, N. W. (1982). Sites of synthesis of chloroplast ribosomal proteins in *Chlamydomonas*. To be submitted to *J. Cell Biol.* [Subsequently published as Schmidt, R. J., Richardson, C. B., Gillham, N. W. and Boynton, J. E. (1983). Sites of Synthesis of Chloroplast Ribosonal Proteins in *Chlamydomonas*. *J. Cell Biol.* **96**, 1451–1463.]

Rick, C. M. (1979). Biosystematic studies in *Lycopersicon* and closely related species of *Solanum*. In J. W. Hawkes, R. N. Lesh and A. D. Skelding (Eds), *The Biology and Taxonomy of the Solanaceae*, Linnean Society Symposium Series No. 7, Academic Press, New York, pp. 667–678.

Ris, H. (1961). The annual invitation lecture. Ultrastructure and molecular organization of genetic systems. *Can. J. Genet. Cytol.*, **3**, 95–120.

Rochaix, J. D. (1972). Cyclization of chloroplast DNA fragments of *Chlamydomonas reinhardi*. *Nature New Biol.*, **238**, 76–78.

Rochaix, J. D. (1978). Restriction endonuclease map of the chloroplast DNA of *Chlamydomonas reinhardii*. *J. Mol. Biol.*, **126**, 597–617.

Roxhaix, J. D. (1981). Organization, function and expression of the chloroplast DNA of *Chlamydomonas reinhardii*. *Experientia*, **37**, 323–332.

Rochaix, J. D. and Darlix, J.-L. (1982). Composite structure of the chloroplast 23S rRNA genes in *Chlamydomonas reinhardii*: Evolutionary and functional implications. *J. Mol. Biol.*, **159**, 383–395.

Rochaix, J.-D. and Malnoë, P. (1978a). Anatomy of the chloroplast ribosomal DNA of *Chlamydomonas reinhardii*. *Cell*, **15**, 661–670.

Rochaix, J.-D. and Malnoë, P. (1978b). Gene localization on the chloroplast DNA of *Chlamydomonas reinhardii*, in G. Akoyunoglou and J. H. Argyroudi-Akoyunoglou (Eds), *Chloroplast Development*, Elsevier/North-Holland Biomedical Press, Amsterdam, pp. 581–586.

Rosenberg, M. and Court, D. (1979). Regulatory sequences involved in the promotion and termination of RNA transcription. *Ann. Rev. Genet.*, **13**, 319–353.

Sager, R. (1972). *Cytoplasmic Genes and Organelles*. Academic Press, New York.

Sager, R. (1977). Genetic analysis of chloroplast DNA in *Chlamydomonas*. *Advances in Genetics*, **19**, 287–340.

Sager, R., Grabowy, C. and Sano, H. (1981). The *mat-1* gene in *Chlamydomonas* regulates DNA methylation during gametogenesis. *Cell*, **24**, 41–47.

Schmidt, G. W., Bartlett, S., Grossman, A. R., Cashmore, A. R. and Chua, N.-H. (1980). *In vitro* synthesis, transport and assembly of the constituent polypeptides of the light-harvesting chlorophyll a/b complex. In C. J. Leaver (Ed), *Genome Organization and Expression in Plants*, Plenum Press, New York, pp. 337–351.

Schmitt, J. M., Bohnert, H.-J., Gordon, K. H. J., Herrmann, R., Bernardi, G. and Crouse, E. J. (1981). Compositional heterogeneity of the chloroplast DNAs from *Euglena gracilis* and *Spinacia oleracea*. *Eur. J. Biochem.*, **117**, 375–382.

Schwarz, Z., Jolly, S. O., Steinmetz, A. A. and Bogorad, L. (1981b). Overlapping divergent genes in the maize chloroplast chromosome and *in vitro* transcription of the gene for tRNA[His]. *Proc. Natl. Acad. Sci. USA*, **78**, 3423–3427.

Schwarz, Zs. and Kössel, H. (1979). Sequencing of the 3′-terminal region of a 16S rRNA gene from *Zea mays* reveals homology with *E. coli* 16S rRNA. *Nature, Lond.*, **279**, 520–522.

Schwarz, Zs. and Kössel, H. (1980). The primary structure of 16S rDNA from *Zea mays* chloroplast is homologous to *E. coli* 16S rRNA. *Nature, Lond.*, **283**, 739–742.

Schwarz, Z., Kössel, H., Schwarz, E. and Bogorad, L. (1981). A gene coding for tRNA[Val] is located near 5′ terminus of 16S rRNA gene in *Zea mays* chloroplast genome. *Proc. Natl. Acad. Sci. USA*, **78**, 4748–4752.

Scowcroft, W. R. (1979). Nucleotide polymorphism in chloroplast DNA of *Nicotiana debneyi*. *Theor. Appl. Genet.*, **55**, 133–137.

Scowcroft, W. R. and Larkin, P. J. (1981). Chloroplast DNA assorts randomly in intraspecific somatic hybrids of *Nicotiana debneyi*. *Theor. Appl. Genet.*, **60**, 179–184.

Sears, B. B. (1980a). Disappearance of the heteroplasmic state for chloroplast markers in zygospores of *Chlamydomonas reinhardtii*. *Plasmid*, **3**, 18–34.

Sears, B. B. (1980b). The elimination of plastids during spermatogenesis and fertilization in the plant kingdom. *Plasmid*, **4**, 233–255.

Seyer, P., Kowallik, K. V. and Herrmann, R. G. (1981). A physical map of *Nicotiana tabacum* plastid DNA including the location of structural genes for ribosomal RNAs and the large subunit of ribulose bisphosphate carboxylase/oxygenase. *Current Genetics*, **3**, 189–204.

Shepherd, H. S., Boynton, J. E. and Gillham, N. W. (1979). Mutations in nine chloroplast loci of *Chlamydomonas* affecting different photosynthetic functions. *Proc. Natl. Acad. Sci. USA*, **76**, 1353–1357.

Shine, J. and Dalgarno, L. (1974). The 3′-terminal sequence of *Escherichia coli* 16S ribosomal RNA: complementarity to nonsense triplets and ribosome binding sites. *Proc. Natl. Acad. Sci. USA*, **71**, 1342–1346.

Sidorov, V. A., Menczel, L., Nagy, F. and Maliga, P. (1981). Chloroplast transfer in *Nicotiana* based on metabolic complementation between irradiated and iodoacetate treated protoplasts. *Planta*, **152**, 341–345.

Spreitzer, R. J. and Mets, L. J. (1980). Non-mendelian mutation affecting ribulose-1,5-bisphosphate carboxylase structure and activity. *Nature, Lond.*, **285**, 114–115.

Spreitzer, R. J. and Mets, L. (1981). Photosynthesis-deficient mutants of *Chlamydomonas reinhardii* with associated light-sensitive phenotypes. *Plant Physiol.*, **67**, 565–569.

Sprouse, H. M., Kashdan, M., Otis, L. and Dudock, B. (1981). Nucleotide sequence of a spinach chloroplast valine tRNA. *Nucl. Acids Res.*, **9**, 2543–2547.

Steege, D. A., Graves, M. C. and Spremulli, L. L. (1982). *Euglena gracilis* small subunit rRNA: Sequence and base pairing potential of the 3' terminus, cleavage by colicin E3. *J. Biol. Chem.*, **257**, 10430–10439.

Steinback, K. E. (1981). Proteins of the chloroplast. In A. Marcus (Ed), *The Biochemistry of Plants*, vol. 6. Academic Press, New York, pp. 303–319.

Steinmetz, A., Mubumbila, M., Keller, M., Burkard, G., Weil, J. H., Driesel, A. J., Crouse, E. J., Gordon, K., Bohnert, H. J. and Herrmann, R. G. (1978). Mapping of tRNA genes on the circular DNA molecule of *Spinacia oleracea* chloroplasts. In G. Akoyunoglou and J. H. Argyroudi-Akoyunoglou (Eds), *Chloroplast Development*, Elsevier/North-Holland Biomedical Press, Amsterdam, pp. 573–580.

Steinmetz, A., Mubumbila, M., Keller, M., Burkard, G., Weil, J. H., Driesel, A. J., Crouse, E. J., Gordon, K., Bohnert, H.-J. and Herrmann, R. G. (1980). Mapping of tRNA genes on the circular DNA molecule of *Spinacia oleracea* chloroplasts. In D. Söll, J. N. Abelson and P. R. Schimmel (Eds), *Transfer RNA: Biological Aspects*, Cold Spring Harbor, New York, pp. 281–286.

Steitz, J. A. (1980). RNA·RNA interactions during polypeptide chain initiation. In G. Chambliss, G. R. Craven, J. Davies, K. Davis, L. Kahan and M. Nomura (Eds), *Ribosomes: Structure, Function and Genetics*, University Park Press, Baltimore, pp. 479–495.

Stubbe, W. (1966). Die Plastiden als Erbträger. In P. Sitte (Ed), *Probleme der Biologischen Reduplication*, Springer, Berlin, pp. 273–288.

Stutz, E., Jenni, B., Knopf, U. C., Graf, L. (1978). Mapping of genes on *Eugleua gracilis* chloroplast DNA. In: G. Akoyunoqlou and J. H. Argyroudi – Akoyunoqlou (Eds.) *Chloroplast Development*. Elsevier/North-Holland Biomedical Press, Amsterdam, pp. 609–618.

Sugiura, M. and Kusuda, J. (1979). Molecular cloning of tobacco chloroplast ribosomal RNA genes. *Molec. Gen. Genet.*, **172**, 137–141.

Tabidze, V. and Beridze, T. (1979). Fine melting curves of chloroplast DNA of higher plants. *Plant Sci. Lett.*, **16**, 157–164.

Takaiwa, F. and Sugiura, M. (1980a). Nucleotide sequences of the 4.5S and 5S ribosomal RNA genes from tobacco chloroplasts. *Molec. Gen. Genet.*, **180**, 1–4.

Takaiwa, F. and Sugiura, M. (1980b). Cloning and characterization of 4.5S and 5S RNA genes in tobacco chloroplasts. *Gene*, **10**, 95–103.

Takaiwa, F. and Sugiura, M. (1981). Heterogeneity of 5S RNA species in tobacco chloroplasts. *Molec. Gen. Genet.*, **182**, 385–389.

Taniguchi, T. and Weissmann, C. (1978). Inhibition of Qβ RNA 70S ribosome initiation complex formation by an oligonucleotide complementary to the 3' terminal region of E. coli 16S ribosomal RNA. *Nature, Lond.*, **275**, 770–772.

Thomas, J. R. and Tewari, K. K. (1974). Conservation of 70S ribosomal RNA genes in the chloroplast DNAs of higher plants. *Proc. Natl. Acad. Sci. USA*, **71**, 3147–3151.

Tilney-Bassett, R. A. E. and Birky, C. W. Jr. (1981). The mechanism of the mixed inheritance of chloroplast genes in *Pelargonium*. *Theor. Appl. Genet.*, **60**, 43–53.

Timothy, D. H., Levings, C. S. III, Pring, D. R., Conde, M. F. and Kermicle, J. L. (1979). Organelle DNA variation and systematic relationships in the genus *Zea*: Teosinte. *Proc. Natl. Acad. Sci. USA*, **76**, 4220–4224.

Tohdoh, N., Shinozaki, K. and Sugiura, M. (1981). Sequence of a putative promoter region for the rRNA genes of tobacco chloroplast DNA. *Nucl. Acids Res.*, **9**, 5399–5406.

Ueda, R. and Wada, M. (1959). Structure and development of plastids in variegated leaves. *Bot. Mag. Tokyo*, **72**, 349–360.

Uzzell, T. and Spolsky, C. (1974). Mitochondria and plastids as endosymbionts: a revival of special creation? *Am. Sci.*, **62**, 334–343.

Uzzell, T. and Spolsky, C. (1981). Two data sets: alternative explanations and interpretations. *Ann. New York Acad. Sci.*, **361**, 481–499.

van Ee, J. H., Man in't Veld, W. A. and Planta, R. J. (1980a). Isolation and characterization of chloroplast DNA from the duckweed *Spirodela oligorrhiza*. *Plant Physiol.*, **66**, 572–575.

van Ee, J. H., Vos, Y. J. and Planta, R. J. (1980b). Physical map of chloroplast DNA of *Spirodela oligorrhiza*; analysis by the restriction endonucleases *Pst*I, *Xho*I and *Sac*I. *Gene*, **12**, 191–200.

VanWinkle-Swift, K. P. (1978). Uniparental inheritance is promoted by delayed division of the zygote in *Chlamydomonas*. *Nature, Lond.*, **275**, 749–751.

VanWinkle-Swift, K. P. (1980). A model for the rapid vegetative segregation of multiple chloroplast genomes in *Chlamydomonas*: Assumptions and predictions of the model. *Current Genetics*, **1**, 113–125.

VanWinkle-Swift, K. P. and Birky, C. W. Jr. (1978). The non-reciprocality of organelle gene recombination in *Chlamydomonas reinhardtii* and *Saccharomyces cerevisiae*. Some new observations and a restatement of some old problems. *Molec. Gen. Genet.*, **166**, 193–209.

Vaughn, K. C. (1981). Plastid fusion as an agent to arrest sorting out. *Curr. Genet.*, **3**, 243–245.

Vaughn, K. C., DeBonte, L. R., Wilson, K. G. and Schaffer, G. W. (1980). Organelle alteration as a mechanism for maternal inheritance. *Science*, **208**, 196–198.

Vaughn, K. C. and Wilson, K. G. (1980). A dominant plastome mutation in *Hosta*. *J. Hered.*, **71**, 203–206.

Vedel, F., Quetier, F. and Bayen, M. (1976). Specific cleavage of chloroplast DNA from higher plants by EcoR1 restriction nuclease. *Nature, Lond.*, **263**, 440–442.

Vedel, F., Quetier, F., Cauderon, Y., Dosba, F. and Doussinault, G. (1981). Studies on maternal inheritance in polyploid wheats with cytoplasmic DNAs as genetic markers. *Theor. Appl. Genet.*, **59**, 239–245.

Vedel, F., Quetier, F., Dosba, F. and Doussinault, G. (1978). Study of wheat phylogeny by EcoR1 analysis of chloroplastic and mitochondrial DNAs. *Plant Sci. Lett.*, **13**, 97–102.

Watson, J. C. and Surzycki, S. J. (1981). Identification of *Chlamydomonas reinhardi* chloroplast genes for transcription and translation using heterologous DNA hybridization probes. *J. Cell Biol.*, **91**, 278a.

Weeden, N. F. (1981). Genetic and biochemical implications of the endosymbiotic origin of the chloroplast. *J. Mol. Evol.*, **17**, 133–139.

Weil, J. H., Guillemaut, P., Burkard, G., Canaday, J., Mubumbila, M., Osorio, M. L., Keller, M., Gloeckler, R., Steinmetz, A., Keith, G., Heiser, D. and Crouse, E. J. (1980). Comparative studies on chloroplast transfer RNAs: tRNA sequences and tRNA gene localization in the rDNA units. 5th International Photosynthesis Congress, Kassandra-Halkidiki, Greece.

Weissbach, H. (1980). Soluble factors in protein synthesis. In G. Chambliss, G. R. Craven, J. Davies, K. Davis, L. Kahan and M. Nomura (Eds), *Ribosomes: Structure, Function and Genetics*, University Park Press, Baltimore, pp. 377–411.

Westhoff, P., Nelson, N., Bünemann, H. and Herrmann, R. G. (1981). Localization of genes for coupling factor subunits on the spinach plastid chromosome. *Current Genetics*, **4**, 109–120.

von Wettstein, D. (1981). Chloroplast and nucleus: Concerted interplay between genomes of different cell organelles. The Emil Heitz Lecture. In H. G. Schweiger (Ed), *International Cell Biology 1980–1981*, Springer-Verlag, Berlin, pp. 250–272.

von Wettstein, D. (1982). Rapporteur's summary: Origin and evolution of plastid proteins. In J. A. Schiff (Ed), *The Origins of the Chloroplasts*, Elsevier/North-Holland Biomedical Press, Amsterdam, pp. 263–273.

von Wettstein, D., Møller, B. L., Høyer-Hansen, G. and Simpson, D. (1982). Mutants in the analysis of the photosynthetic membrane polypeptides. In J. A. Schiff (Ed), *The Origins of the Chloroplasts*, Elsevier/North-Holland Biomedical Press, Amsterdam, pp. 243–255.

Whatley, J. M. (1981). Chloroplast evolution—ancient and modern. *Ann. New York Acad. Sci.*, **361**, 154–165.

Whitfeld, P. R., Herrmann, R. G. and Bottomley, W. (1978). Mapping of the ribosomal RNA genes on spinach chloroplast DNA. *Nucl. Acids Res.*, **5**, 1741–1751.

Wollgiehn, R. and Parthier, B. (1979). RNA synthesis in isolated chloroplasts of *Euglena gracilis*. *Plant Sci. Lett.*, **16**, 203–210.

Wurtz, E. A., Boynton, J. E. and Gillham, N. W. (1977). Perturbation of chloroplast DNA amounts and chloroplast gene transmission in *Chlamydomonas reinhardtii* by 5-fluorodeoxyuridine. *Proc. Natl. Acad. Sci. USA*, **74**, 4552–4556.

Wurtz, E. A. and Buetow, D. E. (1981). Intraspecific variation in the structural organization and redundancy of chloroplast ribosomal DNA cistrons in *Euglena gracilis*. *Curr. Genet.*, **3**, 181–187.

Wurtz, E. A., Sears, B. B., Rabert, D. K., Shepherd, H. S., Gillham, N. W. and Boynton, J. E. (1979). A specific increase in chloroplast gene mutations following growth of *Chlamydomonas* in 5-fluorodeoxyuridine. *Molec. Gen. Genet.*, **170**, 235–242.

Yasui, K. (1929). Studies on the maternal inheritance of plastid characters in *Hosta japonica* Ashers. et Graebn. f. *albomarginata* Mak. and its derivatives. *Cytologia*, **1**, 192–215.

Young, R. A., Macklis, R. and Steitz, J. A. (1979). Sequence of the 16S-23S spacer region in two ribosomal RNA operons of *Escherichia coli*. *J. Biol. Chem.*, **254**, 3264–3271.

Young, R. A. and Steitz, J. A. (1978). Complementary sequences 1700 nucleotides apart form a ribonuclease III cleavage site in *Escherichia coli* ribosomal precursor RNA. *Proc. Natl. Acad. Sci. USA*, **75**, 3593–3597.

Zurawski, G., Perrot, B., Bottomley, W. and Whitfeld, P. R. (1981). The structure of the gene for the large subunit of ribulose 1,5-bisphosphate carboxylase from spinach chloroplast DNA. *Nucl. Acids Res.*, **9**, 3251–3270.

The Evolution of Genome Size
Edited by T. Cavalier-Smith
© 1985 John Wiley & Sons Ltd

CHAPTER 12

Molecular Mechanisms of Duplication, Deletion and Transposition of DNA

PAUL DYSON and DAVID SHERRATT
Department of Genetics, University of Glasgow, Glasgow G11 5JS, UK

INTRODUCTION

The purpose of this chapter is to describe the mechanisms which can give rise to additions and deletions of DNA. We will also indicate some of the situations in which such changes have been described and their biological significance.

Though the occurrence of gross DNA rearrangements has been extensively described and discussed in the past fifty years, it has only recently been appreciated that deletions, duplications, inversions and translocations of DNA comprise a large proportion of 'spontaneous' mutations. Though some of these aberrations are extensive enough to be visualized cytogenetically, many are small enough to be detected only by fine-structure analysis.

Evolutionary success depends on the acquisition of a phenotype better suited to the environment. Traditionally, point mutations have usually been considered the major vehicle by which genotype and phenotype evolve, though the importance of gene duplication in providing the basic material for gene divergence has been frequently acknowledged. The relatively high frequencies of spontaneous gene rearrangements, the realization that gene sequence is much more highly conserved than gene organization in evolution, and the observation that eukaryote structural genes are often separated into functional domains by introns, all reinforce the importance of gene rearrangements in evolution. In discussions of genome structure and evolution, another consideration (discussed in detail elsewhere in this volume) is the idea that genomes may contain 'selfish DNA', derived from self-replicating elements such as viruses, plasmids or transposable elements. The proliferation of such DNA within genomes can have profound effects on their structure, size and activities.

The variation in size and complexity of genomes is enormous. Even closely related eukaryotes show immense differences in DNA content. Indeed, within some organisms, amplification of specific germ-line sequences gives rise to somatic cells with increased DNA content yet having reduced genetic complexity. While other chapters discuss why genomes are so different in their size and complexity, we attempt to discuss the mechanisms that give rise to such variation.

One way that can lead to an increase in genome size is to retain a newly replicated copy of an existing genome or chromosome. This can occur by uncontrolled replication or aberrant segregation and extra copies of DNA molecules derived in this way can recombine with existing molecules to generate larger, physically linked molecules. For example, a number of workers have suggested that single circular chromosomes of bacteria are derived from dimers of an ancestral monomer form (e.g. Hopwood, 1967). In eukaryotes, polyploidy and subsequent chromosome/genome evolution are well established (e.g. Ohno, 1970). But, although duplications of whole chromosomes are clearly important in evolution, increases in DNA content are often the consequence of amplification of only a small fraction of the genome. This produces repetitive sequences which can be divided into those found in heterochromatin that consist of highly repeated tandem arrays, and those that occur in euchromatin and are moderately repeated and dispersed between unique sequences. Selective amplification and maintenance of these and other types of sequence must involve both replication and recombination events.

This chapter, then, concentrates on describing how recombination and replication serve to amplify and delete specific sequences. The next section describes the types of recombination that have been recognized, while subsequent sections discuss the important contributions of unequal exchange, gene conversion and transposition in the spread and maintenance of repeat sequences. The final section gives a brief outline of the mechanisms for generating primary duplications and discusses the concept of the evolution of sequences that can be preferentially replicated in a genome.

RECOMBINATION

Genetic recombination is important to organisms because of its ability to reassort genetic markers in a Mendelian fashion and its role in the generation of genetic rearrangements. Recombination can conveniently be divided into two general types that are distinguished by the extent of DNA homology required for the recombination event. Homologous (or generalized) recombination occurs efficiently only between regions of DNA having at least several hundred base pairs of homology, whereas non-homologous (or illegitimate) recombination can occur at regions of little or no homology. These types of recombination are further distinguished by the proteins required and by their molecular mechanism. Meiotic and mitotic crossing-over, and sister-chromatid exchange is largely the consequence of homologous recombination, which is also the type of recombination measured in traditional genetic crosses. In a genome that consisted entirely of non-repeated unique sequences, homologous recombination would play no role in DNA rearrangements. However, once sequences are duplicated, homologous recombination can act to delete, amplify and rearrange these and other sequences (see later). In contrast, non-homologous recombination is responsible for many types of DNA rearrangement, and is particularly important in the generation of primary duplications of genetic material.

Homologous recombination

Most of our knowledge of the mechanism of homologous recombination is derived from the analysis of fungal and bacteriophage genetic crosses, from the characterization of *E. coli* mutants altered in recombination, and from the study of homologous recombination *in vitro* (Stahl, 1979a; Stahl, 1979b; Clark, 1973; Radding, 1982; Kahn *et al.*, 1981). Early experiments indicated that recombination can occur by an apparent breakage–reunion reaction that involves little or no DNA synthesis. Moreover, the products of recombination can sometimes be reciprocal and sometimes non-reciprocal. These types of observation led Holliday (1964) to formulate a symmetric recombination model (Fig. 12.1(a)) that can easily explain reciprocal recombination but is less satisfactory for some situations in which there is asymmetry in the recovery of recombinant classes. The Aviemore model (Fig. 12.1(b); Meselson and Radding, 1975) was derived to take account of the apparent non-reciprocal formation of heteroduplex DNA in fungal recombination. This model and subsequent derivatives view homologous recombination as an asymmetric process in which a single strand from one of the participating duplexes invades the other duplex to give a structure with a single strand cross-over. Such a structure can now be resolved to give any type of recombinant class.

The credibility of such a scheme has been substantiated by the demonstration *in vitro* that *E. coli* RecA protein, after binding to single stranded DNA, can promote the uptake of that DNA into a homologous or non-homologous duplex (Kahn *et al.*, 1981; West *et al.*, 1981). In the former case, recombinants can be isolated after introduction of the DNA complex into cells. After *rec*A-promoted uptake of single stranded DNA into non-homologous DNA is initiated, the assimilated single strand can be propagated along the duplex, searching for homology. Experiments with the fungus *Neurospora* and bacteriophage lambda have shown that the presence of specific sequences (*cog* and *chi*, respectively) in DNA can promote recombination in the region where they occur. Though the precise role of these sites is not clear it seems likely that if they function early in recombination it is by providing sites in

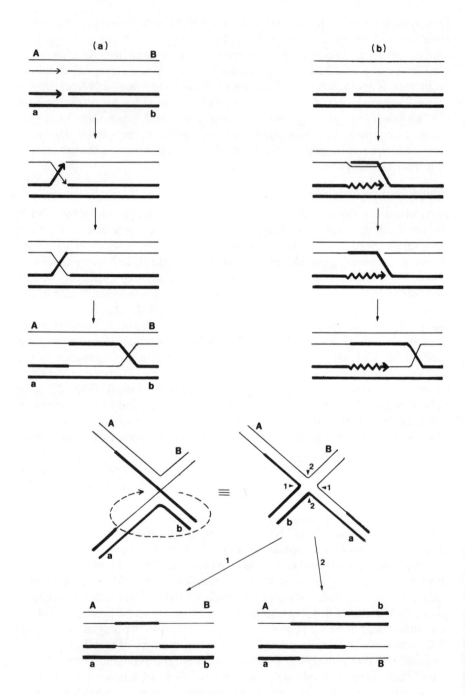

the 'donor' at which single strand aggression can be initiated by either nicking or displacing a single stranded region. If they are required late in recombination, then they could provide sites at which cross-over structures are preferentially resolved. RecBC protein (exonuclease V) of *E. coli* is also involved in homologous recombination, but its precise role *in vivo* remains obscure. Genetic experiments have shown that *rec*BC⁻ mutants of *E. coli* can form recombinant DNA, though very few viable recombinants are formed. This implies that ExoV is used late in recombination, for example, to resolve cross-over structures. However, *in vitro* experiments suggest a possible early role for ExoV in recombination. In the presence of ATP, ExoV can progressively displace a single strand from an end of linear duplex DNA (Taylor and Smith, 1980). The enzyme appears to assimilate a given single strand of a duplex and then at a slower rate release the unwound single strand. Such a process generates single-strand 'tails' and 'loops' which grow in size as the enzyme progresses. After these single strands bind RecA protein, they are potential 'donors' in a homologous recombination reaction. The enzymological properties of ExoV are also consistent with it being able to resolve cross-over recombination intermediates into recombination products.

Non-homologous recombination

Two types of non-homologous recombination have been extensively studied. In both, genes specifying proteins required for the recombination are closely linked to at least one of the recombination sites. Genetic transposition, described in detail later in the chapter, is a replicative process, whereas simple breakage and reunion of DNA strands occurs in site-specific recombination of the type exemplified by bacteriophage lambda integration into and excision from the bacterial chromosome. Such recombination differs from homologous recombination by its obligate reciprocity, apparent symmetry and conceptual simplicity.

Fig. 12.1. Models for recombination. (a) Depicts the Holliday model (Holliday, 1964) which is initiated by synapsis and nicking at the same position on each duplex. Breakage–reunion then occurs to form a 'cross-over'. Branch migration of such a structure will generate hybrid DNA in the region through which the cross-over passes. Isomerization of such molecules (Sigal and Alberts, 1972) gives the structures indicated. Cutting of the 'eye' form at either positions 1 or 2, and subsequent ligation, gives molecules that are non-recombinant or recombinant, respectively, for markers A and B. In both cases the region traversed by branch migration is hybrid. If this contains genetic markers, each duplex will be heteroduplex for the markers. Such heteroduplexes can be corrected (observed as gene conversion), or replicated (observed as post-meiotic segregation). In (b) the Aviemore model (Meselson and Radding, 1975), recombination is initiated asymmetrically and can be driven by replication in the donor. Once a cross-over is formed, isomerization and resolution can occur as in the Holliday model.

Campbell (1962) first postulated that bacteriophage lambda integration into the *E. coli* chromosome was by a reciprocal breakage and reunion of DNA strands at specific sites. Subsequently, a wealth of biochemical and genetic data has elaborated and substantiated his model (Campbell, 1981; Landy and Ross, 1977; Nash *et al.*, 1981). More recently several other site-specific recombination systems, having very different functions, have been described (e.g. see Sherratt *et al.*, 1981). These, like the lambda systems, require a site-specific recombination enzyme, whose gene is adjacent to at least one of the specific recombination sites which vary in size from 14 to several hundred base pairs. In two of the systems, the recombination enzyme appears to make staggered breaks within each of the recombination sites (Mizuuchi *et al.*, 1981; Reed, 1981; Reed and Grindley, 1981), providing in outline a mechanism for the generation of recombinant DNA. The detailed molecular mechanism for such breakage reunion reactions is not known, but two types of mechanism have been suggested. Studying the lambda system, Nash *et al.* (1981) have suggested that the two duplexes in the region of the recombination site come together to form a four-stranded structure stabilized by intermolecular hydrogen bonds, a structure originally proposed by McGavin (1971). Nicking at the same position on each duplex, strand rotation, ligation, followed by nicking, rotations and ligation at the other staggered position on each duplex will generate recombinant DNA. Another type of model, based on that proposed for the action of DNA gyrase and other type II topoisomerases (Gellert, 1981), invokes sequential double-strand cleavage of the two participating duplexes and protein, rather than a four-strand DNA helix, in synapsis and maintenance of the integrity of the recombination intermediates (Fig. 12.2; Sherratt *et al.*, 1981).

The site-specific recombination systems that have been described can in principle act intermolecularly, to join two topologically separate DNA molecules, and intramolecularly, to either invert a DNA sequence (recombination sites inverted) or to delete a sequence (recombination sites directly repeated). In the latter case the 'deleted' sequence may be capable of autonomous existence. In practice, given systems usually show preferred reaction specificities and directions.

Other types of non-homologous or site specific recombination must surely exist. In *E. coli* spontaneous deletions in the *lac*I gene often appear to be derived by recombination across small (5–10 bp) regions of homology (Farabaugh *et al.*, 1978). Similarly the 5 or 9 bp direct repeats bounding transposable elements appear to be substrates for recombination that results in precise excision of the element. Such recombination is *rec*A-independent, does not normally require functional transposable element proteins, but at least in some cases, is dependent on the *him*A protein of *E. coli*. This type of recombination can be distinguished from lambda site-specific recombination, by the apparent non-specificity of the sequences involved (though

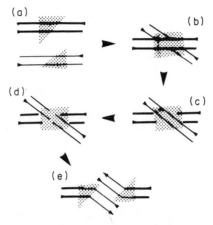

Fig. 12.2. A model for site-specific recombination between double-stranded DNA molecules. This model (Sherratt *et al.*, 1981) was prompted by *in vitro* studies of site-specific recombination mediated by transposable elements (Reed, 1981; Reed and Grindley, 1981). (a) Two regions of DNA each with recombination enzyme bound to the recombination site, (b) protein interactions bring the recombination sites together and now (c) double-strand cleavage of one of the helices occurs and the intact strand is passed between the broken ends. Strand separation and rotation are prevented by bound enzyme. During strand passage a second cleavage occurs (d), and is immediately followed by ligation (e) to produce recombined DNA.

the recombination substrates need to have homology), the low frequency of the reaction, and the absence of a nearby gene for the recombination enzyme. But it should be remembered that lambda site-specific recombination can occur at low frequency between the normal lambda recombination site and secondary bacterial recombination sites that have only partial homology with the normal recombination site.

In conclusion, we feel it likely that most, and possibly all, non-homologous recombination requires some homology at the recombination sites. The nature of the homology, and availability of appropriate recombination enzymes will determine the efficiency of the recombination reaction. Non-homologous recombination of one sort or another is a major mechanism for generating DNA rearrangement.

UNEQUAL EXCHANGE

The consequences of a single cross-over event during recombination of two molecules are outlined in Fig. 12.3. The mechanism to generate the products is envisaged primarily to be homologous recombination, but an efficient site-specific recombination system could also operate, given the right substrate. At its simplest, unequal exchange leads to reciprocal addition and deletion

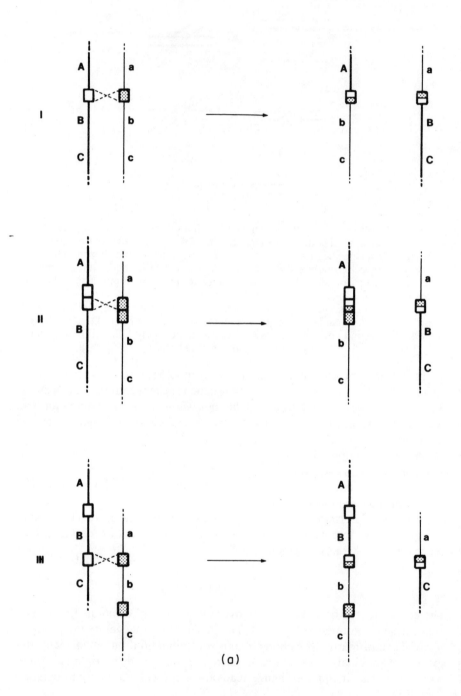

(a)

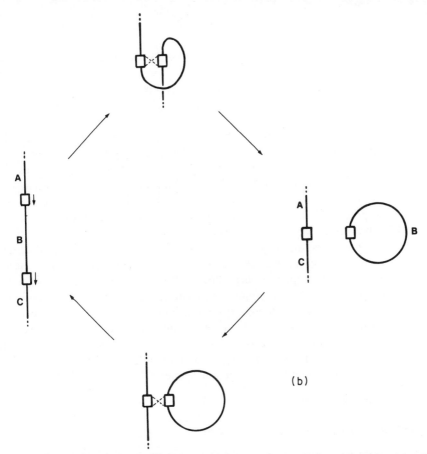

Fig. 12.3. (a) The consequences of a single cross-over during recombination of two DNA molecules. Boxes represent sequences which are efficient substrates for recombination enzymes. I: each participating molecule contains one box: cross-over results in equal exchange between the two molecules. II: each molecule contains two boxes in tandem repeat: this permits unequal exchange between two misaligned molecules resulting in reciprocal addition and deletion in the recombinant molecules of one box. III: each molecule contains two boxes surrounding a unique, non-recombinogenic sequence: unequal exchange results in the reciprocal addition and deletion in the recombinant molecules of one box and the unique sequence (b). If (b) were an essential gene, the chromosome for which it was deleted could not survive in the haploid or homozygous state. A unit of unequal exchange can be envisaged simply as a recombinogenic sequence, as in II, or a recombinogenic sequence plus an additional sequence, as in III. An integral number of units may be exchanged between molecules containing tandem arrays of several units.

(b) Intramolecular recombination and integration of a circular molecule. A single DNA molecule containing directly repeated sequences is a potential substrate for deletion by intramolecular recombination. Recombination across two boxes deletes the intervening sequence, B. A number of fates of the recombinant circular molecule can be envisaged. (1) As illustrated, recombination may reintegrate the circle; integration may occur at dispersed genomic sites if the required homologous substrates for recombination are present at these dispersed sites. (2) An origin of replication contained within the deleted sequence may allow amplification of the circular molecule. A high copy-number, independently replicating circle, besides increasing dosage per cell of sequence B, may increase the potential for integration at dispersed sites. (3) a single-copy deleted circular molecule may simply be diluted out and lost from a population of dividing cells.

in the two participating molecules, and the prerequisite for such exchange is that more than one copy of a recombinogenic sequence is present per molecule.

At this point other factors which may influence unequal exchange frequencies should be considered. An exchanging unit may be an entirely non-coding recombinogenic sequence, or may also include coding sequences. In the latter example, be it via intragenic recombination or recombination across flanking non-coding sequences, unequal exchange can double gene dosage in one segregant, but delete those genes in the other, thereby placing considerable selection pressure on such events.

In the light of this, and given that the enzymology of generalized recombination favours increased frequencies of exchange between longer homologous tracts, it is perhaps not surprising that a typical arrangement of units which may be exchanged unequally is in long tandem arrays. This configuration of repeats allows exchange of several units per cross-over without loss of unique information in the deleted segregant. The result of unequal exchange in tandem arrays is that in a population of DNA molecules a binomial distribution of the numbers of repeated units would be observed, the mean of which can be affected by selection, genetic drift and cellular control mechanisms.

Unequal exchanges between rRNA genes

The predominant products of a cell's transcription are ribosome components, and generally ribosomal genes are present in many tandemly repeated copies in a cell's genome, providing ample opportunities for unequal exchange, giving chromosomes with much reduced and increased doses of these genes.

Prokaryote genome organization is reasonably streamlined, with 'housekeeping' genes largely represented uniquely on a single chromosome. An exception is that rRNA genes may be present in several copies per chromosome, thereby providing a substrate for unequal exchange. Experiments designed to detect unequal exchange in *Salmonella typhimurium* show that such exchanges are much more frequent at rDNA loci than at 37 other loci (Anderson and Roth, 1981). Highest frequencies are observed in the region containing the directly repeated and adjacent A, B and E loci (Fig. 12.4); the arrangement of these loci shows similarities to a repetitive tandem organization. The maximum effect is produced when cells are grown on a rich medium which induces a short generation time. Unequal exchange frequencies are 4 to 10 times lower in cells growing on poor media, implying that rDNA gene dosage may limit growth rate so that cells containing amplified rDNA will have a selective advantage during growth on rich media. It is most likely that unequal exchanges are between daughter molecules following replication, and a short generation time may also increase the number of

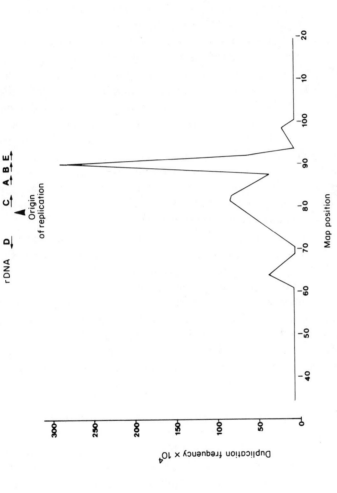

Fig. 12.4. The frequency of tandem duplications at points on the *Salmonella* chromosome with cells grown on rich media. Positions and orientations of five rDNA loci and the origin of replication are indicated. The peak duplication is carried by 3% of the recipient population and occurs between close, directly repeated rDNA cistrons.

replication forks initiated at the origin. The proximity of the origin of replication to the C, A, B and E loci may therefore also contribute to high localized exchange frequencies. Moreover, the adjacent map positions of participating repeat sequences involved in a cross-over event will tend to minimize the degree of misalignment of the two DNA molecules prior to unequal exchange.

Any advantage accruing to amplified dosage is negated if environmental constraints are such that 'excess' DNA constitutes a burden to the cell. Hence a non-reciprocal deletion mechanism involving intramolecular recombination (Fig. 12.3(b)) is important to reduce the length of tandem arrays. Duplications of rDNA, similar to those in *Salmonella*, occur in *E. coli*, and covalently closed circular DNA molecules, which are the presumptive deletion products of intramolecular recombination, have been detected in the latter organism (Hill *et al.*, 1977). If such circles contained an origin of replication they could be autonomously amplified. Moreover, re-integration into homologous chromosomal sites could occur irrespective of autonomous replication. Although amplification of rDNA circles in bacteria has not been observed, a circle derived from intramolecular recombination between the B and E rDNA loci of *E. coli* has been observed to reintegrate at four other dispersed homologous loci (Hill *et al.*, 1982).

While the genome of prokaryote cells usually consists of a single circular chromosome, the nature of unequal exchange in the more complex eukaryote cell appears similar and is a major mechanism of DNA duplication and deletion. A description of unequal exchange of rDNA units in yeast during both mitotic and meiotic division serves to illustrate the major features of such recombination. The rDNA of the yeast, *Saccharomyces cerevisiae*, consists of approximately 140 copies of a 9 kb repeat unit (Schweizer *et al.*, 1969). Each repeat unit codes for a 5S RNA, and for a 35S RNA which is subsequently degraded to mature 25S, 18S and 5.8S ribosomal RNAs (Cramer *et al.*, 1977). Restriction endonuclease analysis has shown that the repeat units are homogeneous and tandemly arranged.

Homogeneity of repeat units is common in eukaryotes and various mechanisms have been proposed to maintain it: master-slave correction (Callan, 1967), saltatory replication (Buongiono-Nardelli *et al.*, 1972), gene conversion and unequal exchange. For yeast rDNA there is direct evidence to oppose the master-slave and saltatory replication hypotheses (Brewer *et al.*, 1980), and the experiments of Szostak and Wu (1980) and Petes (Petes and Botstein, 1977; Petes, 1980) suggests that unequal exchange could serve to maintain rDNA unit homogeneity. These experiments studied the segregation of a LEU2 gene marker, inserted by recombinant DNA techniques into the rDNA. Two types of rDNA, distinguishable because 'form I' has 7 EcoRI restriction sites per unit and 'form II' has only 6, provide an assay to measure exchange between sister and non-sister molecules in a diploid cell heterozygous for the two types, as shown in Fig. 12.5.

Unequal exchange during mitotic growth of haploid strains causes deletions and duplications of 6 to 8 rDNA units, at a frequency of 5×10^{-4} per mitotic division (Szostak and Wu, 1980). The overall frequency of mitotic unequal exchange between sister molecules in the rDNA cluster is about 10^{-2} per generation. Unequal exchange between non-sister molecules in mitotic growth of diploids has not been detected for yeast rDNA, suggesting that it is not a significant mechanism to account for reciprocal deletions/duplications and maintenance of unit homogeneity.

Segregants of a meiotic division in yeast can be monitored by tetrad analysis. As observed for mitosis, meiotic unequal exchange proceeds via sister-molecule exchange, and is detected in approximately 10% of tetrads (Petes, 1980). Recombination between non-sister molecules in the rDNA region is reduced 100-fold compared with elsewhere in the same chromosome, indicating suppression of recombination in that region (Petes, 1979; Petes, 1980). So differing enzymatic pathways may mediate sister-molecule exchanges and meiotic generalized recombination; the former may occur during replication, prior to homologous pairing of chromosomes. This distinction may become clearer when mutants specific to each pathway become characterized.

Yeast rDNA circles resulting from intramolecular recombination have been isolated (Meyerink *et al.*, 1979; Clark-Walker and Azad, 1980). It is estimated that 4 to 5 copies per cell of a 2.8 μm DNA circular molecule are present in *S. cerevisiae*, each circle being equivalent to one rDNA unit. This is indicative of a deletion mechanism but it should be noted that these molecules are extracted from extra-nuclear cytoplasm and are not essential for cell viability. Although it has not been observed, reintegration of such circles may occur.

Several features of the non-transcribed spacers (NTS) which flank rDNA genes suggest that they are recombinogenic, in common with other spacers (Fedoroff, 1979). In general an NTS contains an integral number of short, tandemly repeated sequences similar to highly repetitive DNA (see p. 361). This number may vary from unit to unit, although the number of internal 15 bp repeats of the yeast rDNA spacer is invariable. Moreover, the yeast rDNA NTS contains AT spacers and GC clusters similar to the recombinogenic sequences of the yeast mitochondrial genome (see later), providing a potential substrate for the recombination enzymes promoting sister-molecule exchange.

Xenopus laevis rDNA NTS sequences contain internal repeat sequences of different length which are GC-rich (Bosely *et al.*, 1979). Lengths of these NTS sequences may vary from approximately 2 to 9 kb, probably arising from misalignment of internal NTS sequences during exchange, and such variation provides a characteristic to monitor rDNA amplification in the

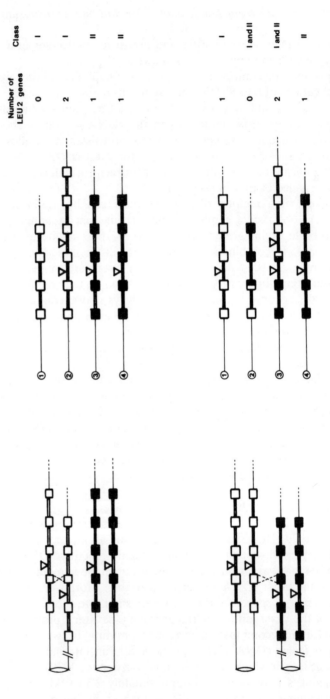

Fig. 12.5. Schematic representation of unequal exchanges of yeast rDNA units resulting from cross-over between sister molecules (top panel), and cross-over between non-sister molecules (bottom panel). The segregation of the LEU 2 insertion (▽) and class I (□)/class II (□) rDNA gives discernibly distinct products from the two cross-overs, as tabulated.

oocyte. Amplification of rDNA is a widespread though not universal feature of oogenesis (Tobler, 1975), providing sufficient DNA templates for the accumulation of ribosomes in oocytes.

The rDNA fraction of *X. laevis* oocytes contains up to 9% of the molecules as circles, varying in size from 12 to 210 kb, equivalent to 1 to 20 rDNA units. Comparison of NTS lengths of amplified units and chromosomal units suggests that amplification starts with the excision of a single unit, possibly by intramolecular cross-over (Wellauer *et al.*, 1976). Subsequent amplification proceeds by rolling circle replication initiated from an origin of replication within the excised circle (Rochaix *et al.*, 1974). Unequivocal evidence for a single excision event before amplification is absent. The issue is complicated as the amplified rDNA is isolated from thousands of oocytes; classes of different length units of total oocyte rDNA could represent the selection of different units for amplification in different oocytes, or the presence of more than one class per oocyte. However, amplified rDNA from a *X. laevis* and *X. borealis* hybrid are always of *X. laevis* chromosome origin (Brown and Blackler, 1972), indicative of a selection process.

Restriction patterns of *X. laevis* rDNA have been used to study its inheritance (Reeder *et al.*, 1976). Forty-eight of 50 matings indicated simple Mendelian inheritance of spacer restriction patterns. The remaining two patterns could be explained by integration of circular rDNA units, though the evidence is not conclusive. Generally, amplified rDNA is a feature of the cell phenotype and not inherited.

Highly-repetitive DNA

Unequal exchange can reciprocally amplify and delete rDNA tandem repeats at high frequency and intramolecular recombination can delete 'excess' units or provide the basis of temporary gross amplification. Moreover, the nature of unequal exchange can contribute to the species-specific homogeneity of the unit sequence. It is easy to conceive of a strong selection pressure for maintenance of coding sequence homogeneity, and this is reflected in the evolutionary conservation of, for example, rDNA genes. But the homogeneity of NTS sequences, which exhibit evolutionary divergence, is not a result of this selection. The key to the short-term maintenance of such homogeneity is that in unequal sister-molecule exchange the adjacent duplicated regions will be identical since they are the sister products of the immediately preceding DNA replication. So the process can maintain a tandemly homogeneous repeated region so long as the exchange rate is several orders greater than the mutation rate. With selective pressure reduced or absent, genetic drift can cause these sequences to diverge.

The NTS sequences strongly resemble highly-repetitive heterochromatic DNA. This DNA is composed of short nucleotide sequences tandemly repeated in long arrays, and exhibits sequence variation from species to species in a similar way to NTS sequences. Similarly, it is envisaged that there is little or no selective force at the nucleotide sequence level, although it is by no means impossible to ascribe function to such DNA. While most of the molecular analyses are consistent with unequal exchange mechanisms for the evolution and maintenance of these tandem arrays, high rates of sister-molecule exchanges in a heterochromatic region have yet to be observed. Nevertheless, highly repetitive tandem sequences may well be recombinogenic for the enzymes of unequal exchange. But while this is a feature of non-coding tandem sequences, intragenic unequal cross-over can create rearrangements of coding sequences. This is a likely origin of human Lepore and anti-Lepore haemoglobins. Generally, though, where it may be of advantage to use unequal exchange to alter gene dosages, it is important that such exchanges do not disrupt coding sequence organization. Increased numbers of repeat sequences in genomes may act as substrates for unequal exchange, yet avoid deleterious disruptions of unique information.

Site-specific deletions and subsequent amplification

During our discussion of recombination types, we distinguished between homologous and site-specific recombination. In practice, it is likely that intermediate types of recombination events occur, involving differing requirements for length of DNA homology and different enzymes. For example, analysis of selected *E. coli* K12 mutants carrying multiple copies of the *amp*C gene in tandem array indicate that flanking sequences of just 12 bp provide a substrate for recombination (Edlund and Normark, 1981). Similarly, deletion events in *E. coli* may involve intramolecular recombination across short homologous sequences as discussed previously.

Site-specific recombination events are less well characterized in eukaryotes. However, two systems provide evidence that deletions can occur between short, specific sequences: spontaneous 'petite' mutants of *S. cerevisiae* and macronuclear DNA amplification of ciliates. Spontaneous cytoplasmic 'petite' mutants of yeast have mitochondrial genomes consisting of tandem repeats of a unit which has been excised from the circular, parental, wild-type genome. A large number of different, though not necessarily exclusive, units varying from 400 to 4000 bp in length can be excised to form the basic repeat unit of a 'petite' genome, and there need be no coding information in the basic unit.

Analysis of a particular spontaneous 'petite' genome indicates that the sequences involved in the primary recombination event are 9 bp direct repeats, generating a 416 bp unit (Gaillard *et al.*, 1980). Generally, repeats of up to 23 bp found in recombinogenic AT-spacer and GC-cluster regions are substrates for excision (Bernardi and Bernardi, 1980). The excised unit contains an origin of replication (de Zamaroczy *et al.*, 1981), allowing subsequent amplification and formation of the 'petite' genome, which is of approximately the same size as the original parental wild-type genome.

The amount of amplification of an excised unit appears to be governed purely on the basis of size; a possible example of nucleotypic selection of non-coding sequences. The cell cycle could be attuned to a temporally regulated DNA replication stage, the length of which is at least partly correlated to the amount of DNA to replicate. This would enable amplification to obtain wild-type genomic size, and because the petite genome bears multiple origins of replication (one per unit) it can replicate more efficiently and hence displace the wild-type genome. This is observed in the diploid progeny of crosses where the 'petite' genome 'suppresses' the wild-type.

Macronuclear DNA in Ciliated Protozoa (Ciliophora = Heterokaryota)

The ciliated protozoa contain two differentiated nuclei: a germinal, diploid micronucleus and a somatic, polyploid macronucleus. Transcription in vegetatively growing cells takes place in the macronucleus; the micronuclear genome is not expressed during the vegetative life of the cell. However, the macronucleus is formed *de novo* from the diploid micronucleus after sexual conjugation: a process involving genome reorganization and amplification.

Macronuclear formation in the order Hypotrichida involves the polytenization of micronuclear chromosomes, subsequent fragmentation of polytene chromosomes and degradation of approximately 95% of the DNA. The remains are small linear molecules, which are extensively replicated to form the mature macronuclear genome. For example, the rDNA is present as discrete linear units of up to a thousand copies per macronucleus, derived from a single unit present in the micronuclear genome.

Each macronuclear linear molecule is flanked by a conserved inverted repeat sequence and a 3' single-stranded tail region of guanosines and ribosylthymines (Klobutcher *et al.*, 1981):

$$5'\text{-}C_4\ A_4\ C_4\ A_4\ C_4\text{---}//\text{---}G_4\ T_4\ G_4\ T_4\ G_4\ T_4\ G_4\ T_4\ G_4\text{-}3'$$
$$3'\text{-}G_4\ T_4\ G_4\ T_4\ G_4\ T_4\ G_4\ T_4\ G_4\text{---}//\text{---}C_4\ A_4\ C_4\ A_4\ C_4\text{-}5'$$

Fragmentation of the polytene chromosome depends on these inverted repeats which are a substrate for looping out and subsequent excision of redundant DNA sequences. Electron microscopy shows that in early macronuclear development, differential chromatinization determines which DNA sequences are excised and eliminated (Meyer and Lipps, 1981). In the course of polytenization 30 nm chromatin fibres become organized in the loop-like structures that are distinct from 12 nm chromatin fibres associated with retained sequences. Excision generates chromatin rings: the DNA in these rings is linear, the structure presumably being linked by a protein involved in excision.

The holotrichous ciliates, including *Tetrahymena* and *Paramecium*, also exhibit site-specific fragmentation. In macronuclear development unpolytenized micronuclear DNA is fragmented at repeats of the hexanucleotide sequence, 5'-CCCCAA-3', generating intermediate linear molecules each containing one unit (Pan and Blackburn, 1981). Subsequent amplification of retained sequences involves about 90% of the micronuclear genome. The amplified linear molecules consist of a pair of units in an inverted repeat arrangement (Fig. 12.6).

Replication of linear molecules poses problems not apparent for circular molecules because all known DNA polymerases are only able to extend existing polynucleotide chains and not start new ones. So various models have been proposed to account for synthesis of 5' ends of linear molecules (Cavalier-Smith, 1974). The single-strand 3w terminal ends of hypotrich macronuclear molecules could enable the binding of RNA primers, or a DNA fragment separated by a nick (Klobutcher *et al.*, 1981). The palindromic arrangement of holotrich macronuclear molecules suggest that terminal cross-linking in the intermediate, single-unit molecule could occur, as proposed for replication of *Paramecium* mitochondrial DNA (Goddard and Cummings, 1977; Fig. 12.7).

Although site-specific DNA-protein interactions may cause genome reorganization in eukaryotes, controlled DNA fragmentation appears to be confined to ciliates, slime moulds, and a very few animals (Chapter 14).

GENE CONVERSION

We have seen how unequal exchange can operate efficiently on families of repetitive sequences organized in tandem arrays to add or delete units. However, a proportion of the genome is composed of dispersed repeat sequences found in mainly unique regions, and these dispersed repeats, in common with tandem repeat families, exhibit sequence homogeneity. From the arguments stated before, it is clear that if unequal exchange were to operate on these sequences, gross deletions of unique information would result. So while genetic transposition may account for the spread of some

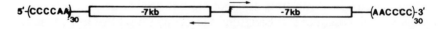

Fig. 12.6.   Structure of a *Tetrahymena* rDNA palindrome. Each molecule consists of a pair of units in inverted repeat. rDNA coding sequences are represented as boxes, flanked by non-coding sequences including the terminal thirty-fold hexanucleotide repeat sequence. About 300 rDNA palindromes are present in the mature macronucleus.

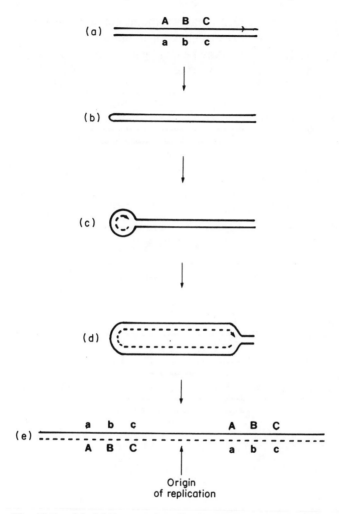

Fig. 12.7.   (a)–(e) Proposed replication pathway to generate palindromic dimers. Terminal cross-linking (b) occurs, allowing replication at the single unit stage to form a palindrome. The site of the origin of replication, as indicated in (e), supports this model; replication of the macronuclear palindrome begins at the centre of the molecule and proceeds as a bidirectional fork to the ends (Truett and Gall, 1977).

dispersed repeats, correction mechanisms such as gene conversion may also contribute to maintenance of homogeneity.

Gene conversion is a term describing the replacement of a sequence at one site in the genome by a related sequence situated elsewhere, with no concurrent change in the 'donor' site information. Evidence for a mechanism to mediate conversion suggests that recombination is intrinsic to the process, giving rise to heteroduplexes: regions of double-stranded DNA containing mispaired bases (Fig. 12.8). Any heterozygosity within the hybrid DNA will segregate with the separation of strands at the next replication; or it may be

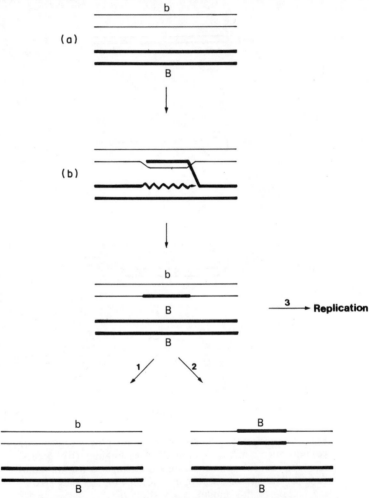

Fig. 12.8.   Schematic representation of proposed events leading to gene conversion. (a) Strand invasion occurs similarly to the Avie-more model for generalized recombination. (b) Nicking and ligation joins strands of the same molecule to the hybrid region, which now has three possible fates. Pathway 1 repairs the heteroduplex to generate the original organization. Pathway 2 repairs the heterod-uplex in favour of the invasive sequence, and in pathway 3 the uncorrected heteroduplex segregates at the next replication. Both pathways 2 and 3 convert sequence b to sequence B.

removed by repair enzymes that correct mispaired bases. Correction may be biased in one direction or the other for any particular mispair, so that heterozygosity is removed predominantly in favour of one parental type, resulting in a deviation from 1:1 allelic ratios.

In eukaryotes, gene conversion can happen during both mitosis and meiosis, and up to half the conversion events in meiosis are associated with positive chiasma interference and reciprocal recombination of flanking markers. This is indicative of conversion accompanying recombination between homologous chromosomes. As discussed before, heteroduplex formation is intrinsic to all models of generalized recombination (Fig. 12.1). However, other meiotic conversion events are not accompanied by reciprocal recombination of flanking markers and exhibit negative chiasma interference. Although the former observation can be explained by interpretation of recombination models (as demonstrated in Fig. 12.1), the latter cannot. For want of better explanations, a mechanism based on initiation, but incompletion of recombination, is proposed to account for the initial formation of heteroduplexes in these other conversion events. This 'abortive' recombination we have termed localized strand invasion (LSI). It can be visualized as occurring by a series of events in which recombination is initiated by formation of a heteroduplex between regions containing related DNA sequences, followed by rejoining of original strands on either side of the heteroduplex (Fig. 12.8), instead of strands of opposite molecules (or sites in intramolecular recombination) as seen in reciprocal recombination.

Events of the LSI category have been studied in yeast during both mitosis (Scherer and Davis, 1980; Jackson and Fink, 1981) and meiosis. Yeast cells are transformed to insert new sequence arrangements into a variety of chromosomal locations by homologous recombination. Gene conversion can be detected if these newly inserted sequences recombine with similar sequences at different locations in the genome in a non-reciprocal manner. The results indicate that the potential for gene conversion is enormous: the postulated events are schematically diagrammed (Fig. 12.9).

The length of sequence converted can vary from a single base pair up to maybe more than 2 kb (Scherer and Davis, 1980), and can encompass both additions and deletions. But there are obvious constraints on both the length and frequency at which a sequence can be converted. For example, the length and nature of flanking homology may contribute to both frequency and length of conversion. Scherer and Davis report that a 350 bp flanking homology may suffice to convert up to 2 kb. The distance between 'donor' and 'recipient' sites may also constrain conversion frequencies. So these constraints may be reflected in the reported frequencies. Mitotic intrachromosomal conversion between duplicated regions may occur at frequencies of 10^{-4} per cell per generation and is at least an order of magnitude greater than mitotic interchromosomal conversion between single copies on homologous

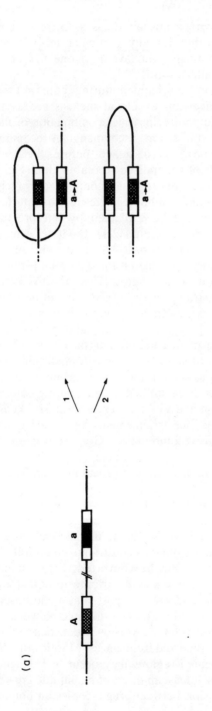

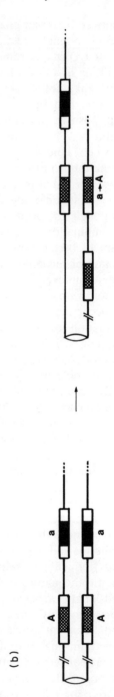

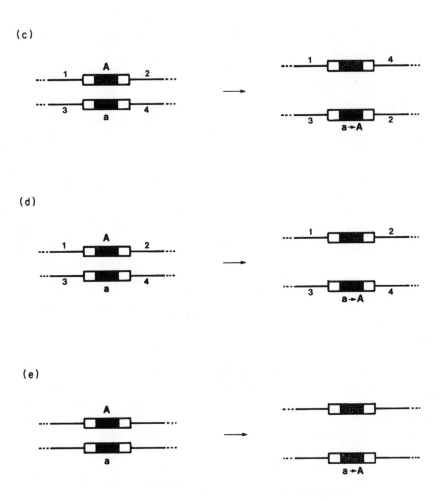

Fig. 12.9. Schematic representation of postulated conversion events. Boxes represent participating sequences; the shaded, A, and black, a, regions are non-homologous, and the white flanking regions are homologous substrates for the enzymes involved. (a) Intramolecular conversion involving (1) directly repeated boxes and (2) repeats in inverted orientation. (b) Intermolecular conversion involving misaligned sister molecules. (c) Intermolecular conversion involving homologous non-sister molecules at meiosis. Conversion is causally related to chiasma formation giving recombination of flanking markers and positive chiama interference. (d) Intermolecular conversion involving homologous non-sister molecules. This conversion is of the LSI type with no accompanying recombination of flanking markers, and negative chiasma interference is exhibited. (e) Intermolecular conversion of the LSI type involving non-homologous molecules.

chromosomes (Jackson and Fink, 1981). Mitotic interchromosomal conversion between non-homologues is reported at 10^{-7} per cell per generation (Scherer and Davis, 1980). Meiotic conversion resulting from LSI or reciprocal recombination have frequencies which are appreciably higher, varying between heterozygous sites from 0.3% to 20% of unselected tetrads (Fogel and Mortimer, 1971).

Genetic regulation, as for any recombination-like event, controls the frequency and direction of conversion. Mating type interconversion in yeast exemplifies this type of control. Each haploid yeast strain contains two unexpressed copies of mating-type alleles at two loci, HML and HMR, that are located away from the MAT locus, where mating-type alleles are expressed. The mating-type of a cell can be changed from MATa to MATalpha or vice versa by conversion of the MAT allele by a copy of one of the 'library' genes (Fig. 12.10). The efficiency of mating-type conversion depends on the *HO* gene, unlinked to MAT or to the silent HML and HMR loci. Heterothallic strains having the recessive *ho* allele switch mating-type rarely, at a frequency of about 10^{-6} per cell per generation (Hicks and Herskowitz, 1977). In contrast, the dominant homothallism allele *HO* promotes very frequent exchanges of mating-type: as often as every cell division (Strathern and Herskowitz, 1979). In a homothallic strain the progeny from a single haploid cell switch mating-type to produce an equal number of MATa and MATalpha cells that can conjugate to form a colony of non-mating MATa/MATalpha diploids. Once a cell expresses both a and alpha information, homothallic mating-type switching ceases (Haber and George, 1979).

The asymmetry of the mating-type interconversions is dependent on the MAT sequence, which normally cannot act as 'donor' to either HML or HMR. Cis-acting mutations within or adjacent to MAT permit the MAT allele to convert HML or HMR (Haber *et al.*, 1980a). This suggests conversion is dependent on an initiation event at the MAT sequence. A prediction from the arrangement of mating type alleles is that a deletion event between two direct repeats might occur. Such deletions occur and at a higher frequency in homothallic strains (1%) than in heterothallic strains, suggesting regulation by *HO* (Haber *et al.*, 1980b). So the mechanism of gene conversion could well share common features with reciprocal recombination, involving interaction of unlinked gene product(s) and a specific DNA sequence.

Convergence versus divergence

Gene conversion acts at significant frequency as a correction mechanism that operates at least in yeast and other fungi. It is conservative in that information at the 'donor' site is retained and there is no appreciable net gain in DNA content in the genome following conversion. The result is the replacement of sequence information at one site by that of another, and this may

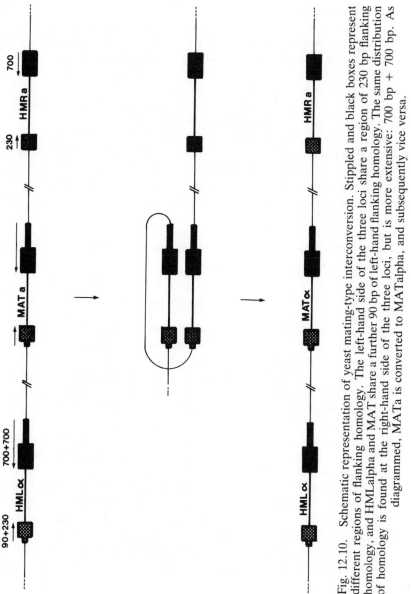

Fig. 12.10. Schematic representation of yeast mating-type interconversion. Stippled and black boxes represent different regions of flanking homology. The left-hand side of the three loci share a region of 230 bp flanking homology, and HMLalpha and MAT share a further 90 bp of left-hand flanking homology. The same distribution of homology is found at the right-hand side of the three loci, but is more extensive: 700 bp + 700 bp. As diagrammed, MATa is converted to MATalpha, and subsequently vice versa.

contribute not only to the maintenance of repeat sequence homogeneity but also to the spread of new or minority repeat varieties. The apparent anomaly of conversion creating either convergence or divergence of repeat sequences in a genome is dependent on how 'aggressive' a potential donor site is, and on the bias in heteroduplex correction. On a short-term basis a directed gene conversion mechanism is likely to contribute to convergence of related sequences, as in the absence of such a mechanism uncorrected mutations can accumulate in the genome. But the potential intrinsic randomness in the choices offered by conversion allows genetic drift acting on populations in the long term to considerably diverge related sequences, especially if the selection pressure on a sequence is reduced or absent.

At the outset our attention was focused on a mechanism to maintain dispersed repeat sequences where obviously unequal exchange is an inviable proposition. Of course correction events will also operate on tandem repeat regions, and conversion and unequal exchange are two mechanisms dependent on generalized (or related) recombination to account for the characteristics of this DNA. But certain dispersed repeat sequences have obviated the requirement for generalized recombination for their maintenance by encoding their own recombination functions: these sequences are known as transposable elements.

GENETIC TRANSPOSITION

Classically, the stability of genomes was deduced both from the relatively non-mutable phenotype of cells and organisms and from genetic analysis that demonstrated the fixed positions of genetic loci within a genome. In essence this deduction still holds water, though it was a mutable phenotype of maize that gave Barbara McClintock in the 1940s the evidence to propose the phenomenon of 'movable elements'. Experiments in the 1960s with *E. coli* gave physical credence to support the idea of specific movable (or transposable) elements (Saedler and Starlinger, 1967; Jordan *et al.*, 1968; Shapiro, 1969) and suggested that such elements were an important cause of spontaneous mutations and genomic rearrangements.

A useful working definition for a transposable element is a specific DNA sequence that can integrate at many loci in a genome. Such integration occurs in the absence of significant DNA homology between the target and the element and, at least in *E. coli*, does not require a functional homologous recombination system. So far such elements have been detected in both gram-negative and gram-positive eubacteria, archaebacteria, blue-green algae and several eukaryotes.

It is the eubacterial transposable elements that have provided the insight to propose the concept that transposition is a replicative process and that transposable elements are replicons. This important concept has provided fuel for much evolutionary debate, as expounded in other chapters of this book, since transposition provides the means for these elements to 'over-replicate' in the host, so that they can be maintained by non-phenotypic selection. The purpose of this section is to describe both the mechanism and biological consequences of transposition.

Prokaryote transposable elements

Physical and genetic characterization of highly polar mutations in the *gal* and *lac* operons of *E. coli* showed that discrete DNA units could integrate at different sites (Saedler and Starlinger, 1967; Jordan *et al.*, 1968; Shapiro, 1969). While originally such mutations were attributed to insertions by a small group of specific sequences (insertion sequences or IS elements) that did not themselves encode phenotypic functions, it was subsequently realized that plasmid-borne determinants for antibiotic resistance were often carried by transposable elements with similar properties. Those elements encoding phenotypic functions in bacteria are termed transposons (Calos and Miller, 1980; Kleckner, 1981).

The essential features of transposition between replicons in bacteria are illustrated in Fig. 12.11. It is duplicative, one copy of the element being inserted into the recipient replicon while one is retained in its original position in the donor. Transposition needs neither a functional homologous recombination system nor sequence homology between the element and the integration site. Preferred integration sites are A + T rich and the frequency with which they occur in DNA depends on the element. Integration of an element invariably results in the duplication of a short sequence adjacent to the insertion site, so that the integrated element is flanked by a short direct repeat. The integrity of this repeat is not required for further transposition, though it is likely to be required for the element's precise excision. Transposable elements invariably have inverted repeats, usually of 15–40 bp at their ends, and usually contain genes whose expression is necessary for transposition.

Currently, it appears that bacterial transposable elements fall into either of two classes, depending on their transposition behaviour (Fig. 12.12). Class I elements include insertion sequences and those composite transposons that contain phenotypic markers bound by insertion sequences. Class II elements belong to the 'Tn3 family'. These are bounded by related 38 bp inverted repeats and make 5 bp duplications on insertion. Inter-replicon transposition of Class II elements has been shown to occur in two sequential separable steps, encoded by separate element functions. The first of these steps, specified by the 'transposase' (*tnp*A product) fuses the donor and recipient replicon between directly repeated copies of the element (Arthur and Sherratt, 1979; Kitts *et al.*, 1982; (Fig. 12.11 and 12.13)).

Such fused replicons or co-integrates, which can be stably maintained in

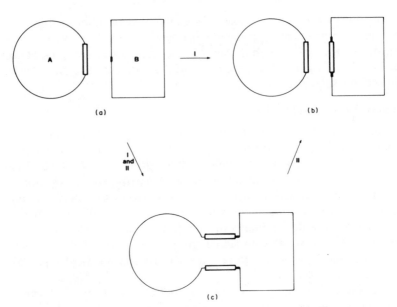

Fig. 12.11.   Intermolecular transposition in bacteria. (a) The transposable element is represented as an unshaded box, residing in replicon A.
There are thought to be two transposition pathways. Class I elements
may transpose directly to give products (b), or may form a cointegrate
molecule (c) which is an intermediate in Class II transposition. Class I
cointegrates are quite stable, whereas Class II cointegrates are rapidly
converted to the products (b) by a site-specific recombination mediated
by the transposon's resolvase protein. As a result of transposition the
element is duplicated, and the target sequence of insertion, represented
as a black box, is duplicated as direct repeats flanking the newly inserted
element in replicon B.

*rec*A⁻ strains, are then converted to the normal transposition end products,
by 'resolvase' (*tnp*R product), a site-specific recombination enzyme that acts
at a specific site (*res*) near the middle of these elements. Class I elements
are distinguished by having no site-specific recombination system to resolve
cointegrate intermediates. Indeed their transposition can apparently proceed
to completion without passing through a cointegrate intermediate. Cointegrates can be formed during their transposition at a variable frequency,
though there appear not to be intermediates in the normal transposition
pathway, and they cannot be converted to normal transposition products in
a *rec*A⁻ strain. Of course, in a *rec*A⁺ strain, homologous recombination can
act to resolve such cointegrates.

Most of the proposed transposition models take account of the apparent
duplicative nature of transposition, and assume that it is accompanied by
semi-conservative replication, one daughter molecule appearing in the target
site and the other remaining in its original position. The models can be

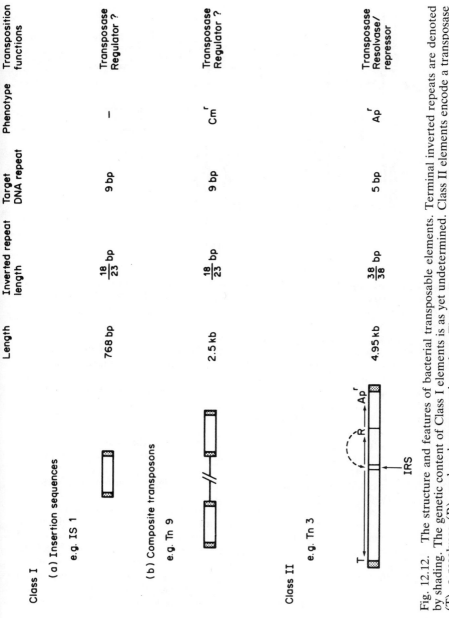

Fig. 12.12. The structure and features of bacterial transposable elements. Terminal inverted repeats are denoted by shading. The genetic content of Class I elements is as yet undetermined. Class II elements encode a transposase (T), a resolvase (R), and a chromosomal marker. The resolvase acts on an internal site (IRS) to both repress transposase gene expression, and mediate site-specific recombination.

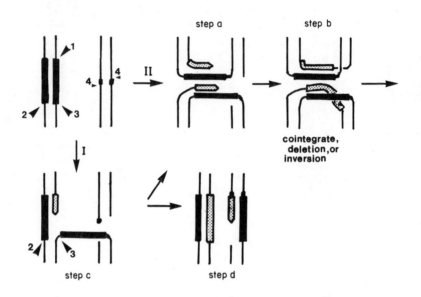

step a   step b

II

cointegrate,
deletion, or
inversion

I

step c   step d

Fig. 12.13. Schematic illustration of the types of transposition model. The symmetrical models (Class II: Shapiro, 1979; Arthur and Sherratt, 1979) assume that the inverted repeats at the ends of the element provide the sequence symmetry for simultaneous or near simultaneous nicking of both element boundaries (cuts marked 1 and 2). Element-encoded transposases provide the best candidates for such nicking and it is presumed that phosphodiester bond energy is conserved by the protein becoming covalently bound to the termini (the 5′ ends of the element in the model of Arthur and Sherratt). In the concerted transposition reaction, staggered breaks are also introduced at the target sites (indicated 4; again with conservation of phosphodiester bond energy) into which the transposon ends are immediately ligated (step a). Semi-conservative replication, most easily initiated from the 3′OH termini (of the donor in the model of Arthur and Sherratt, or alternatively from the recipient, if the original nicks in the donor produce 3′ termini adjacent to the element) followed by ligation complete the first stage of transposition (step b). In inter-replicon transposition, this generates cointegrates, while intramolecular transposition results in either deletion or inversion depending on whether a given donor transposon strand integrates into the same or opposite strand.

Class I models (Sherratt, 1979; Grindley and Sherratt, 1979; Harshey and Bukhari, 1981; Galas and Chandler, 1981) can avoid cointegrates by assuming that the initial nicking is asymmetric: only one end of the element is initially nicked (cut 1) and ligated into its target (step c). Cuts generating either a 3′ or 5′ terminus adjacent to the element are again possible, with subsequent replication occurring most easily from a 3′ in either the recipient or donor, respectively. Complete transposition without cointegration is, in essence, achieved by cleaving the donor element on the same strand at the other end (cut 3; giving d). This site will have no direct sequence relationship with that at which cleavage occurred. Alternatively cointegration can occur by nicking at site 2. For a detailed comparative discussion of these models, see Kleckner (1981).

divided into symmetrical and asymmetrical types, the former easily explaining Class II behaviour (Fig. 12.14) (e.g. Galas and Chandler, 1981). All owe much to the scheme, proposed by Eisenberg *et al.* (1977), for *cis*A protein-mediated replication of øX174 DNA replication.

Transposition is a replicative event, and the evolution of composite elements demonstrates that potentially any chromosomal marker may be amplified by the transposition process. In this respect, phenotypic selection may have determined those transposons which have evolved so far. Transposition from chromosome to plasmid may result both in many-fold amplification: proportionate to the copy-number per cell of the plasmid, and also the subsequent spread throughout a population of the plasmid-borne element. Moreover, the various transposition models propose that intramolecular transposition can generate various genomic rearrangements depending on which transposition pathway is adopted and which strands are involved in the terminating nick-ligation event (Fig. 12.14).

Eukaryote transposable elements

Historically, genetic and cytological aberrations observed in maize suggested the presence of defined elements responsible for translocation, chromosome breakage and inactivation of structural genes (McClintock, 1950; Fincham and Sastry, 1974). The use of more genetically amenable organisms, *Drosophila* and yeast, has provided more direct evidence as to the nature of eukaryote transposable elements.

By analysis of different strains of *Drosophila melanogaster* and yeast cells under continual propagation, evidence for the movement of dispersed repetitive sequences was obtained. These sequences share common characteristics with prokaryote transposable elements: they insert into many different target sequences that share no obvious homology to each other or to the ends of the element; upon insertion direct repeats of a few base pairs of target sequence are found immediately adjacent to the ends of the element; an element has the same terminal sequence when inserted at different chromosomal locations; the elements contain a terminal inverted repetition bounding the main body of the element. These inverted repeats are part of a longer direct repeat (Fig. 12.15; Cameron *et al.*, 1979; Potter *et al.*, 1979), structurally analogous to composite prokaryote transposons such as Tn9.

Typically, the characterized eukaryote elements of *Drosophila* (e.g. *copia*, 412, mdg) and yeast (Ty1) are between 5 and 7 kb in length and are present at about 30 copies per genome. The terminal direct repeats are about 300 bp long (Levis *et al.*, 1980) incorporating 4 bp terminal inverted repeats. Unlike the situation for Tn9, where its terminal long direct repeats of ISI can transpose independently, there is no parallel evidence for independent movement of the direct repeats of a eukaryote element. Moreover, as yet,

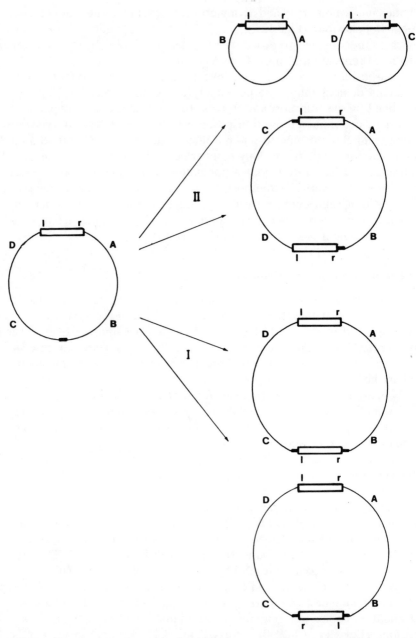

Fig. 12.14. The postulated consequences of intramolecular transposition. l and r denote left and right-hand ends of the element; A, B, C and D the orientations of intervening sequences. The products of Class II-type transposition are shown by the upper arrows, and of Class I direct transposition by the lower arrows.

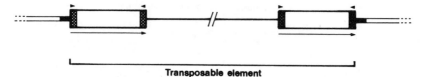

Transposable element

Fig. 12.15.   Structure of *copia*/Ty1-like transposable elements. Shaded regions and arrowheads denote inverted repeats, part of the long terminal direct repeats (boxes). The genetic content and organization of the intervening region has yet to be determined. An integrated element is flanked by direct repeats of the chromosomal target sequence (black).

there is no direct evidence that transposition of eukaryote elements is replicative; but this is suggested by their dispersal in the genome, their structural similarities with prokaryote elements, and comparison with retroviruses, thereby providing a possible transposition mechanism.

Retroviruses are RNA viruses that replicate through DNA intermediates, and have as an obligatory step in their life cycle the integration of a DNA intermediate into the host genome (Temin and Baltimore, 1972). The initial double-stranded DNA reverse transcript is a linear molecule containing a directly repeated sequence (LTR) at each end (Gilboa *et al.*, 1979) which also contains an inverted repeat at each end (Sutcliffe *et al.*, 1980). The linear retroviral DNA molecules can enter the nucleus, whereupon some circles form, containing either one or two LTR sequences (Shank *et al.*, 1978) prior to integration. Integrative recombination with host-cell DNA occurs at the LTR termini, resulting in short flanking target DNA duplications (Majors and Varmus, 1981; Fig. 12.16).

The integrated viral genome, or provirus, is a template for transcription by host RNA polymerase, and the transcripts may be packaged as mature virions. A plausible model for proviral transposition suggests that transcripts may also be the templates for the retroviral reverse transcriptase enzyme, producing DNA reverse transcripts which may integrate elsewhere in the genome.

The retroviral genome may be inherited as part of the host's normal genetic complement. An endogenous provirus is associated with the dilute coat

---

Class II and a proportion of Class I elements adopting the replicon fusion pathway give rise to two indirectly repeated transposons and an inversion of the intervening sequence if the terminal nick-ligation joins opposite DNA strands. If the same strands are joined then the outcome is a deletion of one transposon and the intervening sequence. Class I elements adopting direct transposition will give two indirectly repeated elements with no accompanying sequence inversion if opposite strands are involved. If the same strands are involved, directly repeated elements with no intervening inversion results. If you have trouble following these outcomes then try modelling them with string.

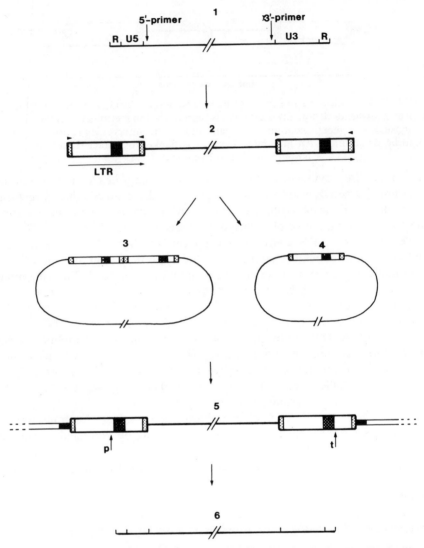

Fig. 12.16.   The retroviral life cycle. (a) An infective single-stranded RNA molecule is reverse transcribed, using the primers as indicated, to generate a linear double-stranded DNA molecule (2). The components of this molecule are indicated: the long-terminal repeats contain (a) U3: unique sequence from the 3' end of the viral RNA; (b) R: sequence repeated at both ends of the viral RNA; (c) U5: unique sequence from the 5' end of the viral RNA. Inverted repeats are shown by arrow-heads.

Circularization may generate either (3), containing two copies of the LTR; or (4), containing just one LTR. Integration of either (3) or (4) generates the proviral structure (5): the target sequence is duplicated as direct repeats (black) flanking the provirus.

Transcription from promotor (p) to terminator (t) sequence generates further viral RNAs (6).

colour mutation of a mouse inbred line (Jenkins *et al.*, 1981). The forward and reverse mutation rates for this locus are about 10^{-6}, suggesting retroviral integration and excision of similar frequencies at this locus.

Retroviruses as a model for eukaryotic transposition

The similarities in structure between proviral DNA and eukaryotic transposable elements, and the isolation of polyadenylated transcripts of these elements, suggest common replication modes. This idea is supported by other evidence: the terminal dinucleotides of a transposable element or provirus are 5'TG...CA3', and in some cases this homology extends further; sequences corresponding to retroviral promoter, polyadenylation signal and primer sequences are implicated for the *Drosophila* transposable element 412; extrachromosomal circular copies of the *Drosophila* element *copia* are found in cultured *Drosophila* cells (Flavell and Ish-Horowicz, 1981).

The notion that circular *copia* molecules, containing either one or two LTR's, are intermediates in transposition is, however, disputed. They may simply be the products of intramolecular recombination. 'Free-ends', residual LTR's left in the chromosome as a result of intramolecular recombination, are detected for proviruses and Tyl elements in yeast, but not for *Drosophila* elements. Reintegration at different sites of circular molecules generated this way could be a replicative process if the circles were capable of autonomous replication; but there is no evidence for this.

So transposition of *copia*/Tyl-like elements could be via a mechanism similar to a retroviral replicative pathway exclusive of an extracellular existence. However, current evidence does not preclude other mechanisms, and also the existence of other dissimilar transposable elements in eukaryotes. Interestingly, a heterogeneous population of circular molecules, with sizes varying from a few hundred nucleotides to 20 kb, have been isolated from *Drosophila* eggs and cultured cells. More than 80% of this DNA is homologous to middle repetitive DNA (Stanfield and Lengyel, 1979).

The rearrangements associated with transposable elements in prokaryotes also appear to be very much a feature in eukaryotes. Ajacent deletion and sequence inversion have been described for Tyl and *copia* elements. Reversion of mutations associated with element insertion occur, indicative of an active excision process.

Recombination between transposable elements

Transposable elements provide 'mobile' sites of homology upon which recombination can act to generate duplications, deletions, inversions and translocations of genetic material. Moreover, since transposition is duplicative, each transposition event provides further opportunities for homologous recombination. The resolvases of Class II elements can, in principle, also promote recombination between transposable elements containing homologous *res* sites. In practice, this recombination is very efficient if the *res* sites are directly repeated in the same molecule, but inefficient if the *res* sites are in inverted orientation or if they are in separate molecules. Homologous recombination, too, appears not to act equally efficiently on all configurations of homology; for example, plasmid/plasmid intermolecular recombination occurs preferentially to intramolecular recombination (e.g. to convert dimers to monomers). This may be because the requirements for the donor (aggressive single strand) and the recipient (superhelicity?) are different and cannot easily be reconciled within the same molecule. Perhaps partly because of this, composite transposable elements like Tn9 appear to undergo homologus recombination only rarely between the two directly repeated IS1s.

Regulation of transposition

Many spontaneous mutations arise as a consequence of transposition. Moreover, transposable element-mediated mutations are of the type that are particularly important in evolution. The frequency of transposition in a cell is almost certainly under both element and non-element control, and in *E. coli*, chromosomal mutations which increase or decrease the transposition rate have been reported (many papers in the *Cold Spring Harbor Symp. Quant. Biol.*, volume 45, 1981 relate to these points and those below). Ways by which transposable elements are able to regulate their own transposition have been described (e.g. Kleckner, 1981). It is likely then that observed transposition frequencies have evolved by natural selection and to some extent reflect a 'compromise' in the symbiosis between the element and its host. Interestingly, the control of transposition appears to be lost in several abnormal situations. For example, it has been reported that transposition of *copia*-like elements in *Drosophila* is higher in cultured embryo cells than in flies themselves. Moreover, the phenomenon of hybrid dysgenesis in *Drosophila melanogaster* has been attributed to uncontrolled transposition (Bregliano *et al.*, 1980). Dysgenic traits such as mutation, chromosomal aberration, distorted segregation and sterility are often present in the offspring of crosses involving parents from two separated populations such as laboratory and wild strains. The basis for these traits may be due to the amplification of transposable elements in the offspring, which could arise as a consequence of the inheritance of a particular element into a regulation deficient background.

The level of transposable elements in an organism is also affected by the rate of the loss, both by exact excision between the direct repeats flanking the element and by 'non-precise' deletions. Neither of these types of event is accompanied by re-transposition. Transposons of the Tn3 family appear to transpose much more readily between plasmids than into the chromosome. Moreover, a specific 'transposon immunity' prevents Tn3 transposition into a plasmid molecule that already contains Tn3 (Wallace *et al.*, 1981).

In conclusion, transposition can be highly controlled by host or element determined regulation, by target selectivity and by events that remove transposable elements from DNA. Very different levels of transposable elements can exist in closely related organisms. For example, the arrangement of structural genes on the chromosome of *Shigella dysenteriae* is extremely similar to that of *E. coli*, though the *Shigella* chromosome contains about 200 copies of IS1 or IS1 like elements (Ohtsubo *et al.*, 1981); this is far more than the six or so of *E. coli*. Whether most of these copies can still transpose is unsure, though they ought to provide many substrates for homologous recombination.

DISCUSSION

Much of this chapter has discussed how unequal exchanges and other homologous recombination events can result in amplification of deletion of specific DNA sequences. We have said little to address the problem of how the primary duplications that are needed to fuel such events occur. Though dimerization or polyploidization of genomes will generate duplications, their nature is such that homologous recombination can generate no new gene arrangements. Transposition, because of its replicative nature (at least in bacteria) is clearly an important source of duplicated sequences. So also is the multiple integration of viral-like sequences. Replicational slippage (Kornberg *et al.*, 1964) and fold-back replication can generate both direct and inverted tandem repeats. Deletion too can occur by replicational slippage, which in some ways is analogous to unequal exchange. Moreover, since repeats as small as 5 bp appear to be able to act as substrates for some type of recombination, mutation will continuously generate and destroy such small recombination sites.

Orgel and Crick (1980) describe two types of DNA in higher organisms: one specific and the other non-specific. The latter originated by the spreading of sequences which had little or no effect on the phenotype and arose by the natural selection of 'preferred replicators' within the genome. Doolittle and Sapienza (1980) have described how this concept can be applied to the spread of transposable elements to account for some middle-repetitive DNA sequences. We too support this idea, arguing that a transposable element can be considered a unique replicon. We would also like to extend this concept in considering tandemly-repeated DNA sequences.

Unequal exchange occurs between newly replicated sister molecules, and it is conceivable that abortive, incomplete replication provides short molecules which can recombine via sister molecule exchange with intact, complete chromosomes. The proximity of a replication initiation point to a tandemly repeated sequence, and especially if initiation is not stringently limited to once per cell division from that point, may provide any amount of sister molecules as substrates for unequal exchange. Moreover, if a 'relaxed' replication initiation point is contained within a repeat unit, the unit may be considered an independent replicator, capable of spreading itself within a genome by unequal exchange. The association between replication initiation points and sequences that might mediate sister-molecule exchanges has been postulated for the spread of the Alu family of sequences in human cells (Jelinek *et al.*, 1980). In the section: unequal exchange, we drew comparisons between non-transcribed spacer (NTS) sequences and highly repeated heterochromatic DNA, and while for the latter there is as yet little evidence for localized replication initiation points, there is in spacer sequences. For example, rDNA amplification in oogenesis is dependent on an origin of replication contained within an excised circle. Analysis of *Xenopus* chromosomal rDNA shows that maybe every unit contains an origin of replication within the NTS (and possibly several per NTS) (Bozzoni *et al.*, 1981). Another developmental strategy indicative of independent replication initiation points is differential replication of germ line sequences in, for example, *Drosophila* (reviewed by Spradling and Rubin, 1981): differential replication of both rDNA and heterochromatic sequences is observed in development. Heterochromatic repetitive DNA has evolved by unequal exchange—this evolution could only be enhanced if a repeat sequence contains an origin of replication.

A handy classification for repeat sequences might be based on their ability to recombine. Transposable elements and some viral genomes can independently recombine with other sequences via functions they specify themselves, while normal cellular recombination pathways are utilized by other repeat sequences in their spread. Assuming an origin of replication is contained within a repeat sequence of the latter type, various forms of spread, other than by unequal exchange, are possible. We have discussed the potential for extrachromosomal replication of circular molecules, and reintegration at various genomic sites. This may contribute to tandem arrays, or alternatively serve to translocate a sequence. If the product of extrachromosomal replication is a circular molecule then only a single cross-over is required to integrate it. Alternatively, if a linear molecule is produced it can integrate via two cross-over events; there is some evidence that integrated tandem duplications of the simian virus 40 genome arise by a double cross-over, although in this case the recombination might be illegitimate (Chia and Rigby, 1981).

In conclusion, we would underline the evolutionary importance of mechanisms responsible for addition and deletion of sequences. The mechanisms described can account for the accumulation of DNA in higher organisms.

REFERENCES

Anderson, P. and Roth, R. (1981). Spontaneous tandem genetic duplications in *Salmonella typhimurium* arise by unequal recombination between rRNA (rrn) cistrons. *Proc. Natl. Acad. Sci. USA*, **78**, 3113–3117.

Arthur, A. and Sherratt, D. (1979). Dissection of the transposition process: A transposon-encoded site-specific recombination system. *Molec. Gen. Genet.*, **175**, 267–274.

Bernardi, G. and Bernardi, G. (1980). Repeated sequences in the mitochondrial genome of yeast. *FEBS Lett.*, **115**, 159–162.

Bosely, P., Moss, T., Machler, M., Portmann, R. and Birnstiel, M. (1979). Sequence organization of the spacer DNA in a ribosomal gene unit of *Xenopus laevis*. *Cell*, **17**, 19–31.

Bozzoni, I., Baldari, C. M., Amaldi, F. and Buongiorno-Nardelli, M. (1981). Replication of ribosomal DNA in *Xenopus laevis*, *Eur. J. Biochem.*, **118**, 585–590.

Bregliano, J. C., Picard, G., Bucheton, A., Pelisson, A., Lavige, J. M. and L'Hentier, P. (1980). Hybrid dysgenesis in *Drosophila melanogaster*. *Science*, **207**, 606–611.

Brewer, B. J., Zakian, V. A. and Fangman, W. L. (1980). Replication and meiotic transmission of yeast ribosomal RNA genes. *Proc. Natl. Acad. Sci. USA*, **77**, 6739–6743.

Brown, D. D. and Blackler, A. W. (1972). Gene amplification proceeds by a chromosome copy mechanism. *J. Mol. Biol.*, **63**, 75–83.

Buongiorno-Nardelli, M., Amaldi, F. and Lava-Sanchez, P. A. (1972).Amplification as a rectification mechanism for the redundant rRNA genes. *Nature New Biol.*, **238**, 134–137.

Callan, H. G. (1967). The organization of genetic units in chromosomes. *J. Cell Sci.*, **2**, 1–7.

Calos, M. P. and Miller, J. H. (1980). Transposable elements. *Cell*, **20**, 579–595.

Cameron, J. R., Loh, E. Y. and David, R. W. (1979). Evidence for transposition of dispersed repetitive families in yeast. *Cell*, **16**, 739–751.

Campbell, A. M. (1962). Episomes. *Adv. Genet.*, **11**, 101–145.

Campbell, A. M. (1981). Evolutionary significance of accessory DNA elements in bacteria. *Ann. Rev. Microbiol.*, **35**, 55–83.

Cavalier-Smith, T. (1974). Palindromic base sequences and the replication of eukaryote chromosome ends. *Nature, Lond.*, **250**, 467–470.

Chia, W. and Rigby, P. W. J. (1981). Fate of viral DNA in nonpermissive cells infected with simian virus 40. *Proc. Natl. Acad. Sci. USA*, **78**, 6638–6642.

Clark, A. J. (1973). Specialized sites in generalized recombination. *Ann. Rev. Genet.*, **7**, 67–86.

Clark-Walker, G. D. and Azad, A. A. (1980). Hybridizable sequences between cytoplasmic ribosomal RNAs and the 3 micron circular DNAs of *Saccharomyces cerevisiae* and *Torulopsis glabrata*. *Nucl. Acid Res.*, **8**, 1009–1022.

Cramer, J. H., Farrelly, F. W., Barnitz, J. T. and Rownd, R. H. (1977). Construction and restriction endonuclease mapping of hybrid plasmids containing *Saccharomyces cerevisiae* ribosomal DNA. *Molec. Gen. Genet.*, **151**, 229–244.

Doolittle, W. F. and Sapienza, C. (1980). Selfish genes, the phenotype paradigm and genome evolution. *Nature, Lond.*, **284**, 601–603.

Edlund, T. and Normark, S. (1981). Recombination between short DNA homologies causes tandem duplication. *Nature, Lond.*, **292**, 269–271.

Eisenberg, S., Griffith, J. and Kornberg, A. (1977). øX174 *cistron A* protein is a multifunctional enzyme in DNA replication. *Proc. Natl. Acad. Sci. USA*, **74**, 3198–3202.

Farabaugh, P. J., Schmeissner, U., Hofer, M. and Miller, J. H. (1978). Genetic studies of the *lac* repressor. VII. On the molecular nature of spontaneous hotspots in the *lacI* gene of *Escherichia coli*. *J. Mol. Biol.*, **126**, 847–863.

Fedoroff, N. V. (1979). On spacers. *Cell*, **16**, 697–710.

Fincham, J. R. S. and Sastry, G. R. K. (1974). Controlling elements in maize. *Ann. Rev. Genet.*, **8**, 15–50.

Flavell, A. J. and Ish-Horowicz, D. (1981). Extrachromosomal circular copies of the eukaryotic transposable element *copia* in cultured *Drosophila melanogaster* cells. *Nature*, **292**, 591–595.

Fogel, S. and Mortimer, R. K. (1971). Recombination in yeast. *Ann. Rev. Genet.*, **5**, 219–236.

Gaillard, C., Strauss, F. and Bernardi, G. (1980). Excision sequences in the mitochondrial genome of yeast. *Nature, Lond.*, **283**, 218–220.

Galas, D. J. and Chandler, M. (1981). On the molecular mechanisms of transposition. *Proc. Natl. Acad. Sci. USA*, **78**, 4858–4862.

Gellert, M. (1981). DNA topoisomerases. *Ann. Rev. Biochem.*, **50**, 879–910.

Gilboa, E., Mitra, S. W., Goff, S. and Baltimore, D. (1979). A detailed model of reverse transcription and tests of crucial aspects. *Cell*, **18**, 93–100.

Goddard, J. M. and Cummings, D. J. (1977). Mitochondrial DNA replication in *Paramecium aurelia*. Cross-linking of the initiation end. *J. Mol. Biol.*, **109**, 327–344.

Grindley, N. D. F. and Sherratt, D. J. (1979). Sequence analysis at IS1 insertion sites: models for transposition. *Cold Spring Harbor Symp. Quant. Biol.*, **43**, 1257–1261.

Haber, J. E. and George, J. P. (1979). A mutation that permits the expression of normally silent copies of mating-type information in *Saccharomyces cerevisiae*. *Genetics*, **93**, 13–25.

Haber, J. E., Mascioli, D. W. and Rogers, D. T. (1980a). Illegal transposition of mating-type genes in yeast. *Cell*, **20**, 519–528.

Haber, J. E., Rogers, D. T. and McCusker, J. H. (1980b). Homothallic conversions of yeast mating-type genes occur by intrachromosomal recombination. *Cell*, **22**, 277–289.

Harshey, R. M. and Bukhari, A. I. (1981). A mechanism of DNA transposition. *Proc. Acad. Sci. USA*, **78**, 1090–1094.

Hicks, J. B. and Herskowitz, I. (1977). Interconversion of yeast mating types. II. Restoration of mating ability to sterile mutants in homothallic and heterothallic strains. *Genetics*, **85**, 373–393.

Hill, C. W., Grafstrom, R. H., Wallis Harnish, B. and Hillman, B. S. (1977). Tandem duplications resulting from the recombination between ribosomal RNA genes in *Escherichia coli*. *J. Mol. Biol.*, **116**, 407–428.

Hill, C. W. and Wallis Harnish, B. (1982). Transposition of a chromosomal segment bounded by redundant rRNA genes into other rRNA genes in *Escherichia coli*. *J. Bacteriol.*, **149**, 449–457.

Holliday, R. (1964). A mechanism for gene conversion in fungi. *Genet. Res.*, **5**, 282–304.

Hopwood, D. A. (1967). Genetic analysis and genome structure in *Streptomyces coelicolor*. *Bacteriol. Rev.*, **31**, 373–403.

Jackson, J. A. and Fink, G. R. (1981). Gene conversion between duplicated genetic elements in yeast. *Nature, Lond.*, **292**, 306–311.

Jelinek, W. R., Toomey, T. P., Leinwand, L., Duncan, C. H., Biro, P. A., Choudary, P. V., Weissman, S. M., Rubin, C. M., Houck, C. M., Deininger, P. L. and

Schmid, C. W. (1980). Ubiquitous, interspersed repeated sequences in mammalian genomes. *Proc. Natl. Acad. Sci. USA*, **77**, 1398–1402.

Jenkins, N. A., Copeland, N. G., Taylor, B. A. and Lee, B. K. (1981). Dilute (d) coat colour mutation of DBA/2J mice is associated with the site of integration of an ecotropic MuLV genome. *Nature, Lond.*, **293**, 370–374.

Jordan, E., Saedler, H. and Starlinger, P. (1968). O° and strong polar mutations in the *gal* operon are insertions. *Molec. Gen. Genet.*, **102**, 353–363.

Kahn, R., Cunningham, R. P., DasGupta, C. and Radding, C. M. (1981). Polarity of heterduplex formation promoted by *Escherichia coli* recA protein. *Proc. Natl. Acad. Sci. USA*, **78**, 4786–4790.

Kilts, P., Lamond, A. and Sherratt, D. J. (1982). Interreplicon transposition of Tn1/3 occur in two sequential genetically separable steps. *Nature, Lond.*, **295**, 626–628.

Kleckner, N. (1981). Transposable elements in prokaryotes. *Ann. Rev. Genet.*, **15**, 341–404.

Klobutcher, L. A., Swanton, M. T., Donini, P. and Prescott, D. M. (1981). All gene-sized DNA molecules in four species of hypotrichs have the same terminal sequence and an unusual 3′ terminus. *Proc. Natl. Acad. Sci. USA*, **78**, 3015–3019.

Kornberg, A., Bertsch, L. L., Jackson, J. F. and Khorana, H. G. (1964). Enzymatic synthesis of deoxyribonucleic acid. XVI. Oligonucleotides as templates and the mechanism of their replication. *Proc. Natl. Acad. Sci. USA*, **51**, 315–323.

Landy, A. and Ross, W. (1977). Viral integration and excision. *Science*, **197**, 1147–1151.

Levis, R., Dunsmuir, P. and Rubin, G. M. (1980). Terminal repeats of the *Drosophila* transposable element *copia*: nucleotide sequence and genomic organization. *Cell*, **21**, 581–588.

Majors, J. E. and Varmus, H. E. (1981). Nucleotide sequence at host-proviral junctions for mouse mammary tumour virus. *Nature, Lond.*, **293**, 370–374.

McClintock, B. (1950). The origin and behaviour of mutable loci in maize. *Proc. Natl. Acad. Sci. USA*, **36**, 344–355.

McGavin, S. (1971). Models of specifically paired like (homologous) nucleic acid structures. *J. Mol. Biol.*, **55**, 293–298.

Meselson, M. and Radding, C. M. (1975). A general model for genetic recombination. *Proc. Natl. Acad. Sci. USA*, **72**, 358–361.

Meyer, G. F. and Lipps, H. J. (1981). The formation of polytene chromosomes during macronuclear development of the hypotrichous ciliate *Stylonychia mytilus*. *Chromosoma*, **82**, 309–314.

Meyerink, J. H., Klootwijk, J. and Planta, R. J. (1979). Extrachromosomal circular ribosomal DNA in the yeast *Saccharomyces carlsbergensis*. *Nucl Acid Res.*, **7**, 69–76.

Mizuuchi, K., Weisberg, R., Enquist, L., Mizuuchi, M., Buraczynska, M., Foeller, C., Hsu, P.-L., Ross, W. and Landy, A. (1981). Structure and function of the phage λ *att* site: size, Int-binding sites, and location of the crossover points. *Cold Spring Harbor Symp. Quant. Biol.*, **45**, 429–437.

Nash, H. A., Mizuuchi, K., Enquist, L. W. and Weisberg, R. A. (1981). Strand exchange in λ integrative recombination: genetics, biochemistry, and models. *Cold Spring Harbor Symp. Quant. Biol.*, **45**, 417–428.

Ohno, S. (1970). *Evolution by Gene Duplication*, Allen and Unwin, London.

Ohtsubo, H., Nyman, K., Doroszkiewicz, W. and Ohtsubo, E. (1981). Multiple copies of iso-insertion sequences of IS1 in *Shigella dysenteriae*, *Nature, Lond.*, **292**, 640–643.

Orgel, L. E. and Crick, F. H. C. (1980). Selfish DNA: the ultimate parasite. *Nature, Lond.*, **284**, 604–607.

Pan, W.-C. and Blackburn, E. H. (1981). Single extrachromosomal ribosomal RNA gene copies are synthesized during amplificaton of the rDNA in *Tetrahymena. Cell*, **23**, 459–466.

Petes, T. D. (1979). Meiotic mapping of yeast ribosomal deoxyribonucleic acid on chromosome XII. *J. Bacteriol.*, **138**, 185–192.

Petes, T. D. (1980). Unequal meiotic recombination within tandem arrays of yeast ribosomal DNA genes. *Cell*, **19**, 765–774.

Petes, T. D. and Botstein, D. (1977). Simple Mendelian inheritance of the reiterated ribosomal DNA of yeast. *Proc. Natl. Acad. Sci. USA*, **74**, 5091–5095.

Potter, S. S., Brorein, W. J., Dunsmuir, P. and Rubin, G. M. (1979).Transposition of elements of the 412, *copia* and 297 dispersed repeated gene families in *Drosophila. Cell*, **17**, 415–427.

Radding, C. M. (1982). Homologous pairing and strand exchange in genetic recombination. *Ann. Rev. Genet.*, **16**, 405–437.

Reed, R. R. (1981). Transposon-mediated site-specific recombination: a defined in vitro system. *Cell*, **25**, 713–720.

Reed, R. R. and Grindley, N. D. F. (1981). Transposon-mediated site-specific recombination *in vitro*: DNA cleavage and protein-DNA linkage at the recombination site. *Cell*, **25**, 721–728.

Reeder, R. H., Brown, D. D., Wellauer, P. K. and Dawid, I. B. (1976). Patterns of ribosomal DNA spacer lengths are inherited. *J. Mol. Biol.*, **105**, 507–516.

Rochaix, J.-D., Bird, A. and Bakkan, A. (1974). Ribosomal RNA gene amplification by rolling circles. *J. Mol. Biol.*, **87**, 473–487.

Saedler, H. and Starlinger, P. (1967). O° mutations in the galactose operon in *E. coli. Molec. Gen. Genet.*, **100**, 178–189.

Scherer, S. and Davis, R. W. (1980). Recombination of dispersed repeated DNA sequences in yeast. *Science*, **209**, 1380–1384.

Schweizer, E., MacKechnie, C. and Halvorson, H. O. (1969). The redundancy of ribosomal and transfer RNA genes in *Saccharomyces cerevisiae. J. Mol. Biol.*, **40**, 261–277.

Shank, P. R., Hughes, S. H., Kung, H. J., Majors, J. E., Quintrell, N., Guntaka, R. V., Bishop, J. M. and Varmus, H. E. (1978). Mapping unintegrated avian sarcoma virus DNA: Termini of linear DNA bear 300 nucleotides present once or twice in two species of circular DNA. *Cell*, **15**, 1383–1395.

Shapiro, J. A. (1969). Mutations caused by the insertion of genetic material into the galactose operon of *Escherichia coli. J. Mol. Biol.*, **40**, 93–105.

Shapiro, J. A. (1979). Molecular models for the transposition and replication of bacteriophage Mu and other transposable elements. *Proc. Natl. Acad. Sci. USA*, **76**, 1933–1937.

Sherratt, D., Arthur, A. and Dyson, P. (1981). Site-specific recombination. *Nature, Lond.*, **294**, 608–610.

Sherratt, D. J. (1979). A model for genetic transposition (round table discussion). *Contr. Microbiol. Immununol.*, **6**, 229–235.

Sigal, N. and Alberts, B. (1972). Genetic recombination: the nature of a crossed strand exchange between two homologous DNA molecules. *J. Mol. Biol.*, **71**, 89–93.

Spradling, A. C. and Rubin, G. M. (1981). *Drosophila* genome organization: conserved and dynamic aspects. *Ann. Rev. Genet.*, **15**, 219–264.

Stahl, F. W. (1979a). *Genetic Recombination: Thinking About It in Phage and Fungi,* Freeman, San Francisco.

Stahl, F. W. (1979b). Specialized sites in generalized recombination. *Ann. Rev. Genet.*, **13**, 7–24.

Stanfield, S. W. and Lengyel, J. A. (1979). Small circular DNA of *Drosophila melanogaster*: chromosomal homology and kinetic complexity. *Proc. Natl. Acad. Sci. USA*, **76**, 6142–6146.

Strathern, J. N. and Herskowitz, I. (1979). Asymmetry and directionality in the production of new cell types during clonal growth: the switching pattern of homothallic yeast. *Cell*, **17**, 371–381.

Sutcliffe, J. G., Shinnick, T. M., Verma, I. M. and Lerner, R. A. (1980). Nucleotide sequence of Moloney leukemia virus: 3' end reveals details of replication, analogy to bacterial transposons, and an unexpected gene. *Proc. Natl. Acad. Sci. USA*, **77**, 3302–3306.

Szostak, J. W. and Wu, R. (1980). Unequal crossing over in the ribosomal DNA of *Saccharomyces cerevisiae*. *Nature, Lond.*, **284**, 426–430.

Taylor, A. and Smith, G. R. (1980). Unwinding and rewinding of DNA by the RecBC enzyme. *Cell*, **22**, 447–457.

Temin, H. M. and Baltimore, D. (1972). RNA-directed DNA synthesis and RNA tumour viruses. *Adv. Virus Res.*, **17**, 129–186.

Tobler, H. (1975). In R. Weber (Ed), *Biochemistry of Animal Development*, vol. 3, Academic Press, New York, pp. 91–143.

Truett, M. A. and Gall, J. G. (1977). The replication of ribosomal DNA in the macronucleus of *Tetrahymena*. *Chromosoma*, **64**, 295–303.

Wallace, L. J., Ward, J. M., Bennett, P. M., Robinson, M. K. and Richmond, M. H. (1981). Transposition immunity. *Cold Spring Harbor Symp. Quant. Biol.*, **45**, 183–188.

Wellauer, P. K., Reeder, R. H., Dawid, I. B. and Brown, D. D. (1976). The arrangement of length heterogeneity in repeating units of amplified and chromosomal DNA from *Xenopus laevis*. *J. Mol. Biol.*, **105**, 487–505.

West, S. C., Cassuto, E. and Howard-Flanders, P. (1981). Homologous pairing can occur before DNA strand separation in general genetic recombination. *Nature, Lond.*, **290**, 29–33.

de Zamaroczy, M., Marotta, R., Faugeron-Fonty, G., Goursot, R., Mangin, M., Baldacci, G. and Bernardi, G. (1981). The origins of replication of the yeast mitochondrial genome and the phenomenon of suppressivity. *Nature, Lond.*, **292**, 75–78.

CHAPTER 13

Are *B* Chromosomes 'Selfish'?

R. N. Jones

*Department of Agricultural Botany, University College of Wales,
Aberystwyth, UK*

INTRODUCTION

The phenotype of eukaryotes reveals little about their chromosome complements. It is impossible by studying the morphology and development of an
animal, or a flowering plant, to say anything about the level of ploidy or
the number and size of the chromosomes. The relationship between the
chromosome karyotype and the 'exophenotype' of an organism is quite
obscure. It is only at the level of the cell and the nuclear phenotype that
relationships begin to emerge, and it becomes possible to equate variation
in chromosome size and nuclear DNA amounts with certain characteristics
such as the cell cycle time and cell size. Even at this level of analysis,
however, we are still bewildered by the fact that up to 90% of the nuclear
DNA may exist in a form which does not code for functional proteins and
does not perform any other known function that we can adequately explain.
This book is concerned with the silent majority of nucleotide base pairs
which inhabit the nucleus, and with the question of why they should inhabit
it at all? The debate generally revolves around the variation that occurs
between species within genera, or larger groupings, rather than between
individuals within species. This is so because the mechanism of heredity
ensures that chromosome complements are self-replicated with high fidelity
and the relationships between individual chromosomes, in terms of number
and structure, are rigidly conserved during cell division: individuals within a
species have karyotypes that are characteristically uniform and stable. There
are certain exceptions. Polyploidy and endopolyploidy annul the usual
constraints because they preserve the numerical relationships between the
chromosomes and do not lead to the kinds of imbalance often associated with
aneuploidy, and with duplications and deletions. There are also widespread

intraspecific structural polymorphisms for certain classes of heterochromatin, such as c-bands (e.g. in *Allium*: Vosa, 1976), knobs (e.g. in maize: Longley, 1938) and supernumerary segments (notably in the Orthoptera: John, 1973), which are all fully integrated into the genome and which exist in apparent harmony with the normal requirements for physiological fitness. These cryptic forms of variation have been the subject of some recent reviews (John and Miklos, 1979; John, 1981) and are briefly discussed in Chapter 4.

Another class of variable DNA, and one that is at extreme variance with the concept of 'balance' within the genome, is that due to the occurrence of supernumerary *B* chromosomes. These *B* chromosomes (*B*s) are non-essential supernumeraries which are not homologous with members of the basic diploid or polyploid complement (*A*s) and which have self-controlling non-mendelian methods of inheritance that permit them to accumulate numerically in populations over generations of outcrossing. They are of interest and significance in the present context because their accumulation system gives rise to widespread polymorphisms for DNA variation, at the intraspecific level, and there is a growing body of evidence and opinion that they may be 'selfish', rather than adaptive, in terms of natural selection and evolution.

CHARACTERISTICS OF *B* CHROMOSOMES

The name *B* chromosomes was allocated by Randolph (1928) to distinguish them from the *A*s in maize. They can only be recognized by cytological screening because their effects are quantitative and not at all obvious in terms of phenotype. During karyotype analysis they stand out as additions which differ in morphology (Fig. 13.1) and which are present in variable numbers in only a proportion of individuals within a population. They are usually smaller than the *A* chromosomes (except in the Orthoptera where they are as large as, and similar in appearance to, the *X*) and often have subterminal centromeres. Contrary to popular belief they are not necessarily heterochromatic (Jones, 1975), and their DNA composition is similar to that of the rest of the complement. *B* chromosomes are never homologous with *A*s, by definition, but will pair among themselves when two or more are present, forming bivalents of multivalents. They are conspicuously lacking in major genes and nucleolus organizers and their origin is unknown.

Another distinctive feature of *B*s is their irregular and non-mendelian mode of inheritance: some show instability at mitosis and may be eliminated altogether from certain tissues. Where the instability takes the form of somatic non-disjunction, as in the spermatagonial cells of some animals, it can constitute an accumulation mechanism (see p. 415). In many plant species mitotic non-disjunction regularly occurs at either the first or second pollen grain mitosis, in the male gametophyte, and an unreduced number of *B*s is *directed* into the germ lines. Pairing of *B*s, and their subsequent distribution

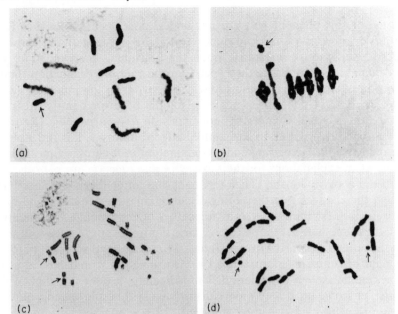

Fig. 13.1. (a) Mitosis in a male embryo of *Pseudococcus obscurus* showing 5 euchromatic chromosomes, 5 heterochromatic darkly stained chromosomes and a heterochromatic *B* (arrowed). (Reproduced with permission from Nur, 1962a, *Chromosoma*, **13**, 249–271.)
    (b) Metaphase I of meiosis in rye showing a single *B* chromosome.
    (c) Two large metacentric *B*s at c-mitosis in *Myrmeleotetrix maculatus*.
        (Photograph by courtesy of Godfrey Hewitt.)
    (d) C-mitosis in chives (*Allium schoenoprasum*) with two small telocentric *B*s. (Photograph by courtesy of John Parker).

at meiosis, is often less efficient than for *A* chromosomes and there is some loss at this stage. In certain species a meiotic drive mechanism favours the movement of *B*s into the germ line cells (see p. 415). *B* chromosomes are thus independent of the *A*s in their inheritance and have modified forms of distribution which enable them to raise their frequencies in populations over generations of sexual reproduction.

As far as their effects are concerned the position can be summarized by saying that their expression is quantitative and additive and is usually undetectable for low numbers. High numbers lead to severe impairment of physiological and reproductive fitness. The cell and nuclear phenotype is more sensitive to their presence and effects have been demonstrated for such characters as cell size, cell cycle time, quantitative variation in nuclear protein and RNA and chromosome behaviour at meiosis. In many cases the effects are not directly proportional to the numbers present, but show a puzzling effect in relation to odd and even numbered combinations, the odds (1*B*, 3*B*) being disproportionately more detrimental than the evens (2*B*, 4*B*). The reasons for this phenomenon are not understood at all (Jones and Rees, 1969). Very few cases are known where the Bs can actually be shown to confer some adaptive advantage upon the individuals or populations that carry them. In some plants they speed up the rate of germination, and in some others they may increase growth and vigour in low numbers (Jones and Rees, 1982). Under experimental conditions of severe density stress they enhance the survival chances of individuals of *Lolium perenne* (Rees and Hutchinson, 1973). Of all their manifold effects, however, it is the aspect of their influence on the pairing and recombination in *A* chromosomes that we turn to when explanations for their existence and function are not otherwise forthcoming or adequate. The argument has often been advanced that through the changes that they bring about in chiasma frequency and distribution in the *A*s, the Bs serve as regulators of the amount and quality of recombination, and that they exercise some control over the release of variability in outbreeders (Hewitt and John, 1967; Jones and Rees, 1967). This argument of course is an attractive one as it involves a fundamental aspect of the genetic system of variation. Unfortunately, it does not explain how the Bs can come to be so widely distributed within and among different species. They have now been recorded in at least a thousand flowering plants and in more than 260 animals (Jones and Rees, 1982). The status of some of these recordings is doubtful though because many have been observed only at mitosis in a few individuals and their homology relationships are untested. Even so there is reason to believe that Bs are a normal part of the genetic system in a significant number of eukaryotes—as many as 10–15%. Within a species it is usual to find them in only a proportion of individuals and for this proportion to vary from one population to another.

   The debate about *B* chromosome polymorphisms revolves around the issue of whether these widespread polymorphisms are simply a direct result of the accumulation mechanisms, or whether they derive from the action of natural selection and evolution, or a combination of both? Throughout the argument runs the complicating factors that the Bs are unnecessary for the normal processes of growth, development and reproduction, and in many cases they are demonstrably harmful to the individuals that carry them.

ADAPTIVE OR 'SELFISH' CHROMOSOMES?

The question of the selfishness of *B* chromosomes was raised as long ago as 1945 by Östergren, in slightly different terms, in his paper on the '. . . parasitic nature of extra fragment chromosomes . . .' (in rye). The question arises now with renewed interest because we are much more aware of the widespread distribution of Bs in natural populations and much more aware too of the redundant, and apparently useless, nature of a large proportion of the nuclear DNA in eukaryotes. A review of some well-documented case histories will assist us in evaluating this question.

Secale cereale: Rye

$2n = 2x = 14 + 0–8$ Bs

The B in rye is half the size of one of the As and is made up of about 850 × 10^6 nuclear base pairs with no known protein-coding function. It has a terminal block of c-heterochromatin at the end of the long arm, and the overall DNA composition similar to that of the A chromosome.

The Bs in rye are stable at mitosis with the same number in every cell and tissue in the sporophyte. At meiosis they pair exclusively among themselves forming univalents (Fig. 13.1), bivalents or multivalents according to their number, and they are distributed to the micro- and megaspores at the end of the reduction division with a high level of regularity. At the first pollen grain mitosis in the male gametophyte they undergo directed non-disjunction and pass undivided into the generative nucleus (Fig. 13.7), and thus into the germ line. An equivalent process takes place in the female gametophyte and the transmission rate of the Bs is thus in excess of that required simply to maintain their level in the progenies. Their numbers increase over the generations, but the build up is counteracted by a certain degree of meiotic elimination and by infertility which accompanies high numbers.

The consequence of this non-mendelian system of heredity can be seen in the experimental results depicted in Fig. 13.2. A sample of seed of Korean

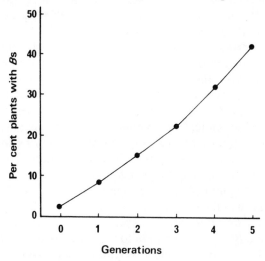

Fig. 13.2.   Accumulating B frequencies in a strain of Korean rye over five generations of open pollination.

rye with an initial *B* frequency of 2% was maintained by the author as an open pollinating population of about 600 plants over five generations of sexual reproduction. At each generation the *B* frequency increased by 8% up to a final level of over 42% when the experiment was ended. At the final sampling the composition of the population was 58.4% 0*B*, 1.3% 1*B*, 37.4% 2*B*, 0.4% 3*B* and 2.5% 4*B* individuals.

Studies on *B* chromosome distribution in natural populations of rye confirm their presence in high frequencies and their widespread occurrence. In 'native' Japanese rye, Kishikawa (1965) found *B*s in a frequency of more than 90% in one of his population samples. Lee and Min (1965) reported the presence of *B*s in all 32 populations which they surveyed in Korea, in frequencies ranging from 2–73%. Müntzing (1957) also sampled Korean rye and he found the *B*s in all seven of his populations in frequencies from 19–90%. In a major survey in Yugoslavian rye Zecević and Paunović (1967) recorded *B*s in all 25 of their populations over the frequency rance 10–55% . The absence of *B*s in primitive rye, in *some* of the population samples, has been reported for strains from Iran (Kranz, 1963) and Afghanistan (Müntzing, 1950).

The major difficulty that arises in any attempt to explain these extensive polymorphisms comes from the fact that the *B*s in rye generally have deleterious effects upon the growth and reproductive potential of individuals that carry them (see Müntzing, 1963) and it is difficult to envisage any conditions under which natural selection could maintain and promote them at such high frequencies. It is tempting to simply ascribe their success to the accumulation mechanism and to suggest that natural selection is of minor importance in determining the patterns of their polymorphism.

One way of tackling this question is to make use of the detailed knowledge available on the cytogenetics of the rye *B* to construct a model which will simulate and predict changes in *B* frequency in randomly mating populations, and which will identify the key factors involved in determining these frequencies. In the model presented by Matthews (1981) and Matthews and Jones (1982), the behaviour of the *B*s was described in mathematical terms and equations devised which included parameters determining their transmission at meiosis, pollen grain (and egg cell) mitosis, and during the development of the gametophyte and sporophyte. The parameters determining transmission rates at meiosis are the *rate of non-pairing* of the *B*s, and the rate of loss (by lagging) of unpaired univalents—the *non-paired loss rate*. Transmission through the pollen (and its equivalent in the egg) is determined by the *non-disjunction rate* and the *direction rate* (i.e. the direction in which the non-disjoined *B*s move in relation to the generative and vegetative pole). Gametic weighting factors and plant weighting factors were used to allow for reduced competitive abilities of pollen grains and plants, respectively.

The equations were transcribed into a computer program into which were put (1) a value for the initial B frequency and distribution of B classes in the population, (2) desired values for the parameters and (3) an input to take account of the fact that plants with four Bs and over are virtually sterile, and to remove them from the calculations at each generation. Sterility is thus a major factor limiting accumulation and in this model places an upper limit on the equilibrium of $4B$/plant. The programme was run for the required number of generations and the output given in the form of computer generated graphs (Figs. 13.3, 13.4, 13.5 and 13.6).

The value of the model lies in the fact that one parameter at a time can be varied, while the others are held constant, and an assessment made of the contribution that each parameter makes to the change in B frequency and to the final equilibrium. Fig. 13.3, for example, displays the output of a simulation showing the effects of different levels of non-pairing of Bs at meiosis. The non-paired loss rate parameter is set at a value of 0.30 (to give a wider separation of the lines on the graph) and the other parameters held at their maximum positive values. The initial B level, as in the other examples, is set at an arbitrary and low level of 0.01 B/plant. The graph shows just how rapidly the Bs can accumulate when directed non-disjunction is given maximum expression, and how the equilibrium frequency can be influenced by the parameters determining transmission at meiosis (variation in the non-paired loss rateparameter has a similar effect). Fig. 13.4 shows that variation in the non-disjunction rate has surprisingly little effect on the final equilibrium level, although it does influence the number of generations required to reach equilibrium, and that it is effective in accumulation even at low rates. Estimates from the literature suggest a mean non-disjunction rate of about 0.86, and a range of variation between 0.76 and 0.95. Direction rate (Fig. 13.5) has a marked effect on final equilibrium levels—a fact not hitherto appreciated. Estimates from the literature give mean values to this parameter of around 0.85. Simulation of the effect of varying selection against gametes (Fig. 13.6), and a similar one for that against plants, reveals that the detrimental effects which Bs have on the competitive fitness of gametes, and of plants, has little consequence to the final equilibrium; the main effect is a delay in the number of generations taken to get there.

As far as can be seen from these simulations, and from what we know from experiments (Fig. 13.2), the Bs in rye are capable of generating and maintaining their polymorphism solely through the agency of their own potent accumulation. Provided newly arisen, or newly introduced, Bs are endowed with the capacity for directed non-disjunction (even at a low rate) they will be able to survive and to distribute themselves throughout populations, even though they impose a fitness burden upon the individuals that carry them. The B chromosomes in rye are 'selfish' because they are harmful to their hosts and no explanations are forthcoming to account for the existence of their polymorphisms other than their accumulation mechanisms. It

Fig. 13.3–6. Simulation graphs showing how parameters controlling transmission of *B*s in rye affect equilibrium frequencies in populations.

EFFECT OF VARYING NONPAIRING RATE

NONPAIRED LOSS RATE = 0.30

NONDISJUNCTION RATE = 1.00

DIRECTION RATE = 1.00

GAMETIC WEIGHTING FACTORS	O	1	2	3	4
	1.00	1.00	1.00	1.00	1.00
PLANT WEIGHTING FACTORS	O	1	2	3	4
	1.00	1.00	1.00	1.00	1.00

INITIAL *B* LEVEL = 0.01

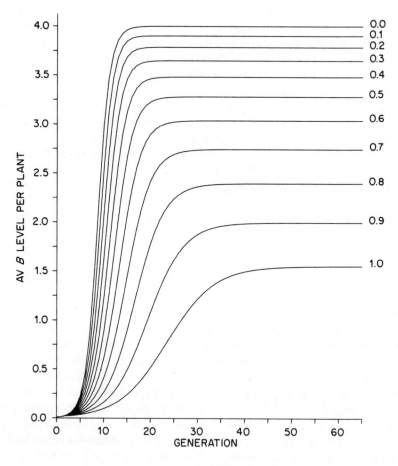

Fig. 13.3.   Effect of varying nonpairing rate.

EFFECT OF VARYING NONDISJUNCTION RATE

NONPAIRING RATE = 0.00
NONPAIRED LOSS RATE = 0.00
DIRECTION RATE = 1.00

GAMETIC WEIGHTING FACTORS	0	1	2	3	4
	1.00	1.00	1.00	1.00	1.00
PLANT WEIGHTING FACTORS	0	1	2	3	4
	1.00	1.00	1.00	1.00	1.00

INITIAL *B* LEVEL = 0.01

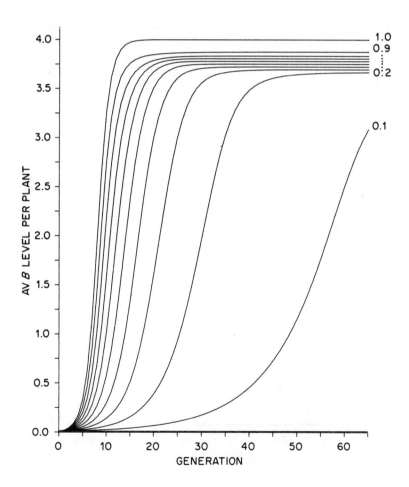

Fig. 13.4. Effect of varying nondisjunction rate.

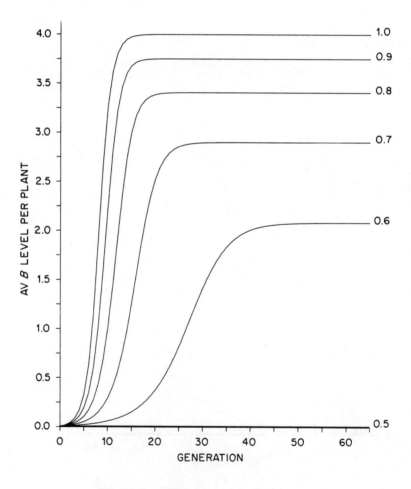

EFFECT OF VARYING DIRECTION RATE

NONPAIRING RATE = 0.00
NONPAIRED LOSS RATE = 0.00
NONDISJUNCTION RATE = 1.00

GAMETIC WEIGHTING FACTORS	0	1	2	3	4
	1.00	1.00	1.00	1.00	1.00
PLANT WEIGHTING FACTORS	0	1	2	3	4
	1.00	1.00	1.00	1.00	1.00

INITIAL *B* LEVEL = 0.01

Fig. 13.5. Effect of varying direction rate.

EFFECT OF VARYING GAMETIC WEIGHTING FACTORS

NONPAIRING RATE = 0.00
NONPAIRED LOSS RATE = 0.00
NONDISJUNCTION RATE = 1.00
DIRECTION RATE = 1.00

PLANT WEIGHTING FACTORS	0	1	2	3	4
	1.00	1.00	1.00	1.00	1.00

INITIAL *B* LEVEL = 0.01

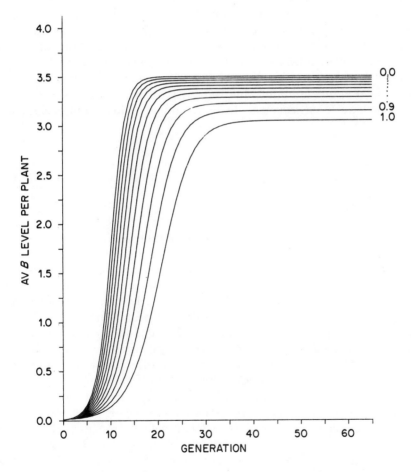

Fig. 13.6.   Effect of varying gametic weighting factors.

does not follow though that natural selection is precluded from playing *any* part in determining their equilibrium frequencies. Different frequencies of *B* chromosomes do exist in different populations. It is possible that the *B*s in some of these populations are young, and have not yet reached equilibrium. It is also likely that many of the populations have reached equilibrium and that they have stabilized at different frequencies and distributions of *B* classes. We expect this situation because the processes of accumulation and elimination which are interacting to give the balance are themselves sensitive to environment. They are physiological processes which involve the pairing, the recombination and the movement of *B* chromosomes within cells, and it would be most surprising if they were excluded from the normal interactions which determine the behaviour and phenotype of other chromosomes. Sensitivity to environment has been demonstrated experimentally. Kishikawa (1970) has shown that transmission of *B*s varies in relation to temperature and soil moisture, and Rees and Ayonoadu (1973) have found a differential survival rate of *B* chromosomes in rye under varying conditions of physiological stress. In addition to this interaction with environment the *B*s are also sensitive to background genotype and to variation arising from evolution in their structural organization (Müntzing, 1951). The 'selfishness' is sensitive and responsive to environment, to selection and to long-term evolutionary change.

Pseudococcus obscurus: Mealy bug

$2n = 2x = 10 + 0 - 6 \ Bs$

The mealy bug has a 'lecanoid' genetic system in which the chromosomes, including the *B*, are holokinetic, with diffuse centromeres, and the meiotic sequence is inverted: the first division is equational and the second reductional. As a further modification the paternal set of chromosomes in the males becomes heterochromatic and inert at an early stage of development in the embryo and remains so throughout development (Nur, 1962a).

In spermatogenesis there are two modified divisions. At the first, both heterochromatic and euchromatic sets of chromosomes divide mitotically: at the second they segregate to opposite poles. Of the four resulting products of meiosis, two are euchromatic and two heterochromatic. The heterochromatic set degenerate and the haploid euchromatic nuclei form the sperm. When *B*s are present in males they are heterochromatic until late prophase I, when they change their status and become euchromatic. They divide mitotically at anaphase I, along with the *A*s, but at anaphase II they segregate together with the maternal euchromatic set into the cells destined to become the sperm, and are thus transmitted in an unreduced number through the males. In females the Bs are transmitted normally except for a slight loss of univalents (Nur, 1962a).

The *transmission rate* (k) of the B chromosome has been defined by Nur (1966a) as 'the ratio of the average number of Bs in the gamete of an individual to the number of Bs of that individual', and on the basis of Mendelian heredity a 1B individual is expected to transmit a single B to half its gametes and to have a k value of 0.5. Controlled crossing experiments under laboratory conditions, using individuals with 1B, 2B, 3B, 4B and 5B classes, show an average transmission rate through the male of 0.88, and through the female of 0.45. In the zygotes therefore the Bs will increase by about 33% at each generation and their frequencies will rise to high levels unless the increment is checked by some form of elimination (Nur, 1962b).

Surveys made on several widely divergent habitats in California, including some experimental populations, showed Bs in mealy bugs to be ubiquitous and present in every population sampled. In the coastal region around Berkeley they were found in frequencies ranging from 74 to 93%, while in the Sacramento Valley their incidence was much lower (approximately 20%). In laboratory populations their frequency reached 100%. Some of the populations have been subjected to repeated sampling and are known to be at equilibrium frequencies. The differences between geographical regions arise from the differences in environmental circumstances under which the equilibria were established.

The population dynamics have been studied by comparing the frequencies and distributions of B classes in female and male parents and in their progenies: the existence of a stable equilibrium situation makes this approach valid and useful. In one population a sample of females were taken from the wild and allowed to lay their eggs in captivity; the B frequencies of both generations were counted cytologically. Frequencies in the male parents were calculated from the data on females and zygotes and from the known facts about transmission rates of the males (and certain other assumptions). The comparisons showed the same frequencies in both the female parents and the zygotes, indicating no difference in fitness of females with or without Bs, but a significantly lower frequency among the males. Evidently Bs are eliminated by selection against males which carry them. Comparisons of B frequencies in the males and their offspring were used to assign relative fitness levels to the males based on a value of 1.00 for the 0B class. The values for the 0B, 1B, 2B, 3B and 4B classes came out at 1.00, 0.52, 0.63, 0.42 and 0.23 respectively (Nur, 1966a).

Experiments on morphology and development later confirmed the findings from the population data. Bs were found to have no detectable effects upon the phenotype or reproductive fitness of females, but a single B in the male reduced the number of sperms and delayed development (Nur, 1966b). The

rate of transmission in the males also decreases with increasing numbers of
Bs.

Nur is firmly of the opinion that the Bs in *Pseudococcus* are maintained
in populations only because of their accumulation mechanism. They appear
to be totally 'selfish' and are kept in check at equilibrium only by the reduced
level of reproductive and developmental fitness which they impose upon the
males that carry them.

Natural selection has a part to play in the population dynamics in as much
as the variation in levels of fitness induced by Bs in the males may find a
different level of expression in different environments. Under laboratory
conditions comparisons between 0B and 1B males show the 1B individuals
to develop more slowly, and the effect to be more pronounced at lower
temperatures. The finding is correlated with population studies which show
the relative fitness of the 0B individuals to fluctuate in relation to seasonal
changes, although it is always higher than that of males with Bs (Nur, 1969).

Myrmeleotettix maculatus: Mottled grasshopper

$2n = 2x = 17\male/18\female + 0 - 3Bs)$

In the mottled grasshopper the standard B is a large mitotically stable and
heterochromatic isochromosome (Fig. 13.1); it is equivalent in size to the X
and has an accumulation mechanism.

Cytological studies have shown that at male meiosis univalent Bs invariably
move undivided to one of the poles at anaphase I and then divide at AII.
In rare cases where they do divide at AI the resulting chromatids attach to
the spindle and move to one of the poles at AII. When two Bs are present
they form bivalents in about half the meiotic cells, and in the rest they are
univalent. There is virtually no elimination of unpaired Bs at male meiosis
(John and Hewitt, 1965a). In crossing experiments involving $0B\female \times 1B\male$,
it came as a surprise therefore to find that the transmission rate was much
lower than the expected level of 0.5. For East Anglian (Foxhole population)
grasshoppers the transmission rates gave a value of $T = 0.381$, and those
from West Wales (Talybont population) were higher at $T = 0.486$ per B. In
$0B\female \times 2B\male$ crosses the corresponding rates per B were $T = 0.264$ for
Foxhole and 0.413 for Talybont. There is thus considerable elimination of
Bs when they are transmitted through the male, which is thought to be due
to defects they cause at the later stages in sperm maturation, or even at
fertilization, rather than at meiosis (Hewitt, 1973a). In reciprocal crossing
experiments the combination of $1B\female \times 0B\male$ gave an enhanced transmission
rate of $T = 0.582$ for Foxhole pairs and $T = 0.887$ for those from Talybont.
In a $2B\female \times 0B\male$ cross (Foxhole) the value was $T = 0.645$. There is thus
an accumulation of Bs when transmitted through the female (see p. 415).

Extensive population surveys covering much of Europe show the *B* chromosome polymorphism in *Myrmeleotettix* to be restricted to the Southern half of Britain. The average frequency of occurrence of *B*s, in the populations where they do occur, is about 40% and the range extends from 7 to 70% (John and Hewitt, 1965b); Hewitt and John, 1967, 1970; Hewitt, 1973a, b, c; Robinson and Hewitt, 1976). In many of these populations the frequencies have remained stable over as long as seven years of sampling, and there is firm evidence for a correlation between the distribution pattern of the *B*s and certain environmental variables such as temperature and rainfall. In general the *B*s are most frequent in warm dry areas in which the species thrives (Hewitt and John, 1967, 1970).

The simplest explanation to account for these established patterns of distribution would be to invoke natural selection and to argue the dynamics of the polymorphism on the grounds of differential fitness and adaptation. As it happens, however, the *B*s have no detectable effects on the morphology of the exophenotype (Hewitt and John, 1970), and this line of argument is difficult to sustain. There are effects at the level of the endophenotype, in terms of changes in chiasma frequency and distribution induced by the *B*s in the *A* chromosomes, which may well be of significance to the long-term survival and evolution of the population, but there is no known means by which these effects could generate and maintain the *B* polymorphisms. On the positive side the accumulation system, and the counter balancing elimination processes likely to be associated with it, seem to offer the most promising line of inquiry so far into an understanding of the population dynamics.

When the frequencies of the 0*B*, 1*B* and 2*B* individuals, found in the Talybont population mentioned above, are used in conjunction with the data on transmission rates to calculate *B* frequencies in the next generation, there is a predicted rise due to accumulation of 28%. Bearing in mind that the population was at equilibrium for the five years prior to the experiment, then the predicted rise must be offset by elimination. The loss of all individuals with 3*B*s or more at each generation (they are rarely found), together with reduction in fitness due to embryo mortality caused by higher levels of *A* chromosome mutation in the 1*B* and 2*B* classes, accounts precisely for the deficit required to maintain the equilibrium (Hewitt, 1973a). The data for the Talybont population are thus consistent and unequivocal in terms of the 'selfish' *B* model.

The Foxhole population data is inconsistent and inconclusive for any model. The problem is that their accumulation, based on the crossing data, is not high enough to maintain them at equilibrium frequencies. When account is taken of loss of embryos due to *A* chromosome mutation and elimination of classes of 3*B*s or more, there is a predicted shortfall of 11% in the next generation. To bring this deficit up to equilibrium requires considerably enhanced levels of fitness in the 1*B* and 2*B* classes, relative to the 0Bs, and there is no firm evidence that such enhanced fitness exists. In fact Nur (1977) has re-analysed Hewitt's data, and that of Robinson and Hewitt (1976), and come to the conclusion that plus B types in the Foxhole population all have reduced, rather than enhanced, levels of fitness. Other evidence also shows that the Bs in *Myrmeleotettix* reduce fitness by slowing down development (Harvey and Hewitt, 1979).

There is also data to show that the incidence of Bs can vary over the annual life cycle of the organism (Robinson and Hewitt, 1976); the frequencies tend to rise in the prediapause autumn eggs, relative to the adult population, and then decline again in the spring embryos. The effects are self-cancelling, however, and there is no change in frequency between one adult generation and the next. The Foxhole population again shows as inconsistency in that a raised B frequency was found in the autumn eggs, when on the basis of prediction it should have been lower. This particular population could well be left out of the argument on 'selfishness' pending further analysis and clarification of its polymorphism.

The evidence for 'selfish' Bs in *Myrmeleotettix* is not as strong as it is in rye, and in the mealy bug, but at the same time it has to be said that we have no convincing alternative explanation either to account for the widespread distribution of Bs in this species.

Melanoplus femur-rubrum

$2n = 2x = 23\sigma/24\female + 0 - 2 \; Bs$

The B is large in size, almost as big as the X, mitotically stable and heterochromatic at prohpase I of meiosis. At male meiosis it moves undivided to one pole at AI and then divides regularly at AII without any elimination. In the female there is a preferential movement at AI of oogenesis into the secondary oocyte, rather than the polar body. Controlled crosses give transmission rates of 0.5 and 0.8 for 1B males and females, respectively; the species thus has an accumulation mechanism (Lucov and Nur, 1973; Nur, 1977). Nur (1977) has presented very convincing data to show that the B in *Melanoplus* is 'parasitic' (i.e. 'selfish') and that equilibrium frequencies are maintained by a balance between accumulation and the reduction in fitness imposed on individuals with 1 or 2Bs.

Lillium callosum

$2n = 2x = 24 + 0 - 2 \; B$

The telocentric B in *Lilium* accumulates during transmission through the female. In $1B\female \times 0B\sigma$ crosses the progeny are in the ratio 80:20 for 1B:0B individuals. The crossing data corresponds with cytological observations which showed that in about 80% of female meioses the *B* is included in the division products at the micropylar end of the embryo sac which includes the egg cell (Kayano, 1957). The effect of the *B*s upon development, particularly in respect of pollen and seed fertility, is markedly deleterious. In natural populations, however, the frequencies of *B*s are high and vary among populations. A mathematical model based on the transmission characters and on the data for distribution of the *B*s led Kimura and Kayano (1961) to the conclusion that the *B*s are 'parasitic' ('selfish') and maintained solely by their meiotic drive mechanism. The forces contributing to the equilibrium were considered to be subject to environmental control, as suggested for the other species described above, and different equilibria can thus be established in different habitats.

More cases of accumulation

Many other cases of polymorphism are known, in both plants and animals, where the *B*s have accumulation mechanisms which lead to enhanced transmission to the progenies, and where the *B*s are capable of raising their numbers in populations until they become stabilized by elimination (Jones, 1975; Jones and Rees, 1982). Examples in animals are mainly found in the Orthoptera, and the accumulation generally results from mitotic non-disjunction leading to numerical variation in *B*s among different follicles of the male testis. Higher numbers of *B*s are found among spermatocytes than among standard somatic tissues, and the accumulation is thought to result because the cells with the increased number are contributed to the germ line in preference to those without (see below).

Accumulation in plants results most frequently from directed non-disjunction at first pollen grain mitosis and is restricted to the male side. The situation in rye, with non-disjunction for both the male and female, is exceptional. When the computer simulation is applied to cases of male accumulation only, as in ryegrass and maize, the outcome is very similar to that described for rye, except that more generations are needed in order to reach equilibrium. The polymorphisms in many plants are extensive, in some cases throughout the range of the species distribution, and no processes are known to account for them other than 'selfish' accumulation.

Cases without accumulation

B chromosome polymorphisms are also known where no accumulation mechanism has been demonstrated: some of them will have to be considered because they are relevant to the debate.

The *B*s in *Centaurea scabiosa* are small and always univalent at meiosis.

They usually divide at AI and then lag and suffer some elimination at AII. Crossing experiments show a small loss during male and female transmission. Fertility and vegetative development are adversely affected by high numbers and there is no indication that they confer any selective advantage upon individuals that may account for their widespread geographic distribution. The polymorphism covers a large part of Europe and there is a correlation between the incidence of Bs and favourable environments for the species (Fröst, 1958).

The 'Nodding Onion', *Allium cernuum*, has an extensive B polymorphism in parts of Pennsylvania (Grun, 1959). The Bs are small, telocentric and unstable at mitosis to such an extent that it is impossible to label plants as belonging to any particular B class. At meiosis they are always univalent: they drift randomly to the poles at AI and then divide equationally at AII. Amazingly, they are transmitted with a high frequency and widely distributed in natural populations. They have no obvious advantage or significance and no ecological preferences.

In *Allium schoenoprasum* (Chives) the Bs are small (Fig. 13.1) and of several morphological types. They are stable at mitosis and have variable degrees of pairing at meiosis. Even when unpaired, however, they are transmitted with a minimum of elimination. There is no accumulation—rather a small loss during transmission as in *Centaurea*—and their effects upon fertility and vigour are deleterious. No explanations have yet been found for the existence of populations with up to 65% of B-containing individuals (Bougourd and Parker, 1979).

In each one of these three cases it has been proposed that some form of natural selection is needed to account for the polymorphism, but, to date, no explanations have been found in any of them. There are no grounds either upon which we may describe them as 'selfish', unless it turns out that they are being generated (rather than accumulated) at higher than mutation frequencies.

CONTROL OF SELFISHNESS

It is quite evident from a consideration of the case histories given above that the 'selfishness' of B chromosomes has its basis in the various methods of preferential distribution and accumulation which direct Bs into the germline. This process is obviously of fundamental importance in relation to the general debate about 'selfish' DNA in eukaryotes and it will be worthwhile inquiring briefly into both the mechanism itself and the basis of its control.

There are essentially two mechanisms of accumulation: one involves directed non-disjunction of unseparated chromatids at mitosis, and the other a preferential segregation of univalent Bs at meiosis. In neither case do the preferences operate with a hundred per cent fidelity.

Preferential segregation at meiosis

In *Myrmeleotettix* the operative stage for this process is MI in the primary oocyte, as described by Hewitt (1976). There is an asymmetrical spindle which is flattened and shortened on the polar body side, elongated and pointed on the egg side. The univalent *B*s in both the 1*B* and 2*B* types (there are relatively few bivalents), are distributed about the spindle with a bias towards the egg side and are preferentially included in the secondary oocyte at anaphase I. Close agreement was found between the frequency of their orientation and the level of preferential transmission in crosses. The precise biophysical or biochemical basis of the preference is open to discussion, but one possibility we may entertain is that the bias in distribution is simply passive and a direct consequence of the unique spindle form, in which case 'selfish' may be too strong a word! A similar system has been proposed to account for enhanced transmission of the *B* in *Melanoplus femur-rubrum* (Lucov and Nur, 1973). As already described, preferential segregation in the mealy bug occurs at AII in males, and in this division, too, there is asymmetry of the spindles, as well as the formation of a double metaphase plate, but the details are not all that well understood (Nur, 1962a).

In *Lilium callosum* there is preferential segregation of univalent *B*s during anaphase I of meiosis in the egg mother cell. At MI the majority of univalent *B*s lie outside the metaphase plate with a preference for the micropylar side of the spindle. They pass undivided to the pole at AI and then divide at AII. In this case there is no asymmetry of the spindle, although the egg mother cell itself is often placed nearer to the micropylar than to the chalazal end of the embryo sac. The mechanics of the preference are thus unknown, and in common with the grasshoppers and the mealy bug we can say little about genetic control.

Non-disjunction at mitosis

Directed non-disjunction at mitosis is the commonest method of *B* chromosome accumulation and it takes place in a variety of situations. In animals, particularly the Orthoptera, non-disjunction is assumed to be the cause of numerical variation of *B*s among different follicles of the testis. This assumption is made because individuals with a standard somatic complement containing 1*B* often have follicle cells with 0*B* and 2*B* as well as with 1*B*. The non-disjunction is also preferential in that those with the raised number (2*B*) are more frequently contributed to the germ line than those without. In 1*B* individuals of *Calliptamus palaestinensis*, for example, the testis follicles with 2*B*s outnumber those with 0*B*s by a ratio of 15:1, and Nur (1963) has suggested that in this species the preferential non-disjunction events occur during embryonic divisions when the cells destined for the germ line are separated from the somatic ones. Boosting of numbers into the germ track by preferential mitotic non-disjunction has also been suggested for some plant species, e.g. *Crepis capillaris*.

Non-disjunction in plant gametophytes is better understood, with the bulk of the information coming from studies on rye and maize (for details of this mechanism in other species, and for some other processes not mentioned here see Jones and Rees, 1982).

In rye there is non-disjunction at the first mitosis of both the microspore and megaspore. On the female side little is known about the mechanism except that it takes place at first division in the embryo sac and the Bs are directed into the micropylar nucleus. In the male gametophyte there is an asymmetrical spindle with a blunt generative pole depressed against the cell wall and a pointed vegetative pole in the central part of the pollen grain, as in *Anthoxanthum aristatum* (Östergren, 1947). The B lags between the separating groups of As at anaphase and then moves undivided to the generative pole. The metaphase plate is closer to the generative pole and the reduced distance and shorter microtubules may favour movement of the non-disjoined Bs in that direction. Centromere function appears to be normal, but disjoining is prevented by sites on either side where the chromatids remain stuck together (Müntzing, 1946). Delayed replication of DNA has been implicated in this process. Control over these events (Fig. 13.7), evidently resides in the Bs themselves because they retain the same capability, and the same timing and sequence of events, even when transferred from rye into a strain of hexaploid wheat (Müntzing, 1970). The controlling elements appear to be located in the terminal heterochromatic segment of the long arm, rather than in the sensitive sticking regions adjacent to the centromere. When the standard subterminal B suffers a deletion of its terminal heterochromatin it also loses its ability to non-disjoin, even though it still has the sensitive sites present. Oddly enough, this ability can be restored by the presence of another standard B within the same cell. The 'gene' controlling non-disjunction can transmit its signal between different Bs, and this mechanism argues against a passive form of 'selfishness' mentioned earlier; indeed it raises further questions about possible genetic variation and the involvement of selection in the 'selfish' process itself.

Zea mays is different again: here the non-disjunction events have been unravelled by genetic analysis and have been shown to take place at the second pollen grain mitosis which gives rise to the sperm nuclei. The non-disjunction occurs with a frequency of 50 to 98%, but it is not directed and the accumulation depends as well on preferential fertilization (at a rate of approximately 65%) by the sperm nucleus carrying the unreduced number of Bs. This preference is lost with high numbers of Bs and this loss limits the accumulation (Carlson, 1969). A detailed account of B chromosome behaviour in the male gametophyte of maize is given in Carlson (1978). The

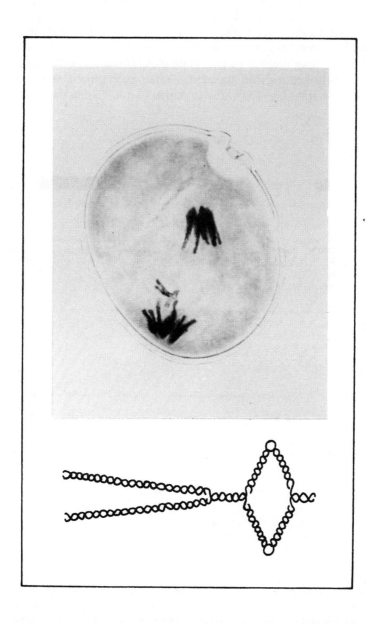

Fig. 13.7. A single *B* undergoing directed nondisjunction towards the generative nucleus at first pollen grain mitosis in rye, together with a diagram of the proposed mechanism by delayed DNA replication.

polymorphism in maize has not been much studied with reference to the population dynamics (although the *B* is widespread in primitive strains), but it has been more thoroughly analysed than in any other species in terms of the elements controlling non-disjunction. The *B* is highly heterochromatic and just larger than the size of the smallest *A*. It has a minute short arm, a block of centric heterochromatin and a small euchromatic segment at the end of the long arm (Fig. 13.8). It is assumed to non-disjoin due to stickiness

(a)

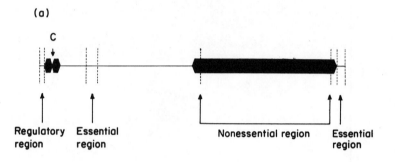

(b)

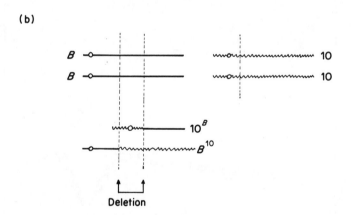

Fig. 13.8. (a) Diagram representing the structure and organization of the maize *B* chromosome in terms of its euchromatin (thin line), heterochromatin (black region), centromere and adjacent sensitive sites (C) and location of controlling elements: based on sources given in the text. (Scale and locations approximate.)

(b) Diagram showing how two of Lin's *B*-10 translocations, with break points at different places in the *B*, are used to give combinations of 10^B and B^{10} with a deletion in the proximal part of the *B*.

and delayed separation of the sensitive centric heterochromatin which responds to controls from other parts of the chromosome.

Translocations between the B and certain of the As (known as TB-As) have proved useful for identifying controlling elements. Roman (1947) working with the TB-4a translocation, which involved chromosome 4 and a break near to the mid point of the long arm of the B, found that non-disjunction of the B^4 (translocated chromosome with the B centromere) could occur in $4^B B^4$ microspores but not in those with $4B^4$. Evidently the region of the B distal to the breakpoint carries a 'gene' controlling the non-disjunction in the centromere region, and the effect is interchromosomal, as in rye. Carlson (1978) found a similar situation for TB-9a. Ward's (1973) analysis using TB-8a, in which the break point is near to the distal tip of the B, showed non-disjunction in $8^B B^8$ pollen (with a frequency of 90%), but not in $8B^8$, and he thus localized the controlling element/'gene' down to the distal euchromatic tip of the B (Fig. 13.8). The controlling 'gene' is thus '*essential*' and its effect can be interchromosomal, i.e. non-disjunction of a translocation B^A chromosome is induced in high frequency by the presence within the same cell of an A^B (or a standard B) carrying the essential 'gene'. Control over non-disjunction, however, is not exclusive to this distal 'gene', as Lin (1979) has shown from a study of TB-10(18) which has a break in the short arm of the B and one in the long arm of chromosome 10. The B^{10} chromosome which carries all of the B except the short arm, and most of the long arm of 10, is capable of non-disjunction, but at a much reduced rate of approximately 20%; the short arm is thus *non-essential*, but its absence reduces the frequency. When the short arm of the B is included in the same cell as B^{10}, in the form 10^B, the non-disjunction of B^{10} is raised from about 20% to about 69%, so the effect of the short arm is *regulating* rather than essential (Fig. 13.8). An earlier analysis by Lin (1978), in which he used a 10^B chromosome from one translocation in combination with a B^{10} from another one, for deletion mapping of proximal parts of the B (Fig. 13.8), showed that while the large heterochromatic region has no influence at all over non-disjunction, there is another *essential* region which can be located in the proximal euchromatin (Fig. 13.8). The analysis requires the use of a whole set of B-10 translocations involving many different break points in the B chromosome (Lin, 1978). It is evident from these analyses (and from more recent work by Carlson and Chou, 1981) that, as summarized in Fig. 13.8, the B in maize has genetic elements that are essential for the induction and the regulation of its own non-disjunction properties, specifically at the second pollen grain mitosis. The controlling elements/'genes' are located in the euchromatin and the large heterochromatic distal segment has no function at all in this respect. The situation is in contrast to that described in rye where the control is vested in the terminal heterochromatic block; although the possibility that this block contains some small euchromatic regions cannot be discounted. A common feature of the system in both rye and maize is that the controlling elements are remote from the sensitive regions they control and the signals may be transmitted between the 'gene' of one *B* and the sensititive region of the same or of a different *B*. As in rye the non-disjoining of the maize chromatids is thought to come about through delayed replication of DNA in the sensitive centromeric regions: evidence for this theory comes from observations by Rhoades and Dempsey (1973) on the effect of *B*s in inducing delayed chromatid separation at knobbed regions of *A* chromosomes, and from the recent discovery that the proximal heterochromatin is the last segment of the *B* to undergo replication (Pryor *et al.*, 1980).

ORIGIN

The way in which *B*s originate is not known for certain in any specific cases, although a great variety of schemes have been proposed from time to time to account for their genesis (see Jones and Rees, 1982, for details and references).

In the Orthoptera they may arise from the *X* chromosomes, with which they show some affinity and similarity at meiosis, or from certain of the small autosomes for which polysomy may be tolerated. In both plants and animals they may be derived from trisomics, tertiary trisomics, small centric fragments produced by inversion heterozygosity or from telocentrics produced by centromere misdivision.

Supernumeraries arising by any of these means all need to fulfil certain conditions in order to secure their survival in the first instance, and then their heredity and integration within the genetic system. Many may arise with mutation frequency and be ephemeral, but those that persist must cease to pair or recombine with *A*s or *A* segments from which they came, must have minimal genetic activity and phenotype effect (unless it be beneficial) and must have an accumulation mechanism. Two of these requirements can be easily fulfilled, as examples of some nascent *B*s will show. In perennial ryegrass a small centric fragment was recovered from a controlled cross between an emasculated trisomic and a normal diploid pollen parent. It had no obvious phenotypic effect and no pairing affinity with any of the basic complement (Karp, 1981). Most probably it derives from centric misdivision of the extra chromosome in the trisomic and it behaves as a typical *B* chromosome, except that its transmission has not yet been determined. It is shown in Fig. 13.9 alongside the standard established *B* of *Festuca pratensis*: the likeness is remarkable and the meiotic behaviour virtually identical. This nascent *B* may well turn out to be ephemeral, but even if it does it shows that duplication of parts of the basic complement is quite feasible without physiological disturbance, and with complete separation and loss of homology from the moment of its birth. The loss of homology is not simply a question

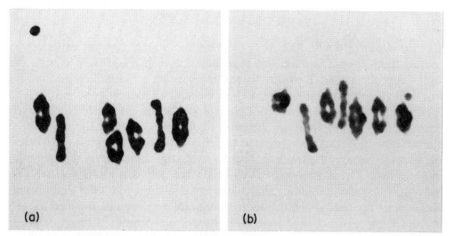

Fig. 13.9. Metaphase I of meiosis showing (a) a univalent standard *B* in *Festuca pratensis* and (b) a univalent nascent *B* of known origin in Perennial ryegrass (*Lolium perenne*).

of size because other fragments have been found in perennial ryegrass, of a similar size, which regularly pair with their homologous regions. It is likely that the fragment lacking homology has lost its telomeres, as happened in another derived fragment that gave rise to an eight-chromosome barley (Fedak, 1976), and its pairing isolation results from a simple structural alteration. At this juncture it is worth mentioning also that *some* trisomics have been found in ryegrass that have no discernible phenotypic consequence, aside from reduced fertility, and one double trisomic has been recovered which has greater vigour than its diploid siblings (Karp, 1981). These examples may be exceptional, but even so they lead us to question whether the general assumption about the imbalance caused by aneuploidy may not be an overstatement. There are evidently 'silent' regions of the genome, or regions capable of being repressed, which are likely candidates for nascent *B*s. Another possibility to consider is that one that involves facultative heterochromatin and gene inactivation. In *Chorthippus paralellus* polysomic for the M4 chromosome, the supernumerary M4s in excess of the diploid euchromatic pair become heterochromatic and pair only among themselves (Hewitt and John, 1968). These are also obvious candidates for nascent *B*s, and they reinforce the idea that loss of homology and lack of gene expression are not serious obstacles to *B* chromosome formation. The question of accumulation remains a problem. Suffice it to say that non-disjunction based on delayed DNA separation in centromeric heterochromatin involves no new phenomenon. In the normal course of the mitotic cycle there is delayed separation of chromatids in the region of the centromere and directed non-disjunction may involve little more than a changes in action of this process contingent upon an altered chromosome form and a particular spindle structure.

CONCLUDING REMARKS

It is far easier to devise ideas and theories than it is to conduct experiments to test them, but if we were to take this liberty it could well lead to the following conclusions. *B* chromosomes that have an accumulation mechanism are 'selfish' and may spread themselves throughout natural populations. The mechanisms appear to have a self-controlling genetic basis which may itself be subject to variation and evolution. In addition to this genetic component the accumulation and counter balancing elimination mechanisms are all based on normal biological processes and as such are subject to environmental modification. The existence of *B* chromosome polymorphisms may well be inevitable for those systems that have accumulation but their equilibrium frequencies are modulated by environment and may even be reduced to zero in circumstances where the burden on fitness is too heavy to bear. Some *B* chromosome systems have no accumulation mechanism and neither are there any alternative explanations to account for their polymorphisms; they remain problematical and open to further inquiry. The origin of *B*s has never been witnessed but the development of a *B* chromosome system, replete with all the necessary characteristics, is not difficult to understand if we accept that there are parts of the genome which lack essential genetic information and for which amplification may be tolerated. The genomes of most eukaryotes certainly have enough DNA to satisfy this requirement.

If the discussion of 'selfish' *B*s were to be carried even further, and to encompass arguments that are more familiar in other areas of host–parasite interactions, it could well lead to speculations about 'gene for gene' relationships and the co-evolution of *A* and *B* genetic systems. Evidence from Nur (unpublished data, cited with permission) suggests that such speculations are not all idle. *Pseudococcus B*s from a high transmission line were placed into F_1s with differing backgrounds determined by euchromatic (active) maternal *A*s deriving from high and low transmission lines. Transmission of *B*s through $1B$ F_1 males carrying a maternal set from a line with low B frequency was significantly lower than that of *B*s through $1B$ F_1 males with a maternal set from a high *B* frequency line. The results point to the existence of 'anti-*B* genes' in the *A* chromosome complements of certain populations, and lead us into an entirely new dimension of research and speculation on the question of 'selfish' *B* chromosomes.

REFERENCES

Bougourd, S. M. and Parker, J. S. (1979). The *B*-chromosome system of *Allium schoenoprasum*. II. Stability, inheritance and phenotypic effects. *Chromosoma*, **75**, 369–383.

Carlson, W. R. (1969). Factors affecting preferential fertilization in maize. *Genetics*, **62**, 543–554.

Carlson, W. R. (1978). The *B* chromosome of corn. *Ann. Rev. Genet.*, **16**, 5–23.

Carlson, W. R. and Chou, T. S. (1981). *B* chromosome nondisjunction in corn: control by factors near the centromere. *Genetics*, **97**, 379–389.

Fedak, G. (1976). Cytogenetics of Wiebe's 16-chromosome barley. *Can. J. Genet. Cytol.*, **18**, 763–768.

Fröst, S. (1958). Studies of the genetical effects of accessory chromosomes in *Centaurea scabiosa*. *Hereditas*, **44**, 112–122.

Grun, P. (1959). Variability of accessory chromosomes in native populations of *Allium cernuum*. *Amer. J. Bot.*, **46**, 218–224.

Harvey, A. W. and Hewitt, G. M. (1979). *B* chromosomes slow development in a grasshopper. *Heredity*, **42**, 397–401.

Hewitt, G. M. (1973a). Variable transmission rates of a *B* chromosome in *Myrmeleotettix maculatus* (Thunb.) Acrididae:Orthoptera. *Chromosoma*, **40**, 83–106.

Hewitt, G. M. (1973b). Evolution and maintenance of *B* chromosomes. *Chromosomes Today*, **4**, 351–369.

Hewitt, G. M. (1973c). The integration of supernumerary chromosomes into the Orthopteran genome. *Cold Spring Harbor Symp. Quant. Biol.*, **38**, 183–194.

Hewitt, G. M. (1976). Meiotic drive for *B* chromosomes in the primary oocytes of *Myrmeleotettix maculatus* (Orthoptera:Acrididae). *Chromosoma*, **56**, 381–391.

Hewitt, G. M. and John, B. (1967). The *B* chromosome system of *Myrmeleotettix maculatus* (Thunb.) III. The statistics. *Chromosoma*, **21**, 140–162.

Hewitt, G. M. and John, B. (1968). Parallel polymorphism for supernumerary segments in *Chorthippus parallelus* (ZETTERSTEDT). I. British populations. *Chromosoma*, **25**, 319–342.

Hewitt, G. M. and John, B. (1970). The *B* chromosome systems of *Myrmeleotettix maculatus* (Thunb.) IV. The dynamics. *Evolution*, **24**, 169–180.

John, B. (1973). The cytogenetic systems of grasshoppers and locusts. II. The origin and evolution of supernumerary segments. *Chromosoma*, **44**, 123–146.

John, B. (1981). Heterochromatin variation in natural populations. *Chromosomes Today*, **7**, 128–137.

John, B. and Hewitt, G. M. (1965a). The *B* chromosome system of *Myrmeleotettix maculatus* (Thunb.) I. The mechanics. *Chromosoma*, **16**, 548–578.

John, B. and Hewitt, G. M. (1965b). The *B* chromosome system of *Myrmeleotettix maculatus* (Thunb.) II. The statistics. *Chromosoma*, **17**, 121–138.

John, B. and Miklos, G. L. G. (1979). Functional aspects of heterochromatin and satellite DNA. *Int. Rev. Cytol.*, **58**, 1–114.

Jones, R. N. (1975). *B* chromosome systems in flowering plants and animal species. *Int. Rev. Cytol.*, **40**, 1–100.

Jones, R. N. and Rees, H. (1967). Genotypic control of chromosome behaviour in rye. XI. The influence of *B* chromosomes on meiosis. *Heredity*, **22**, 333–347.

Jones, R. N. and Rees, H. (1969). An anomalous variation due to *B* chromosomes in rye. *Heredity*, **24**, 265–271.

Jones, R. N. and Rees, H. (1982). B *Chromosomes*, Academic Press, New York.

Karp, A. (1981). The genetic control of meiosis in *Lolium perenne*. Ph.D. Thesis, University of Wales.

Kayano, H. (1957). Cytogenetic studies in *Lillium callosum*. III. Preferential segregation of a supernumerary chromosome in EMC's. *Proc. Jap. Acad.*, **33**, 553–558.

Kimura, M. and Kayano, H. (1961). The maintenance of supernumerary chromosomes in wild populations of *Lilium callosum* by preferential segregation. *Genetics*, **46**, 1699–1712.

Kishikawa, H. (1965). Cytogenetic studies of B chromosomes in rye, *Secale cereale* L., in Japan. *Agric. Bull. Saga Univ.*, **21**, 1–81.

Kishikawa, H. (1970). Effects of temperature and soil moisture on frequency of accessory chromosomes in rye, *Secale cereale* L. *Jap. J. Breeding*, **20**, 269–274.

Kranz, A. R. (1963). Beiträge zur cytologischen und genetischen evolutions-forschung an dem Roggen. *Z. Pflanzenzüchtung*, **50**, 44–58.

Lee, W. J. and Min, B. R. (1965). On accessory chromosomes in *Secale cereale*. I. Frequency and geographical distribution of plants with accessory chromosomes in Korea. *Kor. J. Bot.*, **8**, 1–6.

Lin, B. Y. (1978). Regional control of nondisjunction of the B chromosome in maize. *Genetics*, **90**, 613–627.

Lin, B. Y. (1979). Two new B-10 translocations involved in the control of nondisjunction of the B chromosome in maize. *Genetics*, **92**, 931–945.

Longley, A. E. (1938). Chromosomes of maize from North American Indians. *J. Agric. Res.*, **56**, 177–195.

Lucov, Z. and Nur, U. (1973). Accumulation of B chromosomes by preferential segregation in females of the grasshopper *Melanoplus femur-rubrum*. *Chromosoma*, **42**, 289–306.

Matthews, R. B. (1981). Studies on B chromosomes with particular reference to plant disease. Ph. D. Thesis, University of Wales.

Matthews, R. B. and Jones, R. N. (1982). Dynamics of the B chromosome polymorphism in rye. I. Simulated populations. *Heredity*, **48**, 347–371.

Müntzing, A. (1946). Cytological studies on extra fragment chromosomes in rye. III. The mechanism of non-disjunction at the pollen mitosis. *Hereditas*, **32**, 97–119.

Müntzing, A. (1950). Accessory chromosomes in rye populations from Turkey and Afghanistan. *Hereditas*, **36**, 507–509.

Müntzing, A. (1951). The meiotic pairing of iso-chromosomes in rye. *Portugaliae Acta Biologica Series A. Goldschmidt vol.*, 831–860.

Müntzing, A. (1957). Frequency of accessory chromosomes in rye strains from Iran and Korea. *Hereditas*, **43**, 682–685.

Müntzing, A. (1963). Effects of accessory chromosomes in diploid and tetraploid rye. *Hereditas*, **49**, 371–426.

Müntzing, A. (1970). Chromosomal variation in the Lindström strain of wheat carrying accessory chromosomes of rye. *Hereditas*, **66**, 279–286.

Nur, U. (1962a). A supernumerary chromosome with an accumulation mechanism in the lecanoid genetic system. *Chromosoma*, **13**, 249–271.

Nur, U. (1962b). Population studies of supernumerary chromosomes in a mealy bug. *Genetics*, **47**, 1679–1690.

Nur, U. (1963). A mitotically unstable supernumerary chromosome with an accumulation mechanism in a grasshopper. *Chromosoma*, **14**, 407–422.

Nur, U. (1966a). Harmful supernumerary chromosomes in a mealy bug population. *Genetics*, **54**, 1225–1238.

Nur, U. (1966b). The effect of supernumerary chromosomes on the development of mealy bugs. *Genetics*, **54**, 1239–1249.

Nur, U. (1969). Harmful B chromosomes in a mealy bug: additional evidence. *Chromosoma*, **28**, 280–297.

Nur, U. (1977). Maintenance of a 'parasite' B chromosome in the grasshopper *Melanoplus femur-rubrum*. *Genetics*, **87**, 499–512.

Östergren, G. (1945). Parasitic nature of extra fragment chromosomes. *Botanisker Notiser*, **2**, 157–163.

Östergren, G. (1947). Heterochromatic *B* chromosomes in *Anthoxanthum*. *Hereditas*, **33**, 261–296.

Pryor, A., Faulkner, K., Rhoades, M. M. and Peacock, W. T. (1980). Asynchronous replication of heterochromatin in maize. *Proc. Natl. Acad. Sci. USA*, **77**, 6705–6709.

Randolph, L. F. (1928). Types of supernumerary chromosomes in maize. *Anatomical Record*, **41**, 102.

Rees, H. and Ayonoadu, U. (1973). *B* chromosome selection in rye. *Theoret. Appl. Genet.*, **43**, 162–166.

Rees, H. and Hutchinson, J. (1973). Nuclear DNA variation due to *B* chromosomes. *Cold Spring Harbor Symp. Quant. Biol.*, **38**, 175–182.

Rhoades, M. M. and Dempsey, E. (1973). Chromatin elimination induced by the *B* chromosome of maize. *J. Hered.*, **64**, 13–18.

Robinson, P. M. and Hewitt, G. M. (1976). Annual cycles in the incidence of *B* chromosomes in the grasshopper *Myrmeleotettix maculatus* (Acrididae:Orthoptera). *Heredity*, **36**, 399–412.

Roman, H. (1947). Mitotic nondisjunction in the case of interchanges involving the *B*-type chromosome in maize. *Genetics*, **32**, 391–409.

Vosa, C. G. (1976). Heterochromatic banding patterns in *Allium*. II. Heterochromatin variation in species of the *paniculatum* group. *Chromosoma*, **57**, 119–133.

Ward, E. J. (1973). Nondisjunction: localisation of the controlling site in the maize *B* chromosome. *Genetics*, **73**, 387–391.

Zecĕvić, L. and Paunović, D. (1967). *B* chromosome frequency in Yugoslav rye populations. *Biologia Plantarum (praha)*, **9**, 205–211.

The Evolution of Genome Size
Edited by T. Cavalier-Smith
© 1985 John Wiley & Sons Ltd

CHAPTER 14

Chromatin diminution and chromosome elimination: mechanisms and adaptive significance

DIETER AMMERMANN

Institut fuer Biologie III, Abt. Zellbiologie, Universitaet Tuebingen, Auf der Morgenstelle 28, D-7400 Tuebingen, West Germany

INTRODUCTION

The C-value paradox continues to puzzle cell biologists. Many observations show that most eukaryote cells have more DNA than they need for their well-known functions: inheritance and transcription. One of the earliest observations was made by Boveri (1887) who discovered that *Ascaris* has in its somatic cells only a part of the chromatin which is present in the germ line cell nuclei. The other part is destroyed in a process called chromatin elimination or diminution. Since then this process has also been observed in some other groups of animals. In species which have this elimination three sorts of DNA or chromatin (or chromosomes) can be distinguished:

1. G DNA (germ line DNA): present in the nuclei of the germ line cells.
2. S DNA (soma DNA): a part of G DNA which is present in the soma cell nuclei.
3. E DNA (elimination DNA): a part of G DNA which is found only in germ line cells and is eliminated when soma cells differentiate.

Chromatin diminution seems to be an exception among animals, being found only in some species of the groups Protozoa, Nemathelminthes and Arthropoda. It is, however, connected with the probably 'normal situation' (identity of soma and germ line genomes) by transition stages. In some coccid insects the paternal chromosome set is heterochromatized in males and contributes only little to their phenotypes (Brown and Nur, 1964). Differential replication is another way which leads to differences in DNA between soma and germ line cells (e.g. Hennig *et al.*, 1974; Hess, 1975) and between different soma cells (e.g. Lohmann and Schubert, 1980; Spradling and Mahowald, 1980).

Investigation of the E DNA should provide hints about the role of 'dispensable' DNA in eukaryotes. Unfortunately, there are experimental drawbacks which are the reason for our poor knowledge about E DNA. As parasitic nematodes the Ascaridae are difficult to handle and their egg shells are almost impenetrable. The Copepoda cannot be successfully cultivated in amounts large enough to allow biochemical investigation of their DNA. The same holds true for the few tiny insect species and the mite species which show DNA elimination. Insufficient G and S cell material is obtainable. However, the smallest animals with elimination, the ciliate protozoa, can be grown in large amounts. Some of these species have micronuclei with three times and macronuclei with more than 500 times the DNA amount of a human cell. During the last few years several research groups have investigated the DNA and chromatin of the ciliates. Therefore, this group plays a major role in this chapter which concentrates on the following questions:

1. What is known about G, S and E DNA and about the mechanism of elimination?
2. Does the E DNA have a function? What biological significance might the elimination process have?
3. Can the elimination process give us some hints for the solution of the C-value paradox?

NEMATHELMINTHES

Nematoda

Chromatin diminution (elimination) in *Ascaris* is well known since its first description by Boveri (1887). The elimination of the heterochromatin during the first cleavage divisions has been investigated several times and often described in detail (e.g. White, 1954). Therefore, only a summary of what is known from recent investigations with modern methods of molecular biology is given here.

Four research groups have investigated the DNA and the chromatin elimination of *Ascaris suum* (= *A. lumbricoides*, from the pig) but disagree as to how much DNA is eliminated from the germ line nucleus: 22% (Moritz and Roth, 1976), 27% (Tobler *et al.*, 1972), 34% (Pasternak and Barrell, 1976), 56% (Davies and Carter, 1980). The authors disagree also in the absolute amount of the remaining S DNA: the measured values are between 0.15 and 0.46 pg. For *Parascaris equorum* (Var. *univalens*, from the horse) Moritz and Roth (1976) reported a chromatin elimination of 85% which leads to 0.25 pg remaining S DNA.

The E DNA contains repetitive satellite DNA. Investigations to characterize this DNA fraction gave rather similar results. Roth and Moritz (1981) found that this DNA consists of two families of highly repetitive DNA sequences. One repeated unit is 125 bp, the other 131 bp long. Several repeat unit variants (within both families) could be detected. Mueller *et al.* (1982) characterized the repetitive DNA as a whole set of different but related sequence families, all showing the same repeating unit length of about 120 bp. However, it is not clear whether the E DNA consists entirely of highly repetitive DNA sequences (Moritz and Roth, 1976) or whether approximately 50% of the E DNA consists of repetitive and approximately 50% of unique DNA sequences (Tobler *et al.*, 1972; Goldstein and Straus, 1978).

The elimination process leads to somatic nuclei which contain around 90% unique DNA and 10% repetitive DNA sequences. Only small amounts ('0.01–0.05%': Roth and Moritz, 1981; 'About 1000 copies (or even less)': Mueller *et al.*, 1982) of the highly repeated DNA sequences of the E DNA are left in the somatic nuclei.

It is not known whether the E DNA has any functions. Moritz (1970) found no evidence for an *in vivo* transcription of the E DNA. Mueller *et al.* (1982) did hybridization experiments from which they concluded that *Ascaris* satellite DNA (mainly E DNA) is not transcribed in an amount detectable in any of the tested cells. It is certain that the E DNA is not rDNA (Tobler *et al.*, 1974).

ARTHROPODA

Acarina

Chromatin elimination was found in one species of mite (*Metaseiulus occidentalis*) (Nelson-Rees *et al.*, 1980). In the male embryos half of the chromosomes are eliminated.

Copepoda

Several species of this crustacean group show chromatin elimination during their first cleavage divisions. Beermann (1977) found that during the 4th to 7th cleavage division of three *Cyclops* species around 50% of the G DNA is eliminated. The somatic nuclei then contain the following amounts of DNA: *Cyclops strenuus* 0.9 pg, *C. furcifer* 1.4 pg, *C. divulsus* 1.8 pg.

Besides this, two results of her work seem remarkable:

1. One species has (like *Parascaris*) terminal heterochromatin. Another species has, however, heterochromatin segments which are scattered all along the chromosomes. To eliminate this heterochromatin several chromosome breaks and their controlled rejoining are necessary.
2. Before elimination homologous chromosomes can contain different

amounts of heterochromatin. This is recognizable by the different length of the heterochromatin segments. Nevertheless, they pair. After elimination the homologous chromosomes in the somatic cells have the same size.

The heterochromatin segments seem to be despiralized (lampbrush-like) during diplotene, but it is not known whether there is transcription during this stage or not.

Insecta

Some species of several families of Diptera (Sciaridae, Cecidomyiidae, Chironomidae) of the suborder Nematocera show remarkable differences in chromosome number between germ line and soma nuclei. The germ line cells contain more chromosomes than the soma cells. The elimination of the E chromosomes happens during the first cleavage divisions. The rather complicated elimination process of different species was reviewed by White (1954). No details about the E DNA are known, but its function is better understood than that of any other group of animals with DNA elimination. Bantock (1961) and Geyer-Duszyńska (1966) showed that two species of Cecidomyiidae (gall midges) whose larvae contained only cell nuclei with S chromosomes remained sterile. They were unable to develop sperma, and the oocytes were arrested in the early stages of development. Therefore the E chromosomes are indispensable for the normal course of oogenesis and spermatogenesis. This result was further supported by the findings of Kunz et al. (1970). During the entire oogenesis of the gall midge, *Wachtliella persicariae*, the E chromosomes are despiralized and synthesize RNA. No nucleoli are visible. Therefore the authors suppose that the E chromosomes synthesize mRNA. Unfortunately, biochemical investigations are very difficult with these tiny insects. Therefore it has not yet been possible to determine whether the E chromosomes are just copies of some of the S chromosomes or whether they have genes different from the S chromosome genes.

PROTOZOA

Ciliata

Nuclear dimorphism is typical for ciliates (Ciliophora = Heterokaryota). Each vegetative cell contains a large macronucleus (Ma: sometimes consisting of several parts which are considered here as one Ma) and one or several smaller micronuclei (Mi). During vegetative growth the Ma synthesizes nearly all the RNA of the cell. Without a Ma a cell cannot survive. Before cell division the Ma divides into two daughter nuclei. During sexual reproduction (conjugation) the Ma degenerates. The Ma are the somatic nuclei of the ciliates, because they are responsible for the phenotype of the organism and are discarded during every sexual reproduction.

The diploid Mi are the germ line nuclei of the cells. Their contribution to the phenotype of the cells is small (see below). Many ciliates can live (but not successfully conjugate) without Mi. When two cells start conjugation, the Mi undergo meiosis and haploid gametic nuclei are produced. After exchange and formation of a diploid zygotic nucleus in each cell the partners separate. After mitotic divisions one of the Mi develop into a new Ma.

One of the most interesting questions is: are there besides the striking quantitative differences between Ma and Mi DNA also qualitative differences?

From the five orders of ciliates (which contain around 8000 species) the DNA of only a few species of two orders (Holotricha, Spirotricha, listed in Table 14.1) has been investigated. This small basis does not allow generalizations for all ciliates. Most is known about some species of the suborder Hypotricha (belonging to the Spirotricha). As an example a description of what is known about the Ma and Mi DNA of *Stylonychia mytilus*[1] follows (Ammermann *et al.*, 1974; Steinbrueck *et al.*, 1981). The Mi is a typical eukaryotic nucleus: chromosomes are present and visible during meiosis. The DNA consists of 55% repetitive and 45% unique sequences. Satellite DNA is present. The molecular weight is high ($<14 \times 10^6$ d). The Ma, however, is unique among the nuclei of eukaryotes: chromosomes have never been observed and mitosis does not occur. Reassociation studies showed less than 2% repetitive DNA. Satellite DNA fractions are reduced. The kinetic complexity of the DNA sequences of the Ma is 3×10^{10} d, that is 12 times the size of the *E. coli* genome. This 'Ma unit' is present in approximately 15 000 copies (3×10^{10} d $\times$ 15 000 = Ma DNA content). The size of the DNA molecules is surprisingly low: the average molecular weight is 3×10^6 d. If the Ma DNA is run in an agarose gel, a species-specific banding pattern appears (Swanton *et al.*, 1980; Steinbrueck *et al.*, 1981). Each band consists of DNA molecules of the same size. In the Ma of *Stylonychia mytilus* 12 000–15 000 (Nock, 1981) and in the Ma of *Oxytricha* about 24 000 (Prescott *et al.*, 1979) different genes are present.

These different characteristics of the Ma become established during its development from a Mi after conjugation. The main events (for a detailed discussion see Ammermann *et al.*, 1974) of the development are:

1. Elimination of about 66% of the chromosomes shortly after the beginning of the development of the Ma (first elimination).
2. The remaining chromosomes become giant polytene chromosomes. Shortly after they have reached their final size, they are transected by

---

[1] Now renamed *S. lemnae* (Ammermann, D. and Schlegel, M. (1983). *J. Protozool.*, **30**, 290–4.

membranous material between their bands. Each band is packed into a vesicle. Probably during this step the low molecular weight DNA is produced.

3. Connected with giant chromosome degradation is a loss of DNA. More than 90% of the DNA leaves the nucleus (second elimination).
4. The remaining DNA sequences (less than 2% of the original Mi DNA sequences) are replicated several times until the DNA content of the new Ma is reached. Then the vegetative growth phase of the cell starts.

Is *Stylonychia* a typical example or an exception? It appears that other hypotrich ciliates have the same peculiarities of the Ma and the mode of its development in common (Prescott *et al.*, 1979; Steinbrueck *et al.*, 1981). Table 14.1 shows the data about the DNA of other Hypotricha (i.e. all Spirotricha in this table other than *Stentor*). The comparison with ciliates from other groups shows, however, similarities and differences:

1. The kinetic complexity of the Ma DNA is remarkably similar in all ciliates investigated: about 3×10^{10} d. Besides the species in Table 14.1, Soldo and Brickson (1979) found the same value in several other marine and freshwater species.
2. There are uncertainties about the percentage of repetitive DNA in the Ma. While in the hypotrichs it is below 2%, the occurrence of higher amounts of repetitive DNA in the Ma of species of other groups has been reported.
3. The average molecular weight of Ma DNA which is low in all hypotrich ciliates investigated until now, is apparently higher in other groups. In *Tetrahymena* the DNA in the Ma exists apparently in large molecules although different methods have given differing results (Preer and Preer, 1979). In *Paramecium primaurelia* a low molecular weight Ma DNA was described (McTavish and Sommerville, 1980). However, in the related species *P. tetraurelia* and *P. bursaria* (Steinbrueck *et al.*, 1981) the DNA in the Ma was found to have a high molecular weight. It is interesting that the Ma DNA of *Glaucoma*, a holotrichous ciliate, has a molecular weight which ranges between 1.4 and 66×10^6 d (Katzen *et al.*, 1981). This species therefore, like the hypotrich ciliates, has subchromosomal sized DNA in its Ma.
4. All DNA molecules in the Ma of four hypotrich ciliates (two *Stylonychia* spp., two *Oxytricha* spp.) have at their ends the same terminal inverted repeat sequence which is $5' - C_4 A_4 C_4 A_4 C_4$ . . . and, in addition, a protruding non-complementary single-stranded tail of the sequence $3' - G_4 T_4 G_4 T_4$ . . . (Oka, 1980; Klobutcher *et al.*, 1981). *Euplotes aediculatus*, another hypotrich ciliate, possesses a slightly different terminal sequence arrangement (Klobutcher *et al.*, 1981). The related sequence $5'$ ($C_4 A_2$), 20–70 times repeated in *Tetrahymena* (Blackburn and Gall, 1978) and

Table 14.1. Comparison between Mi and Ma DNA of different ciliates.

	Mi DNA content (1 C) in 10^{12} d	Ma			Chromatin elimination observed during Ma development	References
		DNA content before replication in 10^{12} d	Kinetic complexity (Ma unit) in 10^{10} d	Quantity (in %) of repetitive DNA found		
Holotricha						
Tetrahymena thermophila and *pyriformis*	0.14	6	8	5–10	no	a
Paramecium aurelia (species group)	0.18–0.21	66–176 147*	3–5.2	0–4	no	d, f
Paramecium bursaria	2.3*†	168	1.1	below 2	yes	e
Spirotricha						
Stylonychia mytilus	5.0	473	3.1	below 2	yes	e
Oxytricha sp.	0.4	70	3.6	none	yes	b
Paraurostyla cristata	2.4	168	1.4	below 2	yes	e
Euplotes aediculatus	0.3	228	2.8	below 2	yes	e
Euplotes minuta	0.024–0.04*	51*				
Stentor coeruleus		1650–2250*	6	15		c

* Own unpublished measurements.
† Based on the assumption that the Mi is not polyploid, replicates immediately after division and is therefore 4 C (Just, 1973).

References:
[a] Gorowsky, 1980; [b] Lauth *et al.*, 1976; [c] Pelvat and de Haller, 1976; [d] Sonneborn, 1974; [e] Steinbrueck *et al.*, 1981; [f] McTavish and Sommerville, 1980.

more than 38 times repeated in *Glaucoma* (Katzen *et al.*, 1981), is found at the ends of the free ribosomal gene and also at the ends of many (all?) other DNA molecules in the Ma of these species.

5. Several results in *Tetrahymena* (Gorovsky, 1980) and *Oxytricha* (Boswell *et al.*, 1982; Klobutcher *et al.*, 1984) indicate that the sequences adjacent to the repeated end sequences in the Mi are changed during Ma development. Apparently substantial rearrangements between coding, non-coding and repeated end sequences occur during Ma development. The molecular basis and the details of these rearrangement processes are being investigated in several laboratories.

Chromatin elimination was furthermore found in two species belonging to two groups of ciliates which were not investigated with biochemical methods. In *Chilodonella cucullulus* Radzikowski (1973) found a loss of chromosomal material during the development of a new Ma. In *Trachelonema sulcata*, a species belonging to the Karyorelictida, with a very unusual Ma (it cannot divide), Kovaleva and Raikov (1978) found a DNA loss of 50% during Ma development. The authors suppose that these events lead to Ma which are qualitatively different from the Mi.

In all ciliate species investigated either strong evidence or hints exist which show that during Ma development several processes (e.g. chromatin diminution, differential replication) lead to somatic nuclei which differ from the Mi. The opinion that the Ma is a polyploid edition of the Mi, which was generally accepted even 15 years ago, and which is still found in textbooks, is wrong. However, a new term instead of 'polyploid' has not been coined.

Interspecific variation in the quantity of G, S and E DNA

The DNA content of the Mi and the Ma shows remarkable intraspecific differences (see examples in Table 14.1). Some correlations could be verified recently by spectrophotometric measurements (see Figure 14.1):

1. The amount of Ma DNA (= S DNA) is correlated with the cell size. Because the 'Ma unit' has approximately the same size in all investigated ciliates (= 3×10^{10} d, see Table 14.1), this result means that there is a linear correlation between the number of gene copies in the Ma and the cell size (Soldo *et al.*, 1981; Ammermann and Muenz, 1982). The Ma controls nearly the whole cell metabolism. Therefore, this correlation (gene dosage effect) is not surprising.

2. The Mi DNA content is not correlated either with the cell size or with the Ma DNA content. Even between comparable species (compare, for example, the *Euplotes* species and *Stylonychia mytilus* or *Paramecium aurelia* and *P. bursaria*, see Table 14.1) there are astonishing differences between Mi genome size. The largest known ciliates, e.g. *Stentor coeruleus*

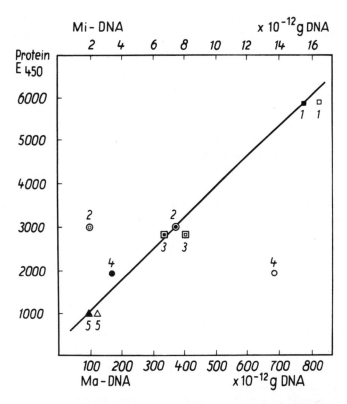

Fig. 14.1. Protein content (in arbitrary units) of whole cells and DNA content of the macro- and micronuclei of differnt ciliates, measured spectrophotometrically. 1: *Stylonychia mytilus* ■ Ma, □ Mi; 2: *Euplotes aediculatus,* ◉ Ma, ◎ Mi; 3: *Paraurostyla weissei* ▣ Ma, ▢ Mi; 4: *Stylonychia putrina* ● Ma, ○ Mi; 5: *Uroleptus caudatus* ▲ Ma, △ Mi. The line shows the correlation between protein content of the cell and DNA content of its Ma. For details see Ammermann and Muenz (1982).

(over 1 mm) and *Spirostomum ambiguum*, have huge Ma but tiny Mi (unpublished measurements). The C-value paradox in ciliates is therefore restricted to the Mi (= G DNA). If we assume that the S DNA sequences are present once in the Mi genomes, then virtually all Mi DNA of *Euplotes minuta*, 10% of the Mi DNA of *Euplotes aediculatus* and 0.6% of the Mi DNA of *Stylonychia mytilus* is S DNA. This fact explains why in species with high Mi DNA content (e.g. *Stylonychia mytilus, Paramecium bursaria*) the chromatin diminution during the development of a new Ma is so easy to demonstrate, while in species with small Mi (*Paramecium aurelia, Tetrahymena thermophila*) no chromatin elimination was found. In species with DNA-poor Mi, chromatin elimination, if it happens, is difficult to demonstrate because these species have only small amounts of

DNA to eliminate and because the elimination process might often be imperceptible: it might be concealed by a simultaneous replication of the remaining Ma DNA sequences.

In *Stylonychia mytilus* and *Oxytricha* it was demonstrated that repetitive DNA is present in the Mi, but most of it is not taken over into the Ma. So a part of the E DNA consists of repetitive DNA sequences. The Mi in *Stylonychia mytilus* contains 45% unique DNA sequences. Because less than 2% of the sequences are taken over into the Ma, the E chromatin must also contain unique DNA sequences. The same is true for *Oxytricha*. The occurrence of repetitive DNA in 'low C-value Mi' was not investigated, because it is too difficult to get enough of them.

Clearly the Mi of ciliates show the same C-value paradox as other eukaryotes. Species with large Mi (= much G DNA) eliminate much DNA (containing repetitive and unique DNA sequences), species with less Mi DNA eliminate less (or perhaps no) DNA. All ciliates so far investigated take over approximately the same amount of Mi DNA (= Ma unit, S DNA) into the Ma.

Possible functions of G, S and E DNA

The function of the *Ma* DNA is obvious: it regulates almost the whole metabolism of the cell. A few Ma DNA sequences of hypotrich ciliates and of *Tetrahymena* were analysed in detail (Gorovsky, 1980; Kaine and Spear, 1982). They consist of a coding region (the largest part of the sequences), a non-coding region on each side and repeated end sequences at each end. The function of the non-coding region and the repeated sequences regions are not known. It would be expected that a nucleus with selected S DNA sequences is transcribed to a high percentage. A clone of *Paramecium primaurelia* contained polyadenylated RNA sequences which were only homologous to 5–8% of the macronuclear DNA (McTavish and Sommerville, 1980). Nock (1981), however, found that 60% of the Ma DNA sequences of *Stylonychia mytilus* are transcribed into nuclear RNA and that the cytoplasmic mRNA sequences are homologous to about 40% of macronuclear DNA. This means that from the calculated total number of 10–20 000 different genes of a Ma more than 50% are active in transcription.

The function of the Mi is much more difficult to characterize. The often heard statement that the Mi has only functions connected with conjugation is too simple and not correct. On the one hand, strains without Mi of nearly all species mentioned in Table 14.1 were rarely found to arise spontaneously in nature or after irradiation in the laboratory. They live successfully. Sometimes their division rate is reduced. The only thing they cannot achieve is successful conjugation (i.e. with surviving exconjugants). On the other hand, the removal of the Mi in all species mentioned above (Table 14.1) has usually fatal consequences for most of the cells. *Stylonychia mytilus* may serve as an example. If the Mi are eliminated, for example with a UV beam, without irradiating the Ma, all cells suffer and most of them die. During the following three-month period some cells gradually 'become used' to living without Mi. Finally, clones without Mi arise, which live, reproduce and age like the cells with Mi. What happens during this period is unknown (Ammermann, 1970). Essentially the same phenomenon can be observed if the Mi of most of the other species of Table 14.1 are removed. In addition, RNA synthesis was demonstrated with autoradiographic methods in the Mi of three *Paramecium* species and *Tetrahymena sp.* (survey: Murti and Prescott, 1970). Nuclear transplantation experiments also showed that the Mi in *Paramecium caudatum* have a function (Fujishima and Watanabe, 1981). The conclusion is that the Mi has hitherto unknown functions during the vegetative growth and division. They can sometimes be taken over by other parts of the cell (the Ma?) if the Mi is eliminated.

Sometimes so-called pseudomicronuclei arise in irradiated cells of *Stylonychia mytilus*. Their origin is unknown. Possibly they are surviving fragments of the Mi after irradiation. They can be very different in size. Often they are much smaller than Mi. Unlike the Ma they contain chromosomes and/ or pieces of chromosomes. Cells suffering after irradiation regain their normal healthy growth immediately after developing pseudo-Mi. If the pseudo-Mi are removed by irradiation, the cells show the same damage as cells whose Mi are removed. Therefore, it is evident that the pseudo-Mi with only a small part of the Mi DNA can substitute for the Mi during vegetative growth and division. However, the cells with the pseudo-Mi cannot conjugate successfully (Ammermann, 1970).

The function of the Mi during conjugation (meiosis, syncaryon formation) is obvious. Whether it has other functions during the conjugation process, is not clear. Sugai *et al.* (1974) found RNA synthesis in the Mi of *Tetrahymena* during the meiotic prophase ('crescent stage'). Most ciliates can start conjugation (i.e. fusion of partner cells, etc.) without Mi.

It is interesting that Steinbrück and Schlegel (1983) found differences in the occurrence of repetitive Mi DNA sequences between clones of different origin (from Germany and East Asia) belonging to the same *Stylonychia* species. They also found differences between different *Stylonychia* species. These differences in the repetitive fraction of the DNA are well known from other eukaryotes (John and Miklos, 1979).

DISCUSSION AND CONCLUSIONS

The elimination of chromatin and chromosomes clearly shows that several animals contain in their germ line cells large chromatin fractions which are not necessary for the function of the soma cells. The obvious question is: for what purpose do the animals have this E chromatin and E DNA? Several possibilities are discussed below, which are not alternatives but which might all be correct:

1. The E DNA could be transcribed into RNA. It is apparent that the E DNA of some insects has functions and is transcribed during oogenesis and spermatogenesis. The question whether the E DNA of Nematoda and Copepoda is transcriptionally active is unanswered. In ciliates there are some observations which point to a low activity of the Mi during vegetative growth and division and/or during conjugation. But whether the S or the E DNA of the Mi is responsible for this activity, is unclear. There seems no doubt that only a small percentage of the E DNA is transcriptionally active.

2. E DNA may be necessary for the control of transcription. E DNA could suppress the activity of most genes in the germ line cells. This hypothesis seems tempting because at least in Nematoda and Ciliata E DNA contains a high percentage of repetitive DNA. It was assumed from several other observations that repetitive DNA may play a role in regulation of gene activity (e.g. Davidson and Britten, 1979). The very variable amount of E DNA in the germ line is, however, not in favour of this hypothesis but does not contradict it.

3. E DNA may have functions during the meiotic processes. These functions would be the most evident for the E DNA. It was often supposed that heterochromatin and repetitive DNA, both present in the E DNA, have important functions during chromosome pairing. A characteristic of both DNA fractions is their great inter- and intraspecific, qualitative and quantitative variation (John and Miklos, 1979). The homologous chromosomes of the Copepoda pair in the meiotic prophase despite differences in the amount of heterochromatin (see Copepoda, p. 430). In *Stylonychia* there are hints that the repetitive DNA sequences of the Mi can be different within strains of one species and between species (see Ciliata, p. 437). But this problem of heterochromatin and repetitive DNA function during meiosis is still unsolved. The reader is referred to the interesting and detailed review of John and Miklos (1979).

   In most ciliates a part of the chromosomes, the centromere, is only needed in the Mi, but not in the Ma. Therefore, it is possible that the centromeric regions of all chromosomes are a part of the E DNA.

4. E DNA may have other functions. Bennett (1971) and Cavalier-Smith (1978) proposed that DNA which does not code for proteins and which is not engaged in regulation of transcription, replication or recombination may have a function as non-genic DNA. Its function could be to increase the DNA amount and the nuclear volume of a cell. One reason to increase

the volume of the nuclei (Cavalier-Smith, 1978) could be to increase the available surface of the nuclei and with this the number of nucleopores. The Mi are, however, transcriptionally rather inactive. As could be expected, the Mi of several ciliate species contain only a few nuclear pores, and it is very unlikely that the volume of the Mi of several species was increased to make space for more nucleopores.

Another reason to increase the volume of the nuclei could be to regulate cell growth and cell size. In many uni- and multicellular organisms cells with large nuclei ('high C-value organisms') are large and have a long generation time while cells with small nuclei ('low C-value organisms') are small and have a short generation time. Therefore, these authors stated that the amount of DNA, and with this the nuclear volume, may be a fundamental physical determinant of cell growth and growth rates. No data about cell size and cell growth of Nematoda, Copepoda and Insecta are available. In ciliates, however, there are some data available but they do not appear to support this hypothesis. The size of the Mi which contain the E DNA is not correlated with the cell size (Fig. 14.1). *Stylonychia mytilus* can live with much smaller pseudo-Mi but their cell size is normal. We have shown for the five hypotrich ciliates from Fig. 14.1 that the generation time is also not correlated with the cell size: *Stylonychia mytilus* with the largest cell size and the largest Mi has the shortest cell cycle while *E. aediculatus*, the second largest cell has the smallest Mi and one of the longest cell cycles (Ammermann and Muenz, 1982). Of course measurements of five species are not sufficient to form an opinion about the hypothesis, but the impression is that they show a trend which might be true for other ciliates, too. It seems certain that generation time and cell size are important parameters which decide the success of the species in their environment. But the results obtained seem to make improbable the hypothesis that the nuclear volume of the Mi (which in most cases contain less than 2% of the Ma DNA content) act as a determinant for cell growth and cell size.

Every hypothesis about the C-value paradox has to find an explanation for the fact that ciliate species which are similar in many aspects have very different amounts of E DNA. *Paramecium aurelia* and *P. bursaria* are rather alike, e.g. in cell size and Ma DNA content (see Table 14.1), but *P. bursaria* has 10 times more Mi DNA than *P. aurelia*. *Stylonychia mytilus* has 8.3 pg E DNA per haploid Mi, *Euplotes aediculatus* only 0.45 pg. To replicate much E DNA before every cell division is energy consuming and the process of sorting out S and E DNA sequences during Ma development is probably a complicated process which might lead to mistakes. Therefore, it seems that these and other disadvantages are equalized by, until now, unknown advantages or additional functions of this E DNA. However, the possible functions which have been so far discussed for this E DNA are not supported by any results of research on animals which show chromatin elimination. Nor do the investigations of E DNA, especially in ciliates, give evidence for or against the hypotheses that this E DNA or most of it is useless 'junk' or 'parasitic' DNA (Orgel *et al.*, 1980). These hypotheses are difficult to test experimentally. One experiment was recently started in *Stylonychia mytilus*. As mentioned earlier its tiny pseudo-Mi (p. 437) can substitute for the normal sized Mi during vegetative growth and division. If most of the Mi DNA is useless, then it should be possible to get cells with pseudo-Mi which contain also the genes necessary for conjugation, but which lack the useless DNA, but this has not yet been achieved. Only further experiments can show whether this means that the 'useless' DNA has a necessary function.

REFERENCES

Ammermann, D. (1970). The micronucleus of the ciliate *Stylonychia mytilus*; its nucleic acid synthesis and its function. *Exp. Cell Res.*, **61**, 6–12.

Ammermann, D. and Muenz, A, (1982). DNA and protein content of different hypotrich ciliates. *Europ. J. Cell Biol.*, **27**, 22–24.

Ammermann, D., Steinbrueck, G., Berger, v. L., Hennig, W. (1974). The Development of the Macronucleus in the Ciliated Protozoan *Stylonychia mytilus*. *Chromosoma*, **45**, 401–429.

Bantock, C. (1961). Chromosome elimination in Cecidomyiidae. *Nature, Lond.*, **190**, 466–467.

Beermann, S. (1977). The diminution of heterochromatic chromosomal segments in *Cyclops* (*Crustacea, Copepoda*). *Chromosoma*, **60**, 297–344.

Bennett, M. D. (1971). The duration of meiosis. *Proc. Roy. Soc. Lond. B.*, **178**, 277–299.

Blackburn, E. H. and Gall, J. G. (1978). A tandemly repeated sequence at the termini of the extrachromosomal ribosomal RNA genes in *Tetrahymena*. *J. Mol. Biol.*, **120**, 33–53.

Boswell, R. E., Klobutcher, L. A. and Prescott, D. M. (1982). Inverted terminal repeats are added to genes during macronuclear development in *Oxytricha nova*. *Proc. Natl. Acad. Sci.*, **79**, 3255–3259.

Boveri, T. (1887). Ueber Differenzierung der Zellkerend waehrend der Furchung des Eies von *Ascaris megalocephala*. *Anat. Anz.*, **2**, 688–693.

Brown, S. W. and Nur, U. (1964). Heterochromatic chromosomes in the coccids. *Science*, **145**, 130–136.

Cavalier-Smith, T. (1978). Nuclear volume control by nucleoskeletal DNA, selection for cell volume and cell growth rate, and the solution of the DNA C-value paradox. *J. Cell Sci.*, **34**, 247–278.

Davies, A. H. and Carter, C. E. (1980). Chromatin diminution in *Ascaris suum*. *Exp. Cell Res.*, **128**, 59–62.

Davidson, E. H. and Britten, R. J. (1979). Regulation of gene expression: possible role of repetitive sequences. *Science*, **204**, 1052–1059.

Fujishima, M. and Watanabe, T. (1981). Transplantation of germ nuclei in *Paramecium caudatum*. *Exp. Cell Res.*, **132**, 47–56.

Geyer-Duszyńska, J. (1966). Genetic factors in oogenesis and spermatogenesis in Cecidomyiidae. In C. D. Darlington and K. R. Lewis (Eds), *Chromosomes Today*, Oliver and Boyd, Edinburgh and London, pp. 174–178.

Goldstein, P. and Straus, N. A. (1978). Molecular characterization of *Ascaris suum* DNA and of chromatin diminution. *Exp. Cell Res.*, **116**, 462–466.

Gorovsky, M. A. (1980). Genome organization and reorganization in *Tetrahymena. Ann. Rev. Genet.*, **14**, 203–239.

Hennig, W., Meyer, G. F., Hennig, J., Leoncini, O. (1974). Structure and function of the y chromosome of *Drosophila hydei. Cold Spring Harbor Symp.*, **38**, 673–683.

Hess, O. (1975). Y-linked factors affecting male fertility in *Drosophila melanogaster* and *Drosophila hydei.* In R. C. King (Ed), *Handbook of Genetics*, Plenum Press, New York, vol. 3, pp. 747–756.

John, B. and Miklos, L. G. (1979). Functional aspects of satellite DNA and heterochromatin. *Int. Rev. Cytol.*, **58**, 1–114.

Just, E. (1973). Untersuchungen zur zeitlichen Ordnung der Syntheseleistungen in der Interphase von *Paramecium bursaria. Arch. Protistenk.*, **115**, 22–68.

Kaine, B. P. and Spear, B. B. (1982). Nucleotide sequence of a macronuclear gene for actin in *Oxytricha fallax. Nature, Lond.*, **295**, 430–432.

Katzen, A. L., Cann, G. M. and Blackburn, E. H. (1981). Sequence-specific fragmentation of macronuclear DNA in a Holotrichous Ciliate. *Cell*, **24**, 313–320.

Klobutcher, L. A., Swanton, M. T., Donini, P. and Prescott, D. M. (1981). All gene-sized DNA molecules in four species of hypotrichs have the same terminal sequence and an unusual 3′ terminus. *Proc. Natl. Acad. Sci. USA*, **78**, 3015–3019.

Klobutcher, L. A., Jahn, C. L., Prescott, D. M,. (1984). Internal sequences are eliminated from genes during macronuclear development in the ciliated protozoan *Oxytricha nova. Cell*, **36**, 1045–1055.

Kovaleva, V. G. and Raikov, J. B. (1978). Diminution and re-synthesis of DNA during development and senescence of the 'diploid' macronuclei of the ciliate *Trachelonema sulcata* (*Gymnostomata, Karyorelictida*). *Chromosoma*, **67**, 177–192.

Kunz, W. (1970). Genetische Aktivität der Keimbahnchromosomen waehrend des Eiwachstums von Gallmücken (*Cecidomyiidae*). *Verh. Dtsch. Zool. Ges.*, **64**, 42–46.

Lauth, M. R., Spear, B. B., Heumann, J. and Prescott, D. M. (1976). DNA of ciliated protozoa: DNA sequence diminution during macronuclear development of *Oxytricha. Cell*, **7**, 67–74.

Lohmann, K. and Schubert, L. (1980). Qualitative changes in DNA indicating differential DNA replication during early embryogenesis of the newt *Triturus vulgaris. J. Embryol. Exp. Morph.*, **57**, 61–70.

McTavish, C. and Sommerville, J. (1980). Macronuclear DNA organization and transcription in *Paramecium primaurelia. Chromosoma*, **78**, 147–164.

Moritz, K. B. (1970). DNA-Variation im keimbahnbegrenzten Chromatin und autoradiographische Befunde zu seiner Funktion bei *Parascaris equorum. Verh. Dtsch. Zool. Ges.*, **64**, 36–42.

Moritz, K. B. and Roth, G. E. (1976). Complexity of germline and somatic DNA in *Ascaris. Nature, Lond.*, **259**, 55–57.

Mueller, F., Walker, P., Aeby, P., Neuhaus, H., Back, E. and Tobler, H. (1982). Molecular cloning and sequence analysis of highly repetitive DNA sequences contained in the eliminated genome of *Ascaris lumbricoides.* In M. M. Burger and R. Weber (Eds), *Progress in Clinical and Biological Research*, vol. 85A, Alan R. Liss Inc., New York, pp. 127–138.

Murti, K. G. and Prescott, D. M. (1970). Micronuclear ribonucleic acid in *Tetrahymena pyriformis. J. Cell Biol.*, **47**, 460–467.

Nelson-Rees, W. A., Hoy, M. A. and Roush, R. T. (1980). Heterochromatization, chromatin elimination and haploidization in the parahaploid mite *Metaseiulus occidentalis* (Nesbitt) (Ascarina:Phytoseiidae). *Chromosoma*, **77**, 263–276.

Nock, A. (1981). RNA and macronuclear transcription in the ciliate *Stylonychia mytilus*. *Chromosoma*, **83**, 209–220.

Oka, Y., Shiota, S., Nakai, S., Nishida, Y. and Okubo, S. (1980). Inverted terminal repeat sequence in the macronuclear DNA of *Stylonychia postulata*. *Gene*, **10**, 301–306.

Orgel, L. E., Crick, F. H. C. and Sapienza, C. (1980). Selfish DNA. *Nature, Lond.*, **288**, 645–646.

Pasternak, J. and Barrell, R. (1976). Quantitation of nuclear DNA in *Ascaris lumbricoides*: DNA constancy and chromatin diminution. *Genet. Res.*, **27**, 339–348.

Pelvat, B. and de Haller, G. (1976). Macronuclear DNA in *Stentor coeruleus*: a first approach to its characterization. *Genet. Res.*, **27**, 277–289.

Preer, J. R. Jr. and Preer, L. B. (1979). The size of macronuclear DNA and its relationship to models for maintaining genetic balance. *J. Protozool.*, **26**, 14–18.

Prescott, D. M., Heumann, J. M., Swanton, M. and Boswell, R. E. (1979). The genome of hypotrichous ciliates. In J. Engberg, H. Klenow and V. Leick (Eds), *Specific Eukaryotic Genes*, *Alfred Benson Symposium*, vol. **13**, Munksgaard, Copenhagen, pp. 85–99.

Radzikowski, S. (1973). Die Entwicklung des Kernapparates und die Nukleinsäuresynthese während der Konjugation von *Chilodonella cucullulus* O. F. Mueller. *Arch. Protistenk.*, **115**, 419–428.

Roth, G. E. and Moritz, K. B. (1981). Restriction enzyme analysis of the germ line limited DNA of *Ascaris suum*. *Chromosoma*, **83**, 169–190.

Soldo, A. T. and Brickson, S. A. (1979). On the nature and size of the DNA genome of several species of marine and fresh water ciliates. *J. Protozool.*, **37**, 11A.

Soldo, A. T., Brickson, S. A. and Lavin, F. (1981). The kinetic and analytical complexities of the DNA genomes of certain marine and fresh-water ciliates. *J. Protozool.*, **28**, 377–383.

Sonneborn, T. M. (1974). *Paramecium aurelia*. In R. C. King (Ed), *Handbook of Genetics*, vol. 2, Plenum Press, New York, pp. 469–594.

Spradling, A. C. and Mahowald, A. P. (1980). Amplification of genes for chorion proteins during oogenesis in *Drosophila melanogaster*. *Proc. Natl. Acad. Sci. USA*, **77**, 1096–1100.

Steinbrueck, G., Haas, I., Hellmer, K. H. and Ammermann, D. (1981). Characterization of macronuclear DNA in five species of ciliates. *Chromosoma*, **83**, 199–208.

Steinbrück, G. and Schlegel, M. (1983). Characterization of two sibling species of the genus *Stylonychia* (Ciliata, Hypotricha): *S. mytilus* Ehrenberg 1838 and *S. lemnae* sp. II. Biochemical characterization. *J. Protozool.*, **30**, 294–300.

Sugai, T. and Hiwatashi, K. (1974). Cytologic and autoradiographic studies of the micronucleus at meiotic prophase in *Tetrahymena pyriformis*. *J. Protozool.*, **21**, 542–548.

Swanton, M. T., Heumann, J. M. and Prescott, D. M. (1980). Gene-sized DNA molecules of the macronuclei in three species of Hypotrichs: size distributions and absence of nicks. *Chromosoma*, **77**, 217–227.

Tobler, H., Smith, K. D. and Ursprung, H. (1972). Molecular aspects of chromatin elimination in *Ascaris lumbricoides*. *Dev. Biol.*, **27**, 190–203.

Tobler, H., Zulauf, E. and Kuhn, O. (1974). Ribosomal RNA genes in germ line and somatic cells of *Ascaris lumbricoides*. *Dev. Biol.*, **41**, 218–223.

White, M. J. D. (1954). *Animal Cytology and Evolution*, Cambridge University Press, Cambridge.

The Evolution of Genome Size
Edited by T. Cavalier-Smith
© 1985 John Wiley & Sons Ltd

CHAPTER 15

The evolutionary significance of middle-repetitive DNAs[1]

W. Ford Doolittle

Department of Biochemistry, Dalhousie University, Halifax, Nova Scotia, B3H 4H7 Canada

MIDDLE-REPETITIVE DNA AS PART OF THE PROBLEM

The kinds of eukaryotic DNA

Eukaryotes show a 10^4- to 10^5-fold range of haploid DNA contents, this variation correlating in no sensible way with anything we might care to call organismal complexity or evolutionary 'advancement' (Cavalier-Smith, 1978, 1982). Only in the protists, fungi and invertebrates does the lowest known DNA content come within even an order of magnitude of that amount conceivably required to code for and regulate the production of all conceivably required proteins, and each of these groups shows within itself at least a 100-fold range of DNA contents (Britten and Davidson, 1971; Hinegardner, 1976; Cavalier-Smith, 1978, 1982; French and Manning, 1980; Herman and Horvitz, 1980; Krumlauf and Marzluf, 1980). Thus when looking at the total DNA of almost any eukaryote, we are not looking at genes; we are looking at non-coding DNA.

Typical of what might be seen, upon slow reannealing of total DNA sufficiently well fragmented, is shown schematically in Fig. 15.1. Some (often very little) of the DNA reanneals at negligible C_ot (DNA concentration in moles of nucleotide/litre × seconds; Britten and Kohne, 1968). This 'zero-time', 'foldback' or 'snapback' DNA consists of sequences which contain inverted repeats and thus renature in a concentration-independent fashion. Some reanneals only at high C_ot. This is considered to comprise sequences present only once in the haploid genome (such as, presumably, most structural genes) and is commonly termed 'single copy' or 'unique-sequence'

[1] The literature review for this chapter was completed in late 1981.

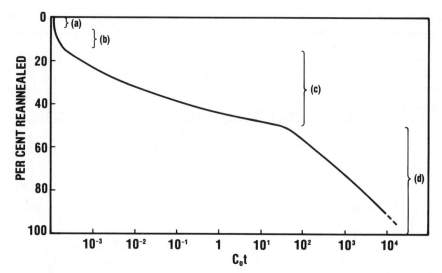

Fig. 15.1.   An imaginary $C_o t$ curve showing % DNA reannealed as a function of $C_o t$ (DNA concentration in moles of nucleotide per liter times seconds) for an imaginary genome with 5% 'foldback DNA': (a), 10% 'highly-repetitive DNA' (b), 35% 'middle-repetitive DNA' (c) and 50% 'unique-sequence DNA'.

DNA. Reannealing between these two more or less well-defined components are DNAs which are repeated from tens to millions of times within the genome, and which can comprise as much as 80–90% of its mass (Britten and Davidson, 1971; Thompson *et al.*, 1980).

In 1971, on the basis of data then available, Britten and Davidson (1971) considered such repetitive DNAs to be of two fundamentally distinct kinds. The first, *'highly-repetitive'* DNA, consisted of sequences which are present in tens of thousands to millions of copies, are often arrayed in long tandem repeats of relatively simple sequence, are often (because of their simple sequences) isolatable by equilibrium density gradient centrifugation as 'satellites' with G + C concent different from that of unique-sequence DNA, are often heterochromatic and are not transcribed. The second, 'moderately', 'intermediate' or *'middle-repetitive'* DNA consisted of sequences which are present in tens to tens of thousands of copies, are more often dispersed as single copies within regions of otherwise unique-sequence DNA, are of more complex sequence (and thus of more average G + C content), are usually euchromatic and are usually transcribed.

The terms highly-repetitive and middle-repetitive persist, but it is not clear to me that they define classes of DNA which are truly distinct. The most highly reiterated family of sequences in the human genome is in fact largely present as single dispersed units of fairly complex sequence (see below). The same is true for certain crabs (Christie and Skinner, 1979). Some tandemly arrayed DNAs are of quite complex sequence (Thayer *et al.*, 1981; Miklos, 1982; Miklos and Gill, 1981); on occasion, some are even transcribed (Varley *et al.*, 1980). Some dispersed elements are of quite simple, satellite-like sequence (Truett *et al.*, 1981). Some repetitive elements exist both in tandem array and in single copies scattered elsewhere (Roiha *et al.*, 1981; Scheller *et al.*, 1981a; Posakony *et al.*, 1981). Worse yet, the distinction between unique-sequence DNAs and repetitive DNAs of high divergence and low copy number may not be as firm as desired. The fractions of the genomes of plants and birds behaving as unique-sequence depends not only on the plant or bird examined but on the stringency of hybridization conditions (Murray *et al.*, 1981; Burr and Schimke, 1980).

Eukaryotic genomes taken together may thus contain DNAs showing a continuous spectrum of copy numbers and intercopy sequence homologies. Only those DNAs which have achieved relatively high copy number and maintain relatively strong homology draw attention to themselves. Among these, we can perhaps most meaningfully distinguish between those which are tandemly arrayed and those which are dispersed within the genome, simply because we can imagine that these have arisen through and are maintained by different genetic mechanisms. Tandem repeats can arise and be maintained relatively homogeneous in sequence by unequal crossing-over (Smith, 1976; Petes, 1980) although gene conversion betweeen misaligned copies of the array may also account for maintenance (Baltimore, 1981; Klein and Petes, 1981; Jackson and Fink, 1981). Dispersed repeats cannot arise without some sort of genomic rearrangement analogous in its consequences to transposition, and cannot (on logical grounds) be maintained by unequal crossing-over. It is to such dispersed repeats, which probably do include much of what has in the past been called middle-repetitive DNA, that this chapter is confined, ignoring that often negligible fraction of such DNA which actually codes for RNAs or proteins of known cellular functions, that is, ignoring 'genes'.

Interspersion patterns

The patterns of interspersion of dispersed repetitive and unique-sequence DNAs shown by most organisms examined up to 1975 (a diverse collection of vertebrates and invertebrates) were similar (Galau *et al.*, 1976); more than 70% of unique-sequence DNA existing in short stretches (1–3 kbp) bracketed by short (200–400 bp), dispersed, repetitive elements of different families (collections of repeats of similar sequence). This pattern—the '*Xenopus*', or 'short-period interspersion' pattern—is *not* universal. In 1975, Manning *et al.*, showed that the *Drosophila* genome contains predominantly long (> 13 kbp) stretches of unique-sequence DNA bracketed by long (> 5 kbp) repetitive elements. Since then, a variety of organisms have been described as having such 'long-period interspersion', and there is no obvious correlation between interspersion pattern and phylogenetic position (see Sapienza and Doolittle, 1981), although there *may* be a correlation with total DNA content (e.g. French and Manning, 1980, but see Smith *et al.*, 1980). It is unclear to what extent the belief that eukaryotic genomes *should* show either short- or long-period interspersion has tempted investigators to categorize what may be intermediate patterns as one or the other. Clearly, individual plant genomes can show a continuum of patterns from short to long (Murray *et al.*, 1979, 1981). The genomes of birds may do this as well (Arthur and Strauss, 1978; Eden and Hendrick, 1978; Burr and Schimke, 1980; Musti *et al.*, 1981).

Families of dispersed middle-repetitive DNAs may thus best be looked at individually, in the genomes in which they live, without premature hope that any broad generalization about their behaviours will emerge. Initial consideration will be given to the properties of those elements which, because of their preponderance, homogeneity or mobility, have drawn attention to themselves in the genomes of *Drosophila*, sea urchins, man and plants, before discussing speculations about their possible cellular functions and evolutionary origin. There are five important parameters defining any family of repetitive elements: family size (number of repeat units), genomic disposition (interspersion pattern), repeat unit length, repeat unit sequence and intrafamilial repeat unit sequence divergence. All are variable. There are really only two important general questions to be asked about a family:

1. At what level, if any, has natural selection operated to give rise to and maintain the family?
2. By what mechanism, if any, is family number and sequence homogeneity maintained?

The two are not unrelated.

DROSOPHILA MIDDLE-REPETITIVE DNAs

Components of the *Drosophila melanogaster* genome

Drosophila melanogaster has one of the smallest genomes found among complex eukaryotes, only ten times as large as that of yeast and forty times as large as that of *Escherichia coli* (Potter and Thomas, 1978). Of its 1.5×10^5 kbp, some 20% is highly-repetitive (average copy number 2×10^5), and almost as much consists of middle-repetitive DNA families of small (10–200 members) family size (Wensink, 1978; Young, 1979). *Drosophila* exhibits 'long-period interspersion', with unique-sequence DNA interrupted at intervals averaging 13 kbp by repetitive elements which are themselves long (5–6 kbp). Indeed, it was the first organism shown to have his pattern (Manning *et al.*, 1975).

The middle-repetitive families of *Drosophila*, examined either individually as cloned repeats, or together as unfractionated DNA reannealed at low $C_o t$, exhibit higher interfamilial sequence homogeneity ($\leqslant$ 5% sequence mismatch) than do middle-repetitive DNAs of organisms showing primarily short-period interspersion (Wensink, 1978; see below). Perhaps three-quarters of this middle-repetitive DNA represents families of relatively long (about 5 kbp), linearly homologous elements dispersed within unique-sequence DNA (Young and Schwartz, 1981). It is elements of this sort about which we know most. The rest may consist largely of quite long arrays of shorter (about 500 bp) elements organized in a way which Wensink *et al.* (1979) described as 'clustered and scrambled'.

Long, linearly homologous copia-like elements

The repetitive elements *copia*, *412* and *297* studied primarily by American workers spawned by Hogness' laboratory (Rubin *et al.*, 1976; Finnegan *et al.*, 1978; Potter *et al.*, 1979; Strobel *et al.*, 1979; Dunsmuir *et al.*, 1980; Levis *et al.*, 1980) and the *mdg* elements studied by Russian workers spawned by Georgiev's laboratory (Ilyin *et al.*, 1978; Bayev *et al.*, 1980; Tchurikov *et al.*, 1981) were initially isolated as 'genes'. Most are abundantly transcribed into polyadenylated [poly(A)$^+$] RNAs which, when used to probe genomic libraries, readily identify clones bearing such *copia*-like elements. Although there are many (perhaps 100) such elements of distinct sequence in the *Drosophila* genome, those which have been characterized exhibit the following common properties, as discussed by Rubin *et al.* (1981), Young and Schwartz (1981), Tchurikov *et al.* (1981) and Finnegan (1981).

1. All appear to comprise families of from 15 to 200 members dispersed throughout the genome, with some tendency for some to be concentrated in chromocentres.
2. Families are usually highly homogeneous, since individually cloned members from the same or different strains or species of *Drosophila* show identical or very similar restriction endonuclease digestion profiles. However, at least one *copia* variant lacking 200 bp of the sequence found in most other copies has been detected (Young and Schwartz, 1981).
3. Among interfertile strains of *D. melanogaster*, families are characterized by constancy of copy number but variability in position. For instance, 15 of 17 elements show more differences than similarities in chromosomal location, as determined by *in situ* hybridization, between the two *melanogaster* strains g-I and g-XII, and yet copy numbers vary by little more than 10% (Young, 1979; Young and Schwartz, 1981).

4. Copy number generally increases (by as much as 15-fold) in *Drosophila* cells grown in culture. This together with the previous observation is conclusive evidence for replicative transposition.

5. *Between-species* comparisons, unlike *between-strain* comparisons, show great differences in copy number. Some *D. melanogaster* elements are totally absent from the genomes of *D. simulans* or *D. virilis*. Shared elements are usually 2–20 times more prevalent in the former species, even though interspecies element homology remains high. Thus, as with dispersed middle-repetitive DNAs of sea urchins (see below), major evolutionary change affects copy number more than copy sequence.

6. *Drosophila copia*-like elements potentially can, and demonstrably do, cause mutations in one or more of the variety of ways in which bacterial transposable elements do this (Rubin *et al.*, 1981; Rasmuson *et al.*, 1981; Ising and Block, 1981; Gehring and Paro, 1980).

Those *copia*-like elements which have been characterized in detail show a common structure, which is similar to that of certain bacterial transposable elements, the yeast transposable element Ty1 and the integrated proviral copies of retroviral genomes (Fig. 15.2). A core of 5–6 kbp of unique

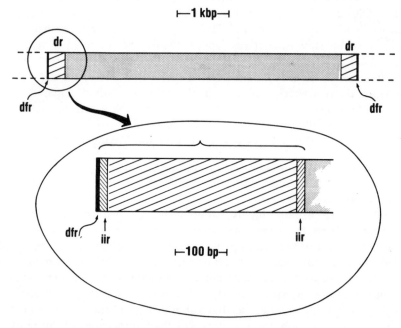

Figure 15.2. Schematic representation of a '*copia*-like' element, showing terminal direct flanking repeats (dfr) of sequences present initially only once at the site of insertion, element specific terminal direct repeats (dr), and the imperfect inverted repeats (iir) at each of *their* termini.

sequence DNA (which is transcribed into translatable mRNAs) is flanked by direct repeats of 300–500 bp. These direct repeats each bear, at each of their termini, short (< 20 bp) regions of imperfect inverted repetition so that the entire element can be viewed as having terminal inverted repeats, like certain bacterial transposons (Calos and Miller, 1980). The 300–500 bp direct repeats of a single element are identical in sequence, but different elements of the same family can have variant sequences, which are again perfectly duplicated at their termini. This suggests that only one copy of the terminal direct repeat is involved in transposition, directing the synthesis of both copies in the resulting product (Rubin *et al.*, 1981). The direct repeats of chromosomal copies are themselves flanked by 4–6 bp short direct repeats whose length is element-specific and whose sequence is specific to the site of insertion. These are, as in the case of bacterial transposable elements, generated by duplication of chromosomal sequences initially present only once at the site of insertion (Dunsmuir *et al.*, 1980; see also Chapter 12).

Most *copia*-like elements examined are abundantly transcribed into several poly(A)+ and poly(A)− RNAs which are present in both cytoplasm and nucleus, but concentrated in the latter. Some of these function as mRNAs and some appear, on the basis of their sizes, to be complete transcripts of entire elements (Young and Schwartz, 1981; Rubin *et al.*, 1981; Georgiev *et al.*, 1981). In this way then, as well as in structure, *copia*-like elements resemble integrated vertebrate retroviral proviruses. It has been proposed that the latter can transpose themselves through reverse transcription into covalently-closed-circular DNAs containing either single or tandemly duplicated copies of the terminal direct repeat, which are then reintegrated into the chromosome by a process similar in its end product to bacterial transposition (Shoemaker *et al.*, 1981; Shimotohno and Temin, 1981). Although intracellular provirus transposition has yet to be demonstrated, the proposed circular intermediate(s) are known intermediates in integration during retroviral infection. Covalently-closed-circular DNAs bearing sequences complementary to much of the middle-repetitive DNA component, and to bulk nuclear poly(A)+ RNA (analogous thus to retroviral RNA genomes) do exist in cultured *Drosophila* cells (Stanfield and Lengyel, 1979, 1980). Recently, Flavell and Ish-Horowicz (1981) have described circular DNAs identical in restriction endonuclease digestion patterns to *copia* elements, and containing either one or two (tandem) copies of the direct repeat, in such cultured cells. It seems perverse not to suppose that *copia*-like elements transpose themselves through cycles of transcription, reverse transcription and reintegration, as Young and Schwartz (1981), Rubin *et al.*, (1981) and Shoemaker *et al.* (1981) suggest, even though this may be difficult to prove and other mechanisms can be devised (Finnegan, 1981).

Copia-like elements are not insignificant components of the *Drosophila* genome. Young (1979) showed that most of *Drosophila's* middle-repetitive

DNA is composed of families of long mobile repeats which presumably behave in just this way. There are, however, at least three other *kinds* of middle-repetitive DNAs in this organism. These, too, appear mobile.

FB elements

Potter *et al.* (1980) isolated cloned *D. melanogaster* fragments which hybridized to total foldback (FB) DNA obtained by reannealing at very low $C_o t$. The cloned unique-sequence DNAs flanking three of these inverted repeats probed different bands in different strains of *D. melanogaster*, as would be expected if the repeated sequences themselves were mobile. These FB elements differed greatly in the size of the repeated region and in the length of the unique-sequence stretch between them. In fact no two of the nine characterized elements appear the same. The inverted repeat sequences themselves are recognized by very few restriction endonucleases, although all show regularly spaced *Taq* I sites and conserved terminal *Hinf* I sites, as if they were made up of tandem repeats of a common simple sequence, with uniquely conserved termini (Truett *et al.*, 1981). This, plus homology demonstrable by hybridization between short fragments, justifies the inclusion of these heterogenous structures within a single family. Sequencing of parts of one element shows imperfectly repetitive, satellite-like structure, with a basic periodicity of 10 bp (GCTTTGCCCA) which is elsewhere amplified to 20 bp and elsewhere to 31 bp, with interruptions, all within an inverted repeat region. The two inverted repeat sequences of a given element are not identical.

Comparisons of sequences around the site where an FB element is present in one strain and absent in another suggest that insertion generates a 9 bp flanking direct repeat of sequences initially present only once at the site of insertion (Truett *et al.*, 1981). *In situ* hybridization experiments with one element (FB5) show it to be present in some 30 copies, whose positions vary greatly between, and even within, strains.

The FB family is the simplest known family of transposable elements. Most of its members are clearly incapable of coding for functions involved in their own transposition and maintenance, and perhaps none of them has that capability. It is difficult to argue that members of this family serve any useful cellular function.

Ribosomal insertion-like elements, clustered and scrambled repeats

The 28s ribosomal RNA genes of *D. melanogaster* are interrupted by two types of insertion, each comprising several size classes (Dawid *et al.*, 1978). Elements homologous to the type I insertion, which interrupts about 50% of all rDNA copies on the X, but not the Y, chromosome, are found elsewhere in the genome, largely in chromocentres (Dawid *et al.*, 1981). Comparisons of the sequences of uninterrupted and interrupted 28S rRNA genes in *Drosophila virilis* are consistent with the notion that type I sequences are indeed transposable elements which generate, upon insertion, direct 14 bp flanking repeats (Rae *et al.*, 1980). In *D. melanogaster*, one of these flanking repeats appears to have been deleted, but copies of the insertion element outside of the rDNA region are flanked by short segments of rDNA sequence, as if they had originated from within that region (Roiha *et al.*, 1981). Although rDNA type I insertions thus appear transposable, they lack the terminal direct or inverted repeats characteristic of *copia*-like or FB elements. This property they share with at least one other transposable element in *Drosophila* (see below) and with members of the human *Alu* I family.

An analysis of cloned type I sequences residing outside the rDNA region shows them to have suffered a variety of substitutions which alter restriction endonuclease sites, as well as larger deletions and insertions (Dawid *et al.*, 1981). Copies of the type I elements are often clustered in non-rDNA regions, both with each other and with inserted or flanking elements which are themselves repeated separately elsewhere in the genome (Kidd and Glover, 1980). This sounds like the 'clustered and scrambled' arrangement of repetitive sequences described by Wensink *et al.* (1979). Regions of this sort consist of quite long arrays of different, short (approximately 500 bp) elements which can be differently associated with different short repeats in different linear patterns in different clusters, and which may also occur as isolated elements flanked by unique-sequence DNA. Young and Schwartz (1981) suggest that as much as one-third of the middle-repetitive DNA of *Drosophila* may consist of such complex rearranged sequences. Similar regions are found in the genomes of chickens (Musti *et al.*, 1981), sea urchins and plants (see below).

Dawid *et al.* (1981) sequenced portions of a repeated element inserted into a type I sequence in such a clustered and scrambled array. It appears to have generated a 13 bp duplication of a sequence normally present only once in the type I element. Like that element itself, it lacks terminal flanking or direct repeats. One of its boundaries, however, comprises an uninterrupted stretch of 18 T (or A) residues separated by one nucleotide from the generated flanking repeat. It thus resembles elements of the human *Alu* I family and related mammalian *Alu* I-like families (see below).

Concluding remarks

Virtually all of the dispersed repetitive sequences in *D. melanogaster* about which we have any certain knowledge appear to be transposable elements, of one of three types. It is hard to argue that elements which are so mobile can play any constant regulatory role in a genome which is otherwise so stable. Many authors have suggested that these elements, and comparable elements in yeast and bacteria, are maintained because they promote genetic variability (e.g. Nevers and Saedler, 1977; Cameron *et al.*, 1979; Strobel *et al.*, 1979; Roeder and Fink, 1980). Others suggest they are maintained because they limit gene flow (Young and Schwartz, 1981), as the demonstrably transposable elements responsible for hybrid dysgenesis demonstrably do (Engels, 1981). Both suggestions require 'group selection' to be responsible for the maintenance of such elements. We (Doolittle and Sapienza, 1980; Sapienza and Doolittle, 1981) and Orgel and Crick (1980) have argued that transposition itself (if it maintains or increases copy number) insures survival of the element transposed, that no other explanation for the existence of such elements is therefore required, that they can thus properly be called 'selfish', and that most dispersed middle-repetitive DNAs which are not members of multigene families may in fact be selfish DNAs. It seems that this is true for the genome of *Drosophila*. The extent to which it is true for more complex genomes, about which we know only a little, is the subject of the rest of this chapter.

SEA URCHIN REPETITIVE DNAs

Characterization of the repetitive fraction of total *S. purpuratus* DNA

The dispersed repetitive sequences of the sea urchin *Strongylocentrotus purpuratus* and its relatives deserve special attention, since it is largely the extensive analysis of their properties which led Britten, Davidson and coworkers to construct the intellectual framework within which much data on eukaryotic repetitive DNAs is interpreted. The *S. purpuratus* haploid genome contains some 8.1×10^5 kbp of DNA. When sheared to a few hundred bp, 25–30% of this DNA reanneals at low C_ot (Graham *et al.*, 1974), and can be shown to comprise a collection of members of some thousand or more non-homologous families. Family sizes (numbers of homologous members) range from as few as three to more than 10^4, and are unimodally distributed around a mean size of between 10^2 and 10^3. The remaining 70–75% of the DNA, under standard conditions of hybridization stringency, is almost entirely 'unique sequence'; there is very little highly-repetitive or 'satellite' DNA (Graham *et al.*, 1974; Klein *et al.*, 1978).

Repetitive and unique-sequence DNAs are interspersed in the genome. Renaturation of larger fragments reveals that more than three-quarters of the unique-sequence DNA is interrupted, at average intervals of about 1 kbp, by repetitive elements (Graham *et al.*, 1974).

Repetitive sequences can be isolated by limited S_1 nuclease digestion of

DNA reannealed at low $C_{o}t$, and sized by gel filtration (Britten *et al.*, 1976; Eden *et al.*, 1977). Two thirds of the duplexes so obtained are 'short' (about 300 bp) and of relatively low thermal stability. Short and long repeats form classes which are distinct not only in size but in sequence; there is little cross-hybridization between them. Although most short repeats appear to be interspersed with unique-sequence DNA, the distribution of long repeats is not easily deduced, and individual family or subfamily structure cannot be readily investigated by kinetic and thermal stability studies of total repetitive DNA.

Characterization of individually cloned repetitive DNAs

In 1977, Britten, Davidson and collaborators began a series of studies on individually cloned repetitive sequences, the results of which have much to tell us about their evolutionary behaviours. DNA of approximately 2 kbp was reannealed to $C_{o}t$ 4 and subjected to limited S_1 digestion to remove non-homologous single-stranded tails of unique-sequence DNA. After ligation to *Eco* RI linkers and digestion with *Eco* RI, fragments were inserted into a plasmid vector, cloned, and amplified in *E. coli*. Since DNAs were in general derived from imperfectly matched family members, the resulting 'repaired', purified and amplified clones bear precise sequences which may not have existed within the family but which are nevertheless no more divergent from typical family members than were the two reannealed parental strands. Clones may also not represent entire repeat units, if one of the parental repeat copies suffered prior shearing or *Eco* RI scission (Scheller *et al.*, 1977). However, the distribution of molecular weights of 26 cloned fragments studied does match that of the S_1 nuclease-derived repeats of total *S. purpuratus* DNA (Klein *et al.*, 1978). Repeat family sizes ranged from 3 to 12 500 members, with a distribution and mode (10^2–10^3) again comparable to that derived from kinetic analyses of unfractionated DNA.

Intrafamilial sequence diversity was measured for 15 of the families represented by individual clones, by determination of the thermal stabilities of hybrids formed between labelled cloned DNA and total genomic DNA at 45° and 55° and subtraction of the former value from the *Tm* of renatured cloned DNA alone ('native *Tm*'). This allowed division of the represented families into three arbitrarily defined classes, of increasing internal sequence divergence (Klein *et al.*, 1978).

Class I families (3 of the 15) are those in which 'native *Tm*' exceeded that of hybrids formed with total DNA at 45°C by no more than 4°C (which corresponds to the amount of sequence heterogeneity observed in unique-sequence DNA from different individuals of a population of *S. purpuratus* [Britten *et al.*, 1978]!).

Class II families (7 of 15) are those in which this difference exceeds 4°, but for which no hybrids with genomic DNA formed at 45° could not also be formed during more stringent hybridization at 55°.

Class III families (3 of 15) are those in which some members have diverged so greatly that they can form some hybrids at 45° which they cannot form at 55°, and thus for which only minimum estimates of family size could be made. There was no correlation between family size and divergence, but there was a general inverse correlation between family member (repeat unit) length and sequence divergence (see below). The average divergence of all families represented by cloned members (10°, corresponding to some 10% primary sequence divergence) again matched that calculated from similar experiments using total *S. purpuratus* repetitive DNA.

The 2034, 2108 and 2109 families

Anderson *et al.* (1981) investigated the genomic dispositions of one long (< 2 kbp) *class I* repeat (*2034*), one long *class II* repeat (*2108*) and one short *class III* repeat (*2109*), using plasmid clones obtained in the earlier studies to identify genomic family members represented in a λ-genomic library of total *S. purpuratus* DNA. The fraction of screened λ clones bearing representatives of the latter two families was that expected if their respective 20 and 1000 members were randomly distributed throughout the genome while the first clone probed only one-third the expected number of plaques, indicating some clustering. That *2108* and *2109* elements indeed usually occur singly was confirmed by Southern hybridization against restriction endonuclease-digested λ clones; only *2034* sequences showed significant clustering. More detailed characterization of a number of genomic members of each family cloned in λ allowed the following conclusions about each (Scheller *et al.*, 1981a; Posakony *et al.*, 1981).

1. The partially clustered *2034* family contains about 2500 repeats of length ≥ 2 kbp. Thermal stabilities of complexes formed between the initially cloned 498 bp probe and each of four individually cloned genomic members are very similar to each other and to thermal stabilities obtained with total genomic DNA. Repeat sequences of this family differ from each other by, on average, only 4%. This is, then, a highly homogeneous family whose members cannot be subdivided into two or more subfamilies, members of each of which show greater homology to each other than to the (nevertheless related) members of other subfamilies. Sequencing of the cloned 498 bp fragment used to characterize this family revealed the presence of numerous translation stop signals in all possible reading frames, and a short region of tandem repetition bearing possible homology

to intron/exon splice junctions, but no real structural clues to function (Posakany *et al.*, 1981).

2. The *2108* family is more complex. Thermal stabilities of hybrids between the plasmid-cloned probe and each of 21 repeat copies cloned in λ fall into two distinct classes, and the melting profile obtained with hybrids between this probe and total genomic DNA is distinctly bimodal. Together with earlier results indicating that, while only 20–50 genomic copies of the *2108* sequence are found by hybridization at 70° and low salt, several hundred more can be detected at lower stringency, this suggests that the *2108* family, unlike the *2034* family, comprises 'subfamilies'. The possibility that there are only 20–50 closely related copies, the remaining *2108* elements being widely divergent from it and each other, was eliminated by using three λ-cloned repeat elements which hybridize only poorly with the initial probe and with each other. Each has other close relatives within the genome so there must be at least four *2108* subfamilies within the *2108* 'superfamily'. The four subfamilies each contain 25 ± 10 members, and since low stringency hybridization reveals at least 400 *2108*-related sequences in the genome, there could well be more than 20 such subfamilies. Heteroduplex and restriction endonuclease mapping of members of each of the few identified subfamilies revealed them to consist of colinear repeat units of about 5 kbp, which differ by only scattered base-pair mismatches. Members of *different* subfamilies, however, are not colinear. They share some common sequence blocks ('subelements'), but these are arranged in different orders in each of the well-characterized subfamilies and show greater primary sequence divergence (base-pair mismatching) between than within subfamilies. What may be a similar 'clustered and scrambled' arrangement has been found for certain repeat units in *Drosophila* (Wensink, *et al.*, 1979) and chickens (Musti *et al.*, 1981). Within a *2108* subfamily, base-pair mismatches between members appear randomly scattered, and members of at least one subfamily cannot, because of frequent termination signals, code for protein.

3. The *class III 2109* family consists of some 10^3 separate and highly divergent repeats a few hundred bp long. Thermal stability profiles of hybrids between the initial plasmid clone and total DNA show a complex pattern, perhaps indicating the existence of a number of subfamilies whose members are less closely related to each other than are those of *2108* subfamilies. Sequencing of several genomic *2109* repeats shows that these too are composed of sub-elements (in this case quite short, 20–40 bp) which may be arranged in different orders in different elements. The low thermal stability of hybrids between members of this family is thus due as much to scrambling of subelements as it is to divergence of sequences within subelements (Posakany *et al.*, 1981).

Evolutionary behaviour of repetitive DNA families

Studies of the type outlined above tell us about the evolutionary behaviour of repetitive sequences within a single genomic lineage. Unfractioned or cloned repeats from *Strongylocentrotus purpuratus* can also be used to detect similar repeats in the genomes of the closely related *S. franscicanus*, or more distant relatives such as *Lytechinus pictus* and *Tripneustes gratilla*, and tell us something about the evolutionary histories of repeated families which have been isolated for 15 (*S. purpuratus/franscicanus*) or even 180 (*S. purpuratus/L. pictus* or *T. gratilla*) million years. Several generalizations can now be made.

1. Repetitive sequences are more strongly conserved than unique sequence DNA. When total repetitive DNAs are used in cross-hybridization experiments of moderately high stringency, *S. purpuratus* is found to share 83% of its repetitive DNA component with *S. franscicanus* and 40% with *L. pictus*, indicating sequence divergence rates (0.04–0.12% divergence per million years) some 2- to 3-fold lower than those shown by unique sequence DNA (Harpold and Craig, 1977). Similar results are obtained with individually cloned repeats (Moore *et al.*, 1978, 1981; Scheller *et al.*, 1981a). The *2108* superfamily for instance, appears to have been well conserved during the 150–200 million years since the divergence of *S. purpuratus* and *L. pictus* or *T. gratilla*, and evidence for the preservation of a similar subfamily structure can be obtained with appropriate subfamily-specific probes (Scheller *et al.*, 1981a).

2. Interspecific repeat hybrids can occasionally be virtually as stable as intraspecific hybrids, but in general it remains true for the sea urchin, as for other eukaryotes, that *within-species sequence homology between members of a given repeat family or subfamily is greater than between-species sequence homology for that family or subfamily* (Moore *et al.*, 1978, 1981). This must reflect either different, sequence-specific, selection in different evolutionary lineages, or the existence of genetic mechanisms which maintain, to a greater or lesser degree, family (subfamily) homogeneity independently of selection pressures operating through phenotype.

3. Short (about 300 bp) repeat elements and long (< 2 kbp) repeat elements show similar *between-species* sequence variation. (In fact, between-species divergence seems not to be correlated with any other readily measurable family property). However, short repeats show much greater *within-species* sequence divergence, a finding established both with total short and long repeats isolated from genomic DNA and with individual cloned members of short and long families (Britten *et al.*, 1976; Eden *et al.*, 1977; Moore *et al.*, 1981). This is most peculiar. It has to mean that the mechanisms by which short and long repeats (which can show the same distribution of family sizes) maintain sequence homogeneity are different

in some way which is not reflected in their 'evolutionary' (*between-species*) behaviour. But this is simply a restatement of the observation and not an explanation.

4. In contrast to the surprisingly strong *between-species* conservation of family (subfamily) sequence, there is very little conservation of family size, and no apparent relationship of this parameter to others. Family sizes for cloned repeats can vary by as much as 20-fold between *S. purpuratus* and *S. franscicanus*, and as much as 350-fold between the former and *L. pictus*. *Thus the major 'evolutionary' events affecting sea urchin repetitive DNAs are massive expansions (and, assuming an overall constancy in repetitive DNA content, contractions) in family size.* This, more than any other result, makes it difficult to argue that most members of any family are maintained in the genome because they have important biochemical or developmental roles to play there.

HUMAN REPETITIVE DNAs

The *Alu* I family: proponderance and structure

The *ca*. 3×10^6 kbp of the human genome consists predominantly (60%) of *ca*. 2 kbp stretches of unique-sequence DNA interrupted by 300 bp-long dispersed middle-repetitive elements, which together comprise some 7% of the genomic mass (Houck *et al.*, 1979; Rinehart *et al.*, 1981). Approximately one-third of these repeats exist in the proper orientation and in sufficient proximity to form 'inverted repeat' structures upon reannealing of denatured large fragments (Houck *et al.*, 1979; Jelinek *et al.*, 1980). Repetitive DNAs (isolated by reannealing total DNA to low $C_o t$ and digestion with S1 nuclease) *and* inverted repeat DNAs (isolated by S1 nuclease digestion of that material which reanneals immediately upon cooling, that is, at $C_o t < 0.0001$) are both predominantly 300 bp long and both predominantly sensitive to the restriction endonuclease *Alu* I, which cleaves them into fragments of 140 and 170 bp.

These observations suggest that human middle-repetitive DNA is largely of one family – the '*Alu* I family'. The estimated number of its members, not all of which retain *Alu* I sites, has climbed from a modest 50 000 in 1979 to 300 000 in 1980, and to a current plateau at 600 000 – 1 000 000 (Deininger and Schmid, 1979; Jelinek *et al.*, 1980; Tashima *et al.*, 1981; Rinehart *et al.*, 1981). It can go no higher, since at this level it represents nearly all of the middle-repetitive component of the human genome. Homologous sequences are nearly as frequent in the DNAs of rats and Chinese hamsters (Haynes *et al.*, 1981).

Renaturation studies using cloned repeats, as well as comparisons of several cloned sequences, reveal the *Alu* I family to be 'homogeneously divergent'; all members show a 10–12% mismatch from the 'consensus sequence' (Deininger and Schmid, 1979; Rubin *et al.*, 1980; Rinehart *et al.*, 1981). If we take 0.5% per million years as the rate of unselected nucleotide substitution, Li *et al.* (1981), and assume that all human *Alu* I elements have diverged without selection from a common ancestral element, then that ancestor should have arisen only 20 million years ago. The major middle-repetitive DNAs of rodents, although differing in certain fundamental structural features from human *Alu* I and thus *independently* divergent, shows approximately 80% sequence identity with it in homologous regions. The last common ancestor of mice and men died about 80 million genes ago. Some kind of mechanism which maintains loose within-species homogeneity while allowing between-species divergence *must* be at work.

With an *Alu* I repeat every 2 kbp, one expects to find copies within characterized human genomic clones, and they are indeed there. There is, for instance, one some 6 kbp 3′ to the human insulin gene (Bell *et al.*, 1980). There are at least eight in a 50 kbp region containing the human β-globin gene cluster (Fritsch *et al.*, 1980, 1981; Coggins *et al.*, 1980; Baralle *et al.*, 1980; Duncan *et al.*, 1981), five in a 20 kbp region containing the human α-globin genes and nine in the introns of a 13 kbp human *onc* gene (Dalla Favera *et al.*, 1981). Several such genomic copies have been sequenced. There is also a 'consensus sequence' available from *Alu* I-digested, unfractioned, low $C_o t$-reannealed, S_1-resistant, total human DNA, and sequences available from several 'BLUR' (Bam-linked-ubiquitous-repeat) clones obtained in a manner analogous to that used by Scheller *et al.* (1977, see above) to obtain clones representative of sea urchin repeat sequences (Rubin *et al.*, 1980).

Together these sequences show that *Alu* I elements, although incompletely homologous, have conserved features and structural peculiarities which must be telling us something about *Alu* I evolution and behaviour. These features

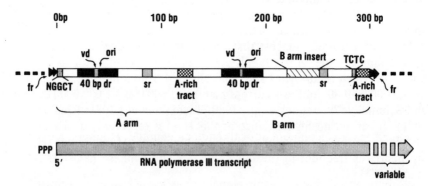

Figure 15.3. Structure of a 'typical' *Alu* I element from human DNA. Symbols explained in text.

have been summarized by Pan *et al.* (1981) and Haynes *et al.* (1981), are presented schematically in Fig. 15.3, and include the following.

1. Most (but not all) *Alu* I elements are flanked by 6–20 bp direct repeats (fr). The sequences repeated are different for each genomic *Alu* I copy. They are thus not 'parts' of *Alu* I elements, but help to define their ends and provide clues to their behaviour (Pan *et al.*, 1981; W. Jelinek and G. F. Saunders, personal communications).
2. The 5' terminus of most *Alu* I elements bears a conserved GGCT one bp removed from the 5'-flanking repeat (5' and 3' being defined by the direction of RNA polymerase III transcription, Fig. 15.3). Some 280–285 bp 3' to that is a conserved TCTC, followed by an A-rich stretch of 8–50 bp. This variable length A-rich region is largely responsible for length variation between individual *Alu* I repeats, and often contains residues other than A. It nevertheless clearly represents the 3' terminus.
3. There are, within *Alu* I sequences, regions of direct internal repetition which allow us to visualize human *Alu* I elements as composed of two incompletely homologous arms, an approximately 130 bp A arm and a B arm which differs from the A by an insertion of 31 bp (residues 220–250). Although there may be (allowing this insertion) significant scattered sequence homologies between the two arms, these are strongest in (a) a 40 bp direct repeat (*dr*) which, except for a 6 bp variable domain (*vd*), is highly conserved among *Alu* I elements and which contains the 14 bp region of homology to viral replication origins (*ori*) discussed below (b) an imperfect 9 bp short repeat (*sr*) 3' to the 40 bp direct repeat and (c) an internal A-rich tract of some dozen bp, which can be considered homologous to the 3'-terminal A-rich tract, and defines the end of the A arm.

Further evidence that human *Alu* I is indeed a 'dimer' is the fact that the predominant repetitive family in mouse and Chinese hamster DNAs shows some 80% primary sequence identity to the human *Alu* I B arm, differing principally in lacking the human-specific 31 bp B arm insertion and in having, in a different location, a rodent-specific 32 bp insertion. The rodent element is clearly the 'monomer'. Although dimeric in nature, human *Alu* I evolves as the 300 bp unit; homologies between all A arms or between all B arms are greater than homologies between A and B arms of the same unit (Krayev *et al.*, 1980; Pan *et al.*, 1981).

'Functions' of the *Alu* I family

Rubin *et al.* (1980) expressed the faith that 'a family of DNA sequences which includes 300 000 highly conserved members interspersed throughout much of the mammalian genome must have an important function'.

Mammalian cells provide, perhaps not surprisingly, a welter of hints about *Alu* I function, none of which actually demonstrate function in the sense defined later, and of which the following is surely an incomplete list.

1. Many *Alu* I family members contain, within the conserved 40 bp direct repeat, the sequence GGAGGCNPuAGPuCPuG, reminiscent of a 14 bp sequence PUGAGGCNGPUGGCGG found within the putative origins of replication of BK, SV40, polyoma and hepatitis B viruses. Some also show, just 5′ to this, the sequence GTACTT, which is found 5′ to the 14 bp sequence in SV40 and BK viruses. But not all do, and rodent *Alu* I-equivalents characteristically do not (Jelinek *et al.*, 1980; Pan *et al.*, 1981; Haynes *et al.*, 1981). The number of *Alu* I sequences is, within an order of magnitude, equivalent to the presumed number of origins of replication in the human genome, and there is some evidence to show that rodent *Alu* I-equivalents can function as origins *in vivo* (Georgiev *et al.*, 1981). However, the existence of structural features in *Alu* I and *Alu* I-like elements which are conserved within species but not between species and the absence of close (cross-hybridizable) relatives in the genomes of non-mammalian vertebrates (W. Jelinek, personal communication) argues against the notion that they are maintained by selection simply to serve as origins of DNA replication. There is no *independent* evidence to indicate that *Alu* I repeats are *the* origins used by cellular replication machinery, or indeed that these origins have long constrained sequences (Laskey and Harland, 1981).

2. *Alu* I transcripts (of either strand) are abundantly represented in heterogeneous nuclear RNA, where they give rise to long base-paired RNA structures. Oligonucleotides identifiable as transcripts of one or the other strand of *Alu* I DNA are the major products of T_1 digestion of isolated nuclear double stranded RNA (Jelinek *et al.*, 1978, 1980). This is the sort of thing predicted by the latest formulation of the Britten-Davidson model for the regulation of gene expression (Davidson and Britten, 1979). But it could also be the inevitable consequence of the fact that mammalian transcriptional units exceed in length, by at least an order of magnitude, the average distance between *Alu* I repeats. Indeed, intronic *Alu* I copies are known (Barta *et al.*, 1981; Dalla Favera *et al.*, 1981). *Alu* I transcripts are also found, but in lesser concentration, in cytoplasmic messenger RNA (Elder *et al.*, 1981a; Calabretta *et al.*, 1981), an observation not easily accommodated by any currently popular model for the regulation of gene expression.

3. Two low-molecular-weight RNAs in man and rodents show homology to the *Alu* I sequence, or parts of that sequence. One is a 100 nucleotide, nonpolyadenylated RNA associated with nuclear RNA in rodents (but not man; Jelinek and Leinwand, 1978), now known to be a '4.5S' RNA,

and an RNA polymerase III product (Haynes *et al.*, 1981). The (homogeneous) 4.5S RNAs of mice and hamsters are clearly homologous in sequence to cloned rodent *Alu* I-like elements, but are shorter than these, and show a ten-nucleotide region of no homology. They must be transcripts of an as yet uncloned subset of such elements. Indeed, only two bands are detected in Southern hybridizations performed at very high stringency between total restriction endonuclease digested DNA and labelled 4.5S RNA (W. Jelinek, personal communication). The second (homogeneous) *Alu* I-related RNA is the 300 nucleotide RNA polymerase III product, '7S RNA', found in human and rodent cells, and incorporated into mammalian retroviruses. Weiner (1980) showed that, although 7S RNA hybridizes to *Alu* I DNA, most or all hybrids have regions of ribonuclease sensitivity, revealing them to be imperfect. Again, this RNA may be the product of a small *Alu* I-like subfamily. One view of these observations is that the vast majority of the 600 000 to 1 million copies of *Alu* I are simply 4.5S or 7S pseudogenes.

4. Most cloned human *Alu* I family members serve as *in vitro* templates for RNA polymerase III (Duncan *et al.*, 1979, 1981; Elder *et al.*, 1981a). *In vitro* transcription appears to initiate at (or at least very near) the 5'-terminal base pair of the repeat sequence and to terminate beyond the 3' border, in regions of unique-sequence DNA having fortuitous T-rich termination signals (Elder *et al.*, 1981a; Duncan *et al.*, 1981). *Alu* I DNA sequences showing homology to the intragenic RNA polymerase III sites of tRNA genes and adenovirus-associated low-molecular-weight RNA genes have been identified (Elder *et al.*, 1981a) and may indeed be initiation sites. Rodent *Alu* I elements, which lack them, are not transcribed by RNA polymerase III *in vitro*, although a variant rodent *Alu* I-like element is (Haynes *et al.*, 1981). My suggestion (Doolittle, 1981) that the strong homology between an *Alu* I region partially overlapping the 40 bp direct repeat and the RNA polymerase III recognition site of *Xenopus* 5S rRNA genes implicated the former as *the* initiation site now seems premature.

Some *Alu* I family members are (or were) transposable elements

The evidence that sequences of the *Alu* I family are, in whole or in part, transposable elements, is entirely indirect. But it is increasing compelling.

1. Human genomic *Alu* I elements (Bell *et al.*, 1980; Pan *et al.*, 1981; Duncan *et al.*, 1981) and most rodent genomic *Alu* I-like elements (Haynes *et al.*, 1981) are flanked by direct repeats which differ from element to element and are not homologous to sequences within the element. The analogy to flanking repeats generated by duplication of (unrelated) target

sequences upon insertion of known prokaryotic and eukaryotic transposable elements is obvious. There are some differences. *Alu* I flanking repeats vary in length from 4 to 20 bp, and do not always abut what would seem to be the ends of the elements. Other known prokaryotic or eukaryotic transposable elements generate repeats which are of characteristic size and directly abutting.

2. Like the *copia*-like elements of *Drosophila*, the Ty1 elements of yeast (Elder *et al.*, 1981b) and proviral copies of vertebrate retroviruses (see above), *Alu* I elements are transcribed, and the 5' terminus of the transcript coincides with one end of the element. The 3' terminus of such transcripts is less clearly defined and may indeed not be coded by most *Alu* I sequences. Transcripts of *copia*-like elements and vertebrate proviruses have been suggested as intermediates in their transposition (see above).

3. Covalently-closed-circular DNAs bearing *Alu* I sequences and varying in size down to that of a single *Alu* I element have been found in human cells (B. Calabretta, *et al.*, in preparation). Similar structures may be intermediates in the intragenomic transposition of *copia*-like elements and retroviral proviruses (see above).

4. Deletions and insertions near resident *Alu* I sequences have been reported (Orkin and Michelson, 1980; Dhruva *et al.*, 1980). Known prokaryotic and eukaryotic transposable elements also promote deletions and insertion adjacent to themselves, although they appear to do so with more precision (Calos and Miller, 1980).

Transposition mechanisms

Alu I family members do not, however, have the internal structure of typical transposable elements. They have no terminal direct or inverted repeats, aside from the short 5' CAGG 3'/3' GTCC 5' found near, but not at, the apparent termini of the consensus rodent *Alu* I-like sequence. There are two ways around this.

1. The first is to assume that the transposing unit comprises two directly repeated copies of *Alu* I, together with the sequences between them, which may or may not carry information related to the transposition process itself (Haynes *et al.*, 1981). Crossing-over between the tandem repeats after transpositional insertion would eliminate the DNA between and generate single *Alu* I copies flanked by direct genomic repeats. An analogous process has been invoked to explain the existence of isolated and non-transposing copies of the direct terminal repeats (δ sequences) of the yeast transposable element Ty1 (Farabaugh and Fink, 1980). *Alu* I-like elements not flanked by direct repeats (of which some are known

in rodents; Haynes *et al.*, 1981) could be the ends of such large transposon-like elements which have not suffered this fate.

2. The second way around is to assume something much more radical, namely that many low-molecular-weight RNAs are, regardless of their functions, potential candidates for insertion into the genome, either directly or as reverse transcripts. Such an assumption has been made by Van Arsdell *et al.* (1981) to explain the existence of multiple dispersed 'pseudogenes' complementary to the small nuclear RNAs (snRNAs) U1, U2 and U3 of human cells. These are flanked by (different) direct repeats and contain the 5'-terminal snRNA sequence, but not all of the 3'-terminal sequence. In at least one case, an A-rich region lies between the truncated 3' end and the 3'-flanking direct repeat.

Other dispersed repetitive sequences in the human genome

I do not wish to leave the impression that the *Alu* I family is the *only* family of dispersed repetitive DNAs in the human genome, although it clearly dominates that genome. There are, for instance, much longer (< 1 kbp) repeats.

Manuelidis (1982) has described a 1.8 kbp fragment, without obvious repetitive internal subelement structure, released from human DNA by digestion with *Hind* III. It comprises some 0.25% of human DNA and hybridizes to a dispersed repetitive 1.5 kbp *Bst* N fragment of the mouse genome. Preliminary sequence analyses show no homology to *Alu* I.

Gillespie *et al.* (1982) report that *Kpn* I digestion of total human DNA releases repeated fragments of 1.1, 1.4 and 1.6 kbp. The first two fragments are also found in other primates, but not other mammals. The repeated elements themselves are substantially larger than the *Kpn* fragments which define them. They can be shown, by Southern hybridization to total DNA digested with other restriction endonucleases, to be dispersed, and do not form hybrids with *Alu* I DNA.

Data on such non-*Alu* I elements in the human genome are too preliminary to allow conclusions about most important family parameters. More complete data are available for some long dispersed repeats in the mouse genome. I will cross the self-imposed species barrier to discuss some of them, with the anticipation that human long repeats will be found to behave similarly. Brown and Dover (1981) showed that cleavage of total *Mus musculus* DNA with *Eco* RI produces a prominent 1.35 kbp fragment. This fragment represents about 1% of the genome and reveals itself, upon further restriction and Southern hybridization analysis, to be part of an approximately 3 kbp dispersed repeat unit. Among all elements of this family (the 'MIF family'), four internal *Hind* III sites can be identified, none of which occurs in all members. These sites are non-randomly distributed among such members;

that is, there are discrete subfamilies of variant sequence. *Mus spretus* contains a similar but not identical family of dispersed repeats. Species of *Apodemus* bear related repeats with restriction endonuclease digestion patterns distinct from those of the *Mus* repeats. Again, there is conservation over considerable evolutionary time, but within-species repeat unit divergence is less than between-species divergence. Some form of concerted evolution is at work. It would be most interesting to know whether mammalian families of long dispersed repeats are more internally homogeneous in sequence than are members of the short *Alu* I family, as has been reported for long (versus short) sea urchin repeat units (see above).

PLANT REPETITIVE DNAs

The comparative approach favoured by many plant molecular biologists and the lack of extensive data on any one species obliges me to treat plants rather shabbily, as if the kingdom were of no greater taxonomic rank or phylogenetic diversity than an individual animal species. The breadth and scattered nature of work on plant repetitive DNAs persuades me to consider primarily the efforts of W. F. Thompson and R. B. Flavell, and their respective collaborators.

Renaturation experiments

Plants (that is, all eukaryotic photosynthesizers which are not algae) show a broad range of haploid DNA contents, and of overall genomic structure, even within taxonomically restricted groups. The leguminous genus, *Vicia*, for instance, exhibits a 7-fold variation in DNA contents (Bennett and Smith, 1976). Peas and mung beans differ not only (by a factor of nearly ten) in DNA content but also in apparent interspersion patterns (Murray *et al.*, 1979, 1981). In flowering plants as a whole genome size ranges from as low as that of *Drosophila* up to 600 times this value! Under standard hybridization conditions, from 75 to more than 85% of the apparent 'excess' behaves as repetitive DNA. That fraction which is operationally defined as unique-sequence increases with total genome mass in both animals and plants, but differently in the two groups (Thompson *et al.*, 1980). Of this unique-sequence DNA, only a small and decreasing fraction *can* be involved in coding for or regulating the production of RNAs and proteins. Thus Murray *et al.* (1979): 'consider most of the available information on genome organization in higher plants to be more directly relevant to the question of genome evolution than to control of gene activity'. This bias, which I share, colours the writings of both Thompson's and Flavell's groups.

The pea (*Pisum sativum*) genome is a large one (approximately 4.7×10^6 kbp) and shows, in standard renaturation kinetic experiments, a repetitive fraction comprising some 85% of the total. The unique-sequence fraction displays short-period interspersion of an extreme sort; only an average of 300–400 bp of unique-sequence DNA separates adjacent units of repetitive DNA (Murray *et al.*, 1979). The mung bean (*Vigna radiata*) has approximately one-tenth as much DNA (4.7×10^5 kbp). Less of it is repetitive, and the unique-sequence DNA is differently disposed; 35% of this is interrupted at intervals between 300 and 1200 bp, 18% is interrupted at intervals between 1.2 and 6.7 kbp and the remaining 46% at intervals of < 6.7 kbp.

The above values for the fractions of total pea and mung bean DNA which are repetitive or unique apply only under standard reassociation conditions ($T = Tm$ - 25°). At Tm - 35°, one-half of the DNA which appears to be unique-sequence in the pea and about one-quarter of that which appears to be unique-sequence in the mung bean reveal themselves to be diverged repetitive sequences. Even the fraction that reanneals as unique-sequence under these less stringent conditions melts as if it contained some 14 and 6% mismatched nucleotides (for pea and mung bean, respectively). Murray *et al.* (1981) conclude that 'virtually all of the sequences of the pea genome which reassociate with single copy kinetics at standard criterion [and recall, this is only 18% of the total DNA] are actually diverged repeats', or '*fossil repeats*' in their terminology. For the mung bean, perhaps as much as 70% of the DNA defined as unique-sequence under standard conditions (and thus about 46% of the total DNA) is indeed that. But since its genome is nearly 10-fold smaller than that of pea, the absolute amount of genuinely unique sequence DNA ($10–20 \times 10^3$ kbp) may be the same for both species.

Thompson and coworkers (Murray *et al.*, 1981; Thompson *et al.*, 1980; Preisler and Thompson, 1981a, b) suggest that most of the excess DNA of plant genomes arises through episodic and recurrent events which amplify and disperse discrete segments of DNA in a manner which may or may not depend on the sequences of these segments. The products of such amplification, which comprise a family of repeats, will drift in sequence and be recognized either as 'true' repetitive (hybridizing at high stringency), 'fossil repeat' (hybridizing at low stringency) or 'unique-sequence' DNAs (no longer capable of forming intrafamilial hybrids), depending on the antiquity of the original amplification event. Some family members will suffer deletion. If deletion balances amplification, genome size will remain constant, although the repetitive sequences which make up most of the genomic mass will be in a state of constant turnover. If the rate of turnover is high, as it may be in pea, then repeats which reanneal under standard conditions, and relatively homogeneous fossil repeats, will dominate the genome. If turnover is low, as it may be in mung bean, perhaps in mammals, and more certainly in some birds (Burr and Schimke, 1980), then diverged fossil repeats and unique-sequence DNA (representing primarily highly diverged repeats) will dominate.

If amplified sequences are dispersed randomly throughout the genome, most of which already consists of previously amplified, dispersed and variously diverged repeats, then interspersion patterns too will reflect turnover rate. If this is high, new elements will (1) be inserted in greater number and (2) more often find themselves near elements still recognizably repetitive-producing 'short-period interspersion'. If turnover rate is low, new elements will (1) be inserted infrequently and (2) more often be inserted into regions which have already drifted into what behaves like unique-sequence DNA—producing 'long-period interspersion'. The question begged, of course, is why some genomes turn over more rapidly than others, but begged questions plague the entire field of genome size evolution.

Thompson and collaborators assume that sequences are chosen randomly for amplification and dispersal. In rapidly turning-over large genomes, chance dictates that the amplified sequences will more often be, or contain, slightly divergent members of as yet recognizable repetitive families ('secondary amplification'), and will thus give rise to new subfamilies, homogeneous within themselves but bearing homology to the rest of the superfamily from which they derive. Superfamilies will therefore appear 'heterogeneous'. In less rapidly turning-over genomes, amplified sequences will more often be unique, or nearly so, and give rise to families which appear homogeneous, having no detectable relatives among existing genomic components. For pea and mung bean, at least, these predictions are borne out (Preisler and Thompson, 1981a, b).

Studies on the positioning and evolution of repetitive elements in the genomes of cereal plants reported by Flavell, Bedbrook and collaborators are consistent with the interpretations outlined above (Flavell *et al.*, 1980, 1981; Bedbrook *et al.*, 1980, 1981). The DNAs of wheat, rye, barley and oats, when renatured at high stringency, appear to contain only 30–40% repetitive sequences, the rest being composed of 'unique-sequence DNA' stretches of > 1 kbp (approximately 20%) or between 1 and 4 kbp (approximately 30%). As stringency is reduced, that fraction which appears repetitive increases (to 80% or more), and the length of apparent unique-sequence stretches decreases (to less than 1 kbp), as expected if the latter consisted primarily of variously interspersed fossil repeats of varying degrees of sequence divergence.

Flavell *et al.* (1980, 1981) describe these cereal genomes as consisting of three types of arrays:

1. Short (200–1200 bp) stretches of what appears under most conditions as unique sequence DNA, interrupted by short (200–600 bp) repetitive elements.
2. Long tandem arrays of related repetitive sequence units.
3. 'Interspersed repeats': regions comprised predominantly of unrelated

short elements each of which is itself a member of a dispersed repeat sequence family, whose other members may be arranged differently (that is, among different elements of different families in different interspersed repeats elsewhere in the genome).
4. Long (< 10 kbp) regions of apparently unique sequence DNA (Rimpau *et al.*, 1978).

The first type of array comprises from 25 to 40% of these genomes, but may be ridden with fossil repeats. The last type comprises from 2 to 9%. The second and third types make up the bulk (50–70%) of cereal DNA but are not easily distinguished kinetically; regions of both types have been cloned and characterized. Where the genes (perhaps 1% of the DNA) are remains unclear.

Wheat, rye, barley and oat genomes all contain some repeated sequences in common, but each has its own unique families of repeated sequences. Even the congeneric *Secale cereale* and *S. silvestre* differ. Of six repeated sequences families characterized by Bedbrook *et al.* (1980), four are unique to the former, and account for most of the differences in telomeric heterochromatin in these species. The four families which are unique are more internally homogeneous in sequence than the two which are shared, indicating that they arose in *S. cereale* since its divergence from *S. silvestre*. In general, repeated sequences families shared between cereal species show greater within-species than between-species sequence homogeneity (Flavell *et al.*, 1980). Such shared families also exhibit species-specific structural features (Bedbrook *et al.*, 1981). Once again, some mechanism of homogenization must be involved. Bedbrook and Flavell consider it to be 'secondary amplification', that is, separate amplification (after species divergence) of individual products of earlier amplifications (which preceded species divergence). There seems to be no other choice.

Cloned repetitive DNAs

Flavell *et al.* (1981) describe a cloned 5.2 kbp fragment of wheat DNA whose properties are those expected of a region of secondarily amplified interspersed repeats. *In situ* hybridization shows that sequences related to all or part of this fragment are scattered throughout the wheat genome. All four subfragments generated by *Hind* III digestion of the cloned insert are repetitive, but show different copy numbers in the wheat genome. These subfragments are also repeated in the genomes of *Triticum monococcum* and *Aegilops squarrosa*, but in different proportions. Thus the 5.2 kbp fragment contains interspersed members of different and independently-evolving repeat families.

The *Secale cereale* (rye) genome contains a 120 bp repeat unit of high copy number and sequence divergence, distributed on all chromosomes but concentrated in telomeres (Bedbrook *et al.*, 1980, 1981). Many of its copies exist in tandem, satellite-like arrays, but others are parts of complex interspersed repeat units. One such unit, represented by a cloned 2.2 kb *Eco* RI fragment, is itself repeated some 4000 times in the genome and is present both as isolated copies and tandem arrays. In addition to the 120 bp repeat, this unit contains other elements which are repeated, in other positions, in the *S. cereale* genome. Some of these other elements, as well as the 120 bp repeat unit, are present, but in different environments, in the genome of *S. silvestre*. Bedbrook *et al.* (1981) propose that 'much of the DNA in the chromosomes of *Secale* species and other cereals is the product of recombination of relatively few diverged sequence families and that various rearrangements are amplified to form high copy-number repeats. After new permutations are amplified at one location, they are distributed by recombination to multiple chromosomal locations'.

INTERPRETATIONS AND SPECULATIONS

Middle-repetitive DNAs are not one kind of thing

It is difficult to present this literature in a coherent and unifying manner, for two reasons. The first is that middle-repetitive DNAs as a class share no common ancestor, perform no common function and exhibit little common behaviour in the eukaryotic genome. The second reason is that each of the four types of genome discussed appears qualitatively different. It is thus easy to see the following:

1. how the relative scarcity and rampant mobility of middle-repetitive DNAs in *Drosophila* leads to the view that they exist only for the purpose of increasing genetic variability (Strobel *et al.*, 1979);
2. how the relatively high abundance, variable family number and complex family structure of sea urchin middle-repetitive DNAs, together with the absence of any direct evidence for transposition, leads to the view that they play complex, orchestrated, developmentally-specific roles in the regulation of gene expression (Davidson and Britten, 1979);
3. how the preponderance and broad dispersal of the Alu I family among human repetitive DNAs leads to the notion that it plays some single essential but developmentally non-specific role in cellular function (Jelinek *et al.*, 1980);
4. how the rapid evolutionary reassortment and frequent species-specific reamplification of repetitive subelements which make up the bulk of the genomes of many plants leads to the notion that they have nothing to do

with anything which matters to the daily life of the cell (Murray *et al.*, 1981).

In spite of all this, two general questions can be asked: are there common reasons for the existence of such DNAs, and are there recurring themes in their evolutionary behaviour? These are, in different form, the same questions posed in the introductory section.

Common reasons for the existence of middle-repetitive DNAs: contributions to fitness?

Until recently, the only kind of natural selection molecular biologists recognized was that which operates on genomic elements indirectly, through their expression in phenotype. Dispersed middle-repetitive DNAs, because they exist, were presumed to make some positive contribution(s) to organismal fitness. In general, such contributions can be viewed as either *sequence-dependent*, requiring that dispersed elements retain a certain consensus sequence so that they can interact with other, independently-coded, sequence-recognizing macromolecules, or *sequence-independent*, requiring only that dispersed elements retain intrafamilial homology so that they can interact with each other.

In either case, some mechanism for maintaining intrafamilial sequence homogeneity is required. It is difficult to see how selection operating individually on each element could ensure that families remain relatively homogeneous around a consensus sequence within species, and yet encourage consensus sequences to vary between species. The difficulty for sequence-dependent element selection is the requirement that different sequences be favoured for what must be a common and fundamental intermolecular recognition process in closely related organisms which do not differ fundamentally in developmental pattern. The difficulty for *sequence-independent* elements is the requirement that a variant sequence be established against selection pressure favouring retention of the original consensus sequence.

Perhaps neither of these difficulties is insurmountable, but it is common to assume that homogeneity is maintained by some sort of correction process; gene conversion, 'conservative' transposition, 'duplicative' transposition (Campbell, 1981), or some sort of saltatory replacement mechanism, any of which could have the effect of replacing many divergent members of the family with newly made (and hence more nearly identical) replicas of just one of these members (Brown and Dover, 1981; Baltimore, 1981; Selker *et al.*, 1981; Britten and Davidson, 1971; Anderson *et al.*, 1981). A form of 'truncation selection' could then operate on families as units, through phenotype, recognizing the fitness of individuals in which the proper sequence-dependent element had been amplified, or in which a greater degree of sequence homogeneity among sequence-independent elements had been established.

Much depends on the nature of these mechanisms. If the process which maintains within-species family homogeneity while encouraging between-species drift can also give rise to a dispersed middle-repetitive family and protect copy number against erosion by deletion (at least to the extent that such protection is observed), then the existence of the family is explained by the homogenizing mechanism itself. The need to invoke function vanishes. The mere existence of dispersed middle-repetitive DNAs does not oblige us to suppose that they have functions.

Contributions to fitness I: Britten-Davidson models

In the late 1960s and early 1970s, Britten and Davidson formulated a model for the control of eukaryotic structural gene expression, *at the transcriptional level*, involving dispersed middle-repetitive elements. Such elements were considered to be control regions ('receptor genes') flanking structural genes. They were thought to be recognized by *trans*-acting regulatory macromolecules (probably RNAs: called 'activator RNAs'). These themselves might be the products of other dispersed or clustered middle-repetitive families of 'integrator genes', whose transcriptions were controlled, *via* operator-like 'sensor genes', by some low-molecular weight effector. Cordinately regulated genes might be flanked by members of the same family of receptor genes. Genes under complex control might be flanked by members of several such families (Britten and Davidson, 1969, 1971).

The model is appealing and flexible, and continues to influence much of the work on middle-repetitive DNAs. Its originators also invoked it to explain the C-value paradox. They argued that there is not likely to be a 100-fold difference in the numbers of structural genes among eukaryotes, but suggested that the *minimum* C-values of many metazoan groups increased with organizational and developmental complexity, considering it 'likely that an ever-growing library of different combinations of groups of producer [structural] genes is needed as more complex organisms evolve' (Britten and Davidson, 1969). The model did not explain the then known variation in C-values within groups of similar apparent complexity, and rested in no small part on the faith that 'the apparently universal occurrence of large quantities of sequence repetition in the genomes of higher organisms suggests strongly that they have an important current function' (Britten and Davidson, 1969).

The discovery that heterogeneous nuclear RNA is the precursor of cytoplasmic mRNA; that much of it never leaves the nucleus; that heterogeneous nuclear RNA compositions are less tissue specific than messenger RNA compositions; that many middle-repetitive DNAs are transcribed either separately or as parts of large, heterogeneous nuclear RNA-producing units;

and that genes are (because of intervening sequences) much larger than the average stretch of unique-sequence DNA—all of which occurred in the last ten years—obliged Britten and Davidson to reformulate their model (Davidson *et al.*, 1977; Davidson and Britten, 1979). Starting from data suggesting that most genes are transcribed at low and constant rates into heterogeneous nuclear RNAs in all nuclei, but that further processing and transport of precursors of specific messenger RNAs is developmentally controlled, they developed a scheme for post-transcriptional control of structural gene expression. The transcription of middle-repetitive DNAs is again a regulated step. Middle-repetitive transcripts in turn regulate the expression of structural genes through the formation of intermolecular RNA: RNA hybrids with internal sequences of constitutively transcribed heterogeneous nuclear RNAs, thus influencing subsequent processing and export. As they note 'The control logic [we] originally postulated is retained . . . The "gene battery", that is, a set of genes under control of a single family of repetitive sequences, is also the unit of regulation we propose here' (Davidson and Britten, 1979).

Evidence consistent with such a scheme derived primarily from work on sea urchins, and includes the following.

1. Transcripts of much of the short and long repetitive fraction of total sea urchin DNA, and of both strands of nine cloned repeats, are found in total sea urchin oocyte RNA, and may represent portions of unprocessed stored messenger RNA precursors. Transcript abundances vary over a 30 to 100-fold range, and are not related to DNA repeat family size. The hybridizing RNAs are themselves much larger than the repeat sequence transcripts they contain (Constantini *et al.*, 1978).
2. In gastrulae and adults, repeat sequence transcripts are largely confined to nuclear RNA. Nuclear transcripts of nine cloned repeats show an abundance variation of at least 100-fold, again not correlated with repeat family size. The relative abundances of transcripts of different repetitive families vary greatly between gastrulae and adult nuclei, that is, repeat sequence transcription *is* developmentally regulated (Scheller *et al.*, 1978).
3. As shown earlier (Moore *et al.*, 1978), *S. purpuratus* repetitive DNAs represented by clones generally show greater family sizes in this genome than in that of *S. franscicanus*, sometimes by factors as great as 20; and yet the relative abundances of the transcripts of these clones vary little (generally less than 2-fold) in the two species. Moore *et al.* (1980) suggest that transcription and processing rates 'have been adjusted during evolution so as to preserve the repeat transcript sequence concentrations despite significant changes in genomic repeat family sizes' and that 'possibly only repetitive sequence elements of given families which are present in the genomes of both species are expressed'. This may be so, but it

leaves begged the question of the *role* of the unexpressed copies which, in the case of the family represented by clone 2109A, comprises 95% of the total.

5. Most of the (relatively unprocessed) transcripts of unique-sequence DNA present in sea urchin eggs are covalently associated with transcripts of repetitive DNAs. These RNAs can be defined as belonging to different sets by the repetitive sequence transcripts they contain (Constantini *et al.*, 1980).

6. Most of the eleven cloned sea urchin actin genes are flanked by members of dispersed repetitive DNA families and similar sequences are found flanking genes of a given type. These repeats belong to families which (like most or all sea urchin repeat families) are transcribed, but are not *themselves* transcribed (Scheller *et al.*, 1981b); a finding consonant only with the earlier Britten–Davidson model.

Most of this evidence is consistent with, but in no way proves, the Britten–Davidson model in any of its formulations. However, it is not so much the difficulty of proving this model with sea urchin data which makes me doubt its validity (Cavalier-Smith, 1978), but the difficulty of generalizing the model to many other eukaryotic cells. If middle-repetitive DNAs play any role in eukaryotic gene expression so fundamental and thus so ancient as that ascribed to them by Britten and Davidson, surely they ought to do so in all eukaryotes. Some eukaryotes such as *Aspergillus* and *Neurospora* (Timberlake, 1978; Krumlauf and Marzluf, 1980) have no detectable middle-repetitive DNA which cannot be assigned to known multigene families (rRNA or tRNA). Virtually all of the middle-repetitive DNA of *Drosophila* is highly mobile and cannot possibly be involved in regulatory interactions of the kind envisioned by Britten and Davidson. Copy number variation for such repeats between *D. melanogaster* and *D. simulans* (very similar to that observed for repeats between *S. purpuratus* and *S. franscicanus*) is observed, and has been used as further evidence against function (Young and Schwartz, 1981). Virtually all of the dispersed middle-repetitive DNA of humans is of the *Alu*I family and yet gene expression in man is surely no less developmentally controlled than it is in sea urchins. *Alu* I family members are some 80–90% homologous to each other. Their apparent homogeneous divergence *might* mask an internal structure consisting of a large number of small subfamilies of entirely homogeneous and different sequences playing different regulatory roles. However, inter-subfamily homology would still be so high that most RNA:RNA hybrids should be mismatched. No specific regulatory mechanism could work efficiently with such an error frequency. Plant repetitive sequence organization is more complex, but it is so far only this complexity and our consequent ignorance of detail which allows us to say that regulatory mechanisms of the sort postulated by Britten and Davidson *could* operate within plant genomes.

Contributions to fitness II: chromosomal domains and superstructure

Transcriptional units mapped so far seem to include little more than introns, exons and 5'- and 3'-terminal sequences a few hundred bp on either side (Darnell, 1979). However, this does not mean that they are not parts of very much larger functional *domains*, including non-transcribed regions which, although perhaps not regulatory in the sense that operators, promotors and potential terminators can be seen to be in bacteria, are nevertheless essential for function. The extraordinary sequence conservation of the DNA which separates the genes of either the α or β human globin gene clusters and which comprises more than 80% of the mass of those clusters suggests that it is not simply 'junk' (Jeffreys, 1982). These, and other developmentally regulated and clustered genes of related families may lie within domains whose function(s) can only be guessed at (Van der Ploeg *et al.*, 1980). Scheller *et al.* (1981b) have suggested, for instance, that the multiple sea urchin actin genes are differentially regulated because they have been inserted into different 'genomic domains[s] already involved in specification of an extant state of differentiation or developmental structure'. Repetitive sequences may play roles in the definition or maintenance of such domains. These might well be sequence-independent roles, and cannot be further defined until the very existence and significance of domains becomes clear.

Manuelidis (1982) and Bennett (1982) have separately suggested that repetitive DNAs play what would have to be sequence-independent roles in a yet even higher order of chromosome structure and function. Manuelidis, for instance, notes that 'studies of prophase, metaphase and anaphase nuclei have indicated that there are discrete positions for each chromosome and that these ultimately play a significant role in defining the shape of the interphase nucleus and the exact position of the nucleolus. Alignment of centromeric and Giemsa-like bands on chromosome arms could be facilitated by some repeated DNA subsets . . . and indeed such sequences could participate in the orderly alignment and association of chromosomes, as well as three dimensional recruitment of different adjacent chromosome segments for orderly transcription and heterochromatinization'. Bennett finds, in plant nuclei, similar evidence for ordered disposition of non-homologous chromosomes, and that this is governed primarily by chromosomal arm length, which in turn reflects largely the (sequence-independent?) amplification of repetitive elements, most of which may be telomeric. Moreau *et al.* (1981) suggest that repeated (although not necessarily homologous) A + T-rich regions punctuate the eukaryotic genome in a systematic way which may be related to higher-order chromosome recognition functions in expression, meiosis and mitosis.

Those who can think in several dimensions are grappling with a problem of immense importance, and it may have something to do with repeated sequences, either tandem or dispersed. However, that cannot yet be determined, and the strong experimental evidence showing that tandemly-repeated 'satellite' DNAs do not have many of the important and quite similar functions in somatic cell chromosome behaviour previously assigned to them (although they may affect recombination frequency in the germ line) suggests caution (Miklos and Gill, 1981; Miklos, 1982).

Contributions to fitness III: the promotion of genetic variability

The recognition that the frequent mobility, interspecies variation in copy number and transposon-like structure of many middle-repetitive DNAs in small-genomed eukaryotes like yeast and *Drosophila* is incompatible with any regular role in cellular function prompted many investigators to suggest that these DNAs perform an 'evolutionary function' (Nevers and Saedler, 1977; Cameron *et al.*, 1979; Strobel *et al.*, 1979; Roeder and Fink, 1980; Jain, 1980; Shapiro, 1982). Their movements create, through deletions, insertions and other rearrangements, genetic variability on which natural selection can act. Although the middle-repetitive DNAs of most large-genomed organisms like sea urchins or plants have not directly been shown to be independently transposable, rearrangements affecting or catalysed by them are frequently invoked as major determinants in the evolutionary process (Britten and Davidson, 1971; Davidson and Britten, 1979; Gillespie *et al.*, 1982; Flavell *et al.*, 1981; Bedbrook *et al.*, 1981).

This may be so, but it does not mean that the promotion of genetic variability is the 'function' of middle-repetitive DNAs; it does not mean that such an evolutionary 'function' is the reason for existence of middle-repetitive DNAs. Surely when we talk about the reason for existence, or *function*, of a biological structure, we are referring to that phenotypically expressed effect(s) whose contribution to the fitness of the organism is (are) responsible for its maintenance and/or origin through natural selection. Structures may have *effects* incidental to those defining their *function*, and these need not be of immediate, or even ultimate, selective advantage. The genetic variability which transposable elements promote is most easily seen as such an incidental effect (Doolittle and Sapienza, 1980; Orgel and Crick, 1980; Sapienza and Doolittle, 1981). The alternative, otherwise, must be to assume that natural selection has favoured mechanisms which generate variability at the level of the population, even though they bestow no selective advantage, and impose some disadvantage, upon the individual, that is, we must invoke 'group selection' (Maynard Smith, 1978). As pointed out earlier (in the discussion of transposable elements in *Drosophila*), it is precisely for elements of this sort that such an invocation is unnecessary (see also Chapter 16).

Recurring themes in the evolution of middle-repetitive DNAs

If it is true that the middle-repetitive DNAs as a class, and most middle-repetitive DNAs as individual families, have no common sequence-dependent function which is the product of natural selection operating through pheno-type, and (unless speculations about their roles in higher order chromosome structuring and functioning prove correct) have no common sequence-inde-pendent function either, might we still expect them to exhibit common behav-iours? We might if these behaviours reflect similarities in the genetic mechan-isms and/or (non-phenotypic or intragenomic) selection pressures which give rise to and maintain them. Among common behaviours, the following might be included.

1. *Family size varies more than family sequence in separate evolutionary lineages.* Data obtained by examining different strains or species of *Droso-phila* and sea urchins leaves the impression that variation of copy number, rather than variation in concensus sequence, is the dominant feature which differentiates the middle-repetitive components of the genomes of related species. If we include as special cases the appearance of entirely new families resulting from the amplification of sequences not initially detect-able as repetitive, and the amplification in one lineage of repeat sequence subelements which are differently arrayed or amplified in another lineage, then additional data from sea urchins and much plant data can be inter-preted in these same terms.

2. *Between-species variation is greater than within-species variation.* This may be generally true for middle-repetitive DNAs. It has obvious important implications. If species B and C each have *n* numbers of a repeat family, and derive from an ancestor A which had *n* numbers, and if all that happened to those *n* numbers since the B:C divergence was neutral sequence drift, then within-species homogeneity should equal between-species homogeneity. That it does not must mean that mechanisms which homogenize family member sequences are generally operating in euka-ryotic genomes.

3. *Much middle-repetitive DNA evolution may involve 'saltatory' events.* The *Drosophila*, sea urchin and plant data must mean that the sizes of some repetitive sequence families have increased while the sizes of others have decreased, since the divergence of evolutionary lineages represented by different species. Whether this reflects sequence-specific selection (at either the level of the genome or the individual) or results from neutral drift processes (Ohta, 1981) remains unclear. Much of the data seems to

indicate that amplification is a sporadic or saltatory process. The homogeneous divergence of the human *Alu* I family, and the apparent absence of any substantial subfamily of *highly* homologous *Alu* I repeats, seems to suggest that it is not now growing as actively as it did in the past, or at least is now less subject to homogenization than it was in the past. The variation in intrafamilial sequence homology observed between individual repeat sequence families in a single sea urchin species, which is independent of family size, would seem to indicate that various families have experienced sudden spurts of growth at different times in the evolution of that species. Not all of them are now growing or homogenizing themselves at the same rate. The existence in plants of subfamilies of highly homologous sequences within superfamilies of high divergence, and the existence there of species-specific rearrangements of repetitive elements into now composite elements which themselves can comprise large internally homogeneous subfamilies, seems to reflect sporadic amplification events.

It may well be true that family growth *in general* involves the amplification and dispersal, over a short period of time, of single members of a previously quiescent and diverging population of repeats, themselves the product of an earlier amplification, and that continual slow growth by the duplication of randomly chosen family members at a rate slightly higher than that of loss of members through deletion is not the predominant mode of middle-repetitive DNA evolution. However, we cannot be sure of this until we have developed sophisticated mathematical models which predict the outcome, in terms of family substructure and intrafamilial sequence homogeneity, of both modes of growth. These models will have to consider all mechanisms currently entertained for maintaining family homogeneity ('gene conversion', 'duplicative transposition' and 'conservative transposition'). They will have to take into account both the population biology of elements of a family within a genome and the spread, either selective or neutral, of elements within the genomes of a population of sexually reproducing organisms. They must also recognize the possibility that the existence of subfamilies may not reflect saltatory amplification of individual variants, but instead gradual divergence within element subpopulations which are partially isolated from each other because, for instance, they reside on different chromosomes. These models will be complex and attempts to develop them have only just begun (Ohta and Kimura, 1981; Ohta, 1981).

4. *Short elements may differ qualitatively from long ones.* Sea urchin data show rather clearly that families of long ($\geq$ 2 kbp) repeat elements exhibit greater intrafamilial sequence homology than do families of short (approximately 300 bp) repeat elements, regardless of family size. There is no comparable body of data from any other single organism, although

such a statement *might* be made for long versus short (*Alu* I-like) repeats in mammalian genomes (Brown and Dover, 1981), and the latter are clearly more divergent in sequence than are the best characterized long repeats—the *copia*-like elements of *Drosophila* and yeast. If this is generally true, then the ways in which short repetitive elements preserve sequence homogeneity must differ from those in which long repetitive elements do this, either qualitatively (in terms of mechanisms) or quantitatively (in terms of frequencies of events).

We do not know what those mechanisms are. Brown and Dover (1981), Dover (1980) and Dover *et al.* (1981, 1982) have proposed that something analogous to gene conversion maintains within-species homogeneity while allowing between-species concensus sequence variation. It would do that, but it is not immediately clear why it should affect short repeats differently from long repeats, nor why different families of the same size (copy number) do not all show the same degree of sequence homogeneity. They clearly do not do so in sea urchins. Unless we assume that 'gene conversion' mechanisms prefer certain repeat sequences or repeat lengths, or that different repeat families characteristically occupy regions of chromosomes which differ in the readiness with which they will allow 'gene conversion' to occur, other mechanisms must be invoked to explain the data.

5. *Most middle-repetitive DNAs may be transposable.* We know that the vast majority of middle-repetitive DNAs in yeast and *Drosophila* are transposable elements analogous in structure to bacterial transposable elements, and that they generate repeats of genomic sequences at the site of insertion. The mechanisms by which eukaryotic elements transpose themselves may well be different from the mechanism used by most bacterial elements, although the final outcome could be the same: creation of a new daughter element without loss of the multiplying maternal element. We have strong reasons to believe that the *Alu* I family is, or was, a family of transposable elements. These are the middle-repetitive DNAs whose structures and behaviours we know the best. Dare we make the assumption that most eukaryotic middle-repetitive DNAs are, or were, transposable?

Duplicative transposition *will* account for the construction of complex repeat families in plant genomes, although conservative transposition and amplification through unequal crossing-over will do so as well. Duplicative transposition (in which transposition does not destroy the material elements so that copy number is increased) coupled with random deletion occurring at lesser, equal, or greater frequency can account for repeat family size amplification, repeat family sequence homogenization, and repeat family size diminution. Sequence-specific transposition mechanisms can account for observed variations in intrafamilial sequence homology.

Gene conversion or conservative transposition (in which there is no increase in copy number) will account for some but not all of these behaviours. Neither can account for the creation of middle-repetitive DNA families, but duplicative transposition can.

Middle-repetitive DNAs as the product of intragenomic evolution

Dover (1980) has proposed that repetitive DNAs are the inevitable by-product of cellular mechanisms which lead, perhaps unwittingly, to the amplification of randomly chosen sequences, and that these processes, coupled with random deletion, lead to continual turnover of repeated sequences. Homology between repeated sequences is maintained, he believes, by something like gene conversion, which is again a by-product of normal cellular mechanisms selected to perform other functions. To the extent that amplification, dispersal and homogenization are independent of the sequences of the elements affected by them, these elements can properly be called 'ignorant' (Dover, 1980). However, some repetitive elements are clearly 'designed' to perpetuate themselves within genomes, the transposable elements of bacteria, yeast and *Drosophila* being the most obvious examples. For elements such as these, for which amplification, dispersal and homogenization mechanisms are at least in part element-determined, and clearly element sequence-dependent, the term 'selfish' *is* appropriate (Doolittle and Sapienza, 1980; Orgel and Crick, 1980). Surely in any genome which does turnover, there exists a selection pressure which gives rise to and maintains sequences of this sort.

However, we and Dover and Thompson and Flavell and their collaborators all agree on what may be the most important point. The generation of repeated sequences is an inevitable consequence of genome turnover and intragenomic selection, and recurring themes in middle-repetitive DNA behaviours need only reflect the rules of the game. The repeats may have *effects*. They may be involved in chromosomal 'resetting', a term coined by Gillespie *et al.* (1982) in a Goldschmidtian context to describe rearrangements which many people (Dover *et al.*, 1982; Brown and Dover, 1981; Stanley, 1979; Gould, 1980) now like to believe are the major determinants of major evolutionary change. They may be involved in generating the excess DNA required by those who believe that C-value itself is selectively significant (Cavalier-Smith, 1978, 1982; Bennett, 1982). These are the possible *effects* of middle-repetitive DNAs; they need not be their *functions*.

ACKNOWLEDGEMENTS

I thank W. Jelinek, G. F. Saunders, L. Manuelidis, M. Bennett, D. Gillespie, G. Dover, A. Jeffrey, E. Davidson and G. Miklos for providing me with information prior to publication, the Medical Research Council and Natural Sciences and Engineering Research Councils of Canada for support of research which stimulated my interest in repetitive DNAs, and C. Sapienza and C. W. Helleiner, and H. Doolittle for advice and criticism.

REFERENCES

Anderson, D. M., Scheller, R. H., Posakony, J. W., McAllister, L. B., Trabert, S. G., Beall, C., Britten, R. J. and Davidson, E. H. (1981). Repetitive sequences of the sea urchin genome. Distribution of members of specific repetitive families. *J. Mol. Biol.*, **145**, 5–28.

Arthur, R. R. and Strauss, N. A. (1978). DNA sequence organization in the genome of the domestic chicken. *Can. J. Biochem.*, **56**, 257–264.

Baltimore, D. (1981). Gene conversion: some implications for immunoglobulin genes. *Cell*, **24**, 592–594.

Barelle, F. E., Shoulders, C. C., Goudbourn, S., Jeffreys, A. and Proudfoot, N. J. (1980). The 5' flanking region of human epsilon-globin gene. *Nucleic Acids Res.*, **8**, 4393–4404.

Barta, A., Richard, R. I., Baxter, J. D. and Shine, J. (1981). Primary structure and evolution of rat growth hormone gene. *Proc. Natl. Acad. Sci. USA*, **78**, 4867–4871.

Bayev, A. A., Jr., Krayev, A. J., Lyubormirskaya, N. V., Ilyin, Y. V., Skryabin, K. G. and Georgiev, G. P. (1980). The transposable element Mdg 3 in *Drosophila melanogaster* is flanked with the perfect direct and mismatched inverted repeats. *Nucleic Acids Res.*, **8**, 3263–3273.

Bedbrook, J., Jones, J. and Flavell, R. (1981). Evidence for the involvement of recombination and amplification events in the evolution of *Secale* chromosomes. *Cold Spring Harbor Symp. Quant. Biol.*, **45**, 755–760.

Bedbrook, J. R., Jones, J., O'Dell, M., Thompson, R. D. and Flavell, R. B. (1980). A molecular description of telomeric heterochromatin in Secale species. *Cell*, **19**, 545–560.

Bell, G. I., Pictet, R. and Rutter, W. J. (1980). Analysis of the regions flanking the human insulin gene and sequence of an Alu family member. *Nucleic Acids Res.*, **8**, 4091–4109.

Bendich, A. J. and Anderson, R. J. (1977). Characterization of families of repeated DNA from four vascular plants. *Biochem.*, **16**, 4655–4663.

Bennett, M. D. (1982). The nucleotypic basis of the spacial ordering of chromosomes in eukaryotes and the implications of the order for genome evolution and phenotypic variation. In G. A. Dover and R. B. Flavell (Eds), *Genome Evolution*, Academic Press, London, pp. 239–260.

Bennett, M. D. and Smith, J. B. (1976). Nuclear DNA amounts in angiosperms. *Phil. Trans. Roy. Soc. Lond. B.*, **274**, 227–273.

Britten, R. J., Cetta, A. and Davidson, E. H. (1978). The single copy sequence polymorphisms of the sea urchin *Strongylocentrotus purpuratus*. *Cell*, **15**, 1175–1186.

Britten, R. J. and Davidson, E. H. (1969). Gene regulation for higher cells: a theory. *Science*, **165**, 349–357.

Britten, R. J. and Davidson, E. H. (1971). Repetitive and nonrepetitive DNA sequences and a speculation on the origins of evolutionary novelty. *Quart. Rev. Biol.*, **46**, 111–137.

Britten, R. J., Graham, D. E., Eden, F. C., Painchaud, D. M. and Davidson, E.

H. (1976). Evolutionary divergence and length of repetitive sequences in sea urchin DNA. *J. Mol. Evol.*, **9**, 1–23.

Britten, R. J. and Kohne, D. E. (1968). Repeated sequences in DNA. *Science*, **161**, 529–540.

Brown, S. D. M. and Dover, G. (1981). The organization and evolutionary progress of a dispersed repetitive family of sequences in widely separated rodent genomes. *J. Mol. Biol.*, **150**, 441–466.

Burr, H. E. and Schimke, R. T. (1980). Intragenomic DNA sequence homologies in the chicken and other members of the class Aves: DNA re-association under reduced stringency conditions. *J. Mol. Evol.*, **15**, 291–307.

Calabretta, B., Robberson, D. L., Maizel, A. L. and Saunders, G. F. (1981). mRNA in human cells contains sequences complementary to the Alu family of repeated DNA. *Proc. Natl. Acad. Sci. USA*, **78**, 6003–6007.

Calos, M. P. and Miller, J. H. (1980). Transposable elements. *Cell*, **20**, 579–595.

Cameron, J. R., Loh, E. Y. and Davis, R. W. (1979). Evidence for transposition of dispersed repetitive DNA families in yeast. *Cell*, **16**, 739–751.

Campbell, A. (1981). Evolutionary significance of accessory DNA elements in bacteria. *Ann. Rev. Microbiol.*, **35**, 55–83.

Cavalier-Smith, T. (1978). Nuclear volume control by nucleoskeletal DNA, selection for cell volume and cell growth rate, and the solution of the DNA C-value paradox. *J. Cell Sci.*, **34**, 247–278.

Cavalier-Smith, T. (1982). Skeletal DNA and the evolution of genome size. *Ann. Rev. Biophys. Bioengineering*, **11**, 273–302.

Christie, N. T. and Skinner, D. M. (1979). Interspersion of highly repetitive DNA with single copy DNA in the genome of the red crab *Geryon quinquedens*. *Nucleic Acids Res.*, **6**, 781–796.

Coggins, L. W., Grindlay, G. J., Vass, K. J., Slater, A. A., Montague, P., Stinson, M. A. and Paul, J. (1980). Repetitive DNA sequences near three human-type globin genes. *Nucleic Acids Res.*, **8**, 3319–3333.

Constantini, F. D., Britten, R. J. and Davidson, E. H. (1980). Message sequences and short repetitive sequences are interspersed in sea urchin egg poly (A)$^+$ RNAs. *Nature, Lond.*, **287**, 111–117.

Constantini, F. D., Scheller, R. H., Britten, R. J. and Davidson, E. H. (1978). Repetitive sequence transcripts in the mature sea urchin oocyte. *Cell*, **15**, 173–187.

Dalla Favera, R., Gelmann, E. P., Gallo, R. C. and Wong-Staal, F. (1981). A human *onc* gene homologous to the transforming gene (*v-sis*) of simian sarcoma virus. *Nature, Lond.*, **292**, 31–35.

Darnell, J. E. Jr.(1979). Transcription units for mRNA production in eukaryotic cells and their DNA viruses. *Prog. Nucleic Acids Res. Mol. Biol.*, **22**, 327–353.

Davidson, E. H. and Britten, R. J. (1979). Regulation of gene expression; possible role of repetitive sequences. *Science*, **204**, 1052–1059.

Davidson, E. H., Klein, W. H. and Britten, R. J. (1977). Sequence organization in animal DNA and a speculation on hnRNA as a co-ordinate regulatory transcript. *Dev. Biol.*, **55**, 69–84.

Dawid, I. B., Long, E. O., DiNocera, P. P. and Pardue, M. L. (1981). Ribosomal insertion-like elements in *Drosophila melanogaster* are interspersed with mobile sequences. *Cell*, **25**, 297–408.

Dawid, I. B., Wellauer, P. K. and Long, E. O. (1978). Ribosomal DNA in *Drosophila melanogaster*. I. Isolation and characterization of cloned fragments. *J. Mol. Biol.*, **126**, 749–768.

Deininger, P. L. and Schmid, C. W. (1979). A study of the evolution of repeated

DNA sequences in primates and the existence of a new class of repetitive sequences in primates. *J. Mol. Biol.*, **127**, 437–460.

Dhruva, B. R., Schenk, T. and Subramanian, K. N. (1980). Integration *in vivo* into simian virus 40 DNA of a sequence that resembles a certain family of genomic interspersed repeated sequences. *Proc. Natl. Acad. Sci. USA*, **77**, 4514–4518.

Doolittle, W. F. (1981). 5S ribosomal RNA genes and the *AluI* family; evolutionary and functional significance of a region of strong homology. *FEBS Lett.*, **126**, 147–149.

Doolittle, W. F. and Sapienza, C. (1980). Selfish genes, the phenotype paradigm and gene evolution. *Nature, Lond.*, **284**, 601–603.

Dover, G. A. (1980). Ignorant DNA? *Nature, Lond.*, **285**, 618–620.

Dover, G. A., Brown, S. D. M., Coen, E. S., Strachan, T. and Rick, M. (1982). Concerted sequence evolution and species isolation. In G. A. Dover and R. B. Flavell (Eds), *Genome Evolution*, Academic Press, London, pp. 343–372.

Dover, G. A., Strachan, T. and Brown, S. D. M. (1981). The evolution of genomes in closely-related species. In G. G. E. Scudder and J. L. Reveal (Eds), *Evolution Today*, Hunt Institute for Botanical Documentation, Pittsburgh, pp. 337–350.

Duncan, C., Biro, P. A., Choudary, P. V., Elder, J. T., Wang, R. R. C., Forget, B. G., DeRiel, J. K. and Weissman, S. M. (1979). RNA polymerase III transcriptional units are interspersed among human non-alpha globin genes. *Proc. Natl. Acad. Sci. USA*, **76**, 5095–5099.

Duncan, J. H., Jagadeeswaran, P., Wang, R. R. C. and Weissman, S. M. (1981). Structural analysis of templates and RNA polymerase III transcripts of *Alu* family sequences interspersed among the human β-like globin genes. *Gene*, **13**, 185–196.

Dunsmuir, P., Brorein, W. J., Jr., Simon, M. A. and Rubin, G. M. (1980). Insertion of the *Drosophila* transposable element *copia* generates a 5 base pair duplication. *Cell*, **21**, 575–579.

Eden, F. C., Graham, D. E., Davidson, E. H. and Britten, R. J. (1977). Exploration of long and short repetitive sequence relationships in the sea urchin genome. *Nucleic Acids Res.*, **4**, 1553–1567.

Eden, F. C. and Hendrick, J. P. (1978). Unusual organization of DNA sequences in the chicken. *Biochemistry*, **17**, 5838–5844.

Elder, J. T., Pan, J., Duncan, C. H. and Weissman, S. M. (1981a). Transcriptional analysis of interspersed repetitive polymerase III transcription units in human DNA. *Nucleic Acids Res.*, **9**, 1171–1189.

Elder, R. T., St John, T. P., Stinchcomb, D. T. and Davis, R. W. (1981b). Studies on the transposable element Ty 1 of yeast. I. RNA homologous to Ty 1. *Cold Spring Harbor Symp. Quant. Biol.*, **45**, 581–584.

Engels, W. R. (1981). Hybrid dysgenesis in *Drosophila* and the stochastic hypothesis. *Cold Spring Harbor Symp. Quant. Biol.*, **45**, 561–565.

Farabaugh, P. J. and Fink, G. R. (1980). Insertion of the eukaryotic transposable element Ty1 creates a 5 base pair duplication. *Nature, Lond.*, **286**, 352–356.

Finnegan, D. J. (1981). Transposable elements and proviruses. *Nature, Lond.*, **292**, 800–801.

Finnegan, D. J., Rubin, G. M., Young, M. W. and Hogness, D. S. (1978). Repeated gene families in *Drosophila melanogaster*. *Cold Spring Harbor Symp. Quant. Biol.*, **42**, 1053–1063.

Flavell, A. J. and Ish-Horowicz, D. (1981). Extrachromosomal circular copies of the eukaryotic transposable element *copia* in cultured *Drosophila* cells. *Nature, Lond.*, **292**, 591–595.

Flavell, R. B., O'Dell, M. and Hutchinson, J. (1981). Nucleotide sequence organization in plant chromosomes and evidence for sequence translocation during evolution. *Cold Spring Harbor Symp. Quant. Biol.*, **45**, 501–508.

Flavell, R., Rimpau, J., Smith, D. B., O'Dell, M. and Bedbrook, J. R. (1980). The evolution of plant genome structure. In C. J. Leaver (Ed), *Genome Organization and Expression in Plants*, Plenum Press, New York, pp. 35–47.

French, C. K. and Manning, J. E. (1980). DNA sequence organization in the thysanuran *Thermobia domestica*. *J. Mol. Evol.*, **15**, 277–289.

Fritsch, E. F., Lawn, R. M. and Maniatis, T. (1980). Molecular cloning and characterization of the human β-like globin gene cluster. *Cell*, **19**, 959–972.

Fritsch, E. F., Shen, C. K. J., Lawn, R. M. and Maniatis, T. (1981). Re-organization of repetitive sequences in mammalian globin gene clusters. *Cold Spring Harbor Symp. Quant. Biol.*, **45**, 761–775.

Galau, G. A., Chamberlin, M. E., Hough, B. R., Britten, R. J. and Davidson, E. H. (1976). Evolution of repetitive and non-repetitive DNA. In F. J. Ayala (Ed), *Molecular Evolution*, Sinauer Associates, Sunderland, Massachusetts, pp. 200–224.

Gehring, W. J. and Paro, R. (1980). Isolation of a hybrid plasmid with homologous sequences to a transposing element of *Drosophila melanogaster*. *Cell*, **19**, 897–904.

Georgiev, G. P., Ilyin, Y. V., Chmeliauskaite, V. G., Ryskov, A. P., Kramerov, D. A., Skryabin, K. G., Krayev, A. S., Lukanidin, E. M. and Grigoryan, M. S. (1981). Mobile dispersed genetic elements and other middle repetitive DNA sequences in the genomes of *Drosophila* and mouse; transcription and biological significance. *Cold Spring Harbor Symp. Quant. Biol.*, **45**, 641–654.

Gillespie, D., Donehower, L. and Strayer, D. (1982). Evolution of primate DNA organization. In G. A. Dover and R. B. Flavell (Eds), *Genome Evolution*, Academic Press, London, pp. 113–133.

Gould, S. J. (1980). Is a new and general theory of evolution emerging? *Paleobiology*, **6**, 119–130.

Graham, D. E., Neufeld, B. R., Davidson, E. H. and Britten, R. J. (1974). Interspersion of repetitive and non-repetitive DNA sequences in the sea urchin genome. *Cell*, **1**, 127–137.

Harpold, M. M. and Craig, S. P. (1977). The evolution of repetitive DNA sequences in sea urchins. *Nucleic Acids Res.*, **4**, 4425–4437.

Haynes, S. R., Toomey, T. P., Leinwand, L. and Jelinek, W. R. (1981). The chinese hamster Alu-equivalent sequence: a conserved, highly repetitious interspersed deoxyribonucleic acid sequence in mammals has a structure suggestive of a transposable element. *Mol. Cell. Biol.*, **1**, 573–583.

Herman, R. K. and Horvitz, R. H. (1980). Genetic analysis of *Caenorhabditis elegans*. In B. M. Zuckerman (Ed), *Nematodes as Biological Models,* vol. 1, Academic Press, New York, pp. 227–261.

Hinegardner, R. (1976). Evolution of genome size. In F. J. Ayala (Ed), *Molecular Evolution*, Sinauer Associates, Sunderland, Massachusetts, pp. 179–199.

Houck, C. M., Rinehart, F. P. and Schmid, C. W. (1979). A ubiquitous family of repeated DNA sequences in the human genome. *J. Mol. Biol.*, **132**, 289–306.

Ilyin, Y. V., Tchurikov, N. A., Ananiev, E. V., Ryskov, A. P., Yenikolopov, G. N., Limborska, S. A., Maleeva, N. E., Gvozdev, V. A. and Georgiev, G. P. (1978). Studies on the DNA fragments of mammals and *Drosophila* containing structural genes and adjacent sequences. *Cold Spring Harbor Symp. Quant. Biol.*, **42**, 959–969.

Ising, G. and Block, K. (1981). Derivation-dependent distribution of insertion sites for a Drosophila transpososon. *Cold Spring Harbor Symp. Quant. Biol.*, **45**, 527–544.

Jackson, J. A. and Fink, G. (1981). Gene conversion between duplicated genetic elements in yeast. *Nature, Lond.*, **292**, 306–307.

Jain, H. K. (1980). Incidental DNA. *Nature, Lond.*, **288**, 647–678.

Jeffreys, A. J. (1982). Evolution of globin genes. In G. A. Dover and R. B. Flavell (Eds), *Genome Evolution*, Academic Press, London, pp. 157–176.

Jelinek, W. R., Evans, M., Wilson, M., Salditt-Georgieff, M. and Darnell, J. E. (1978). Oligonucleotides in heterogeneous nuclear RNA: similarity of inverted repeats and RNA from repetitious DNA sites. *Biochemistry*, **17**, 2776–2783.

Jelinek, W. R. and Leinwand, L. (1978). Low molecular weight RNAs hydrogen-bonded to nuclear and cytoplasmic poly (A) – terminated RNA from cultured Chinese hamster ovary cells. *Cell*, **15**, 205–214.

Jelinek, W. R., Toomey, T. P., Leinwald, L., Duncan, C. H., Biro, P. A., Choudary, P. V., Weissman, S. M., Rubin, C. M., Houck, C. M., Deininger, P. L. and Schmid, C. W. (1980). Ubiquitous, interspersed repeated sequences in mammalian genomes. *Proc. Natl. Acad. Sci. USA*, **77**, 1398–1402.

Kidd, S. J. and Glover, D. M. (1980). A DNA segment from *D. melanogaster* which contains five tandemly repeating units homologous to the major DNA insertion. *Cell*, **19**, 103–119.

Klein, H. L. and Petes, J. D. (1981). Intrachromosomal gene conversion in yeast: a new type of genetic exchange. *Nature, Lond.*, **289**, 144–148.

Klein, W. H., Thomas, T. L., Lai, C., Scheller, R. H., Britten, R. T. and Davidson, E. H. (1978). Characteristics of individual repetitive sequence families in the sea urchin genome studied with cloned repeats. *Cell*, **14**, 889–900.

Krayev, A. S., Kramerov, D. A., Skryabin, K. G., Ryskov, A. P., Bayev, A. A. and Georgiev, G. P. (1980). The nucleotide sequence of the ubiquitous repetitive DNA sequence B₁ complementary to the most abundant class of mouse fold-back RNA. *Nucleic Acids Res.*, **8**, 1201–1215.

Krumlauf, R. and Marzluf, G. A. (1980). Genome organization and characterization of the repetitive and inverted repeat DNA sequences in *Neurospora crassa. J. Biol. Chem.*, **255**, 1136–1145.

Laskey, R. A. and Harland, R. M. (1981). Replication origins in the eukaryotic chromosome. *Cell*, **24**, 283–284.

Levis, R., Dunsmuir, P. and Rubin, G. M. (1980). Terminal repeats of the Drosophila transposable element *copia*: nucleotide sequence and genomic organization. *Cell*, **21**, 581–588.

Li, W.-H., Gojobori, T. and Nei, M. (1981). Pseudogenes as a paradigm of neutral evolution. *Nature, Lond.*, **292**, 237–289.

Manning, J. E., Schmid, L. W. and Davidson, E. H. (1975). Interspersion of repetitive and nonrepetitive DNA sequences in the *Drosophila melanogaster* genome. *Cell*, **4**, 144–155.

Manuelidis, L. (1982). Repeated DNA sequences and nuclear structure. In G. A. Dover and R. B. Flavell (Eds), *Genome Evolution*, Academic Press, London, pp. 263–285.

Maynard Smith, J. (1978). *The Evolution of Sex*, Cambridge University Press, Cambridge.

Miklos, G. L. G. (1982). Sequencing and manipulating highly repeated DNAs. In G. A. Dover and R. B. Flavell (Eds), *Genome Evolution*, Academic Press, London, pp. 41–68.

Miklos, G. L. G. and Gill, A. C. (1981). Nucleotide sequences of highly repeated DNAs; compilation and comments. *Genet. Res. Camb.*, **39**, 1–30.

Moore, G. P., Constantini, F. D., Posakony, J. W., Davidson, E. H. and Britten,

R. J. (1980). Evolutionary conservation of repetitive sequence expression in sea urchin egg RNAs. *Science*, **208**, 1046–1048.

Moore, G. P., Pearson, W. R., Davidson, E. H. and Britten, R. J. (1981). Long and short repeats of sea urchin DNA and their evolution. *Chromosoma*, in press.

Moore, G. P., Scheller, R. H., Davidson, E. H. and Britten, R. J. (1978). Evolutionary change in the repetition frequency of sea urchin DNA sequences. *Cell*, **15**, 649–660.

Moreau, J., Matyash-Simirniaguina, L. and Scherrer, K. (1981). Systematic punctuation of eukaryotic DNA by A + T-rich sequences. *Proc. Natl. Acad. Sci. USA*, **78**, 1341–1345.

Murray, M. G., Palmer, J. D., Cuellar, R. E. and Thompson, W. F. (1979). Deoxyribonucleic acid sequence organization in the mung bean genome. *Biochem*, **18**, 5259–5266.

Murray, M. G., Peters, D. L. and Thompson, W. F. (1981). Ancient repeated sequences in the pea and mung bean genomes and implications for genome evolution. *J. Mol. Evol.*, **17**, 31–42.

Musti, A. M., Sobieski, D. A., Chen, B. B. and Eden, F. C. (1981). Repeated deoxyribonucleic acid clusters in the chicken genome contain homologous sequence elements in scrambled order. *Biochemistry*, **20**, 2989–2999.

Nevers, P. and Saedler, H. (1977). Transposable genetic elements as agents of gene instability and chromosomal rearrangements. *Nature, Lond.*, **268**, 109–115.

Ohta, T. (1981). Population genetics of selfish DNA. *Nature,.Lond.*, **292**, 648–649.

Ohta, T. and Kimura, M. (1981). Some calculations on the amount of selfish DNA. *Proc. Natl. Acad. Sci. USA*, **78**, 1129–1132.

Orgel, L. E. and Crick, F. H. C. (1980). Selfish DNA: the ultimate parasite. *Nature, Lond.*, **284**, 604–607.

Orkin, S. H. and Michelson, A. (1980). Partial deletion of the α globin structural gene in human α-thalassaemia. *Nature, Lond.*, **286**, 538–541.

Pan, J., Elder, J. T., Duncan, C. H. and Weissman, S. M. (1981). Structural analysis of interspersed repetitive polymerase III transcription units in human DNA. *Nucleic Acids Res.*, **9**, 1151–1170.

Petes, T. D. (1980). Unequal meiotic recombination within tandem arrays of yeast DNA genes. *Cell*, **19**, 765–774.

Posakony, J. W., Scheller, R. H., Anderson, D. M., Britten, R. J. and Davidson, E. H. (1981). Repetitive sequences of the sea urchin genome. III. Nucleotide sequences of cloned repeat elements. *J. Mol. Biol.*, **149**, 41–76.

Potter, S. S., Brorein, W. J. Jr., Dunsmuir, P. and Rubin, G. M. (1979). Transposition of elements of the *412*, copia and *297* dispersed repeated gene families in *Drosophila*. *Cell*, **17**, 415–427.

Potter, S. S. and Thomas, C. A. Jr., (1978). The two-dimensional fractionation of *Drosophila* DNA. *Cold Spring Harbor Symp. Quant. Biol.*, **42**, 1023–1031.

Potter, S., Truett, M., Phillips, M. and Maher, A. (1980). Eukaryotic transposable genetic elements with inverted terminal repeats. *Cell*, **20**, 639–647.

Preisler, R. S. and Thompson, W. F. (1981a). Evolutionary sequence divergence within repeated DNA families of higher plant genomes. I. Analysis of reassociation kinetics. *J. Mol. Evol.*, **17**, 78–84.

Preisler, R. S. and Thompson, W. F. (1981b). Evolutionary sequence divergence within repeated DNA families of higher plant genomes. II. Analysis of thermal denaturation. *J. Mol. Evol.*, **17**, 85–93.

Rae, P. M. M., Kohorn, B. D. and Wade, R. P. (1980). The 10 kb *Drosophila virilis*

28s rDNA intervening sequence is flanked by a direct repeat of 14 base pairs of coding sequence. *Nucleic Acids Res.*, **8**, 3491–3504.

Rasmuson, B., Westerberg, B. M., Rasmuson, A., Gvozdev, V. A., Belyaera, E. S. and Ilyin, Y. V. (1981). Transpositions, mutable genes, and the dispersed gene family Dm 225 in *Drosophila melanogaster. Cold Spring Harbor Symp. Quant. Biol.*, **45**, 545–551.

Rimpau, J., Smith, D. and Flavell, R. (1978). Sequence organization analysis of wheat and rye genomes by interspecies DNA/DNA hybridization. *J. Mol. Biol.*, **123**, 327–359.

Rinehart, F. P., Ritch, T. G., Deininger, P. L. and Schmid, P. W. (1981). Renaturation rate studies of a single family of interspersed repeated sequences in human deoxyribonucleic acid. *Biochemistry*, **20**, 3003–3010.

Roeder, G. S. and Fink, G. R. (1980). DNA rearrangements associated with a transposable element in yeast. *Cell*, **21**, 239–249.

Roiha, H., Miller, J. R., Woods, L. C. and Glover, D. M. (1981). Arrangements and rearrangements of sequences flankings the two types of DNA insertion in *D. melanogaster. Nature, Lond.*, **290**, 749–753.

Rubin, G. M., Brorein, W. J. Jr., Dunsmuir, P., Flavell, A. J., Levis, R., Strobel, E., Toole, J. J. and Young, E. (1981). *Copia*-like transposable elements in the *Drosophila* genome. *Cold Spring Harbor Symp. Quant. Biol.*, **45**, 619–628.

Rubin, G. M., Finnegan, D. J. and Hogness, D. J. (1976). The chromosomal arrangement of coding sequences in a family of repeated genes. *Progr. Nucleic Acid Res. Mol. Biol.*, **19**, 221–226.

Rubin, C. M., Houck, C. M., Deininger, P. L., Friedmann, T. and Schmid, C. W. (1980). Partial nucleotide sequence of the 300-nucleotide interspersed repeated human DNA sequences. *Nature, Lond.*, **284**, 372–374.

Sapienza, C. and Doolittle, W. F. (1981). Genes are things you have whether you want them or not. *Cold Spring Harbor Symp. Quant. Biol.*, **45**, 117–125.

Scheller, R. H., Anderson, D. M., Posakony, J. W., McAllister, L. B., Britten, R. J. and Davidson, E. H. (1981a). Repetitive sequences of the sea urchin genome. II. Subfamily structure and evolutionary conservation. *J. Mol. Biol.*, **149**, 15–39.

Scheller, R. H., Constantini, F. D., Kozlowski, M. R., Britten, R. J. and Davidson, E. H. (1978). Specific representation of cloned repetitive DNA sequences in sea urchin RNAs. *Cell*, **15**, 189–203.

Scheller, R. H., McAllister, L. B., Crain, W. R. Jr., Durica, D. S., Posakony, J. W., Thomas, T. L., Britten, R. J. and Davidson, E. H. (1981b). Organization and expression of multiple actin genes in the sea urchin. *Mol. Cell Biol.*, **1**, 609–628.

Scheller, R. H., Thomas, T. L., Lee, A. S., Klein, W. J., Niles, W. D., Britten, R. J. and Davidson, E. H. (1977). Clones of individual repetitive sequences from sea urchin DNA constructed with synthetic *Eco RI* sites. *Science*, **196**, 197–200.

Selker, E. U., Yanofsky, C. Y., Driftmier, K., Metzenberg, R. L., Alzner, DeWeerd, B. and RajBhandary, U. L. (1981). Dispersed 5S RNA genes in *N. crassa*: structure, expression and evolution. *Cell*, **24**, 819–828.

Shapiro, J. A. (1982). Changes in gene order and gene expression. *Nat'l Cancer Inst. Monographs.*, **60**, 87–110.

Shimotohno, K. and Temin, H. M. (1981). Evolution of retroviruses from cellular movable genetic elements. *Cold Spring Harbor Symp. Quant. Biol.*, **45**, 719–730.

Shoemaker, C., Goff, Gilbou, E., Paskind, M., Mitra, S. M. and Baltimore, D. (1981). Structure of cloned retroviral circular DNAs: implications for virus integration. *Cold Spring Harbor Symp. Quant. Biol.*, **45**, 711–717.

Smith, G. P. (1976). Evolution of repeated DNA sequences by unequal cross-over. *Science*, **191**, 528–535.

Smith, M. J., Lui, A., Gibson, K. K. and Etzkorn, J. K. (1980). DNA sequence organization in the common Pacific starfish *Pisaster ochraceous*. *Can. J. Biochem.*, **56**, 1048–1053.

Stanfield, S. W. and Lengyel, J. A. (1979). Small circular DNA of *Drosophila melanogaster*: chromosomal homology and kinetic complexity. *Proc. Natl. Acad. Sci. USA*, **76**, 6142–6147.

Stanfield, S. W. and Lengyel, J. A. (1980). Small circular deoxyribonucleic acid of *Drosophila melanogaster*: homologous transcripts in nucleus and cytoplasm. *Biochemistry*, **19**, 3873–3877.

Stanley, S. M. (1979). *Macroevolution: Pattern and Process*, San Francisco, Freeman.

Strobel, E., Dunsmuir, P. and Rubin, G. M. (1979). Polymorphisms in the chromosomal locations of elements of the *412*, *copia* and *297* dispersed repeated gene families in *Drosophila*. *Cell*, **17**, 429–439.

Tashima, M., Calabretta, B., Torelli, G., Scofield, M., Maizel, A. and Saunders, G. F. (1981). Presence of a highly repetitive and widely dispersed DNA sequence in the human genome. *Proc. Natl. Acad. Sci. USA*, **78**, 1508–1512.

Tchurikov, N. A., Ilyin, Y. V., Skryabin, K. G., Ananiev, K. V., Bayev, A. A. Jr., Krayev, A. S., Zelentsova, E. S., Kulguskin, V. V., Lyubomirskaya, N. V. and Georgiev, G. P. (1981). General properties of mobile dispersed genetic elements in *Drosophila melanogaster*. *Cold Spring Harbor Symp. Quant. Biol.*, **45**, 655–665.

Thayer, R. E., Singer, M. F. and McCutchey (1981). Sequence relationships between single repeat units of highly reiterated African Green monkey DNA. *Nucleic Acids Res.*, **9**, 169–181.

Thompson, W. F., Murray, M. G. and Cuellar, R. E. (1980). Contrasting patterns of DNA sequence organization in plants. In C. J. Leaver (Ed), *Genome Organization and Expression in Plants*, Plenum Press, New York, pp. 1–15.

Timberlake, W. E. (1978). Low repetitive DNA content in *Aspergillus nidulans*. *Science*, **202**, 973–974.

Truett, M. A., Jones, R. S. and Potter, S. S. (1981). Unusual structure of the FB family of transposable elements in *Drosophila*. *Cell*, **24**, 753–764.

Van Arsdell, S. W., Denison, R. A., Bernstein, L. B., Weiner, A. M., Manser, T. and Gesteland, R. F. (1981). Direct repeats flank three small nuclear RNA pseudogenes in the human genome. *Cell*, **216**, 11–17.

Van der Ploeg, L. H. T., Konings, A., Oort, M., Roos, D., Bernini, L. and Flavell, R. A. (1980). γ-β-thalassaemia studies showing that deletion of the γ and δ-genes influences β-globin gene expression in man. *Nature*, **283**, 637–642.

Varley, J. M., MacGregor, H. C. and Erba, H. P. (1980). Satellite DNA is transcribed on lampbrush chromosomes. *Nature, Lond.*, **283** 686–688.

Weiner, A. M. (1980). An abundant cytoplasmic 7S RNA is complementary to the dominant interspersed middle repetitive DNA sequence family in the human genome. *Cell*, **22**, 209–218.

Wensink, P. C. (1978). Sequence homology within families of *Drosophila melanogaster* middle-repetitive DNA. *Cold Spring Harbor Symp. Quant. Biol.*, **42**, 1033–1039.

Wensink, P. C., Tabata, S. and Pachl, C. (1979). The clustered and scrambled arrangement of moderately repetitive elements in *Drosophila* DNA. *Cell*, **18**, 1231–1246.

Young, M. W. (1979). Middle-repetitive DNA: a fluid component of the *Drosophila* genome. *Proc. Natl. Acad. Sci. USA*, **76**, 6274–6278.

Young, M. W. and Schwartz, H. E. (1981). Nomadic gene families in *Drosophila*. *Cold Spring Harbor Symp. Quant. Biol.*, **45**, 629–640.

The Evolution of Genome Size
Edited by T. Cavalier-Smith
© 1985 John Wiley & Sons Ltd

CHAPTER 16

Recombination, Genome Size and Chromosome Number

BRIAN CHARLESWORTH

Department of Biology, University of Chicago, 1103 E. 57th Street, Chicago, Illinois 60637, USA.

SUMMARY

The processes likely to be important in genome evolution are examined from the point of view of population genetic theory. The evolution of recombination rates, chromosome number and chromosome rearrangements is first discussed. The evolutionary modification of genome size, by means of polyploidy, gene duplication and deletion and transposable elements is then considered. For each of these aspects of genome evolution it is possible to provide a list of well-defined processes that are mechanistically plausible. It appears likely, however, that several of these processes are usually causally involved; it is difficult in such cases to determine which of them plays the predominant part.

INTRODUCTION

The last few years have witnessed a revolution in our knowledge of the structure of the genome of higher organisms, due to the introduction of techniques for DNA sequencing and gene cloning. Among other findings, it is now apparent that much of the genome consists of repeated sequences of DNA which are frequently not transcribed, whose functional significance (if any) is obscure, and which often change rapidly in evolutionary time in sequence, organization and quantity (Dover and Flavell, 1982). Evolutionary biologists have only just started to consider the implications of these discoveries, but a good deal of theoretical work has been carried out in recent years on more classical problems of genome evolution, such as the

489

evolutionary control of rates of genetic recombination, chromosome arrangements and gene duplication (e.g. Maynard Smith, 1978; Ohta, 1980).

The purpose of this chapter is to survey the population processes that seem likely to be most important in genome evolution, and which have been explored theoretically by well-formulated models. The point of view consistently adopted here is that most evolutionary change is the consequence of alterations of gene and chromosome frequencies within natural populations, as a result of the interaction of mutation, selection, genetic drift and recombination. Some unconventional forces, such as meiotic drive and the replication of transposable genetic elements, will also be discussed. Arguments that explain evolutionary phenomena in terms of long-term advantages to the species will be avoided; these have been unduly prevalent in this field, and still appear in writings on the new discoveries in molecular genetics (e.g. Nevers and Sadler, 1977; Strobel *et al.*, 1979). Reasons for doing this are given by Fisher (1958, pp. 49–50), Williams (1966), Lewontin (1970) and Maynard Smith (1964) among others. Also omitted is discussion of directly functional interpretations of such phenomena as repeated DNA (e.g. Cavalier-Smith, 1978; Davidson and Britten, 1979). This is not because they are of no interest or importance but simply because such functional interpretations present no new problems to the population geneticist. The aim here is to discuss possible evolutionary mechanisms rather than to arrive at definitive explanations for the data. The method of presentation is to describe the assumptions and results of the models in largely verbal and non-technical terms. This inevitably leads to over-simplification, but should help an exchange of ideas between theorists and experimentalists in this area. Questions relating to evolutionary aspects of recombination and chromosome number will be dealt with first, followed by the evolution of genome size, with particular reference to gene duplication and transposable elements.

EVOLUTIONARY FORCES AFFECTING RECOMBINATION AND CHROMOSOME NUMBER

This section examines various evolutionary mechanisms that can influence the frequency of genetic recombination and the number of chromosomes in the genome. Since the expected amount of recombination between a pair of loci sampled at random from a genome is a function of both these quantities, it may be anticipated that they will respond in a similar way to a given selective situation. Nevertheless, each has certain properties that warrant separate study, and these will be pointed out in what follows.

Selection for reduced recombination

The section starts by presenting reasons for believing that there is a general and widespread selective force favouring reduction in the level of genetic recombination to near zero. Since recombination is an almost universal phenomenon in the living world, this raises the question of 'why does the genome not congeal?' (Turner, 1967); some attempts to answer this question will be described under the heading: Selection for increased recombination. Certain basic concepts of population genetics used throughout the chapter will also be introduced here.

There is a large amount of literature dealing with the population genetics of systems in which a selection pressure exists for reducing the rate of recombination between loci (reviewed by Maynard Smith, 1978, Chapter 5; see also Feldman *et al.*, 1981). The case usually studied is that of a random-mating, diploid population segregating for a pair of loci with alleles A,a and B,b, respectively, which are under the control of natural selection. With non-overlapping generations (as in organisms such as annual plants), the effects of selection can be studied by assigning a fitness to each genotype, defined as the product of survival to adulthood and expected fecundity when adult. Since only relative fitnesses affect the course of selection, it is usual to choose one genotype as a standard and express the fitnesses of the other genotypes as fractions of the fitness of the standard (Table 16.1). The extent to which the relative fitness of a genotype deviates from unity can be measured by its *selection coefficient* which can be negative or positive.

Table 16.1. Selection in a two-locus system.

	AA	Aa	aa
BB	$1-s_1-s_2-e_1$	$1-s_2$	$1-t_1-s_2-e_2$
Bb	$1-s_1$	1	$1-t_1$
bb	$1-s_1-t_2-e_3$	$1-t_2$	$1-t_1-t_2-e_4$

The entries in the table represent the fitnesses of the nine possible genotypes. The coefficients s_1 and t_1 measure selection at the A locus when B is heterozygous; similarly, s_2 and t_2 measure selection at the B locus when A is heterozygous. The coefficients e_1-e_4 measure the strength of epistatic interactions in fitness between the loci, and are all zero only if fitness effects are purely additive across loci. Note that none of the eight coefficients is constrained to be positive, but that all the fitnesses must be positive.

The simplest representation of the joint effects of two loci on fitness is the *additive* model, in which the fitness of each genotype is obtained by addition of separate contributions from each locus (Table 16.1). If the joint effects of the loci on fitness deviate from additivity, it can be shown that, at least for some values of the frequency of recombination between the loci, an equilibrium population polymorphic for both loci will display *linkage disequilibrium*, such that the frequencies of the four classes of gamete AB, Ab, aB and ab deviate from the frequencies expected if the alleles were combined at random into gametes (Ewens, 1979, Chapter 6). If the gamete frequencies are written

x_1, x_2, x_3, x_4, respectively, and the allele frequencies are p_1 and q_1 for A and a, and p_2 and q_2 for B and b, it can be shown that

$$x_1 = p_1p_2 + D, \; x_2 = p_1q_2 - D,$$
$$x_3 = q_1p_2 - D, \; x_4 = q_1q_2 + D \tag{1}$$

where D is the coefficient of linkage disequilibrium ($D = x_1x_4 - x_2x_3$).

From equation (1) it is seen that $D > 0$ if AB and ab exceed the frequencies expected under random combination of alleles at the two loci, and $D < 0$ if they are in deficiency. If the frequency of recombination between the loci is $R > 0$, a population with no selection will eventually come to a state with $D = 0$, since the value of D in one generation is $(1 - R)$ of its previous value. Selection under the additive model leads to a similar final state, but when there is *epistasis* in fitness effects (deviation from additivity), selection tends to favour some combinations of genes over others, and hence opposes the breakdown of linkage disequilibrium by recombination. If epistatic selection is sufficiently powerful in relation to recombination, D will be non-zero at equilibrium.

In such circumstances, theoretical work has shown that there is a selection pressure favouring a reduction in recombination between A and B, as originally suggested by Fisher (1930, Chapter 5). If we postulate a *modifier* locus with alleles C and c, that itself has no effect on fitness but merely modifies the frequency of recombination between A and B, then it appears that whichever allele reduces recombination will tend to spread if the initial population is at equilibrium with non-zero D between A and B (Feldman *et al.*, 1981). If D is zero, then the spread of a modifier either does not occur or occurs at a significant rate only if the modifier is present at above some threshold frequency (Charlesworth and Charlesworth, 1973). The speed of spread of a modifier reducing recombination depends on the tightness of linkage of the modifier to the loci it controls; it should be greatest for chromosome rearrangements, such as inversions, that completely suppress recombination between the loci concerned, at least when heterozygous over the standard arrangement, and which are themselves absolutely linked to the selected loci. There is evidence that the widespread inversion polymorphisms of *Drosophila* species involve this mode of selection, although the individual loci under selection cannot be identified (Dobzhansky, 1970, Chapter 5; Charlesworth, 1974). Genic modifiers of recombination rates have also been identified in a variety of species (Maynard Smith, 1978, Chapter 5; Turner, 1979), and could cause a gradual evolutionary shift towards reduced recombination between loci involved in epistatic selection. Chiasma localization (White, 1973, pp. 171–172) is another extremely effective means of restricting recombination to small portions of the genome.

If epistatic selection between polymorphic loci is a widespread phenomenon, it seems likely that the mechanism outlined above would favour genomes in which recombination is virtually absent. In general, it is improbable that interactions between different loci always generate additive fitness effects, and it is known that an extremely important form of selection, stabilizing selection favouring the intermediate phenotypic values of metrical traits, generates strong epistasis in fitness (Bulmer, 1980, Chapter 10). Direct evidence for epistatic selection is hard to come by, however, except in *Drosophila* where, in addition to inversion polymorphism, a variety of experiments on variation in fitness components testify to its operation (Spiess, 1958; Charlesworth and Charlesworth, 1976). But recombination occurs at a substantial rate in female (but not male) *Drosophila*. This exemplifies the problem mentioned at the start of this section, and suggests the need to seek for mechanisms favouring increased recombination.

Selection for increased recombination

It has sometimes been argued that the close interrelationship between DNA repair mechanisms and recombination provides a sufficient explanation for the evolutionary maintenance of the latter (e.g. Bernstein *et al.*, 1981). This seems unlikely, however, because certain organisms such as male *Drosophila* have a meiotic system lacking recombination, yet do not seem to suffer any greater incidence of mutational defects in consequence. Furthermore, it is difficult to see how this type of effect could explain the known species differences in recombination rates; for example, *D. subobscura* has map lengths two or three times those of *D. melanogaster*, despite genetic homology of the chromosome arms (Loukas *et al.*, 1979). Similar considerations apply to the argument that at least one chiasma must be formed per bivalent in order to ensure regular disjunction of the homologues (White, 1973, p. 169).

For these reasons, a good deal of effort has been put into the investigation of theoretical models which generate selection for increased recombination. This topic has been reviewed in detail by Maynard Smith (1978, Chapters 6 and 7), and there is little point in presenting an exhaustive account here. Instead a classification of the various types of model will be proposed, and their main features briefly described. Alternative classifications have been proposed (Felsenstein, 1974; Maynard Smith, 1978) and no special merit is claimed for the one below.

Hitch-hiking models

Hitch-hiking is the process by which the frequency of an allele at one locus is changed as a result of a change in allele frequency, caused by selection at a linked locus (Kojima and Schaffer, 1967; Maynard Smith and Haigh, 1974). The change in frequency at the first locus is caused by its being in linkage disequilibrium with the second locus as a result of historical accident or the effects of genetic drift in a finite population. An important source of linkage disequilibrium due to historical accident is when a selectively favourable mutation becomes established in a population as a result of a single mutational event. It will, of course, occur initially in a gamete with a unique complement of alleles at other polymorphic loci. As it rises in frequency, these alleles will tend to be dragged along with it; the closer their linkage, the longer this perturbation effect lasts (Maynard Smith and Haigh, 1974; Thomson, 1977). An example of this phenomenon is the association between the sickle-cell haemoglobin allele at the human β-globin locus and a closely-linked restriction site (Kan and Dozy, 1978).

The importance of hitch-hiking from the present viewpoint is that it implies that selection at one locus can interfere with the operation of selection at a second, linked locus, the Hill-Robertson effect (Hill and Robertson, 1966). Hence, recombination may be favoured, since it reduces the opposition of the selective forces. A number of different processes may be classed under the heading of hitch-hiking.

Firstly, mutants at two or more separate loci may be increasing in frequency. In the absence of recombination, only one of these can be fixed by selection, unless a double mutation occurs. Recombination may therefore be favoured. This process was postulated by Fisher (1930, p. 133), and has since been studied quantitatively by Felsenstein and Yokoyama (1976) and Charlesworth *et al.* (1977). A related model, in which one of the loci is maintained polymorphic by selection while the other experiences a change in allele frequency, was studied by Strobeck *et al.* (1976).

A second type of model assumes recurrent mutation at a pair of loci, generating unfavourable alleles which are kept rare by natural selection. Under certain conditions on the joint fitness effect of the loci concerned, increased recombination can be favoured, presumably because it allows a more effective reduction in the frequency of the mutant allele at each locus (Feldman *et al.*, 1981). This model has been extended to an arbitrary number of loci by Kondrashov (1982), who has shown that a population with free recombination always has a higher mean fitness at equilibrium under mutation–selection balance than a non-recombining population, provided that the adverse effect on fitness of a mutation increases with the number of mutations already present at other loci (c.f. Crow, 1970, pp. 140–148). There is empirical evidence for such a relation between fitness and number of mutations in *Drosophila* (Mukai, 1969); since mutational loads are known to be large in higher organisms (Crow, 1970), this model provides an attractive explanation of non-zero recombination rates.

A model that is related closely to this one is 'Muller's ratchet' (Muller, 1964; Felsenstein, 1974; Haigh, 1978; Maynard Smith, 1978). This model also envisages recurrent mutation to deleterious alleles at a large number of loci. In a finite population, there is a chance that all the mutant-free chromosomes will eventually be lost from the population by genetic drift. In the absence of recombination, they can never be regenerated, unless the rate of back mutation is improbably high. The fittest class that remains now carries one deleterious mutation per chromosome, and this is exposed to the same hazard of random loss. A ratchet-like mechanism thus operates, leading to a gradual increase in the mean number of mutations carried by the chromosomes of the population. But if recombination takes place, the process is arrested, since mutation-free chromosomes can be generated by cross-overs between chromosomes with mutations at different loci. A modifier inducing recombination can thus experience a selective advantage, since it will be carried up in frequency in association with the mutant-free chromosome (Felsenstein and Yokoyama, 1976).

Another type of hitch-hiking model was proposed by Slatkin (1975), who envisaged two loci responding to two different environmental gradients affecting fitness. It seemed attractive to suppose that increased recombination would be favoured in such a system, since it would allow greater freedom to each locus to respond to selection. A detailed study by Charlesworth and Charlesworth (1979) showed that this could happen, but under rather restricted circumstances, and that decreased recombination was often favoured instead.

Temporally fluctuating environments

If conditions change from time to time in such a way that the *combinations* of genes favoured by selection vary across generations, it is intuitively reasonable to suppose that increased recombination would facilitate the passage of the population into a new state when the environment changes (Sturtevant and Mather, 1938; Maynard Smith, 1971), and hence would be favoured by selection. This is indeed found to be the case when a formal model is constructed with a locus controlling the degree of recombination between two selected loci subject to fluctuating selection pressures, such that a state with $D < 0$ is favoured in some generations, and $D > 0$ in others. It has been suggested by several authors (Hamilton, 1980; Hutson and Law, 1981; Bell, 1982, pp. 153–159) that host–parasite or predator–prey interactions could drive cycles of D of this sort. May and Anderson (1983) describe a formal model for the case of a host–parasite system.

A closely related process was proposed by Slatkin and Lande (1976) on the basis of a non-genetic model, and investigated by Maynard Smith (1980) using a genetical formulation. It involves a metrical trait subject to stabilizing selection. As mentioned on p. 493, if the optimal value of the phenotype remains constant, there will tend to be linkage disequilibrium among the loci controlling the trait, and selection for reduced recombination. But if the optimum fluctuates between generations, or shows a directional trend, there may be an advantage to a modifier gene that increases the rate of recombination, since increased recombination permits a more rapid response to the directional component of selection (cf. Mather, 1943).

Spatially heterogeneous environments (sib competition)

This model was originally proposed by Williams and Mitton (1973) and Williams (1975, Chapters 2–4), and investigated more formally by Maynard Smith (1976). The environment is imagined to be divided into patches, such that different genotypes at a set of loci have fitnesses that vary according to the patch in which they are born. At least some of the individuals who develop in a single patch are siblings or half siblings, so that selection operates to some extent through competition between the offspring of the same parent. Recombination means that a parent is able to produce a wider range of progeny than in the absence of recombination, and therefore increases its chance of producing at least some progeny capable of surviving in a randomly selected patch.

We have little information to help us to decide which of the mechanisms described above may be important in controlling the evolution of recombination rates in nature. Each has certain *a priori* advantages and disadvantages. The hitch-hiking models, particularly the ones involving deleterious mutations, have the advantage that they operate through processes that must occur universally in natural populations. They have the disadvantage that they mostly provide a selection pressure in favour of only a very low level of recombination between the loci under selective control, and hence are not attractive candidates for explaining high rates of recombination and large species differences in recombination rates.

Fluctuating environment models are attractive in that they can select for almost any level of recombination (Charlesworth, 1976; Maynard Smith, 1980) and, at least in principle, could explain species differences. There are some doubts about their plausibility, however, since the first model requires a rather special type of constraint on the way in which fitnesses change with time (Maynard Smith, 1971; Charlesworth, 1976; May and Anderson, 1983). The second requires an appropriate balance to be struck between the strength of stabilizing selection and the rate of change of the optimum with time.

The spatial heterogeneity model suffers from a number of drawbacks discussed by Maynard Smith (1978, pp. 107–108). In particular, it is clear that many organisms in which there is little opportunity for sib competition have appreciable rates of genetic recombination.

Changes in chromosome number

Chromosome number can be reduced by *centric fusion*, in which two acrocentric chromosomes undergo a translocation with breakpoints in the centromeric heterochromatin. This produces one metacentric chromosome with two large arms, and another with two small arms, consisting of hetero-chromatin which is subsequently lost (White, 1973, p. 224). Subsequent pericentric inversions can convert the large metacentric into an acrocentric. Evolutionary reduction in chromosome number due to centric fusions has been well documented in both plants and animals (Stebbins, 1971, Chapter 4; White, 1973, Chapter 12). The reverse process, centric fission or dissociation, is generally thought to be less common, although recent evidence suggests its importance in mammalian karyotype evolution (Imai and Crozier, 1980). Formally, a centric fusion behaves as a dominant gene that reduces recombination between loci on the two different arms involved in the fusion (Fig. 16.1); a fission behaves like a recessive gene increasing recombination. Since a selectively favoured recessive gene has a lower chance than a dominant of establishing itself in a random-mating population (Haldane, 1927), it is to be expected that centric fusions would tend to outnumber centric fissions, whatever the mode of selection involved. This would not, of course, apply to predominantly inbreeding species.

Selective forces favouring a reduction in recombination will favour a centric fusion between two chromosomes, provided that there are suitable fitness interactions between loci on the chromosomes concerned, sufficiently strong to generate linkage disequilibrium between unlinked genes. Evidence for such a linkage disequilibrium has, for example, been obtained in the grass-hopper *Moraba scurra* (Lewontin and White, 1960). This type of effect could, therefore, be very important in selecting for centric fusions (e.g. Lewis and John, 1968). A different selective force was suggested by White (1973, p. 613), and analysed theoretically in modified form by Charlesworth and Charlesworth (1980). This involves the existence of a selectively maintained polymorphism on an autosome, such that selection affects males and females differently, with the result that the equilibrium allele frequencies are different in the two sexes. In such a situation, there is effectively linkage disequilibrium

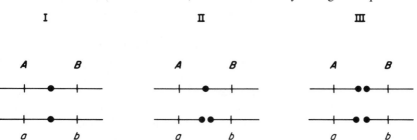

Figure 16.1. This displays the linkage relations between two loci on different chromosome arms, for the three possible karyotypes in a population segregating for a centric fusion. In karyotypes I and II, *A* and *B* behave as linked loci, whereas they are unlinked in karyotype III.

between the sex chromosomes and the autosomal locus, with the result that a centric fusion between the X or Y chromosome and the autosome is favoured. Since in general it is likely that selection will act to some extent differently on the two sexes, this force could well be a causative agent in the frequently observed fusions between autosomes and sex chromosomes (see White, 1973, Chapter 17 for examples of these). Selection for centric fissions could be generated by any of the mechanisms discussed in the previous section: selection for increased recombination.

A completely different mechanism that can promote the spread of chromosome rearrangements is *meiotic drive*, whereby a genetic element causes distorted segregation in heterozygotes, such that more than 50% of the gametes contain the element in question. Examples of cytologically detectable chromosome changes that are associated with meiotic drive are known in a variety of organisms (Zimmering et al., 1970). White (1978, pp. 212–213) has argued for the importance of meiotic drive in causing species differences in karyotype, but quotes only one example in which there is definite evidence for the involvement of drive. Drive is, however, undoubtedly of major significance in maintaining supernumerary (B) chromosomes in species of both plants and animals (Jones and Rees, 1982). These chromosomes are usually heterochromatic, with deleterious effects on the fitness of their carriers, and are not homologous with members of the normal complement. They display a variety of drive mechanisms by which they can maintain their presence in populations despite counter-selection at the individual level. In plants, for instance, preferential disjunction in meiosis I may be found, with an increased frequency of inclusion of the B chromosomes into the functional egg nucleus.

In contrast to these modes of evolution based on individual or gamete selection, many authors have discussed the possibility that chromosome rearrangements may be fixed by genetic drift in small populations (Wright, 1941; Bengtsson and Bodmer, 1976; Lande, 1979; Hedrick, 1981; Slatkin, 1981; Walsh, 1982). The problem is complicated by the fact that most classes of chromosome rearrangement cause a loss of fertility when heterozygous, due to the production of unbalanced gametes as a result of disturbances to normal segregation. Since a new mutation in a random mating population is present predominantly in heterozygotes, the chance that a rearrangement associated with such a fertility loss can rise to a high frequency in a local population is small, except with very restricted population size. Migration between populations lowers the chance of establishment of new arrangements within a local population, but increases the rate at which they can spread through the species (Lande, 1979; Slatkin, 1981). The most favourable population structure for rapid chromosomal evolution by this mechanism is when there is a high degree of subdivision into small, partially isolated populations, with a high rate of extinction of populations and recolonization from adjacent areas. As a result of the recolonization process, there is a finite chance that, at some future date, each local population will be descended from just one of the populations present in a given generation, by analogy with the process of random fixation of genes by drift in a single population. An alternative pathway is the origination of a new species from a small isolated population; if the isolate becomes fixed for a new arrangement, this will obviously come to characterize the species as a whole (Wright, 1940).

It seems likely that this process of random fixation of chromosomal arrangements has played a major role in the alteration of chromosome number by centric fusions or fissions (Lande, 1979; Bengtsson, 1980). The apparent preponderance of fusions over fissions in many groups could be explained by the greater ease with which fusions originate by mutation (White, 1973, p. 224). Extreme inbreeding, such as self-fertilization, will accelerate this mode of karyotypic evolution, since fewer heterozygotes occur, and some of the classic botanical examples of rapid chromosomal evolution are associated with high rates of selfing (e.g. *Clarkia*, Lewis, 1973). (It should be noted, however, that selective factors influencing recombination rates and chromosome numbers may be affected by inbreeding, e.g. Charlesworth *et al.* (1977, 1979); Malefijt and Charlesworth (1979). Associations between inbreeding and karyotypic evolution do not, therefore, constitute strong evidence for the role of drift.)

A related model of chromosome evolution has been proposed by Wilson *et al.* (1974, 1975) and Bush *et al.* (1977), who suggest that rearrangements may be associated with favourable mutations caused by the effects of the rearrangement on the regulation of gene expression. Because of the fertility loss to heterozygotes, they postulate that such rearrangements can only be established in small populations. The difficulty with this view is the scanty evidence that chromosomal variants which get established in natural populations are directly connected with favourable phenotypic effects. The most carefully studied case, that of the inversion polymorphisms of *Drosophila* spp., has not yielded any examples of such a direct influence of arrangements on phenotype. The argument that there is a general correlation between morphological and karyotypic evolution (Wilson *et al.*, 1974; Bush *et al.*, 1977) confuses correlation with causation, and does not provide unequivocal support for the proposed mechanism.

CHANGES IN GENE NUMBER

The mechanisms discussed so far contribute to changes in the arrangement of the genetic material as opposed to changes in its quantity. Alterations of quantity can proceed by a number of processes: polyploidy, aneuploidy, duplication, deletion and the activities of transposable elements. These will each be discussed in turn.

Polyploidy and aneuploidy

Population genetics has little of interest to say about changes in chromosome number by polyploidy or aneuploidy. Allopolyploidy, the multiplication of the entire set of chromosomes in a species hybrid, has clearly played a major role in the evolution of flowering plants (Stebbins, 1971, Chapter 5). One selective advantage of allopolyploidy lies in its ability to regularize pairing in a species hybrid, since a doubling of chromosome number allows preferential pairing of chromosomes derived from the same species. The increased cell size of polyploids may also play a role. Exactly why species hybrids should be so successful relative to the progenitor species under some circumstances is obscure.

Once a genome has become polyploid, all loci are of course present in duplicate. This provides the potential for evolutionary divergence of the homologous pairs, either by the acquisition of new functions, or by degeneration of one of the pair. This latter process has been much studied by population geneticists (Li, 1980; Maruyama and Takahata, 1981). It occurs when the relationship between dose of gene product and fitness is such that the fixation at one out of a pair of loci of a mutation inactivating the gene or gene product has a negligible effect on fitness. If the population size is large, mutations will tend to accumulate at similar frequencies at both members of the pair, so that fixation is retarded (Fisher, 1935). In a finite population, however, there is a chance that genetic drift can cause a mutation to become fixed before anything happens at the other locus. An alternative model has been suggested by Allendorf (1979), who proposed that selection may often favour an intermediate level of gene product, so that inactivation of one of the pair would be advantageous in a polyploid. There is a considerable amount of evidence that a large proportion of structural genes has become secondarily diploid in old polyploid groups such as the Salmonidae (Ferris and Whitt, 1979), presumably as the result of one or other of the processes.

Aneuploidy, a change in chromosome number involving less than the whole complement, has no obvious positive selective significance, although species comparisons (especially in plants: Stebbins, 1971, pp. 163–168) indicate that it can play an evolutionary role. In diploids, aneuploidy is generally lethal or highly deleterious, so that it is not surprising that aneuploidy is commonly associated with polyploidy, where the disturbance of genic balance is obviously less serious. Processes analogous to those described above for individual loci may well play a role in the evolution of aneuploid series.

Gene duplication

As emphasized long ago by Bridges (1935) and Muller (1936), duplication of genes as a result of accidents of chromosome replication is the only way in which the number of copies of a gene obeying the ordinary Mendelian rules can be increased in a diploid genome. Tandem duplications, in which the original DNA sequence and its duplicate remain at adjacent sites and in the same orientation seem to be the most common form of duplication preserved in evolution (Ohno, 1970). Once a duplication has become fixed in a population, the way is open for evolutionary divergence of the two duplicates, if an advantageous mutation conferring a new function becomes fixed in one of the two copies. Evolutionary degeneration of one of the two copies by similar processes to those described above for polyploids may also occur. This may account for the apparently non-functional pseudogenes that have been described in several cases of duplicate gene complexes, such as the globin genes (e.g. Zimmer *et al.*, 1980; Nishioka *et al.*, 1980). As shown by Maruyama and Takahata (1981), in large populations degeneration of a duplicated locus by random fixation of deleterious alleles proceeds most rapidly when there is close linkage between the loci involved in the duplication.

There are several processes by which a tandem duplication can become fixed in a population.

Random genetic drift

A duplication, like any other class of genetic change, may become fixed by change in a finite population. The action of drift is, however, complicated by the occurrence of unequal crossing-over in individuals homozygous for a duplication. As shown in Fig. 16.2, this results in the production of one recombinant chromosome carrying a triplication and one carrying a single copy (Sturtevant, 1925; Sturtevant and Morgan, 1923). Unequal cross-overs including the triplication can then generate quadruplication, and so on. Unequal crossing-over in itself cannot change the mean number of gene copies per chromosome in the population, but a population that has gone to fixation for a duplication has the potentiality to produce 'mutants' carrying single copies (the high rate of reversion of the *Bar* duplication in *D. melanogaster* motivated the classic experiments of Sturtevant and Morgan that uncovered the process of unequal crossing-over). In contrast, a population fixed for the single copy chromosome class is effectively trapped, since the rate of origination of new duplications is much lower than the rate of unequal crossing-over. The consequence of this is that genetic drift over a long period of time will result in an ever-increasing probability that the population will carry a single copy; since the expected copy number is not changed by drift or unequal crossing-over, this is counterbalanced by a low probability of populations carrying very high copy numbers. This effect has been studied by Ohta and Kimura (1981) and Ohta (1981).

In summary, although duplications can be fixed by genetic drift, the occur-

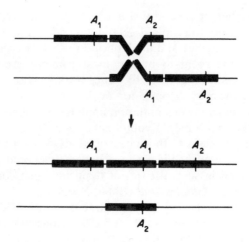

Fig. 16.2.   This displays the consequences of unequal crossing-over for the number of gene copies and for the genetic similarity of adjacent members of the gene family. A_1 and A_2 are alternative alleles at the same site within the gene involved in the duplication.

rence of unequal crossing-over means that the most likely state is eventual loss of the duplication, although there is a low probability that the population achieves a high number of copies. Since the probability of loss is lowest when there is little recombination, repeated systems of genes are most likely to be observed in regions of restricted recombination, such as centromeric regions. This may account for the occurrence of families of apparently non-functional, highly repeated DNA sequences in such regions (John and Miklos, 1979).

Selection for increased copy number

Abundant cellular components, such as ribosomes, may require a rate of synthesis beyond the capacity of single structural genes, creating a selective advantage for duplicating the relevant loci. This is clearly the explanation for the existence of many gene clusters, such as the highly duplicated ribosomal RNA genes (Ohno, 1970). But unequal crossing-over in such a system will tend to generate variability in copy number, whereas selection would presumably tend to favour some optimal copy number, with fitness falling off as a function of deviation from the optimum number. Depending on the strength of this stabilizing selection relative to the rate of generation of variance in copy number by unequal crossing-over, there will be more or less variability in copy number in an equilibrium population (Crow and Kimura, 1970, pp. 294–296). Population studies of duplicated esterase loci in various *Drosophila* species have provided direct evidence on the form of the copy number distribution in natural populations (Roberts and Baker, 1973; Loukas and Krimbas, 1975).

In such systems, naturally, there will be a selective advantage to devices for restricting the amount of crossing-over in the region of the duplication system, on the lines discussed in the previous section: polyploidy and aneuploidy. This may help to explain the frequent association of ribosomal RNA genes with heterochromatic chromosome regions, in which crossing-over is restricted. On the other hand, the occurence of unequal cross-overs tends to ensure genetic similarity between neighbouring loci in a system of duplications, as suggested by Smith (1974) and extensively analysed by Ohta and colleagues (Ohta, 1980). This process of homogenization is shown in Fig. 16.2. A similar effect is produced by gene conversion events between unequally paired members of a duplicated system, for which there is increasing evidence (e.g. Leigh Brown and Ish-Horowicz, 1981; Jackson and Fink, 1981). Since crossing-over and gene conversion are both resultants of the same process of hybrid DNA formation (Stahl, 1979), it is to be expected that both will occur in duplication systems. The occurrence of unequal crossing-over and unequal gene conversion will tend to retard the process described above (p. 501) of degeneration of components of a duplicated system by random fixation of deleterious alleles. Evidence for such degeneration, apparently resulting from the insertion of a transposable element, has been described for the ribosomal RNA genes of *D. melanogaster* (Roija *et al.*, 1981). The short-term advantage of restricting crossing-over in such systems may therefore have deleterious long-term consequences for the functional efficiency of highly duplicated systems.

Selection for duplication of a polymorphic locus

It has been suggested by several authors (e.g. Partridge and Giles, 1963; Fincham, 1966) that the maintenance of a balanced polymorphism by heterozygote advantage creates a selection pressure in favour of a duplication, since it allows the production of homozygous individuals carrying both alleles involved in the polymorphism. If a new duplication arises in a system segregating for two alleles A_1 and A_2, it will of course initially contain the same allele as the locus which was duplicated, A_1 say. No especial advantage would normally accrue to such a duplication gamete, unless there was an advantage in increasing the amount of gene product as discussed above. It would therefore behave as a selectively neutral variant. If it persisted for a sufficiently long time, a duplication gamete of the type A_1A_2 could eventually be generated by crossing-over or conversion. This might have a selective advantage, since (under suitable conditions), the A_1A_2 gamete would exhibit heterozygote advantage over both A_1 and A_2 non-duplicated gametes. The duplication would then spread to fixation. Detailed models of this have been studied by Spofford (1969). As pointed out by Lewis and Wolpert (1979), the process can only occur in sexually reproducing species, where recombination generates the appropriate gamete types. Forms of balancing selection other than heterozygote advantage could have the same effect on the evolution of duplications, for example in situations where the two alleles are adapted to different ecological niches.

Other mechanisms

Partial duplication of a gene may result in a gene coding for a new amino acid sequence, as in the human haptoglobin 2 variant which apparently resulted from an unequal cross-over in a heterozygote for the Hp^{1F} and Hp^{1S} variants (Smithies, 1964). A new function associated with such a partial duplication might be selectively advantageous. Such a process might have been important in the evolution of the highly internally duplicated structure of immunoglobulin heavy chains (Ohno, 1970, p. 79). Similarly, if a whole set of loci is duplicated instead of a single gene, a new phenotype could be caused by the resulting disturbance in genic balance, as in the *Bar* duplication. In rare instances, such a change could be advantageous.

Deletions

Loss of genic material due to deletions is generally likely to be deleterious, so such deletions will usually only occur in natural populations as rare variants. As discussed above, gene loss may be neutral or even advantageous in polyploid genomes and in tandem duplications, so that deletions may become established in such systems. Deletions of non-coding DNA, as in the loss of centric heterochromatin associated with centric fusions, may also be selectively neutral, and hence could be established by genetic drift.

Transposable elements and repetitive DNA

Recent work on the molecular biology of the eukaryote genome has revealed the existence of a large fraction of DNA which is present in multiple copies, and has no obvious functional significance for the organism (e.g. Finnegan *et al.*, 1978; John and Miklos, 1979; Dover and Flavell, 1982). Differences in DNA amount and sequence between related species are frequently due to differences in the repeated DNA. This is usually divided into two classes: (1) middle-repetitive DNA, present in copy numbers in the tens to thousands, according to family, and dispersed throughout the genome; (2) highly repeated, satellite DNA present in thousands to millions of tandemly duplicated copies. The highly repeated DNA usually consists of relatively short sequences. Variation in its quantity may, if it really has no functional significance, fit the neutral model of evolution of tandemly duplicated arrays described in the previous section: gene duplication (Ohta and Kimura, 1981; Ohta, 1981). In contrast, the middle-repetitive DNA consists of sequences of a few hundred to a few thousand loci, frequently with terminal repeats. Their structure is reminiscent of that of bacterial transposons or vertebrate retroviruses. It is therefore attractive to suppose that they may be capable of self-replication and of increasing the number of copies within the genome, by integration of replicates of pre-existing copies into new sites in the genome (Doolittle and Sapienza, 1980; Orgel and Crick, 1980). Direct evidence for such 'transposition' and increase in copy number has been obtained in the case of the element responsible for the hybrid dysgenesis syndrome in *Drosophila melanogaster* (Bingham *et al.*, 1982), which is caused by chromosomal insertions of the element.

As suggested by Doolittle and Sapienza (1980) and Orgel and Crick (1980), this component of the genome may therefore be maintained by its own power of replication within the genome, rather than because of any advantage conferred on the individuals who carry it. There is an important difference between sexual and asexual reproduction in this respect, pointed out by Hickey (1982). In an asexual population, increase in copy number within a genome has no consequences except for members of the clone within which the increase occurs. Thus, introduction of a single individual carrying a transposable element into a population lacking it cannot result in the spread of the element throughout the population, except by the action of genetic drift. With sexual reproduction, however, there is the possibility of segregation and recombination, so that a transposable element that integrates itself into a new genomic site may be passed to a different progeny individual from its parental element. Providing that there is a net increase every generation in the number of copies per individual genome, the element will eventually spread throughout the population. This spread does not require the occurrence of genetic recombination; segregation of non-recombining chromosomes is sufficient. This difference between sexual and asexual reproduction may explain the larger proportion of middle-repetitive DNA in eukaryotes than in prokaryotes.

This raises the question of what prevents an indefinite increase in the number of copies of a given family of transposable elements within the genome. Two possibilities present themselves. The first is that the replication is self-limiting, i.e. the transposition rate per element declines with the number of elements already present. There is evidence in prokaryotes that the presence of elements in the genome can inhibit the transposition of other elements of the same class (Sherratt *et al.*, 1983; Johnson and Reznikoff, 1984). In *D. melanogaster* the reciprocal effects observed in the hybrid dysgenesis syndrome (which only occurs when a genome containing the transposable element concerned is placed over a genome lacking it, in a cytoplasmic background derived from a strain lacking the element) provide evidence for self-limitation of copy number (Engels, 1981). The population genetics of a selectively neutral transposable element with self-limited replication has been studied by Charlesworth and Charlesworth (1983) and Langley *et al.* (1983), who have determined some of the properties of the steady-state distribution of copy number per individual that results from the interaction of transposition with genetic drift in a finite population.

The alternative possibility is that the presence of transposable genetic elements in the genome is selectively disadvantageous to the individuals who carry them, perhaps because of mutations induced when they insert into the genome. If the fitness of individuals declines with the number of copies of a given family of transposable elements that they carry, it is attractive to suppose that a balance might be reached between the increase in copy number generated by transposition, and the decrease in copy number due to selection, resulting in a stable distribution of numbers of copies per genome (Orgel and Crick, 1980). This possibility has been investigated quantitatively by Brookfield (1982) and Charlesworth and Charlesworth (1983), using computer and analytical models. It appears that it is indeed possible for such a stable distribution of copy number to be set up, but that this requires the logarithm of fitness to fall off more steeply than linearly as a function of copy number (Fig. 16.3). If the fitness function is not sufficiently downwardly curved (curves B and C in Fig. 16.3), then the pressure of transposition overcomes that of selection, and there is an indefinite built-up of the number

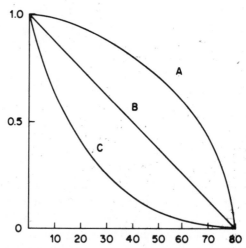

Fig. 16.3.   Possible forms of the relationship between the logarithm of fitness of an individual and the number of transposable elements of a given family present in its genome. Stability of copy number is maintained only if the curve is of type A, with a sufficiently strong curvature, in the absence of self-regulation of transposition rates.

of copies per genome of the element. As mentioned in the section: selection for increased recombination, there is evidence for fitness curves of type A (Mukai, 1969).

There is an interesting relationship between recombination and the ability of transposable elements to spread. The absence of genetic recombination in a chromosome region allows the accumulation of deleterious mutations through the operation of Muller's ratchet, as discussed in the section: selection for increased recombination. If transposable elements are subject to counter-selection, it is evident that, when once integrated into the genome, they would behave like deleterious mutations. It would therefore be expected that regions where crossing-over is absent would tend to accumulate transposable elements, and hence be rich in repetitive DNA. MacGregor and Horner (1980) and MacGregor (1982) describe such a situation in the crested newt, *Triturus cristatus*, where chromosome 1 is permanently heterozygous for a chromosome arm heteromorphism in which the alternative morphs are both lethal as homozygotes, and in which crossing-over is suppresssed in the heterozygote. This region has accumulated repetitive DNA and genes from other chromosomes, presumably as the result of the action of transposable elements. If the harmful effects of transposable elements are recessive, then random fixation by genetic drift of an element at a particular site on one of the two homologues could also contribute to this accumulation process, by the mechanism suggested by Nei (1970). Similar effects could also be involved in the heterochromatinization of the Y chromosome (cf. Jones, 1977; Charlesworth, 1978) and in the accumulation of repetitive sequences in regions of restricted crossing-over, such as areas proximal to the centromeres (John and Miklos, 1979).

DISCUSSION

The processes discussed above suggest that genomic evolution is probably under the control of a number of interacting factors. As is frequently found in evolutionary biology, it is difficult in many cases to distinguish experimentally between alternative explanations of the same phenomena. For example, there seems to be no possibility at present of deciding which of the factors discussed in the section: selection for increased recombination, are likely to be most significant in the evolutionary maintenance of genetic recombination and in causing species differences in recombination frequencies. Similarly, the respective roles of genetic drift and selection in karyotypic evolution are unclear. It seems indeed likely that many aspects of genomic evolution involve more than one causal factor. The role of population genetics theory in such cases is to determine which processes are both mechanistically plausible, and have evolutionary consequences consistent with the data. At present, it would seem that there is a set of well-characterized population processes capable of explaining, at least in principle, the main features of the evolution of recombination rates, chromosome number and arrangement, and the evolution of gene duplications and multigene families. The study of the evolutionary dynamics of transposable elements is still in its infancy, but will undoubtedly become an important interface between population genetics and molecular genetics.

REFERENCES

Allendorf, F. W. (1979). Rapid loss of duplicate gene expression by natural selection. *Heredity*, **43**, 247–258.

Bell, G. (1982). *The Masterpiece of Nature*, Croom Helm, London.

Bengtsson, B. O. (1980). Rates of karyotype evolution in placental mammals. *Hereditas*, **92**, 37–47.

Bengtsson, B. O. and Bodmer, W. F. (1976). On the increase of chromosome mutations under random mating. *Theor. Pop. Biol.*, **9**, 260–281.

Bernstein, H., Byers, G. S. and Michod, R. E. (1981). Evolution of sexual reproduction: importance of DNA repair, complementation, and variation. *Amer. Nat.*, **117**, 537–549.

Bingham, P. M., Kidwell, M. G. and Rubin, G. M. (1982). The molecular basis of *P-M* hybrid dysgenesis: the role of the *P* element, a *P*-strain specific transposon family. *Cell*, **29**, 995–1004.

Bridges, C. B. (1935). Salivary chromosome maps. *J. Hered.*, **26**, 60–64.

Brookfield, J. F. Y. (1982). Interspersed repetitive DNA sequences are unlikely to be parasitic. *J. Theor. Biol.*, **94**, 281–299.

Bush, G., Case, S. M., Wilson, A. C. and Patton, J. (1977). Rapid speciation and chromosomal evolution in mammals. *Proc. Natl. Acad. Sci. USA*, **74**, 3942–3946.

Bulmer, M. G. (1980). *The Mathematical Theory of Quantitative Genetics*, Oxford University Press, Oxford.

Cavalier-Smith, T. (1978). Nuclear volume control by nucleoskeletal DNA, selection for cell volume and cell growth rate, and the solution of the C-value paradox. *J. Cell Sci.*, **34**, 247–278.

Charlesworth, B. (1974). Inversion polymorphism in a two-locus genetic system. *Genet. Res.*, **23**, 259–280.

Charlesworth, B. (1976). Recombination modification in a fluctuating environment. *Genetics*, **83**, 181–195.

Charlesworth, B. (1978). Model for the evolution of Y-chromosomes and dosage compensation. *Proc. Natl. Acad. Sci. USA*, **75**, 5618–5622.

Charlesworth, B. and Charlesworth, D. (1973). Selection of new inversions in multilocus genetic systems. *Genet. Res.*, **21**, 167–183.

Charlesworth, B. and Charlesworth, D. (1976). An experiment on recombination load in *Drosophila melanogaster. Genet. Res.*, **25**, 267–274.

Charlesworth, D. and Charlesworth B. (1979). Selection on recombination in clines. *Genetics*, **91**, 581–589.

Charlesworth, D. and Charlesworth, B. (1980). Sex differences in fitness and selection for centric fusions between sex chromosomes and autosomes. *Genet. Res.*, **35**, 205–214.

Charlesworth, B. and Charlesworth, D. (1983). The population dynamics of transposable elements. *Genet. Res.*, **42**, 1–27.

Charlesworth, D., Charlesworth, B. and Strobeck, C. (1977). Effects of selfing on selection for recombination. *Genetics*, **86**, 213–226.

Charlesworth, D., Charlesworth, B. and Strobeck, C. (1979). Selection for recombination in self-fertilizing species. *Genetics*, **93**, 237–244.

Crow, J. F. (1970). Genetic loads and the cost of natural selection. In K. Kojima (Ed), *Mathematical Topics in Population Genetics*, Springer-Verlag, Berlin.

Crow, J. F. and Kimura, M. (1970). *An Introduction to Population Genetics Theory*, Harper and Row, New York.

Davidson, E. H. and Britten, R. J. (1979). Regulation of gene expression: possible role of repetitive sequences. *Science*, **204**, 1052–1056.

Dobzhansky, T. (1970). *Genetics of the Evolutionary Process,* Columbia University Press, New York.

Doolittle, W. F. and Sapienza, C. (1980). Selfish genes, the phenotype paradigm, and genome evolution. *Nature, Lond.*, **284**, 601–607.

Dover, G. A. and Flavell, R. B. (1982). *Genome Evolution and Phenotypic Variation*, Academic Press, New York and London.

Engels, W. R. (1981). Hybrid dysgenesis in *Drosophila* and the stochastic loss hypothesis. *Cold Spring Harbor Symp. Quant. Biol.*, **45**, 561–566.

Ewens, W. J. (1979). *Mathematical Population Genetics*, Springer-Verlag, Berlin.

Feldman, M. W., Christiansen, F. B. and Brooks, L. D. (1981). Evolution of recombination in a constant environment. *Proc. Natl. Acad. Sci. USA*, **47**, 4838–4841.

Felsenstein, J. (1974). The evolutionary advantage of recombination. *Genetics*, **78**, 737–756.

Felsenstein, J. and Yokoyama, S. (1976). The evolutionary advantage of recombination. II. Individual selection for recombination. *Genetics*, **83**, 845–859.

Ferris, S. D. and Whitt, G. S. (1979). The role of speciation and divergence time in the loss of duplicate gene expression. *J. Mol. Evol.*, **12**, 267–317.

Fincham, J. R. S. (1966). *Genetic Complementation*, Benjamin, New York.

Finnegan, D. J., Rubin, G. M., Young, M. W. and Hogness, D. S. (1978). Repeated gene families in *Drosophila melanogaster*. *Cold Spring Harbor Symp. Quant. Biol.*, **42**, 1053–1063.

Fisher, R. A. (1930). *The Genetical Theory of Natural Selection*, 1st edn, Oxford University Press, Oxford.

Fisher, R. A. (1935). The sheltering of lethals. *Amer. Nat.*, **69**, 446–455.

Fisher, R. A. (1958). *The Genetical Theory of Natural Selection*, 2nd edn, Dover, New York.

Haigh, J. (1978). The accumulation of deleterious genes in a population—Muller's ratchet. *Theor. Pop. Biol.*, **14**, 251–267.

Haldane, J. B. S. (1927). A mathematical theory of natural and artificial selection. Part V. Selection and mutation. *Proc. Camb. Phil. Soc.*, **23**, 838–844.

Hamilton, W. D. (1980). Sex versus non-sex versus parasite. *Oikos*, **35**, 282–290.

Hedrick, P. W. (1981). The establishment of chromosomal variants. *Evolution*, **35**, 322–332.

Hickey, D. H. (1982). Selfish DNA: a sexually-transmitted nuclear parasite. *Genetics*, **101**, 519–531.

Hill, W. G. and Robertson, A. (1966). The effect of linkage on limits to artificial selection. *Genet. Res.*, **8**, 269–294.

Hutson, V. and Law, R. (1981). Evolution of recombination in populations experiencing frequency-dependent selection with time delay. *Proc. Roy. Soc. Lond. B*, **213**, 345–359.

Imai, H. T. and Crozier, R. H. (1980). Quantitative analysis of directionality in mammalian karyotype evolution. *Amer. Nat.*, **116**, 537–569.

Jackson, J. A. and Fink, G. R. (1981). Gene conversion between duplicated genetic elements in yeast. *Nature, Lond.*, **292**, 306–311.

John, B. and Miklos, G. L. G. (1979). Functional aspects of satellite DNA and heterochromatin. *Int. Rev. Cytol.*, **58**, 1–114.

Johnson, R. C. and Reznikoff, W. S. (1984). Copy number control of Tn 5 transpositions. *Genetics*, **107**, 9–18.

Jones, K. W. (1977). Snake's eye view of Adam and Eve. *Nature, Lond.*, **268**, 107–108.

Jones, R. N. and Rees, H. (1982). *B-chromosomes*. Academic Press, New York.

Kan, Y. W. and Dozy, A. M. (1978). Polymorphism of DNA sequence adjacent to human β-globin structural gene: relationship to sickle mutation. *Proc. Natl. Acad. Sci. USA*, **75**, 5631–5636.

Kojima, K. I. and Schaffer, H. E. (1967). Survival process of linked mutant genes. *Evolution*, **21**, 518–531.

Kondrashov, A. S. (1982). Selection against harmful mutations in large sexual and asexual populations. *Genet. Res.*, **40**, 325–332.

Lande, R. (1979) Effective deme sizes during long-term evolution estimated from rates of chromosomal evolution. *Evolution*, **33**, 234–251.

Langley, C. H., Brookfield, J. F. Y. and Kaplan, N. (1983). Transposable elements in Mendelian populations. I. Theory. *Genetics*, **104**, 457–471.

Leigh Brown, A. J. and Ish-Horowicz, D. (1981). Evolution of the 87A and 87C heat shock loci in *Drosophila*. *Nature*, **290**, 677–682.

Lewis, H. (1973). The origin of diploid neospecies in *Clarkia*. *Amer. Nat.*, **107**, 161–170.

Lewis, J. and Wolpert, L. (1979). Diploidy, evolution and sex. *J. Theor. Biol.*, **78**, 425–438.

Lewis, K. R. and John, B. (1968). The chromosomal basis of sex determination. *Int. Rev. Cytol.*, **23**, 277–379.

Lewontin, R. C. (1970). The units of selection. *Ann. Rev. Ecol. Syst.*, **1**, 1–18.

Lewontin, R. C. and White, M. J. D. (1960). Interaction between inversion polymorphisms of two chromosome pairs in the grasshopper, *Moraba scurra*. *Evolution*, **14**, 116–129.

Li, W. H. (1980). Rate of gene silencing at duplicate loci: a theoretical study and interpretation of data from tetraploid fishes. *Genetics*, **95**, 237–258.

Loukas, M. and Krimbas, C. B. (1975). The genetics of *Drosophila subobscura* populations. V. A study of linkage disequilibrium in natural populations between genes and inversions of the E chromosome. *Genetics*, **80**, 331–347.

Loukas, M., Krimbas, C. B., Mavragani-Tsipidou, P. and Kastritsis, C. D. (1979). Genetics of *Drosophila subobscura* populations. VIII. Allozyme loci and their chromosome map. *J. Hered.*, **70**, 17–26.

MacGregor, H. C. (1982). In G. A. Dover and R. B. Flavell (Eds), *Genome Evolution and Phenotypic Evolution*, Academic Press, New York and London.

MacGregor, H. C. and Horner, H. (1980). Heteromorphism for chromosome 1, a requirement for normal development in crested newts. *Chromosoma*, **76**, 111–122.

Malefijt, M. de Waal and Charlesworth, B. (1979). A model for the evolution of translocation heterozygosity. *Heredity*, **43**, 265–281.

Maruyama, T. and Takahata, N. (1981). Numerical studies of the frequency trajectories in the process of fixation of null genes at duplicated loci. *Heredity*, **46**, 49–58.

Mather, K. (1943). Polygenic balance and natural selection. *Biol. Rev.*, **18**, 32–64.

May, R. M. and Anderson, R. M. (1983). Epidemiology and genetics in the coevolution of parasites and hosts. *Proc. Roy. Soc. Lond. B*, **219**, 281–313.

Maynard Smith, J. (1964). Kin selection and group selection. *Nature, Lond.*, **201**, 1145–1147.

Maynard Smith, J. (1971). What use is sex? *J. Theor. Biol.*, **30**, 319–335.

Maynard Smith, J. (1976). A short-term advantage for sex and recombination through sib-competition. *J. Theor. Biol.*, **57**, 239–242.

Maynard Smith, J. (1978). *The Evolution of Sex*. Cambridge University Press, Cambridge.

Maynard Smith, J. (1980). Selection for recombination in a polygenic model. *Genet. Res.*, **35**, 269–277.

Maynard Smith, J. and Haigh, J. (1974). The hitch-hiking effect of a favourable gene. *Genet. Res.*, **23**, 23–35.

Mukai, T. (1969). The genetic structure of natural populations of *Drosophila melanogaster*. VII. Synergistic interaction of spontaneous mutant polygenes controlling viability. *Genetics*, **61**, 149–161.

Muller, H. J. (1936). Bar duplication. *Science*, **83**, 528–530.

Muller, H. J. (1964). The relation of recombination to mutational advance. *Mut. Res.*, **1**, 2–9.

Nei, M. (1970). Accumulation of non-functional genes on sheltered chromosomes. *Amer. Nat.*, **104**, 311–321.

Nevers, P. and Sadler, H. (1977). Transposable genetic elements as agents of instability and chromosomal rearrangements. *Nature, Lond.*, **268**, 109–115.

Nishioka, Y., Leder, A. and Leder, P. (1980). An unusual alpha-globin-like gene that has clearly lost both globin intervening sequences. *Proc. Natl. Acad. Sci. USA*, **77**, 2806–2809.

Ohno, S. (1970). *Evolution by Gene Duplication*, Springer-Verlag, Berlin.

Ohta, T. (1980). Evolution and variation of multigene families. In *Lecture Notes in Biomathematics*, 37, Springer-Verlag, Berlin.

Ohta, T. (1981). Population genetics of selfish DNA. *Nature, Lond.*, **292**, 648–649.

Ohta, T. and Kimura, M. (1981). Some calculations on the amount of selfish DNA. *Proc. Natl. Acad. Sci. USA*, **78**, 1129–1132.

Orgel, L. E. and Crick, F. H. C. (1980). Selfish DNA: the ultimate parasite. *Nature, Lond.*, **284**, 604–607.

Partridge, C. W. H. and Giles, N. H. (1963). Sedimentation behaviour of adenylosuccinase formed by interallelic complementation in *Neurospora crassa*. *Nature, Lond.*, **199**, 304–305.

Roberts, R. M. and Baker, W. K. (1973). Frequency distribution and linkage disequilibrium of active and null esterase isozymes in natural populations of *Drosophila montana*. *Amer. Nat.*, **107**, 709–726.

Roija, H., Miller, J. R., Woods, L. C. and Glover, D. M. (1981). Arrangements and rearrangements of sequences flanking the two types of rDNA insertion in *Drosophila melanogaster*. *Nature, Lond.*, **290**, 749–753.

Sherratt, D. J., Arthur, A., Bishop, R., Dyson, P., Kitts, P. and Symington, L. (1983). Genetic transposition in bacteria. In K. F. Chater, C. A. Cullis, D. A. Hopwood, A. W. R. Johnston and H. W. Woodhouse (Eds), *Genetic Rearrangement*, Croom Helm, London.

Slatkin, M. (1975). Gene flow and selection in a two-locus system. *Genetics*, **81**, 787–802.

Slatkin, M. (1981). Fixation probabilities and fixation time in a subdivided population. *Evolution*, **35**, 477–488.

Slatkin, M. and Lande, R. (1976). Niche width in a fluctuating environment – density-independent model. *Amer. Nat.*, **110**, 31–35.

Smith, G. P. (1974). Unequal crossover and the evolution of multi-gene families. *Cold Spring Harbor Symp. Quant. Biol.*, **38**, 507–513.

Smithies, O. (1964). Chromosomal rearrangements and protein structure. *Cold Spring Harbor Symp. Quant. Biol.*, **29**, 309–319.

Spiess, E. B. (1958). Effects of recombination on viability in *Drosophila. Cold Spring Harbor Symp. Quant. Biol.*, **23**, 239–250.

Spofford, J. B. (1969). Heterosis and the evolution of duplications. *Amer. Nat.*, **103**, 407–432.

Stahl, F. W. (1979). *Genetic Recombination*, Freeman, San Francisco.

Stebbins, G. L. (1971). *Chromosomal Evolution in Higher Plants*, Edward Arnold, London.

Strobeck, C., Maynard Smith, J. and Charlesworth, B. (1976). The effects of hitch-hiking on a gene for recombination. *Genetics*, **82**, 567–548.

Strobel, E., Dunsmuir, P. and Rubin, G. M. (1979). Polymorphisms in the chromosomal locations of elements of the 412, *copia* and 297 dispersed repeated families in *Drosophila. Cell*, **17**, 429–439.

Sturtevant, A. H. (1925). The effects of unequal crossing over at the *Bar* locus. *Genetics*, **10**, 117–147.

Sturtevant, A. H. and Mather, K. (1938). The interrelations of inversions, heterosis and recombination. *Amer. Nat.*, **72**, 447–452.

Sturtevant, A. H. and Morgan, T. H. (1923). Reverse mutation of the *Bar* gene correlated with crossing over. *Science*, **57**, 746–747.

Thomson, G. (1977). The effect of a selected locus on linked neutral loci. *Genetics*, **85**, 753–788.

Turner, J. R. G. (1967). Why does the genome not congeal? *Evolution*, **21**, 645–656.

Turner, J. R. G. (1979). Genetic control of recombination in the silkworm. I. Multigenic control of chromosome 2. *Heredity*, **43**, 273–293.

Walsh, J. B. (1982). Rate of accumulation of reproductive isolation by chromosome rearrangements. *Amer. Nat.*, **120**, 510–532.

White, M. J. D. (1973). *Animal Cytology and Evolution*, 3rd edn, Cambridge University Presss, Cambridge.

White, M. J. D. (1978). *Modes of Speciation*, Freeman, San Francisco.

Williams, G. C. (1966). *Adaptation and Natural Selection*, Princeton University Press, Princeton.

Williams, G. C. (1975). *Sex and Evolution*, Princeton University Press, Princeton.

Williams, G. C. and Mitton, J. B. (1973). Why reproduce sexually? *J. Theor. Biol.*, **39**, 545–554.

Wilson, A. C., Bush, G. L., Case, S. M. and King, M. C. (1975). Social structuring of mammalian populations and rate of chromosomal evolution. *Proc. Natl. Acad. Sci. USA*, **72**, 5061–5065.

Wilson, A. C., Sarich, V. M. and Maxson, L. R. (1974). The importance of gene rearrangement in evolution: evidence from studies on rates of chromosomal, protein and anatomical evolution. *Proc. Natl. Acad. Sci. USA*, **71**, 3028–3030.

Wright, S. (1940). Breeding structure of populations in relation to speciation. *Amer. Nat.*, **74**, 232–248.

Wright, S. (1941). On the probability of fixation of reciprocal translocations. *Amer. Nat.*, **75**, 513–522.

Zimmer, E. A., Martin, S. L., Beverley, S. M., Kan, Y. W. and Wilson, A. C.

(1980). Rapid duplication and loss of genes coding for the α chains of hemoglobin. *Proc. Natl. Acad. Sci. USA*, **77**, 2158–2162.

Zimmering, S., Sandler, L. and Nicoletti, B. (1970). Mechanisms of meiotic drive. *Ann. Rev. Genet.*, **4**, 409–463.

Index